Heat Transfer and Spacecraft Thermal Control

Progress in Astronautics and Aeronautics

Martin Summerfield, Series Editor
PRINCETON UNIVERSITY

VOLUMES	EDITORS
1. Solid Propellant Rocket Research. 1960	Martin Summerfield PRINCETON UNIVERSITY
2. Liquid Rockets and Propellants. 1960	Loren E. Bollinger THE OHIO STATE UNIVERSITY Martin Goldsmith THE RAND CORPORATION Alexis W. Lemmon Jr. BATTELLE MEMORIAL INSTITUTE
3. Energy Conversion for Space Power. 1961	Nathan W. Snyder INSTITUTE FOR DEFENSE ANALYSES
4. Space Power Systems. 1961	Nathan W. Snyder INSTITUTE FOR DEFENSE ANALYSES
5. Electrostatic Propulsion. 1961	David B. Langmuir SPACE TECHNOLOGY LABORATORIES, INC. Ernst Stuhlinger NASA GEORGE C. MARSHALL SPACE FLIGHT CENTER J. M. Sellen Jr. SPACE TECHNOLOGY LABORATORIES
6. Detonation and Two-Phase Flow. 1962	S. S. Penner CALIFORNIA INSTITUTE OF TECHNOLOGY F. A. Williams HARVARD UNIVERSITY
7. Hypersonic Flow Research. 1962	Frederick R. Riddell AVCO CORPORATION
8. Guidance and Control. 1962	Robert E. Roberson CONSULTANT James S. Farrior LOCKHEED MISSILES AND SPACE COMPANY
9. Electric Propulsion Development. 1963	Ernst Stuhlinger NASA GEORGE C. MARSHALL SPACE FLIGHT CENTER

10. Technology of Lunar Exploration. 1963	Clifford I. Cummings and Harold R. Lawrence JET PROPULSION LABORATORY
11. Power Systems for Space Flight. 1963	Morris A. Zipkin and Russell N. Edwards GENERAL ELECTRIC COMPANY
12. Ionization in High-Temperature Gases. 1963	Kurt E. Shuler, Editor NATIONAL BUREAU OF STANDARDS John B. Fenn, Associate Editor PRINCETON UNIVERSITY
13. Guidance and Control—II. 1964	Robert C. Langford GENERAL PRECISION INC. Charles J. Mundo INSTITUTE OF NAVAL STUDIES
14. Celestial Mechanics and Astrodynamics. 1964	Victor G. Szebehely YALE UNIVERSITY OBSERVATORY
15. Heterogeneous Combustion. 1964	Hans G. Wolfhard INSTITUTE FOR DEFENSE ANALYSES Irvin Glassman PRINCETON UNIVERSITY Leon Green Jr. AIR FORCE SYSTEMS COMMAND
16. Space Power Systems Engineering. 1966	George C. Szego INSTITUTE FOR DEFENSE ANALYSES J. Edward Taylor TRW INC.
17. Methods in Astrodynamics and Celestial Mechanics. 1966	Raynor L. Duncombe U.S. NAVAL OBSERVATORY Victor G. Szebehely YALE UNIVERSITY OBSERVATORY
18. Thermophysics and Temperature Control of Spacecraft and Entry Vehicles. 1966	Gerhard B. Heller NASA GEORGE C. MARSHALL SPACE FLIGHT CENTER
19. Communication Satellite Systems Technology. 1966	Richard B. Marsten RADIO CORPORATION OF AMERICA

20. Thermophysics of Spacecraft and Planetary Bodies Radiation Properties of Solids and the Electromagnetic Radiation Environment in Space. **1967**	Gerhard B. Heller NASA GEORGE C. MARSHALL SPACE FLIGHT CENTER
21. Thermal Design Principles of Spacecraft and Entry Bodies. 1969	Jerry T. Bevans TRW SYSTEMS
22. Stratospheric Circulation. 1969	Willis L. Webb ATMOSPHERIC SCIENCES LABORATORY, WHITE SANDS, AND UNIVERSITY OF TEXAS AT EL PASO
23. Thermophysics: Applications to Thermal Design of Spacecraft. 1970	Jerry T. Bevans TRW SYSTEMS
24. Heat Transfer and Spacecraft Thermal Control. 1970	John W. Lucas JET PROPULSION LABORATORY

(Other volumes are planned.)

The MIT Press
Cambridge, Massachusetts,
and London, England

Progress in
Astronautics and Aeronautics

An American Institute of Aeronautics
and Astronautics Series

Martin Summerfield, Series Editor

Volume 24

Heat Transfer and Spacecraft Thermal Control

Technical papers selected from the AIAA 8th Aerospace Sciences Meeting, January 1970, and the AIAA 5th Thermophysics Conference, June-July 1970, subsequently revised for this volume.

Edited by
John W. Lucas

Jet Propulsion Laboratory
Pasadena, California

Printed by The Alpine Press Inc.
Bound in the United States of America by
The Colonial Press Inc.

ISBN 0 262 12042 9 (hardcover)

Library of Congress catalog card number:
70-147076

III.
Thermal Joint Conductance

IV.
Heat Transfer Analysis

V. Multilayer Insulation

VI. Thermal Control Devices

Preface

Many requirements must be satisfied in order to achieve a successful spacecraft mission. One very pervasive requirement is that every spacecraft component must be maintained within either operating or survival temperature limits at all times. This is a very difficult requirement to satisfy because of the many boundary conditions.

First it is necessary to satisfy a number of external and internal conditions. Clearly, the thermal control system must operate satisfactorily in the high vacuum of space and in the presence of many types of radiation. Of the latter, the thermal and visible radiation from the sun is of critical importance. For example, the decreasing solar intensity must be compensated for on a mission to Mars or the outer planets. While the spacecraft is near Earth or another planet, the thermal radiation and reflected solar energy from the planet must also be taken into account. Throughout a mission the orientation of the spacecraft with respect to incoming radiation is a determining factor. Ultraviolet and particulate (that is, protons, electrons) radiations can modify the spacecraft surface thermal radiation properties. These changes and their effects must be minimized, and compensations must be made for any remaining effects.

In addition to the external conditions just mentioned, there is also a set of internal conditions. There is internal heat from operating electronic equipment on board the spacecraft; this heat dissipation varies according to the phase of the mission. There are, of course, pulses of heating from the rocket motor during midcourse maneuvers. Heat paths and heat capacity inside the spacecraft vary during the mission due to, for example, the gradual expenditure of propellants.

Another category of constraints has to do with schedules and ground handling. Thermal designs must be such that they can be converted into hardware in a dependable, reproducible, and timely manner. During various phases of assembly, testing, and ground transportation the thermal system hardware must resist degradation and, if degraded, must be amenable to expeditious repair. During the boost phase the system must withstand the structural vibration and aerodynamic heating loads. Once operating successfully in space, the thermal system must operate remotely for periods up to several years.

A final category of constraints includes the necessity to develop and supply thermal systems that are very low in weight and require low or zero power for operation, all at minimal cost.

These boundary conditions have presented a tremendous challenge. As a matter of fact, it has been absolutely necessary to develop much knowledge and many techniques which never existed before in order to do the job. It became necessary:

1. to understand optical (thermal radiation) properties of surfaces with far more accuracy and in greater detail than heretofore, first in order to know what values of properties (for example, thermal emittance) to use, and second, to be able to tailor the properties and maintain them;

2. to understand radiation heat transfer far more thoroughly than previously so as to be able to calculate with the new higher precision that is required in the frequent absence of convection heat transfer;

3. To understand thermal conductance in order, for example, to predict and control heat transfer across joints in vacuum;

4. to understand combined radiation and conduction on internal and external portions of spacecraft in free flight and in multilayer "super" insulations; and

5. to develop devices to control heat transfer to and from various portions of spacecraft in order to maintain necessary temperatures under varying conditions.

The progress of thermophysics to date on these tasks has provided a wealth of knowledge and understanding which has far-reaching consequences in a number of areas. For one, the foundation has been set for far more difficult space missions: missions to the planet Mercury and even closer to the sun, landings on Mars and Venus, and missions to the outer planets and the far reaches of the solar system. A second area is increasingly more detailed information about our planet Earth. This information includes solar heat balance of the atmosphere and Earth surface (weather); spectral characteristics of various surfaces such as plant leaves (crop damage detection); and local water, ground, and atmosphere temperatures (thermal pollution). Yet a third area is thermal aspects of the moon and the planets. For example, thermal behavior of the interior and surface of the moon is related to the evolution of the moon. A fourth area is the application of thermophysics technology to civil systems on Earth. Some of these are lighter and more efficient thermal insulations, smaller thermal sensors for medical applications, and thermal radiation sensors for skin-temperature diagnosis.

Representative examples of the type of heat transfer and advanced thermophysical work that is currently being done in response to space mission needs are contained in this volume. As an exercise, it is left to the reader, as he peruses the volume, to extrapolate the work reported here to a number of applications beyond those mentioned in the preceding paragraphs or in the respective papers.

The first paper, by Heller, discusses the recent past, status, and future directions of thermophysics. It will be noted that the area of thermophysics is broad and that the remaining sections in the book discuss some of its many aspects.

Chapter II is a major one in that it covers various aspects of the critically important area of thermal and visible radiation properties of solid/vacuum interfaces. The first subsection of Chapter II contains papers on the fundamental aspects of surface radiation. The first two papers, by Kroes, Kulshreshtha, Wegner, Mookherji, and Hayes and by Morrison and Sancier, discuss changes in radiative properties in terms of solid-state parameters. The paper by Gilligan and Brzuskiewicz includes the application of the scanning electron microscope, and Grimm brings up the interesting question of the validity of Kirchoff's law.

The second subsection of Chapter II deals with the application of radiation heat transfer analysis to surface radiation properties. The paper by Look and Love discusses the effects of surface roughness while that of Edwards and Bertak suggests a quick method of measuring the degree of diffuseness or, conversely, specularity of a surface.

The third subsection of Chapter II is on the engineering measurement of surface radiation properties. Sherrell and Shahroki's paper presents an indirect method for measuring hemispherical thermal emittance. Reid and Coon deal with measurements of the emittance of copper while the paper by Bartsch, Hudgins, and Geyer covers coated refractory metals. Loehrlein, Winter, and Viskanta present a photographic method for measuring bi-directional reflectance. The paper by Smith, Müller, Frost, and Hsia describes measurements of super- and subspecular maxima of reflected polarized radiation.

Chapter III covers thermal conductance across joints. Fletcher and Gyorog describe dimensionless parameters which correlate experimental data on conductance between similar metals while McKinzie takes up cyclic behavior. The paper by Gyorog reports on measurements of the effect of interstitial materials. Yovanovich applies thermal conductance concepts to ball bearings.

Chapter IV is on heat transfer analysis. The paper by Pfahl discusses heat transfer solely by the radiant transfer mode. Ishimoto and Pan add conduction while the paper by Novotny and Olsofka also adds an intervening gas. Finally, Larkin combines external convection with conduction in sandwich panels.

Chapter V deals with the rapidly developing area of multilayer or "super" insulation. Chau and Moy present the relative magnitudes of radiation, gas conduction, and joint conduction heat transfer in the perpendicular direction in multilayer insulation. The first paper by Pogson and MacGregor is on heat transfer in the parallel direction, while their second relates heat transfer in the parallel and perpendicular directions and suggests a means for decreasing the perpendicular heat flow. MacGregor, Pogson, and Russell discuss the numerical prediction of heat transfer through multilayer insulation, while the paper by Richardson, Ruccia, and French is on the application of multilayer insulation to space suits.

Chapter VI is on thermal control devices. The paper by Grodzka discusses the fundamental behavior of a number of phase-change materials. El-Saden deals with combined thermal-galvano-magnetic effects. The paper by Doughty and Jones presents experimental data on a solar shield for cryogenic propellant tanks. Cockfield and Killen discuss a radiator for rejecting heat at high temperatures, while Hulett and Zierman describe a staged radiator design for providing low temperatures for infrared detector systems. The concluding paper by Wilbers, Schelden, and Conti is on insulation materials for a Mars lander.

A number of individuals played key roles in the preparation of this book. Vernon Klockzien was responsible for the thermophysics sessions at the AIAA 8th Aerospace Sciences Meeting in New York. Richard Bobco, Technical Program Chairman of the AIAA 5th Thermophysics Conference, was of great assistance to me, as General Chairman, in organizing the technical portion of the Conference. The Session Chairmen, in turn, selected the papers for their sessions and later prepared a review of each paper in their respective sessions. Members of the 1970 Thermophysics Committee, with the assistance of several of their colleagues, prepared additional reviews of the papers. Dr. Martin Summerfield, Editor in Chief, and Miss Ruth Bryans, Director, Scientific Publications, provided guidance and assistance throughout the entire preparation and editing of this volume. My gratitude is also due Miss Bonnie Blair, especially for maintaining accurate files and correspondence on the continuous status of all papers.

John W. Lucas

November 1970

The Thermophysics Committee of the American Institute of Aeronautics and Astronautics

Membership in 1969 and 1970

Jerry T. Bevans (Chairman, 1969)
TRW SYSTEMS INC.
REDONDO BEACH

Richard P. Bobco (1969 and 1970)
HUGHES AIRCRAFT COMPANY
LOS ANGELES

Donald G. Burkhard (1969 and 1970)
UNIVERSITY OF GEORGIA
ATHENS

Walter G. Camack (1969 and 1970)
MARTIN MARIETTA CORPORATION
DENVER

Robert P. Caren (1969 and 1970)
LOCKHEED MISSILES & SPACE COMPANY
PALO ALTO

William F. Carroll (1969 and 1970)
JET PROPULSION LABORATORY
PASADENA

Edward T. Chimenti (1970)
NASA MANNED SPACECRAFT CENTER
HOUSTON

Jerry W. Craig (1969)
NASA MANNED SPACECRAFT CENTER
HOUSTON

Gary L. Denman (1970)
AIR FORCE MATERIELS LABORATORY
WRIGHT PATTERSON AIR FORCE BASE

Donald K. Edwards (1969)
UNIVERSITY OF CALIFORNIA
LOS ANGELES

Erwin Fried (1970)
GENERAL ELECTRIC COMPANY
PHILADELPHIA

Robert G. Hering (1970)
UNIVERSITY OF ILLINOIS AT URBANA-CHAMPAIGN

Billy P. Jones (1969 and 1970)
NASA MARSHALL SPACE FLIGHT CENTER
HUNTSVILLE

Warren Keller (1969 and 1970)
NASA HEADQUARTERS
WASHINGTON, D.C.

Robert M. Kendall (1970)
AEROTHERM CORPORATION
MOUNTAIN VIEW

Robert Kidwell (1970)
NASA GODDARD SPACE FLIGHT CENTER
GREENBELT

Vernon G. Klockzien (1969 and 1970)
THE BOEING COMPANY
SEATTLE

Tom J. Love (1970)
UNIVERSITY OF OKLAHOMA
NORMAN

Donald S. Lowe (1969)
UNIVERSITY OF MICHIGAN
ANN ARBOR

John W. Lucas (1969; Chairman, 1970)
JET PROPULSION LABORATORY
PASADENA

Hyman Marcus (1969 and 1970)
AIR FORCE MATERIELS LABORATORY
WRIGHT PATTERSON AIR FORCE BASE

J. T. Neu (1969)
GENERAL DYNAMICS/CONVAIR
SAN DIEGO

Simon Ostrach (1969)
CASE WESTERN RESERVE UNIVERSITY
CLEVELAND

Sinclaire M. Scala (1969)
GENERAL ELECTRIC COMPANY
PHILADELPHIA

Donald L. Stierwalt (1969)
NAVAL WEAPONS CENTER
CORONA

Elmer R. Streed (1969)
NASA AMES RESEARCH CENTER
MOFFETT FIELD

Chang Lin Tien (1970)
UNIVERSITY OF CALIFORNIA
BERKELEY

Alfred E. Wechsler (1970)
ARTHUR D. LITTLE, INC.
CAMBRIDGE

J. A. Wiebelt (1969)
OKLAHOMA STATE UNIVERSITY
STILLWATER

Editorial Committee for Volume 24

I. INTRODUCTION

THE STATUS OF THERMOPHYSICS AS A MULTIDISCIPLINE AREA IN ASTRONAUTICS AND AERONAUTICS

Gerhard B. Heller*

NASA Marshall Space Flight Center, Huntsville, Ala.

Abstract

A survey and critical review of the field of thermophysics are presented. Problems are highlighted in the areas of technology and applications to space vehicles and to hypervelocity vehicles, such as the Space Shuttle. Specific attention is given to spacecraft thermal design, active and semiactive thermal control, optical and radiative properties of thermal control coatings, the electromagnetic radiation environment, effects of the space environment on thermal control surfaces, thermal problems of hypervelocity vehicles, contamination of thermal control surfaces and of space optics, the optical and radiation properties of natural planetary surfaces, and remote sensing. In a number of areas problems still outstanding are pointed out.

Introduction

Summaries on the status of thermophysics were presented by the author at several prior AIAA meetings, such as the Annual Meeting in 1964, and as introductions to the Specialist Conferences in 1965 and 1967.[1,2] Also, the author and several associates have recently prepared a report delineating various aspects of thermophysics research at the NASA Marshall Space Flight Center (MSFC).[3]

Since this is the first paper in this book, it might be worthwhile to discuss the total field of thermophysics. Thermophysics acquired

---

Presented as Paper 70-812 at the AIAA 5th Thermophysics Conference, Los Angeles, Calif., June 29-July 1, 1970.

*Director, Space Sciences Laboratory.

considerable impetus by the launching of Explorer I and the ensuing space program achievements of the last 12 years. An additional widening of interests has occurred because of the initiation of the supersonic transport airplane and the inception of studies for the Space Shuttle.

The charter of the Thermophysics Committee of the AIAA is as follows:

> "Study and applications of the properties and mechanisms involved in thermal energy transfer within and between solids, and between an object and its environment, particularly by radiation. Study of environmental effects on such properties and mechanisms."

It is evident from this charter that emphasis is placed on radiative transfer, and rightly so, because of the importance of the role it assumes in the thermal balance of all space vehicle and hypersonic aerodynamic problems, such as entry vehicles and transports of very high Mach numbers.

In this framework the "thermophysics community" is concerned with a better understanding of the thermal environment of these vehicles and space projects and the interaction of the environment with the surfaces of the vehicles and with the thermal equilibrium at the surfaces and within various subsystems and components. Aspects of thermophysics include the following: spacecraft thermal design; active and passive thermal control; semiactive devices, such as heat pipes; the optical and radiative properties of thermal control coatings and surfaces; the contamination of optical and thermal control surfaces by outgassing from spacecraft or the exhaust of rockets; advanced laboratory techniques for optical measurements; analytical methods to arrive at computer models of the thermal control; the electromagnetic radiation environment, such as solar "constant," planetary albedo and IR radiation; the thermal balance of the earth; the effect of the space environment on thermal control surfaces; and the thermophysics of high speed entry into atmospheres, including the ablation process of protective materials. All of these areas are important either as straightforward engineering tasks for which solutions are known and have to be applied to each project, or as advanced research and technology tasks for which solutions are still needed.

Areas which have been initiated in the last 5 years are a logical extension of the basic interdisciplinary activities of thermophysics. These include the thermophysical properties of natural planetary surfaces: the surface of the moon and, of course, the optical and radiation characteristics of the earth. The knowledge we have acquired for the solution of space problems can be applied to our home planet Earth. Foremost among these applications are the study of earth resources by remote sensing from orbit and the pollution of the earth environment.

Spacecraft Thermal Design

Thermal design is now an essential engineering activity in the development of all space vehicles, and especially the interfacing with the total spacecraft design has been a problem of serious consequences. In many cases a space vehicle has been designed and manufacturing started before the thermal design was initiated. Thermal considerations often require a combination of active and passive thermal control techniques that have led to costly redesign. This is avoidable if the thermal design is an essential part of a system from the very beginning. Thus a systems approach to thermal design is mandatory.

For the thermal design of each new project a number of new technological challenges have to be met. In the ideal case the technology should be on hand when a new project starts. However, in many cases technology problems have to be solved after the project is already under way to avoid delay in starting a project. It is believed that more emphasis should be given to technology in the preparation of such projects. The thermal design problems of past or existing projects are discussed in the proceedings of the joint NASA-DOD Conference of the Thermal Control Working Group, August 16-17, 1967, Dayton, Ohio.[4] More discussion in the open literature and general documentation would be desirable so that all can learn from past experiences, even if this means talking about some failure or troubles caused by thermal problems with space vehicles. In many cases it is not necessarily a clear-cut failure. Design experiences might include "fixes" because of thermal problems leading to an over-design which could be avoided were more known about each project experience; thus, technology could be advanced to provide quicker and better solutions. An example of such hardware experience follows.

Figure 1 is an artist's concept of a Pegasus satellite. Three of these spacecraft were launched in 1965. The S-IVB stage was coated with S-13, a paint composed of ZnO as the pigment and methyl silicone resin as the binder. The α_S/ϵ_T ratio was expected to start with 0.22 and increase within a year's effective lifetime of the spacecraft to not more than 0.27. The first measurements in orbit showed a value of 0.52,[5,6] which increased with time in orbit. If it were not for the active thermal control provided by a louver assembly between the Pegasus electronic canister and the S-IVB, the electronics would have exceeded their thermal limits. The reason for the high values of the ratios of solar absorptance to infrared emittance was not simply the degradation caused by the space environment, but a combination of the contamination from the rocket firings and the subsequent effect of the space environment on the contaminated surfaces.

The experience from the Pegasus contamination was taken into consideration in other projects, such as Apollo. Better solutions are needed for future projects. Thermal control surfaces less subject to contamination are needed, as are solutions to the contamination problem, which is becoming more severe for optical surfaces of astronomical telescopes and will also become more severe because of the extended stay times in space which are contemplated for future space projects. It is not possible to cover all the thermal design problems that are expected in future projects which are part of the national space program as prepared by the Space Task Group under the chairmanship of Vice President Agnew during 1969.[7] However, some general comments can be made. There is still the problem of finding better space-stable thermal control surfaces. If ZnO is eliminated as the pigment and other pigments or organic polymers coated by metal are considered, the possible effects of autoacceleration of degradation should be analyzed. This phenomenon is not as evident in ZnO as in other materials. A knowledge of the difference between ZnO and other materials could provide guidance in the selection of new pigments or new concepts. The second surface mirror coatings proved to be very successful on Vanguard I; however, the problems connected with the application of this method to large vehicles have not been solved. In addition to the applicability to large surfaces and an ϵ_T of 0.8 or more, a reduction in cost is necessary. The entire area of second surface mirrors has been largely neglected, despite its early success. A very important aspect of the development of thermal control surfaces is the long duration of planned future missions. These surfaces should be space stable or have a predictable degradation value for 10 or 20 years.

A promising area of thermal control is the use of fusible materials in all cases where strong temperature excursions occur for only short periods of time up to an hour;[8,9] for example, flight experiments which are outside the thermally controlled envelope of a spacecraft, or surfaces exposed to peak heating during planetary entry or peak heating on surfaces of the earth-to-orbit Shuttle. A heat sink of fusible material could be used to supply a constant temperature level for sensitive equipment, such as optical components or detectors. A thermocouple ice bath is an example of a laboratory application.

It is gratifying to note that many tools resulting from thermal design efforts have been perfected in the past years and, as such, are no longer classified in the area of critical technology requirements. Some examples are the sophisticated computer programs which allow analysis of the thermal design of a space vehicle and its time variations with 1000 nodes, or more if needed, and the simulation of the space environment in thermal vacuum tests. Large chambers have been built for full-scale testing of space vehicles. The problem here still seems to be that albedo cannot be simulated, the real need for exact spectral simulation of the solar radiation has not been proven, and the serious concern about possible contamination of optical surfaces of space experiments in such chambers. However, progress is being made with respect to these problems.

Active and Semiactive Thermal Control

Active thermal control devices have not been developed as a technology program with the goal to have such devices when needed. However, some excellent devices have been developed as integral components of satellite projects. A few examples are 1) the pneumatically controlled louvers on Nimbus and the internal pneumatic controls on Telstar, 2) the thermal switches on Surveyor, 3) the bimetallic-strip-controlled louvers on Pegasus, Mariner, OGO, OAO, Nimbus D, and other spacecraft, 4) the Maltese cross on the later Pioneers and on Relay, and 5) electric heaters which have been widely used for active thermal control on spacecraft, such as OGO, OAO, and ATM.

For many years there has been discussion of the heat pipe and its potential for thermal control. The problem is not whether a heat pipe "works" in zero g. Since its principle is based on capillary forces and on the high speed of gas flow, it "works" in the laboratory even at minus 1 g. The question is how well it will work and for how long. One problem is connected with the ground testing of a complicated three-dimensional heat pipe system under 1 g conditions and, then, to extrapolate to a zero-gravity environment. For extended periods, i.e., 10 to 20 years for future space bases, the corrosiveness of heat pipe fluids could limit the applicability. It is well known that heat pipes have been successfully flown on Explorer 36 (GEOS II)[10] and PAC I.[11] They equalized the temperatures between the sun-exposed and shadowed side.

Optical and Radiative Properties of Thermal Control Coatings

The specific properties of thermal control coatings are the absorptance for solar radiation α_s and the emittance of thermal infrared radiation ϵ_T. Figure 2, adapted from a recent study by Gates, Zerlaut, and Carroll,[12] shows the range of presently available thermal control surfaces for a mission in space not to exceed 1 year. In an earlier study,[13] Snoddy and Miller discussed the availabili of such coatings under the constraints of easy applicability to large space vehicles, stability under space conditions, and low cost. Selective black solar absorbers, such as Tabor surfaces, are often used as quantity heat absorbers around solar cell arrays. The metallic paints shown in Fig. 2 are more often used for their corrosion inhibiting properties on internal structures than for thermal control. Black paints, such as the Cat-A-Lac black coatings, combine high absorptance with high emittance. Zinc oxide pigments in silicone or silicate polymer bindings, or other white paints, have been of primary importance in thermal control of large spacecraft surfaces and are planned for extensive use on the Skylab configuration.

Thermal control surfaces for stability requirements up to 5 years are shown in Fig. 3, adapted from the previously mentioned paper by Gates et al. While laboratory testing is still in progress to evaluate the stability of these surfaces for missions of this length, this projection of available surfaces is the author's estimate of the state-of-the-art.

Coatings or surfaces for thermal control usage for mission lifetimes exceeding 5 years are currently not available for application because of the scarcity of stability tests of these lengths under combined solar uv and proton irradiation. Any of the surfaces listed in Figs. 2 or 3 are candidates for such a mission, but particularly likely candidates are metallized polymers,[14] sandblasted metals and, for its lower absorptance and stability to uv and proton irradiation, Zn_2TiO_4.[15,16] Extensive laboratory testing of these materials is required before a selection is available for missions, such as the Space Station, with lifetimes of 10 to 20 years.

Much progress has been made on measurement techniques for thermophysical properties during the first decade of space exploration. Further refinement is required in a number of areas, including specific problems in connection with the earth-to-orbit shuttle and the optical degradation of optical surfaces in the neighborhood of space stations. In the past, the thermophysicists have perfected the analytical techniques for thermal design. The already-mentioned multinode computation codes with 1000 or more nodes allow one to analyze thermal behavior under varying external and internal heat load conditions. Of course, specific codes and specific inputs of design and environmental constraints to a thermal control computer program are needed for each specific project as part of the project activity. The author has been a strong advocate of the computer approach. This should precede and supplement all thermal vacuum testing and should be the preparation for the evaluation of results of flight measurements. Considerable progress has been made in the area of radiative transport, which includes engineering applications for multiple reflections between surfaces. Solutions are available for simple geometries if the surfaces are all specularly or all diffusely reflecting. Individual cases with specular and diffuse reflectors in some arbitrary configurations have been solved.[17] However, solutions for the general case, where each surface interfacing with others has mixed specular-diffuse reflectance characteristics, are still lacking. Besides analytical solutions for such cases, better methods for experimental determination of the bidirectional reflectance with both specular and diffuse characteristics have to be found. The task is formidable, since a hemisphere has about 20,000 square degrees; therefore, for a modest angular resolution of 1°, 20,000 data points are needed to describe just one surface with only one incident angle of illumination and at only one wavelength. This calls for a fully automated system using a test setup directly connected to the terminal of a computer. More direct radiative measurements of thermal control surfaces in space are needed. Laboratory and computation techniques can be carried only to a certain point. Presently there are still discrepancies between predictions and the actual thermal behavior in space, which only direct space experiments in a measured-defined space environment will resolve.

Electromagnetic Radiation Environment

The exact determination of the electromagnetic environment continues to be an important task. Recent measurements from airplanes have contributed to this area.[18,19] The value of the solar constant which has been used for thermal design efforts up to now is being revised. Measurements to increase our knowledge of the statistical variations and absolute levels of planetary albedo and IR radiation are equally important. Besides their importance for the thermal radiative equilibrium of space vehicles, they are important for an understanding of the thermal balance of our planet Earth.

Effects of the Space Environment on Thermal Control Surfaces

A great deal of attention has been focused on this area of thermophysical activity because of its importance to the thermal design of space vehicles. It was realized that all earlier laboratory measurements of the degradation of certain commonly used coatings were wrong because of the bleaching effects of the oxygen in the atmosphere on the color centers when the degraded sample was removed from the vacuum chamber. Figure 4 shows this bleaching effect on S-13, a thermal control coating based on ZnO as pigment and a methyl silicone resin as binder. The well-recognized effect is dependent both on the atmospheric pressure and on time; all measurements of the degradation of coatings are now made in situ. This means that the exposure to uv radiation and to simulated solar wind is made in vacuum and that the measurements of the optical properties are made in the same chamber without breaking the vacuum. The same precaution should be taken for flight experiments where thermal control samples are exposed to the space environment and then brought back for laboratory measurements. The samples should be kept under a vacuum of 10^{-6} torr or better, or, if leakage cannot be prevented, under an inert atmosphere with ≤ 1 ppb oxygen (which corresponds to the 10^{-6} torr vacuum). The best method is again the in situ measurement which could be accomplished by an astronaut with a portable spectroreflectometer.[20]

Despite the many efforts aimed at an analysis of the interaction mechanisms, the effect of electromagnetic radiation or particle radiation on thermal control coatings is not completely understood. This understanding becomes more important if we want to extend the stability of thermal control coatings to 10 or more years, as already discussed in connection with the requirements for thermal control design.

Another important aspect for a space base will be the question whether thermal control surfaces can be repaired or overcoated in space by an astronaut. Proper cleaning techniques should be developed to remove the effects of contamination.

Thermal Problems of Hypervelocity Vehicles

With the advent of ballistic missile projects and of satellite entry, a great number of interesting thermophysical problems had to be solved. Ballistic missile reentry was proved to be feasible when, in 1956, the first ablative type reentry nose cone was recovered from the ocean. The technology of ablative materials also proved to be a good solution for satellite entry from orbit or for entry of space vehicles with parabolic and hyperbolic velocities, such as those returning from the moon or from other planets. The reasons for this are well known, especially the fact that the total heating periods are short and that the unablated material provides sufficient insulation to protect the structural shell of the vehicle from the heat for this relatively short period.

The initiation of studies for an earth-to-orbit-and-return shuttle has posed additional and very interesting problems in the area of thermophysics. Figure 5 shows a typical heat flux-time curve for the ascent and reentry of a shuttle booster vehicle prepared by members of the Aerophysics Division of the Aero-Astrodynamics Laboratory at NASA MSFC. The first peak occurs at staging and the second occurs during the gliding reentry in the atmosphere. Figure 6 shows a similar typical curve for the entry into the atmosphere of a shuttle orbiter on its return flight.[21,22] This curve shows two peaks which are typical for the entry of a high L/D vehicle. The heat flux is applicable to the point on the leading edge shown in the sketch included in Fig. 6. Initial braking is performed at a high angle of attack. The second peak occurs during the glide phase. Some trade-off can be made to increase the first peak and thus reduce the second one. The total heat transfer to the vehicle is dependent on the total energy that has to be dissipated. The most advantageous method is to dissipate the maximum possible amount by wave drag rather than by frictional drag. High heat fluxes for extended periods call for entirely different approaches. Ablation might still be a good solution for the peaks; however, the required refurbishment after each flight is an undesirable feature. It can be expected that the challenges connected with the hypersonic vehicles such as the shuttle will provide new impetus to thermophysics.

Contamination of Thermal Control Surfaces and Space Optics

Contamination of thermal control surfaces has three aspects: on the ground, during launch, and in space. Contamination on the ground occurs during assembly, thermal vacuum testing, and handling prior to and during launch. In many cases, more than a year passes between the application of coatings and the launch of a spacecraft. An example of an approach to this problem was in connection with the Pegasus project where a portable spectroreflectometer was used to measure the thermal radiation properties of all thermal control coatings up to 1 day before launch at NASA Kennedy Space Center. In this way, the optical properties were known just before launch, and one could determine whether degradation had occurred on the ground. The requirements for cleanroom assembly of spacecraft and special procedures for handling during shipment, etc., have greatly improved the ground-based contamination for many satellite projects. However, the contamination in space is still a very serious problem. Exhaust jets from reaction control engines have a wide expansion cone and produce a contamination cloud surrounding the whole space vehicle. During simulated tests in chamber A at NASA Manned Spacecraft Center[23] contaminants from the Reaction Control System engines were found on surfaces which were behind barriers and 180° away from the jets (Fig. 7). The dark surfaces are specularly reflecting, and no light is reflected in the direction of the camera. On the right side are the same surfaces after exposure to the exhaust gases from the jets. The illuminating light is now partially diffusely reflected, and the surfaces become visible.

The effect of such contamination is more serious in space because of the simultaneous effect of uv radiation. Without this irradiation the contaminants would probably leave the surface again after a short time, depending on the accommodation coefficient. However, with the presence of uv radiation, and the solar wind for deep space probes, the contaminants undergo reactions which prevent them from vaporizing and which lead to high solar absorptance. Figure 8 shows the effect of the increase of the solar absorptance on the average radiator outlet temperature of a space station.[24] With the docked experiment modules, which is the operational mode of the station, very little margin is left between the initial α_s of 0.2 and its degradation to high values.

Another aspect of the contamination problem is connected with the optics of space telescopes, cameras, gratings, etc., which have requirements for optical quality that are more stringent than for thermal control surfaces. The effect of contamination can be a flat reduction of reflection or transmission, depending upon the type of optics, or the occurrence of diffuse reflection with severe effect on the imaging quality or on spectral resolution. Selective absorption in specific wavelengths or wave bands can affect the quantitative evaluation of spectral lines. This can be caused by absorption edges in the contaminants or by interference effects if the thickness of the contaminant reaches a quarter wavelength of a specific value of λ. Efforts are under way to fully understand the sources of contamination, the physical processes, and the effects on optics. Figure 9 shows, for example, the effects of Reaction Control System engine contamination on the reflectivity of grazing incidence X-ray optics. The wavelength is $\lambda = 1.79$ Å. The angle of incidence varies from 9 to 50 minutes of arc.[25]

The Optical and Radiation Properties of Natural Planetary Surfaces

A very interesting subject of thermophysical research is the study and the measurement of the optical and radiation properties of the moon and other natural planetary surfaces. In addition to the scientific value of such studies, the results are needed for the thermal design of space vehicles which land on the moon or other planetary bodies. Until the actual lunar landing of Surveyor and Apollo, all of our knowledge of the moon was based on the analysis of information obtained via electromagnetic radiation, such as photographs from the earth and from the Lunar Orbiter, IR radiation, thermal mm waves, and radio reflection signals from gound-based observatories. These measurements, combined with investigations in the laboratory with simulated lunar materials, allowed us to formulate lunar thermophysical models. The scientific returns from the Apollo 11 and Apollo 12 lunar landings are tremendous and have resulted in a new understanding of our earth-moon relationship and of the formation of the solar system. The results in the thermophysical area are somewhat disappointing. We still have not analyzed the lunar materials with sufficient thoroughness and accuracy to be able to distinguish between the various models which were developed. The question may arise whether we still need these models, since we have direct information from the moon. The answer is that the need is greater than ever, since we do not have a complete understanding of the thermophysical properties of the lunar soil. We still do not know why the thermal and radiative properties of the lunar surface derived from IR measurements do not match those obtained from thermal mm waves, or those from radar echoes. Figure 10 shows thermal conductance of simulated lunar soil (basalt powder)[26] and some recent data points obtained by Birkebak from the analysis of lunar soil from Tranquillity Base.[27] At low temperature the data points fall between those for densities of 1.3 and 1.5 g/cm^3 of the "simulated lunar soil." The deviation at higher temperatures and the scatter are not completely understood. These first results constitute a good start, but more sophisticated work in this area would be highly desirable.

With the strong interest in the planetary exploration, more work will probably be extended toward a better understanding of the thermal and radiative properties of the terrestrial planets — Mars, Venus, and Mercury — the asteroids, and the outer planets which provide additional challenges because of their entirely different composition.

Remote Sensing

There is much interest in the subject of earth resources and their determination by remote sensing techniques. The vast knowledge and experimental results obtained in connection with the characterization of the surfaces of spacecraft are applicable to the problems of characterization of natural surfaces on the earth. The spectral dependence of bidirectional reflectance could be of vital importance to characterize plants and animals, geological formations, or water surfaces. The exploration of earth resources has a considerable appeal to the man in the street who expects such direct tie-ins of the space effort to everyday problems.

Conclusions

This survey has shown the field of thermophysics as a very active multidisciplinary area which plays a vital role in space projects and hypervelocity vehicles. The field has matured to a well-established area of theoretical and experimental research with many engineering applications to space projects. New projects which are now in the study phase, like the Space Station and the Space Shuttle, impose new challenging tasks for thermophysicists. New developments to meet these challenges are needed in thermal control surfaces which are space stable for extended time in space of 10 to 20 years. The Space Shuttle requires coatings which provide oxidation protection for high-temperature materials and also act as thermal control coatings during the orbital phase of a mission. Active systems or unfoldable radiators might be needed to meet such a diverse thermal requirement.

In the well-established type of space projects more work is needed to obtain a thorough understanding of the underlying physical processes. Examples of these are: the degradation process; the contamination of thermal control surfaces and space optics caused by outgassing, waste dumps, or rocket firings; measuring techniques for bidirectional reflectance; and heat transfer between mixed diffuse-specular surfaces, etc. A very broad and very promising area is that of the lunar thermophysical properties, those of other planetary bodies, and those of our home planet Earth. The area of remote sensing from earth orbit for the purpose of providing information on natural resources is an area of great importance.

References

[1]Heller, G. B., "Preface," AIAA Progress in Astronautics and Aeronautics: Thermophysics and Temperature Control of Spacecraft and Entry Vehicles," edited by G. B. Heller, Vol. 18, Academic Press, New York, 1966, pp. xi-xv.

[2]Heller, G. B., "Preface," AIAA Progress in Astronautics and Aeronautics: Thermophysics of Spacecraft and Planetary Bodies," edited by G. B. Heller, Vol. 20, Academic Press, New York, 1967, pp. ix-xiii.

[3]"Thermophysics Research at Marshall Space Flight Center," Research Achievements Review, Vol. III, No. 6, NASA TMX-53820, 1969.

[4]Proceedings of the Joint Air Force-NASA Thermal Control Working Group, Johnson and Boebel, eds., AFML-TR-68-198, Wright-Patterson Air Force Base, Ohio, 1968.

[5]Schafer, C. F. and Bannister, T. C., "Thermal Control Coating Degradation Data from the Pegasus Experiment Packages," AIAA Progress in Astronautics and Aeronautics: Thermophysics of Spacecraft and Planetary Bodies, Vol. 20, edited by G. B. Heller, Academic Press, New York, 1967, pp. 457-473.

[6]Linton, R., Thermal Design Evaluation of Pegasus, NASA TN D-3642, Dec. 1966.

[7]"The Post-Apollo Space Program: Directions for the Future," Space Task Group Report to the President, Sept. 1969.

[8]Bannister, T. C., "Space Thermal Control Using Phase Change," NASA TMX-53402, March 1966.

[9]Bannister, T. C. and Richard, B. E., "Microscopic Observation of Interfacial Phenomena," AIAA Progress in Astronautics and Aeronautics: Thermophysics Applications to Thermal Design of Spacecraft, Vol. 23, edited by J. T. Bevans, Academic Press, New York, 1970, pp. 175-188.

[10]Harkness, R. E., "The GEOS-II Heat Pipe System and Its Performance in Test and Orbit," Report TG-1049, April 1969, Applied Physics Lab., Johns Hopkins University, Baltimore, Md.

[11]Stitandic, E. A. and Powers, E. I., "The Thermal Design and Flight Performance of PAC Spacecraft Utilizing Heat Pipes" (to be published as a NASA Goddard Space Flight Center report, Greenbelt, Md.).

[12]Gates, D. W., Zerlaut, G. A., and Carroll, W. F., "Recent Advances in Spacecraft Thermal-Control Materials Research," presented at the 20th IAF Congress, Mar del Plata, Argentina, Oct. 10, 1969 (to be published in the IAF Proceedings).

[13]Snoddy, W. C. and Miller, E., "Areas of Research on Surfaces for Thermal Control," Symposium on Thermal Radiation of Solids, S. Katzoff, ed., NASA SP-55, 1965, Washington, D. C.

[14]Mookherji, T., "A Review of the Stability of Metallized Polymers and Dielectrics Exposed to the Degrading Influences of Space Environment," Summary Report, Contract NAS8-30148, RL-SSL-1192, Jan. 1969-July 1970, Teledyne-Brown Engineering Company, Huntsville, Ala.

[15]Zerlaut, G. A. and Roger, F. O., "The Behavior of Several White Pigments as Determined by In Situ Reflectance Measurements of Irradiated Specimens," Proceedings of the Joint Air Force-NASA Thermal Control Working Group, 1968, AFML-TR-68-198, Air Force Materials Laboratory, Wright-Patterson Air Force Base, Ohio.

[16]Zerlaut, G. A. and Ashford, N. A., "Development of Space-Stable Thermal-Control Coatings," Interim Report U6002-85, NASA-MSFC Contract NAS8-5379, Feb. 1970.

[17]Howell, J. R. and Ziegler, R., "Thermal Radiative Heat Transfer," Vols. I and II, NASA SP-164, 1969.

[18]Kaplin J., "Near-Earth Electromagnetic Radiation Environment," Rept. 4, Under Contract NAS8-21420, Dec. 1968, Space Physics Lab., General Electric.

[19]"The Solar Constant and the Solar Spectrum Measured from a Research Aircraft at 38,000 Feet," TMX-322 68-304, NASA-GSFC, 1968.

[20]Wilkes, D. R., "Preliminary Results from a Space Compatible Portable Reflectometer," presented at the Open Forum Session of the AIAA Thermophysics Specialist Conference, June 1969, San Francisco, Calif.

[21]Heneidi, F. and Vaniman, J. T., "Reentry and Ascent Environment Determination for MSFC Space Shuttle," S&E-ASTN-PT-70-M-20, Jan. 24, 1970, Marshall Space Flight Center, Huntsville, Ala.

[22]Forney, J. A., "Re-entry and Ascent Environment Determination for MSFC Space Shuttle,"S&E-AERO-AT-70-2, Jan. 28, 1970, Marshall Space Flight Center, Huntsville, Ala.

[23]Arnett, G. M., "Lunar Excursion Module RCS Engine Vacuum Chamber Contamination Study," NASA-MSFC TM X-53859, July 1969.

[24]Private communication from J. T. Vaniman, S&E-Astronautics Lab, Marshall Space Flight Center, June 1970, Huntsville, Ala.

[25]Reynolds, J. M., "Reflection Efficiency of Various Materials at X-Ray Wavelengths," presented at the 47th Annual Meeting of the Alabama Academy of Science, April 1970, Auburn University, Auburn, Ala.

[26]Fountain, J. A. and West, E. A., "Thermal Conductivity of Particulate Basalt as a Function of Density in Simulated Lunar and Martian Environments," Journal of Geophysical Research, Vol. 75, No. 20, July 10, 1970, pp. 4063-4069.

[27]Birkebak, R. C., Cremers, C. J., and Dawson, P. J., "Thermal Radiation Properties and Thermal Conductivity of Lunar Materials," Science, Vol. 167, No. 3918, 1970, p. 724.

Fig. 1 Artist's concept of Pegasus satellite.

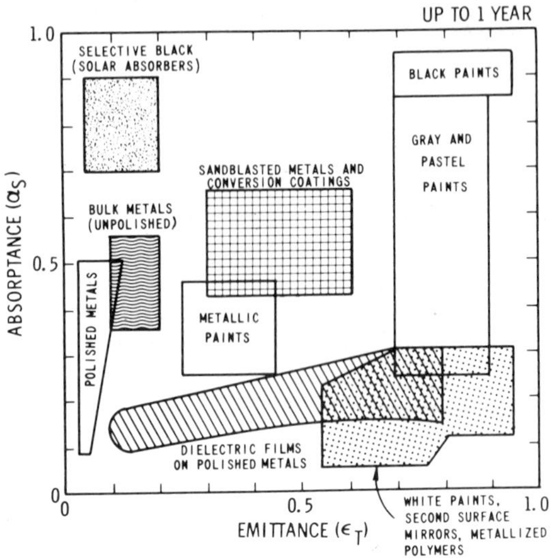

Fig. 2 Available coatings and surfaces based on the state of the art.

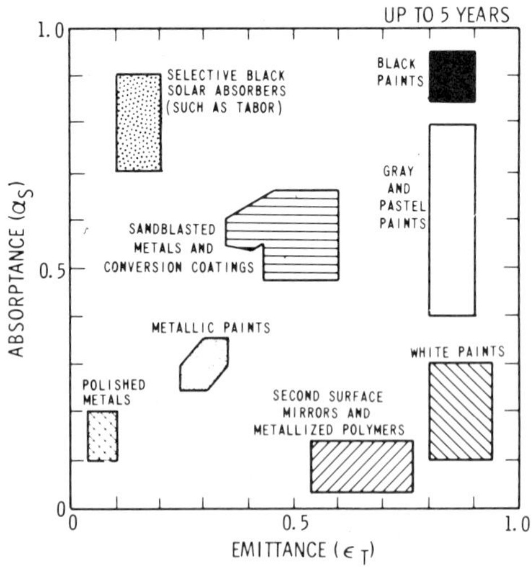

Fig. 3 Available coatings and surfaces based on the state of the art.

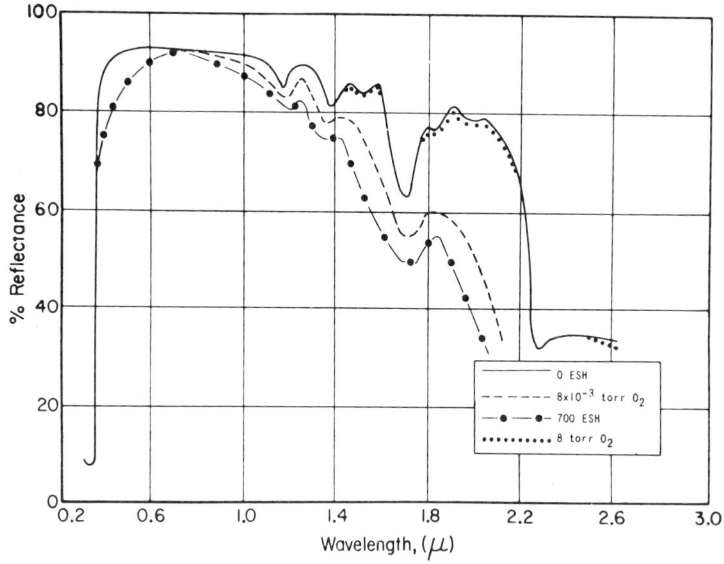

Fig. 4 uv degradation and oxygen recovery of S-13.

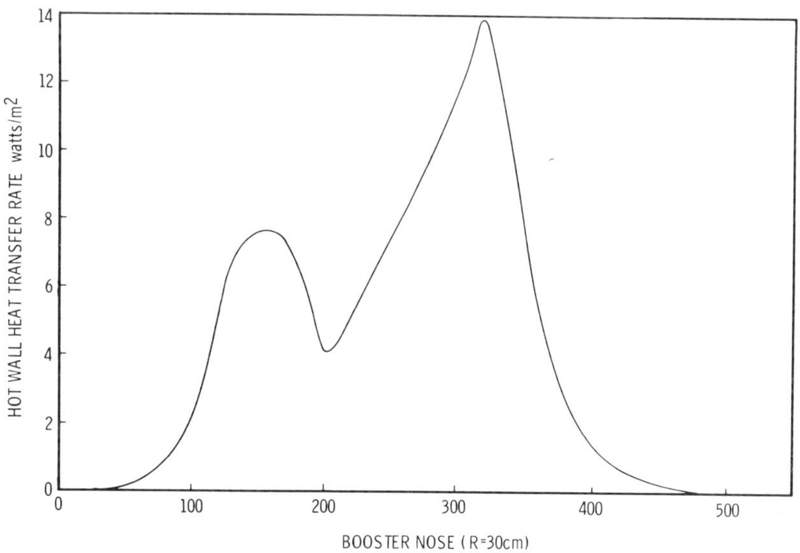

Fig. 5 Baseline booster of shuttle — nose heat-transfer rate vs time.

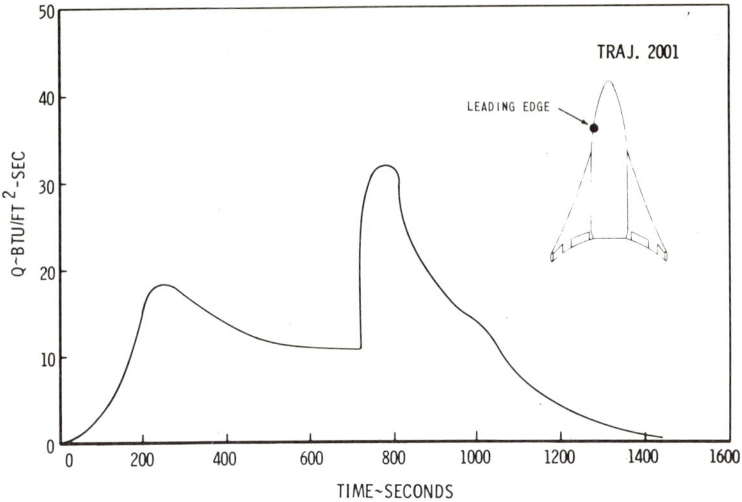

Fig. 6 Reentry heating rates for point on shuttle orbiter.

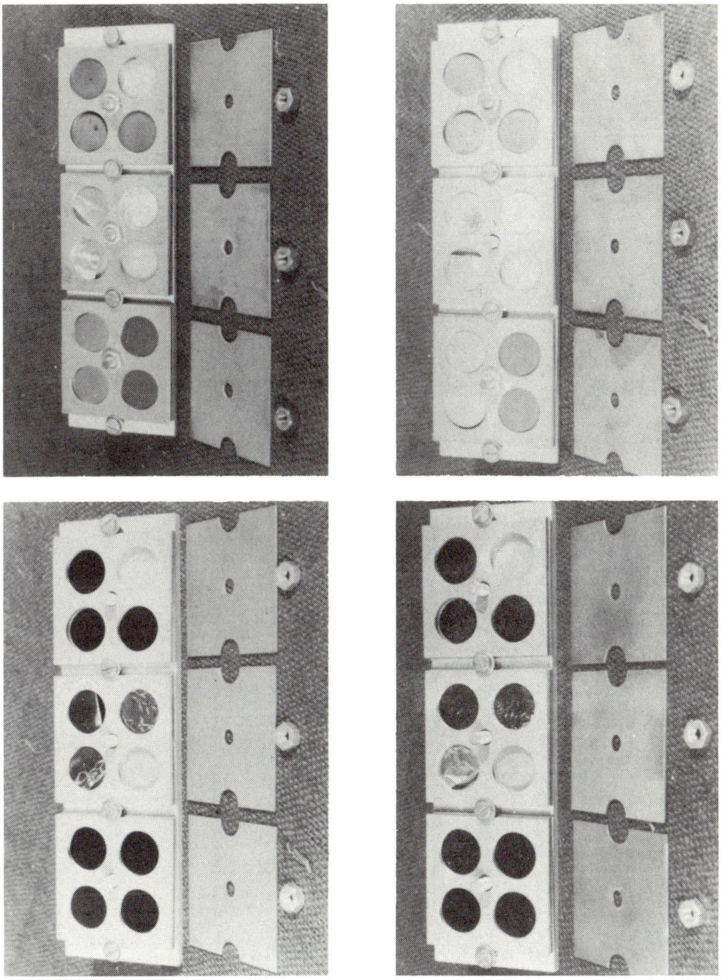

Fig. 7 RCS engine contamination test — sample test beds.

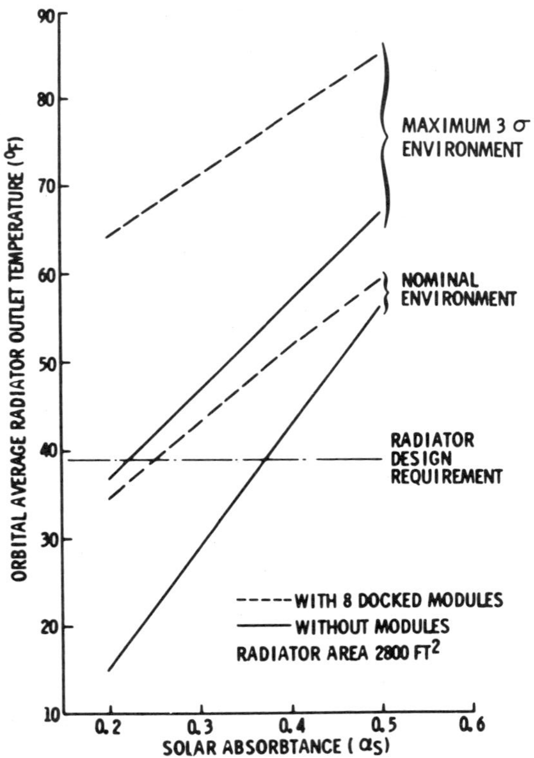

Fig. 8 Space Station radiator performance.

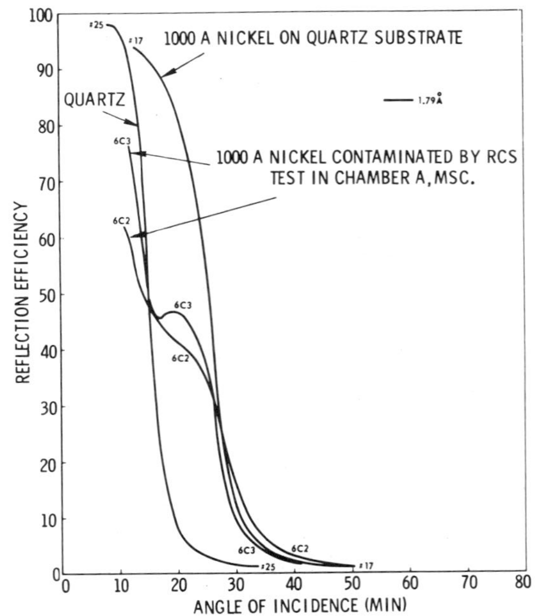

Fig. 9 Reflection efficiency curves at 1.79 Å.

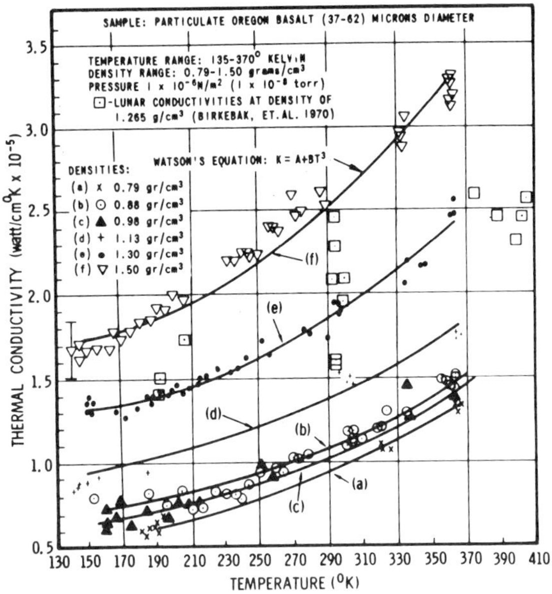

Fig. 10 Thermal conductivity of particulate basalt as a function of temperature and density compared to measured lunar conductivities.

II. SURFACE RADIATION PROPERTIES

II.1 Fundamental Aspects

EFFECTS OF ULTRAVIOLET IRRADIATION ON ZINC OXIDE

Roger L. Kroes,* Arun P. Kulshreshtha,† and U. E. Wegner†
NASA George C. Marshall Space Flight Center, Huntsville, Ala.
and
T. Mookherji‡ and J. D. Hayes‡
Brown Engineering Company, Huntsville, Ala.

Abstract

Zinc oxide has been used extensively as a paint pigment for thermal control coatings of spacecraft. In a space environment, this coating degrades and absorbs a high percentage of incident solar energy. A program was initiated by our laboratories to characterize the degradation in terms of solid-state parameters. In these studies a major effort was undertaken to detect and identify the defect centers produced by the ultraviolet irradiation. The results obtained from different types of samples have been correlated to obtain information about the importance of bulk and surface phenomena in the ultraviolet degradation.

I. Introduction

The degradation of white oxide thermal control coatings by ultraviolet (UV) radiation has been a recognized problem for several

---

Presented as Paper 70-829 at the AIAA 5th Thermophysics Conference, Los Angeles, Calif., June 29-July 1, 1970. Part of this study was performed by Teledyne-Brown Engineering Company, Huntsville, Ala. under contract NAS8-30148.

*Physicist, Thermal Environment Physics Branch, Space Thermophysics Division, Space Sciences Lab.

†Postdoctoral Resident Research Associate of National Academy of Sciences — National Research Council, U. S. A.; assigned to Thermal Environment Physics Branch, Space Thermophysics Division, Space Sciences Lab.

‡Physicist, Brown Engineering Company.

years. A great amount of effort has been exerted to determine the mechanisms by which the UV radiation damages the oxides. Most of these studies have been performed on zinc oxide. Several models have been offered to describe the degradation that manifests itself by an increase in the optical absorption coefficient in two spectral regions, the 0.39- to 0.8-μ range and the 1.0- to 2.4-μ range.

The general model that has been advanced to describe the degradation of zinc oxide is described as follows. The UV photons, which are absorbed near the surface, produce free electrons and holes. The photo-produced holes that diffuse to the surface recombine with electrons at surface oxygen, thereby neutralizing the surface oxygen. The neutralized surface oxygen is then evolved from the ZnO surface if the zinc oxide is in a vacuum environment. The first oxygen to be evolved is chemisorbed oxygen, but as the irradiation is continued, surface lattice oxygen is also evolved. The evolution of oxygen leaves the surface zinc rich, and the excess zinc diffuses into the bulk of the zinc oxide. Thus, the net result of the UV irradiation is the generation of excess zinc and an increase in the concentration of free electrons.

The mechanisms by which the above actions cause the increased visible and infrared region absorption are not clearly defined. Some workers believe that the enhanced infrared (IR) absorption is a result of additional free carrier absorption which is caused by the increase in the free electron concentration. Others believe that the enhanced IR absorption is a result of an increase in the density and population of defect levels lying near the conduction band.

The increased visible absorption is likewise not clearly understood. It has been explained by some workers that this is the result of the excess zinc precipitating out at dislocations. Precipitation of lithium has been shown[1] to occur in ZnO and to cause severe lattice strain in the neighborhood of the dislocation. Such strain could result in a decrease in the separation between the conduction and valence band extrema and, in effect, decrease the band gap in the neighborhood of the precipitation. This would produce a low energy tail on the fundamental absorption edge, similar to the visible degradation observed. Another explanation to the increased visible absorption is that it is a result of defect centers whose energy levels lie just above the valence band.

The studies described in this report were undertaken to provide more data to aid in the further establishment or refinement of the proposed degradation models[2,3] by providing more information concerning the generation and population of defect levels by UV radiation and by determining the energy levels and the magnetic properties of the defect states.

These studies were conducted by the authors in five separate experiments. The work was correlated, and in some cases, different experiments were made on the same samples. It is believed that a separate description of the different studies followed by a collective summary and conclusion will provide the clearest presentation of the results.

II. Electrical Properties of Thin Films

Objectives

The principal objective of the electrical measurements was to determine the mechanisms that cause a large conductivity increase in zinc oxide when it is UV irradiated and to gain a better understanding of the electronic energy levels and transport processes that may be associated with UV-induced IR absorption.

The particular objectives of these studies were to 1) study the effects of UV irradiation on the free carrier concentration, 2) look for defect centers sensitive to UV irradiation by means of mobility measurements, 3) determine if UV irradiation creates and populates defect states with energy levels near the conduction band, and 4) determine to what extent surface band-bending affects the population of defect centers with energy levels near the conduction band.

To achieve these objectives, the basic electrical measurements, that is the temperature dependence of conductivity, the Hall mobility, and the carrier concentration were made on zinc oxide films at various levels of degradation. The variation of the Hall mobility and the carrier concentration with photoexcitation was measured, and IR photoconductivity measurements were performed.

Experimental Procedures

The general experimental arrangement for the Hall measurements consisted of an ion-pumped vacuum dewar, the tail assembly of which projected into the pole gap of a Magnion UFS-3 magnet with a 1-in.-diameter axial hole in one pole piece. The sample was mounted on the brass cryostat tip with standard four contact electrical connections. The electrical contacts were made by soldering thin copper wires directly to the zinc oxide, to vacuum-evaporated chromium lands, or to vacuum-evaporated indium-aluminum lands using indium as the solder. All three types of contacts provided good ohmic connections. The Hall signal was processed by conventional direct current techniques utilizing a high input impedance operational amplifier as a preamp.

Transparent, thin films of zinc oxide were prepared from vacuum-evaporated zinc films by maintaining the zinc films at 600° C in a pure oxygen atmosphere for 24 to 48 hours.

In making the temperature measurements, a fine (0.005 in.) chromel-alumel thermocouple was affixed to the top surface of the sample substrate approximately 3 mm from the zinc oxide film. All measurements were made in a high vacuum environment of at least 3×10^{-6} torr.

The experimental arrangement for the spectral photoconductivity measurements included a Xenon flashlamp, a set of narrow-bandpass interference filters and the vacuum cryostat described above. The experiment consists of measuring the voltage change, v, across the load resistor, R, when the sample is flashed with light. It can be shown[4] that the change in conductivity, $\Delta\sigma$, of the film is related to the measured voltage change, v, by

$$\Delta\sigma = v(R + r_o)^2/[r_o^2 BR - vr_o R(r_o + R)] \tag{1}$$

where $r_o = 1/\sigma_o$ is the sample resistance before illumination and B is the sample bias voltage. The load resistance, R, was chosen to be nearly equal to the sample resistance so that the photovoltage was related to the photoconductance by

$$v = B/4 \cdot \Delta\sigma/\sigma_o \tag{2}$$

The incident intensity of the exciting light was determined using a blackened thermocouple whose response had been calibrated with a 1000° C blackbody. The spectral photoconductivity was then corrected for intensity and energy differences, and the normalized photovoltage vs wavelength of exciting light was plotted.

The spectral photoconductivity of a thin ZnO film was first measured in vacuum at room temperature with no prior treatment. The film showed no photoconductivity in the range of 0.8 to 1.7μ. Next, the sample was cooled to 100° K, and curve A of Fig. 1 was obtained for the spectral photoconductivity. The sample was then irradiated 20 ESH and curve B was obtained at 100° K. Curve C shows the spectral photoconductivity at 300° K after 20 ESH irradiation.

The sample was then exposed to air at room temperature and, after recovery to its original resistivity, was reevacuated, and the experiment was repeated. Curve D illustrates the photoconductivity in vacuum immediately after recovery, and curve E shows the spectral photoconductivity after 80 ESH irradiation. Both curves were obtained for a sample temperature of 100° K.

Temperature Variation of Conductivity

After the sample chamber was evacuated to 10^{-6} torr, the sample was cooled to ~100° K and maintained at this temperature for approximately 15 min. The sample was then slowly warmed to a maximum temperature of about 400° K, and the sample resistance was measured as a function of temperature using a General Radio Model 1230-A electrometer. Usually, two or more runs were made without additional treatment to check the repeatability of the experiment. During the warming cycle of the first run, the chamber pressure usually increased to about 5 or 6×10^{-6} torr, but this pressure increase was not observed during successive runs. Consequently, the second run was usually taken as the control. The sample was again cooled to 100° K and irradiated for a few seconds with low intensity ultraviolet light. Such slight irradiation was sufficient because of the high sensitivity of the zinc oxide film. In general, an order of magnitude decrease in sample resistance at 100° K was attempted for each successive measurement cycle. After irradiation, the sample was held at 100° K in the dark for about 15 min to allow the resistance to stabilize. The sample was then warmed and the resistance measured as a function of temperature.

Figure 2 and curve A of Fig. 3 illustrate the temperature dependence of the conductivity of typical zinc oxide films in vacuum before irradiation. Figure 2 indicates that the conductivity is extrinsic with an activation energy, ΔE, of the order of 0.1 ev, as defined by $\sigma = \sigma \exp(-\Delta E/kT)$. Curve A of Fig. 3 indicates the presence of some type of trapping action, as evidenced by the shoulder in curve A for sample 726SH-1. Since the resistance of this film was so high, it was believed that this film may be amenable to thermally stimulated conductivity (TSC) measurements. Consequently, curves B, C, and D of Fig. 3 were taken after sample 726SH-1 was irradiated at 100° K for increasingly greater successive doses of ultraviolet radiation.

Temperature Variation of Mobility

The temperature variation of the mobility of a zinc oxide film prior to irradiation was not obtained because of the low mobility and excessive noise of the unirradiated film. The noise was a function of the sample resistance; by reducing the resistance with illumination, the noise was reduced to an acceptable level. The initial experimental procedure involved evacuating the sample chamber, cooling the sample to 100° K, and irradiating the sample to a resistance of about 10^7 ohm. The sample resistance was allowed to stabilize in the dark (15 min), after which the sample was slowly warmed and Hall measurements were made at selected temperatures. The magnetic field was reversed three times at each selected temperature, and the average of these readings was taken as the Hall voltage for that temperature. The voltage drop across the Hall probes was continuously recorded on a strip chart recorder. After attaining room temperature, the sample was again cooled to 100° K, irradiated to further reduce the resistance, and the above procedure performed. This cycling was continued until the resistance could not be reduced further at 100° K by irradiation with the 250-w mercury arc lamp.

Figure 4 shows the Hall mobility, μ_H, vs temperature, T, measured by this initial procedure, while Fig. 5 shows the electron concentration, n, vs 1000/T for the same data.

Figures 4 and 5 indicate that surface scattering and ionized impurity scattering are the dominant scattering mechanisms and that irradiation damage recovery rate increases as the sample is warmed.

Figure 6 shows the temperature variation of μ_H for the same film after irradiation and a 15-min recovery at 100° K (curve A), followed by heating to room temperature and recooling to 100° K once (curve B), and then heating to room temperature and recooling to 100° K twice (curve C). The recovery with temperature is evident in curve A.

Variation of Mobility with Photoexcitation

The Hall voltage measurements for this experiment were made in the manner as described in the previous paragraph, except that in this case the sample temperature was not varied. The sample resistance was varied by irradiating the sample with increasingly higher intensities of illumination. The Hall voltage was measured with the film in a quasisteady-state condition under illumination, whereas the temperature dependent data were taken with the sample in the dark. The intensity of irradiation was varied by using neutral density filters and by varying the focus of the ultraviolet light.

Figure 7 shows the variation of mobility and electron concentration with sample current as varied by irradiation. Both the mobility and the carrier concentration increased with irradiation as reported previously[5]. Figure 8 shows the variation of mobility with electron concentration and indicates a saturation of mobility with increasing electron concentration.

The photoconductivity data show that there exists an absorption at 1.0μ prior to UV irradiation, but that this absorption is greatly enhanced by the UV irradiation. Absorptions near 1.4μ and 1.7μ are induced by UV irradiation. The energy states associated with these three absorptions are believed to be donor states with energy levels approximately 1.2, 0.9, and 0.7 ev below the conduction band.

The photo-Hall effect studies show that both the mobility and the free electron concentration increase because of UV irradiation. The increase in electron concentration is believed to be due to the desorption of chemisorbed oxygen with the subsequent decrease in electronic energy band-bending and to the creation of additional zinc donors by lattice photolysis at the surface. The temperature dependence of the electron concentration in irradiated films indicated that the Fermi level for the irradiated films was either above the bottom of the conduction band or that the donor centers were so dense that impurity band conduction was significant.

The temperature dependence of the mobility indicated that the dominant scattering mechanisms for thin films in the temperature range of 100 to 300° K were surface scattering and ionized impurity scattering. The variation of the mobility with photoexcitation indicated that the UV radiation neutralized scattering centers. It is believed that this neutralization was accomplished by the elevation of the Fermi level and the subsequent population of ionized donor centers.

III. Electrical Properties of Crystals

Objectives

The three primary objectives of this work were 1) to study the effects of UV radiation on the free carrier concentration and mobility in zinc oxide crystals, 2) to look for energy levels created by or sensitive to UV radiation, and 3) to determine to what extent bulk electrical properties were affected by UV irradiation.

Experiment Procedures

Single, colorless crystals of ZnO were obtained from Minnesota Mining and Manufacturing Company in the form of a parallelopiped with typical dimensions of 10 × 2 × 2 mm with the long dimension along the hexagonal c-axis. Some of the crystals were oriented with the a-axis, but in this case, the maximum length was 5 mm. The specific electrical conductivity at room temperature was about 10^{-2} mho cm^{-1}. These crystals were grown by the vapor phase technique, and the emission sepctrographic data revealed that the major possible contamination could be that of copper and tin; however, the total impurity contents from these data were found to be less than about 10 ppm.

Before making electrical contacts, the crystals were lightly etched in 85% phosphoric acid. Good ohmic contacts were obtained by using indium or silver paste. No appreciable difference was observed in the two cases, though at low temperatures the latter type of contacts sometimes became noisy and nonohmic. Ultraviolet excitation of the samples in a vacuum of 10^{-6} torr was accomplished using a 200-w, high pressure mercury arc lamp without any filters.

For the Hall effect measurements, the conventional dc technique was employed using a Cary Model 3810 vibrating reed electrometer. A magnetic field of 5000 gauss was applied in each direction of the Hall points. The current through the crystal was always kept at a low value (in this case, less than 10^{-3} amps) to avoid the joule heating. The direction of current was frequently reversed. Sapphire substrates were used to give good thermal contact between the crystal and the cold finger of the cryostat for low-temperature measurements. A copper-constantan thermocouple was affixed to the top of the substrate, and the temperature was measured by digital equipment. The temperature range for our cryostat was from -170 to 120° C. For measurements at still higher temperatures (up to about 700° C), a special system was used that enabled us to keep the crystal in vacuum or in a controlled gaseous environment. The experiment was so arranged that the temperature variation of conductivity, volt-ampere characteristics, and time dependences of various parameters could be directly recorded on an X-Y plotter.

Temperature Variation of Electrical Parameters in Dark

From studies of the effect of vacuum, it was observed that in most cases when the surface of the crystal was not polished (or was deliberately damaged) or when the crystal possessed relatively larger room temperature resistivity (as in the case of some lithium-doped samples), the conductivity generally increased with evacuation but could be quickly recovered by passing air over the samples. However, quantitative reproduction of these results with respect to the ambient vacuum or the time of evacuation could not be obtained. Flow of nitrogen into the vacuum system did not influence the conductivity. It may be mentioned that Royal et al.[6] tried to probe into the feasibility of using ZnO thin films for monitoring the partial pressure of oxygen. In our case, we see no such possibility because pure, fairly polished crystals showed conductivity to be independent of the ambient pressure.

The temperature variation of conductivity for a number of specimens in dark is shown in Fig. 9. The dashed curve represents an impure sample and has been shown for the sake of comparison. From the Hall and thermoelectric measurements, we concluded that the conductivity was always the n-type.

Carrier concentration as a function of temperature is shown in Fig. 10. We have chosen the ratio of the Hall mobility to the true microscopic mobility as $3\pi/8$ for calculating the electron concentration from the Hall data, as is true in the case of acoustical mode scattering. Although, for a uniaxial crystal of type ZnO, the electrical conductivity tensor should have two components with differing values, parallel and perpendicular to the axis, our results on the electrical parameters were almost similar for c- and a-axis oriented crystals. This agrees with Hutson's[7] observations that the conductivity of ZnO is isotropic within the experimental uncertainty. The broken-line curve in Fig. 10 represents the typical electron concentration plot of Hutson's purest crystal.[8] The calculated average slope from $n \sim \exp(-E/kT)$ yields an activation energy of 0.18 ev.

It has been suggested, on the basis of the variation of the intensity of the luminescence and of the dipolar Debye absorption with the copper doping,[9] that a level of about 0.20 ev could be introduced by copper doping. As mentioned earlier, though our samples are relatively pure, they have some copper impurity, which may be generating a broad energy level in the neighborhood of 0.18 to 0.20 ev by entering the ZnO lattice, substitutionally as well as interstitially. However, further experimentation is necessary before a final assignment to the observed level can be made.

A typical curve for the dark Hall mobility as a function of temperature is shown in Fig. 11. Also, in the case of mobility, we did not see any appreciable crystal operational effect. It is interesting to note that Seitz and Whitmore[10] have observed a larger mobility in the a-direction than for the c-direction. This directional dependence has been attributed to the effective mass anisotropy for ZnO. However, in the absence of any quantitative data from the authors, we are unable to discuss these crystallographic orientation effects.

Analysis of our data shows that the mobility varies as $T^{-1.6}$ in the experimental range of temperature. This is in agreement with the values obtained earlier[11] on as-grown and heat-treated ZnO specimens. Although one expects a relation of the type $\mu \sim T^{-1.5}$ for pure acoustical mode scattering, it could be safely concluded that, in case of pure ZnO single crystals, lattice vibrations play a dominant role in the scattering process.

Temperature Variation of Electrical Parameters with Photo-excitation

The effects of ultraviolet irradiation upon the temperature variation of conductivity, electron concentration, and mobility are shown in Figs. 12, 13, and 14, respectively. Curves A were plotted from crystals in the dark, while curves B were recorded under constant UV irradiation in a vacuum of 10^{-6} torr. It is evident that the electrical parameters remain nearly unchanged in the higher temperature region. Upon lowering the temperature one observes increasing deviation between the values obtained in the dark and under excitation. The temperature at which such deviation starts to build up is largely dependent upon the excitation intensity, but the nature of the curves remains the same as shown.

Curves C were plotted after cutting off the UV excitation (after reaching a near saturation condition for the excitation) and keeping the crystals in the dark and under vacuum for about 1 hr. These results remain the same even when the crystals are placed in dry nitrogen. Electron concentration is found to be reversible at all temperatures, but mobility requires considerable time (sometimes as much as a few days) to repeat the initial temperature-dependence data. However, if the samples are opened to an oxygen or air environment after UV irradiation, curves A in all the cases could be readily traced back.

Discussion

The observed effect of photoexcitation upon the electrical parameters can be explained by suitably extending a model proposed earlier by us.[5] In this model, we considered that there is chemisorbed oxygen on the surface of untreated ZnO acts as an acceptor. Because of the trapping of electrons from the conduction band by the adsorbed oxygen, the electronic energy bands near the surface are bent upward.

Ultraviolet excitation generates free-hole pairs in ZnO at a rate strongly dependent upon their depth within the crystal. The generation function at the depth x, is given by

$$G(x) = \int_{\nu_i}^{\infty} (1 - R) \times \alpha(\nu)\, I(\nu) \times \exp(-\alpha x) \times d\nu$$

where α is the absorption coefficient for frequency ν, corresponding to the excitation wavelength; and I is the incident intensity at this frequency. The limiting condition for this relation is that the size of the crystal should be large compared to the penetration depth.

This condition is satisfied for zinc oxide for wavelengths less than 3800 Å. From the relation for generation function, it can be found that G(x) will be appreciable only within a surface layer of about 10^{-4} cm,[2] which approximates the diameters of particles of finely divided powder and the thickness of thin films, but is small compared to the dimensions of a single crystal.

The large negative surface in ZnO causes the photo-produced holes to migrate to the surface, causing neutralization of the absorbed oxygen. Band-bending is thereby quickly reduced, and results in a greater population of electrons in the conduction band as well as in the defect levels situated near the conduction band. However, the chances for detecting any small change in the carrier concentration by UV excitation of single crystals at toom temperature are fairly remote because, as mentioned earlier, the change takes place only in a surface layer of thickness, which is small compared to the bulk, and because the bulk electron concentration is comparatively larger at this temperature. The situation is different in the case of thin films and powders where electrical and other properties such as luminescence and electron paramagnetic resonance acquired after prolonged UV exposure should remain practically irreversible without opening the specimen to air. Our earlier results confirm this viewpoint.

At lower temperatures, the bulk electron concentration drops a few orders, as is evident from Fig. 10, and the concentration of the photogenerated carriers becomes comparable to it, or even larger. Therefore, upon lowering the temperature under constant photoexcitation, the net carrier concentration does not change as much as it would have by making measurements in the dark.

The mobility of carriers in UV-irradiated ZnO consists of two parts, one depending upon phonon scattering and the other as a result of the scattering from oxygen ion vacancy centers, which are created when neutralized oxygen is evolved from the surface under vacuum environment. This complex scattering gives rise to curve B of Fig. 14.

Upon cessation of the excitation, the oxygen ion vacancy centers are slowly filled by the oxygen available in the system, and the generation of the new electron-hole pairs in the surface layer of ZnO stops. However, the time of relaxation for the filling up of vacancy centers will be very extensive compared to the carrier recombination time, depending upon the ambient vacuum condition. This explains the typical temperature-dependence behavior of electron concentration and mobility (curves C of Figs. 13 and 14, respectively) when the measurements are made some time after cutting off the excitation. The carrier concentration is seen to return immediately to its initial value, whereas the mobility requires considerable time to repeat its behavior in the dark.

IV. Electron Paramagnetic Resonance and Magnetic Susceptibility Studies

Objectives

The objectives of these studies were to 1) determine the species responsible for paramagnetism developed by UV irradiation of zinc oxide powders, 2) determine the type of paramagnetism developed, 3) determine the symmetry of the developed paramagnetism, 4) determine the interaction of the developed paramagnetism with the lattice, and 5) determine whether or not the developed paramagnetism is a result of color center formation.

Experimental Procedure

Electron paramagnetic resonance (EPR) measurements were made on SP-500 ZnO powder using a Varian V4500-15 X-band spectrometer equipped with a 100-kilohertz modulation unit. The resonant frequency of the loaded cavity was in the range of 9.4 to 9.5 gigahertz and was measured with a Hewlett Packard 5245L electronic frequency counter having a 5255A plug-in unit. The static magnetic field was calibrated using a Varian F8-A nuclear magnetic resonance flux meter. Measurements at various temperatures between room and liquid nitrogen temperature were initiated employing a gas flow type cryostat in conjunction with an electronic temperature controller. Temperature was measured and monitored with a copper-constantan thermocouple and a Honeywell Electronik 15 temperature recorder. Calibration of the thermocouple and the recorder was made using standard methods. The results of measurements are summarized in the following paragraphs.

Results and Discussion

Figure 15 shows the room temperature first derivative of the absorption spectrum of the powder samples at atmospheric pressure, under vacuum, and upon irradiation in vacuum. The ZnO powder sample as such has a very small signal. The total integrated absorption intensity (hereafter called only intensity) of this signal increases upon evacuation to about 10^{-6} torr. This signal goes back to its original intensity level when oxygen is again admitted to the system. The peak-to-peak line width ΔH_{pp} of the signal decreases to 3.9 gauss in vacuum from about 5.1 gauss in air. The effect of the removal of the surface oxygen by evacuation seems to be twofold; it increases the concentration of the paramagnetic defect centers because of the reduction in the degree of band-bending and narrows the line because of the change in the interaction of the oxygen with the surface unpaired electrons. This interaction is probably dipolar in nature.

When the sample in vacuum is irradiated with UV light, the signal intensity increases considerably depending on the total amount of dose. The increase is rapid during the initial stage of irradiation and then slowly reaches a saturation for the particular radiation source. This EPR absorption spectrum consists of one line with no structure, as can be seen from Fig. 15c. The line is narrow having a peak-to-peak width of about 4.2 gauss and has a Lorentian shape, which is revealed by line-shape analysis. The symmetrical shape of this resonance line indicates that there is practically no g-factor anisotropy and is in agreement with that reported by others[12,13].

The lack of observable hyperfine splitting because of Zn^{67} ($I = 5/2$) merely indicates that the wavefunctions of the unpaired electrons overlap with several defect centers and are thus delocalized. The absence of hyperfine structure in our spectra verifies the result of others.[12,13] The effect of over-modulation as the cause of the absence of hyperfine structure was investigated. The modulation amplitude in our experiments was about 0.50 gauss, which is about 1/10 that of the line width. The effect of over-modulation was experienced beyond a modulation amplitude of about 2.0 gauss.

The g-value of this UV-irradiated signal is 1.956. In general it agrees with the results reported by others[12–19] and not with Kasai,[20] since the line is UV sensitive.

The temperature variation of the g-value and the line width for the UV-irradiated sample are presented in Figs. 16 and 17. Both g-value and line width are found to show some decrease with the lowering of temperature. The change in g-value is very small, and the peak-to-peak line width is about 4.3 gauss at 300° K and changes to 3.4 gauss at 93° K.

Since the line shape essentially remains the same throughout the whole temperature range of investigation, an estimate of temperature dependence of signal intensity can be made. A plot of the ratio of the signal intensity at different temperature to that at room temperature (I_T/I_{RT}) against temperature is shown in Fig. 18. The nature of variation of the ratio I_T/I_{RT} with temperature shows that the defect centers obey I/T law. This clearly shows that the paramagnetism is neither Pauli type[21] nor Van Vleck high frequency type.[22] This means that neither do the paramagnetic electrons behave like a degenerate gas nor do they have a diamagnetic ground state.

Absence of hyperfine structure caused by the overlap of the electronic wave functions strongly suggests the formation of a shallow donor band, and we believe that the signal at $g \simeq 1.956$ is because of delocalized electrons at a donor level.

The mean magnetic susceptibility for powders, and the susceptibility along and perpendicular to the crystallographic c-axis for single crystals were measured using a very sensitive electrodynamic microbalance.[23] The susceptibility of $CuSO_4 \cdot 5H_2O$ along its c-axis was taken as standard for these measurements.

Powders were packed in small quartz ampules, evacuated and sealed, and the susceptibility was measured before and after UV irradiation. The same procedure was adopted for the single crystals.

Our measurements on unirradiated and UV-irradiated ZnO powder and single crystals show the development of paramagnetism because of UV irradiation. Measurements on UV-irradiated single crystals show no observable anisotropy of the developed paramagnetism and support our observation symmetrical EPR resonance.

V. Luminescence Studies

Objectives

The primary objective of these studies was to determine the effect of UV radiation on the density and population of defect states that can be excited to produce luminescence and thus to gain information about the processes by which UV radiation degrades ZnO.

Experimental Procedure and Results

The photoluminescence of ZnO powder (SP-500, New Jersey Zinc Company) was studied for untreated, UV-irradiated, and carbon monoxide (CO)-reduced samples. Two different experimental configurations were used (Fig. 19). With the setup shown in Fig. 19a, 3660-angstrom light from an AH-6 high pressure mercury lamp was passed through a McPherson Model 225 monochromator and focused on the sample to stimulate luminescence. The luminous flux emitted by the sample was focused on the entrance slit of a second McPherson Model 225 monochromator. The luminescence spectrum was scanned by this second monochromator, detected by a sodium salicylate-coated EMI 9514 photomultiplier tube, and recorded on a strip chart recorder. Utilizing this first setup, no luminescence was measureable for untreated SP-500 in vacuum or in air.

Next, a sample of SP-500 was irradiated with the unfiltered radiation from an AH-6 lamp for 14 hr while the sample was in a dynamic vacuum of approximately 10^{-6} torr. Then, the in situ photoluminescence spectrum was recorded using the setup shown in Fig. 19a. Figure 20A shows this spectrum.

Next, the sample was exposed to air, and the photoluminescence spectrum was again recorded (Fig. 20B). It was found that the luminescence bleached to about one-half its original value in approximately 100 minutes.

Samples of SP-500 were also given a reducing treatment by heating them in an atmosphere of CO at 1000° C for approximately 1 hr. This treatment produces the reaction:

$$2\,ZnO + CO \rightarrow ZnO{:}Zn + CO_2$$

The luminescence spectrum of a typical sample after the reducing treatment is shown in Fig. 21. No quantitave bleaching data were taken for the reduced samples, but qualitatively, there was no pronounced bleaching.

It is seen that both UV irradiation in vacuum and high temperature CO reduction can produce green band luminescence in SP-500 ZnO. UV irradiation also produces a luminescence band in the near-ultraviolet region. Air was found to quench both the ultraviolet and the green luminescence.

Since the first experimental setup was not sensitive enough to detect luminescence in untreated SP-500 ZnO, measurements were made using the arrangement depicted in Fig. 19b. The increase in sensitivity was gained by using a Spectra Physics He-Cd UV-laser as the excitation source ($\lambda = 3250$ Å). Figure 22 shows the luminescence spectra of untreated ZnO under various conditions taken with this arrangement.

In the upper part of Fig. 22, the temperature of zinc oxide in a vacuum of 5×10^{-5} torr was varied from 293° K to 80° K, whereas in the lower part the pressure of zinc oxide at 293° K was changed from 760 torr to 5×10^{-5} torr. The figure reveals that decreasing the temperature causes a decrease in the luminescence intensity in all bands. Especially the luminescence intensity of the green band is strongly reduced. By decreasing the air pressure, the intensity of only the green band decreases, whereas the intensity of the ultra-ultraviolet band becomes enhanced.

Our results are in good agreement with measurements by W. Maenhout-Van der Vorst and F. Van Crayenest.[24] For their ZnO_B powder, they found such a weak emission that it was barely detectable by their instruments. However, at temperatures above 773° K, the luminescence intensity increased very strongly with temperature. We have found the same behavior for a temperature variation from 80° K up to room temperature (Fig. 22, upper part). The use of the CW-mode UV-laser in our experiments enabled us to measure this extremely low intensity luminescence at low temperatures.

Our results of the pressure dependence of the luminescence intensity can be explained by the luminescence mechanism proposed by W. Maenhout-Van der Vorst and Crayenest. In this model (Fig. 23), irradiation with ultraviolet light excites an electron from the valence to the conduction band (process 1). This electron is trapped by some defect having an energy level about 0. 8 ev under the conduction band. This electron may recombine with a positive hole in or near the valence band and the energy emitted is light of wavelength 500 nanometers (process 2). At an energy of 3. 11 ev under the conduction band, another level exists that is denoted as b-center. Ultraviolet light can also excite an electron from the b-center to the conduction band (process 3). A recombination of an electron from the a-center with a positive hole in the b-center results in light emission of energy 3. 11 ev-0. 8 ev = 2. 31 ev or a wavelength of 530 nanometers (process 4). The edge emission in ZnO is caused by electron transitions from levels near the conduction band to the valence band (process 5).

VI. Conclusions

A summary of the most important conclusions of this work follows:

1) IR photoconductivity indicates that IR sensitive defect centers are created by UV irradiation.

2) Thin film Hall studies show that UV irradiation raises the electron Fermi level above the bottom of the conduction band and that UV irradiation neutralizes surface scattering.

3) Crystal mobility studies demonstrate that UV irradiation produces ionized scattering centers.

4) EPR studies associate paramagnetism with delocalized, bound electrons.

5) Luminescence measurements demonstrate that luminescent defect levels are present in untreated SP-500 ZnO and that UV irradiation greatly enhances the luminescence and that air quenches the enhanced luminescence.

Our studies have shown that UV irradiation of ZnO results in the production and population of defect centers with energy levels near the conduction band and that these centers are sensitive to infrared radiation. UV irradiation also increases the free electron concentration to such a density that free carrier absorption in the near infrared region should become appreciable. The luminescence studies demonstrated that luminescent defect levels were present in untreated SP-500 ZnO and that UV irradiation enhanced the population and density of those levels. Any attempt to assoicate the UV and green bands with specific defects would be only speculation from our data. However, Dingle[25] has associated a green band in ZnO crystals with divalent copper impurities situated at zinc lattice sites. The EPR studies showed that the paramagnetic defect centers produced by UV irradiation are donor-type centers, with trapped delocalized electrons as the paramagnetic species.

References

[1]Lander, J. J., "Reactions of Lithium as a Donor and an Acceptor in ZnO," Journal of Physical and Chemical Solids, Vol.15, 1969, 11. 324-334.

[2]Blakemore, J. S., "A Model for Extraterrestrial Solar Degradation of ZnO," Transactions of the IEEE, AES-2, 1966, pp. 332-336.

[3]Sklensky, A. F., MacMillan, H. F., and Greenberg, S. A., "Solar Radiation Induced Damage to Optical Properties of ZnO-Type Pigments," TR 4-17-68-1, Feb. 1968, Lockheed Missiles & Space Co.

[4]Ryvkin, S. M., Photoelectric Effects in Semiconductors, Consultants Bureau, N. Y., 1964.

[5]Kroes, R. L. et al., "Studies of Ultraviolet Irradiation Effects in ZnO-Type Thermal Control Coating Pigments," AIAA Paper 69-639, San Francisco, Calif., June 1969.

[6]Royal, T. M., Wortman, J. J., and Monteith, L.K., "An Investigation of Thin Film Oxygen Partial Pressure Sensors," Rept. CR-1182. Sept. 1968, NASA.

[7]Hutson, A. R., "Electronic Properties of ZnO," Journal of the Physics and Chemical of Solids, Vol. 8, 1959, pp. 461-471.

[8]Hutson, A. R., "Hall Effect Studies of Doped Zinc Oxide Single Crystals," The Physical Review, Vol. 108, 1957, pp. 222-230.

[9]Pfahn, A., "Properties of ZnO Phosphors Doped with Li, Ni, and Cu," Journal of the Electrochemical Society, Vol. 110, 1963, pp. 381-384.

[10]Seitz, M. A. and Whitmore, D. H., "Electronic Drift Mobilities and Space-Charge-Limited Currents in Lithium-Doped Zinc Oxide," Journal of the Physics and Chemistry of Solids, Vol. 29, 1968, pp. 1033.

[11]Bogner, G., "Messunger der Elektrischen Leitfahigheit und des Halleffekts an ZnO-Kristallen und ihre deutung durch stirbander," Journal of the Physics and Chemistry of Solids, Vol. 19, 1961, pp. 235-250.

[12]Geisler, C. H. and Simmons, G. L., "High Temperature Induced EPR Signals in Zinc Oxide," Physics Letters, Vol. 11, 1964, pp. 111-112.

[13]Müller, K. A. and Schneider, J., "Conduction Electron Spin Resonance in Group II-IV Semiconductors and Phosphors," Physics Letters, Vol. 4, 1963, pp. 288-291.

[14]Schneider, J. and Rauber, A. "Paramagnetische resonanz von Donatoren in Zinkoxyd," Zeitschrift für Naturiorschung, Vol. 16a, 1961, pp. 712-714.

[15]Kokes, R. J., "The Influence of Chemisorption of Oxygen on the Electron Spin Resonance of Zinc Oxide," Journal of Physical Chemistry, Vol. 66, 1962, pp. 99-103.

[16]Baranov, E. V., Kholmogorov, V. E., and Terenin, A. N., "Photoinduced EPR Signals in Zinc Oxide," Doklady Akademii Nauk SSSR, Vol. 146, 1962, pp. 125-128.

[17]Sedaka, M. and Kivan, T., "The Adsorped State of Oxygen on Zinc Oxide as Revealed by the Electron Spin Resonance Method," Bulletin of the Chemical Society of Japan, Vol. 38, 1965, pp. 1414-1415.

[18]Fujita, Y. and Turkevich, J., "General Discussion of Adsorption and Chemisorption," Discussions of the Faraday Society, Vol. 41, 1966, pp. 394-412.

[19]Lal, R. B. and Arnett, G. M., "Electron Paramagnetic Resonance of Photosensitive Donors in ZnO," Journal of the Physical Society of Japan, Vol. 21, 1966, pp. 2743-2744.

[20]Kasai, P. H., "Electron Spin Resonance Studies of Donors and Acceptors in ZnO," The Physical Review, Vol. 130, 1963, pp. 989-995.

[21]Seitz, F., Modern Theory of Solids, McGraw-Hill, New York, 1940, p. 159.

[22]Van Vleck, J. H., Electric and Magnetic Susceptibilities, Oxford University Press, 1932, p. 145.

[23]Neogy, D. and Lal, R. B., "A Sensitive Magnetic Balance for Measuring Absolute Susceptibilities of Single Crystals and Dilute Solutions," Journal of Scientific and Industrial Research, Vol. 21B, 1962, pp. 103-107.

[24]Van der Vorst, W. M. and Van Crayenest, F., "The Green Luminescence of ZnO," Physica Status Solidi, Vol. 9, 1969, pp. 749-752.

[25]Dingle, R., "Luminescent Transitions Associated with Divalent Copper Impurities and the Green Emission from Semiconducting Zinc Oxide," Physical Review Letters, Vol. 23, 1969, pp. 579-581.

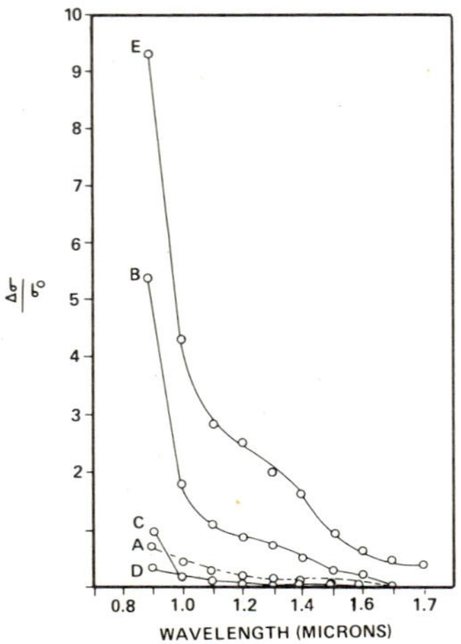

Fig. 1 Induced impurity photoconductivity.

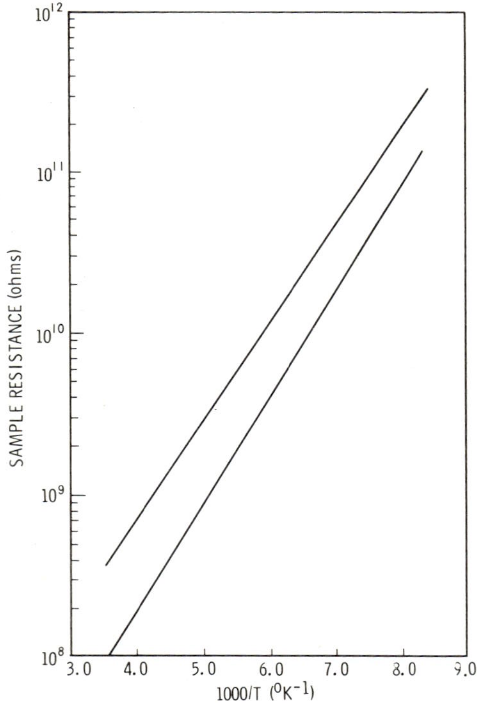

Fig. 2 Temperature dependence of conductivity of ZnO samples.

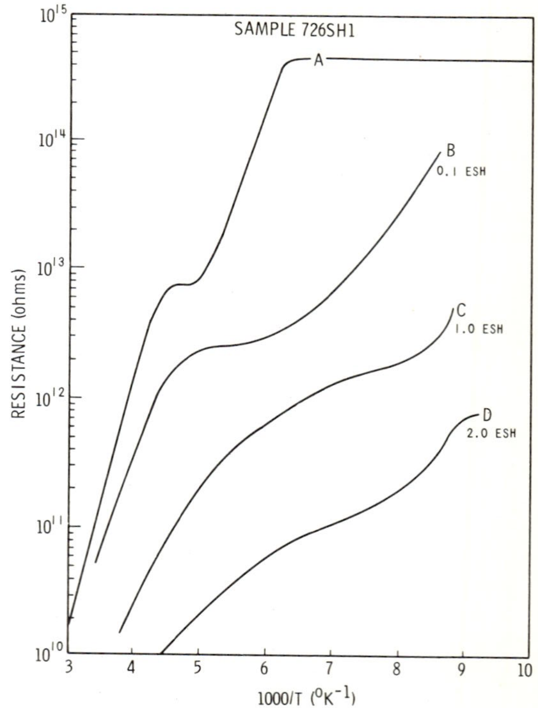

Fig. 3 Temperature dependence of conductivity after various doses of irradiation.

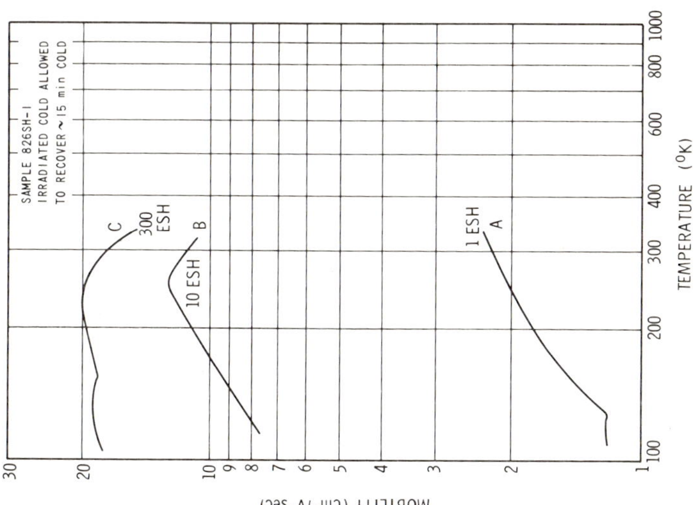

Fig. 4 Temperature dependence of Hall mobility.

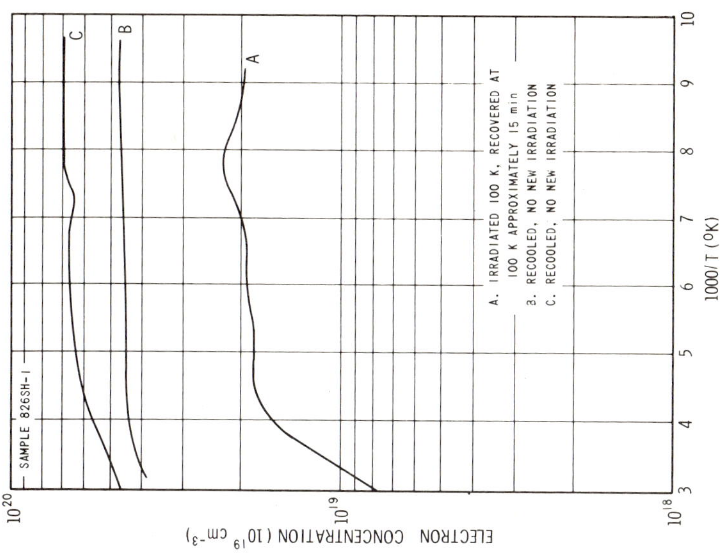

Fig. 5 Temperature dependence of electron concentration.

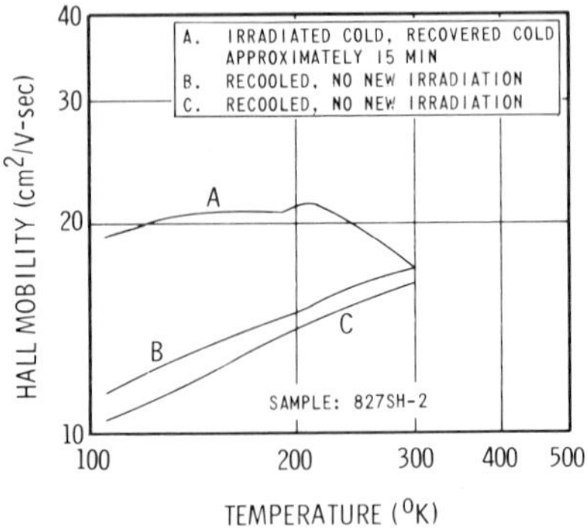

Fig. 6 Effect of temperature cycling on the temperature dependence of Hall mobility.

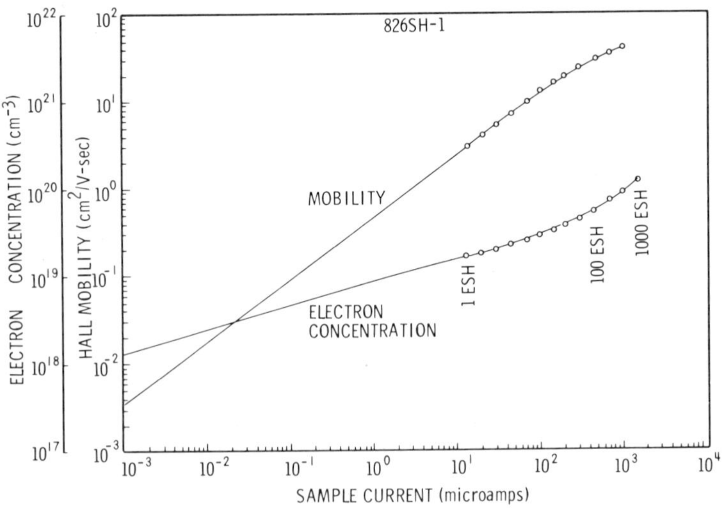

Fig. 7 Variation of Hall mobility and carrier concentration with photoexcitation current.

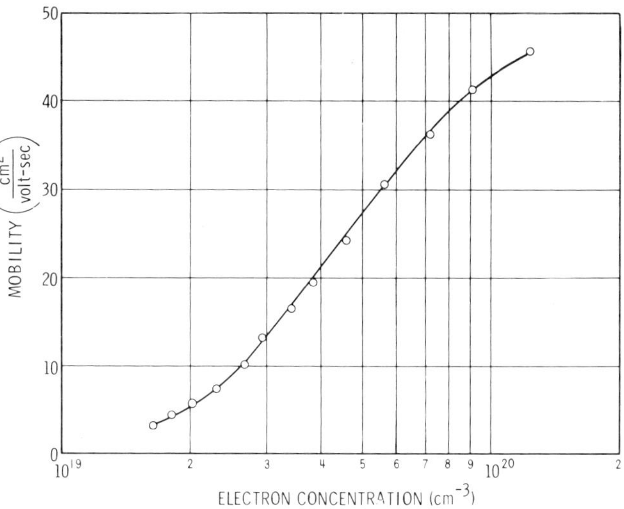

Fig. 8 Variation of Hall mobility with carrier concentration during photoexcitation.

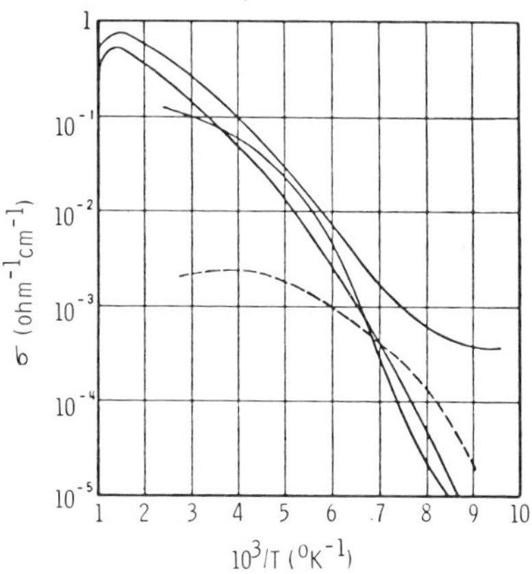

Fig. 9 Dark conductivity of pure ZnO crystals as a function or reciprocal temperature (dashed curve is for an impure crystal).

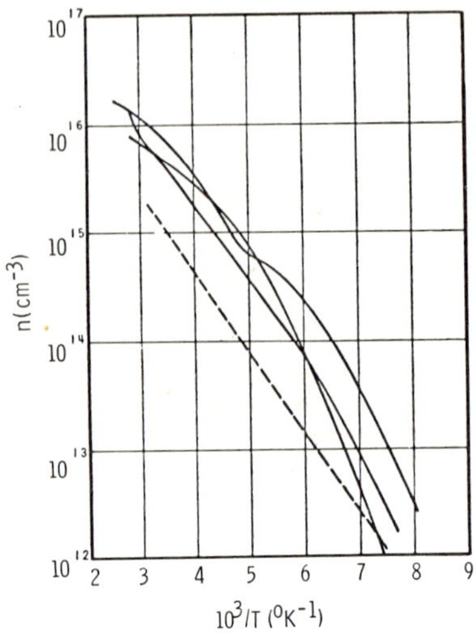

Fig. 10 Electron concentration vs T^{-1} for three pure ZnO samples in dark (dashed curve is after Hutson[7] and represents the temperature dependence for n for his purest crystal).

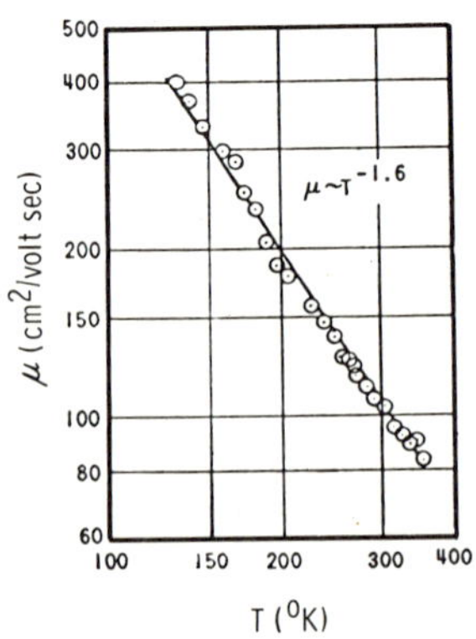

Fig. 11 Temperature variation of dark Hall mobility for a typical ZnO specimen.

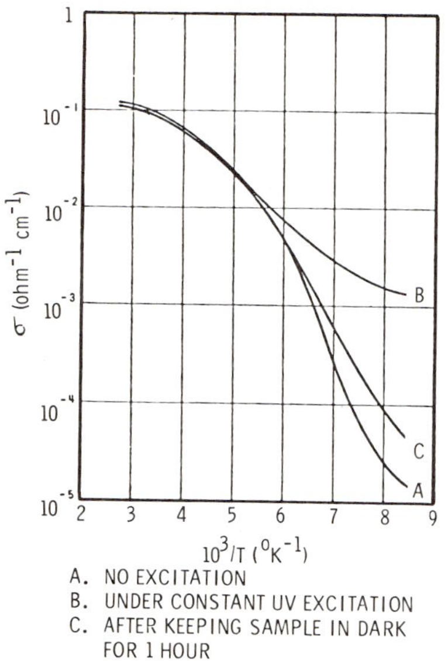

Fig. 12 Temperature variation of electrical conductivity. Curve A, no excitation. Curve B, under constant UV excitation. Curve C, after keeping sample in dark for 1 hr.

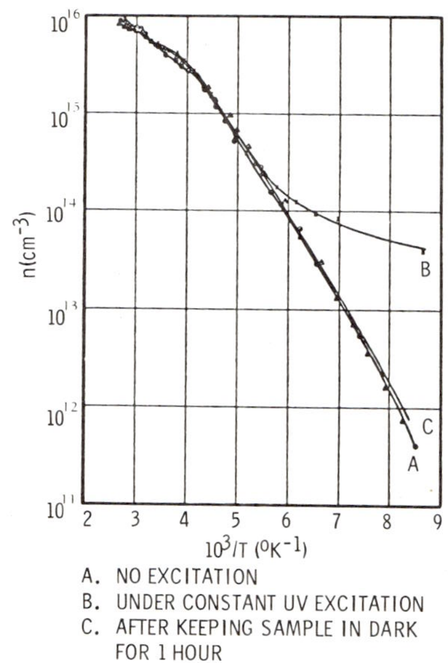

Fig. 13 Temperature variation of carrier concentration. Curve A, no excitation. Curve B, under constant UV excitation. Curve C, after keeping sample in dark for 1 hr.

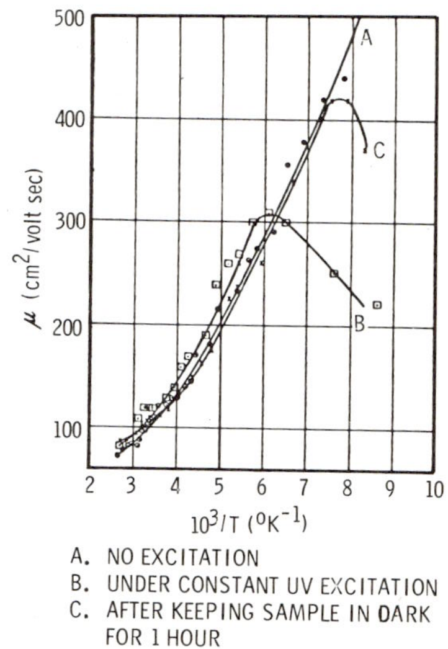

Fig. 14 Temperature variation of Hall mobility. Curve A no excitation. Curve B, under constant UV excitation. Curve C, after keeping sample in dark for 1 hr.

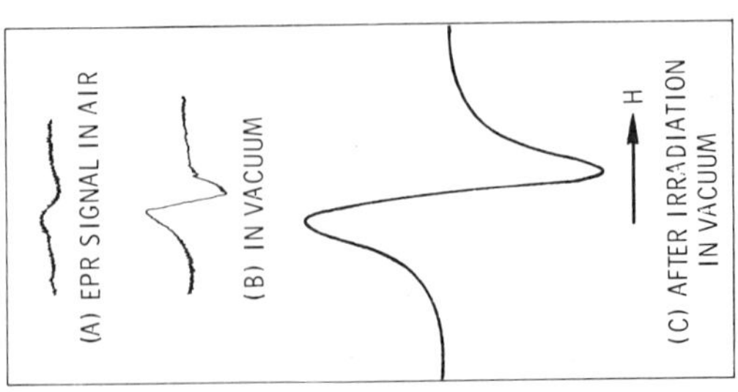

Fig. 15 EPR signal. a) EPR signal in air; b) In vacuum; c) After irradiation in vacuum.

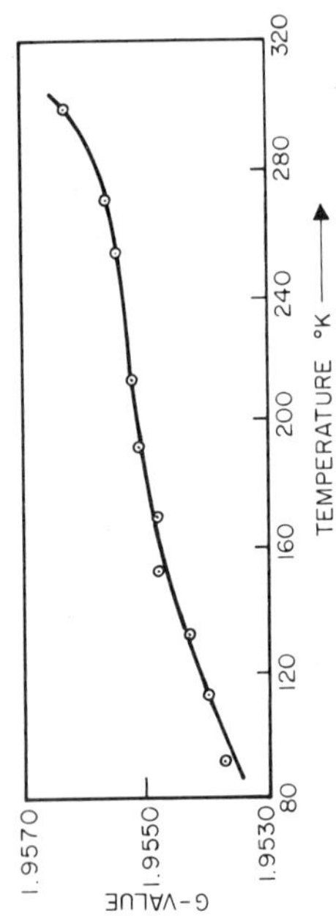

Fig. 16 Temperature variation of g-factor.

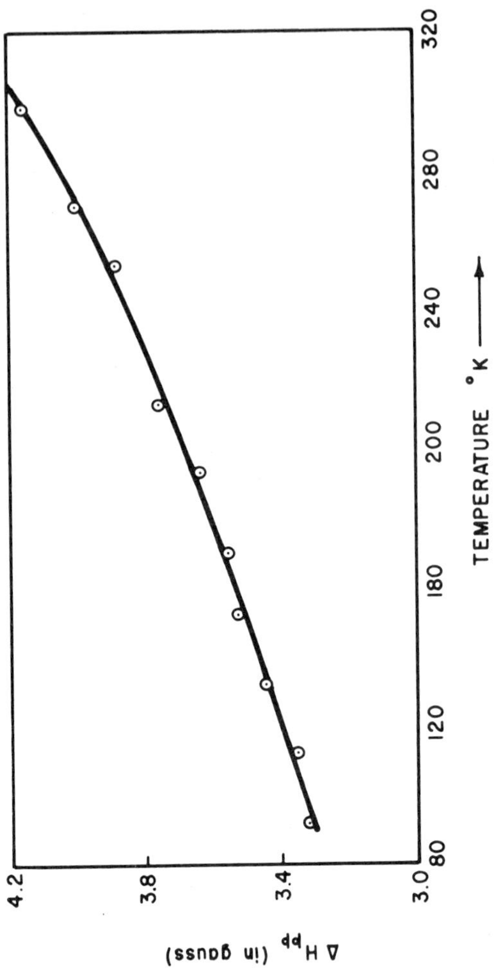

Fig. 17 Temperature variation of line width.

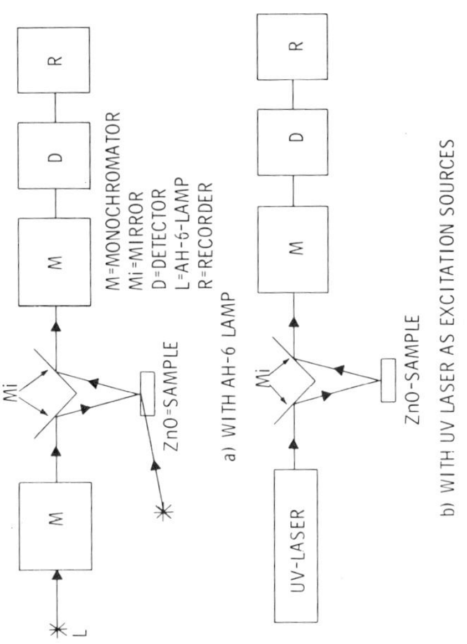

Fig. 19 Experimental setup for luminescence measurements. a) With AH-6 lamp; b) With UV laser as excitation sources.

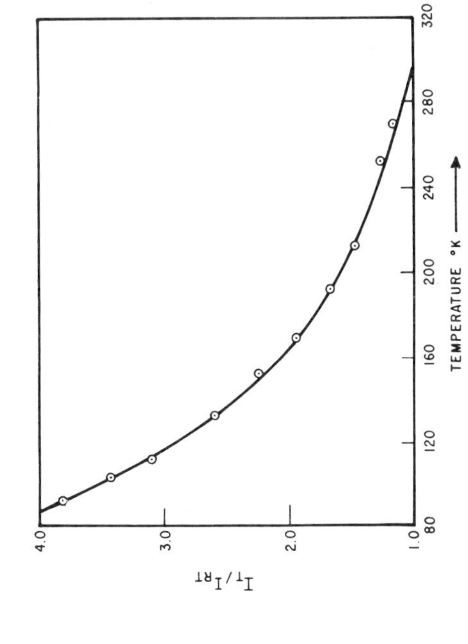

Fig. 18 Temperature variation of signal intensity.

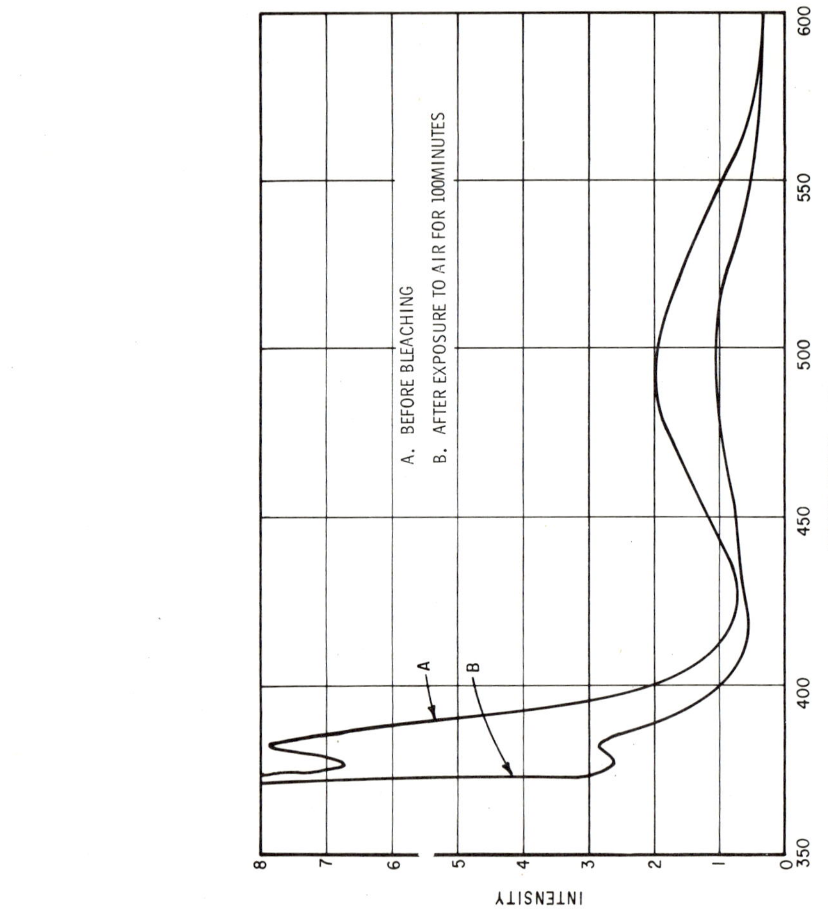

Fig. 20 Luminescence spectra of UV activated ZnO.

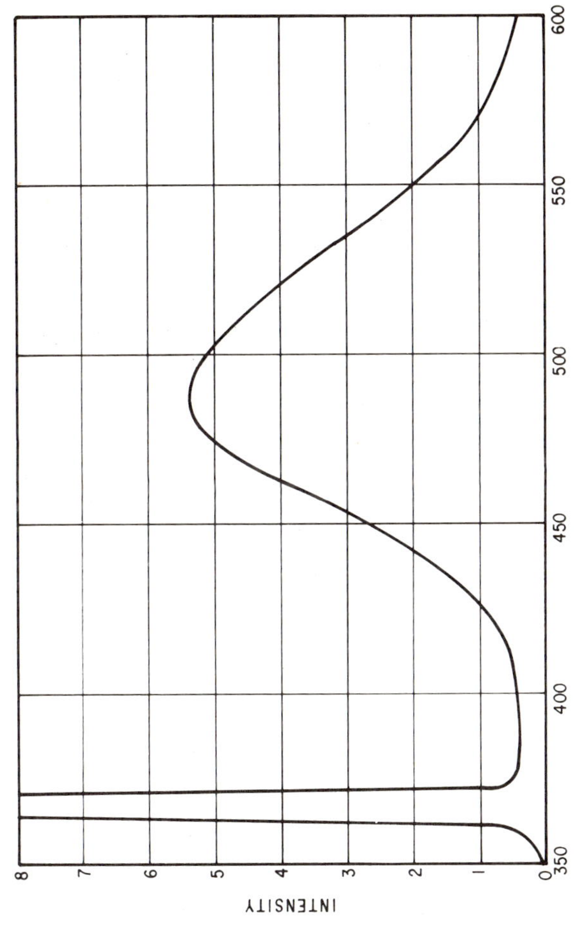

Fig. 21 Luminescence spectrum of CO reduced ZnO.

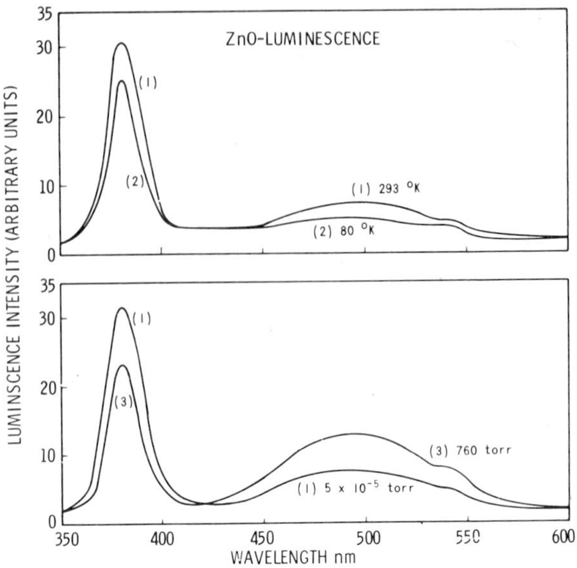

Fig. 22 ZnO luminescence after UV laser excitation. a) Temperature dependence; b) Air-pressure dependence.

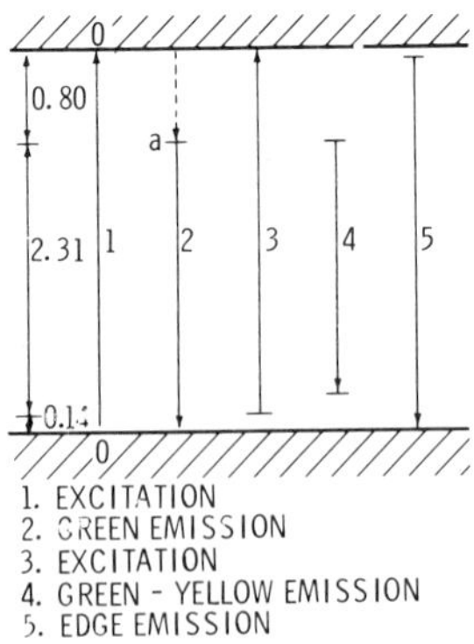

Fig. 23 Luminescence model for ZnO. 1 Excitation. 2 Green emission. 3 Excitation. 4 Green-yellow emission. 5. Edge emission.

SURFACE RECOMBINATION CENTERS AS PROTECTION AGAINST VACUUM PHOTOLYSIS OF THERMAL CONTROL COATINGS

S. Roy Morrison* and K. M. Sancier†

Stanford Research Institute, Menlo Park, Calif.

Abstract

Photoproduced holes and electrons in an oxide pigment can cause chemical changes in the pigment or the vehicle and thus cause discoloration of a coating. This paper reports studies of recombination center additives to prevent such discoloration. Surface recombination centers are chemical species (energy levels) provided at the semiconductor surface, which can recombine the energetic hole-electron pairs before they can cause chemical changes. Phases of the program briefly discussed are: 1) identification of chemical redox couples having good characteristics for recombination centers on ZnO; 2) tests with the ferro/ferricyanide couple for photostabilization of ZnO as detected by conductance changes on single crystals; 3) tests with this redox couple of photostabilization of ZnO powders as detected by electron spin resonance; and 4) preliminary tests of photostabilization as detected by optical degradation. It is observed that the degradation rate can be decreased many orders of magnitude by an appropriate surface additive.

Presented as Paper 70-830 at the AIAA 5th Thermophysics Conference, Los Angeles, Calif., June 29-July 1, 1970. This research was supported by the Jet Propulsion Laboratory, California Institute of Technology, sponsored by NASA under Contract NAS 7-100. We acknowledge valuable discussions with W. Carroll of Jet Propulsion Laboratory concerning optical degradation of thermal control coatings and in the planning of this program. We acknowledge also the assistance of J. Schutt of NASA Goddard Space Flight Center, and M. J. Brown of Electromechanical Research Company in performing the preliminary optical degradation tests.

* Senior Physicist, Surface Physics and Chemistry Group.

† Senior Physical Chemist, Surface Physics and Chemistry Group.

Introduction

When ultraviolet is incident on oxide pigments, particularly in a vacuum environment, chemical changes occur that affect the optical absorption or the coloration of the coating. We have been engaged in a research program with the objective of using surface recombination centers to retard these chemical changes. This contribution summarizes the program to date.

The only direct action of incident bandgap light on a semiconductor is to create hole-electron pairs. If such photoproduced electrons and holes are mobile, they can migrate to the particle surface and chemically react with surface species. Surface species thus may be reduced by the photoproduced electrons or oxidized by the photoproduced holes[1] and, particularly if the product species is volatile in its new oxidation state, e.g., oxygen, the chemical change can be permanent. The function of surface recombination centers is to intercept and recombine hole-electron pairs before they can cause such chemical changes. In order to fulfill this function, a recombination center by our requirement is an intentional additive M which has a high capture cross section for holes and electrons, and which can undergo the reactions

$$M^n + p \rightarrow M^{(n+1)} \quad (1)$$

$$M^{(n+1)} + e \rightarrow M^n \quad (2)$$

(where p represents a hole, e an electron, and where n is the valence of the additive ion) where both M^n and M^{n+1} are stable nonvolatile species. It is seen from Eqs. (1) and (2) that after capture of an equal number of holes and electrons there is no net chemical change at the surface.

Specifically, our objectives have been to find a satisfactory additive for ZnO and to develop a theoretical understanding so that suitable additives for other pigments can be selected. In order to pursue these objectives, the program was divided into several parts: 1) studies of hole capture by additives, 2) studies of electron capture by additives, and 3) studies of hole-electron recombination and of photolysis. Each of these will be discussed briefly.

Hole Capture Cross Section of Various Species M^n

Studies have been made of hole capture by reducing agents M^n dissolved in aqueous solution, using an electrochemical technique. The technique depends on an observation termed "current doubling."

The current-doubling effect originates as follows. A single crystal wafer of n-type ZnO is used as the anode of an electrochemical cell and the wafer is illuminated to produce holes. These holes migrating to the surface produce a certain anodic current. It is found that if a two-equivalent species such as methanol is added to the solution, the photocurrent almost doubles. A reaction of the form

$$MeOH + p \rightarrow MeOH^{+} \tag{3}$$

$$MeOH^{+} \rightarrow \text{products} + e \tag{4}$$

is presumed to occur. Thus for every photoproduced hole reaching the surface and oxidizing the methanol by one equivalent, an electron is injected by the unstable intermediate to provide the second equivalent of oxidation, and the current doubles. One-equivalent species do not cause a current increase, as hole capture forms a stable species.

The phenomenon of current doubling can be used to compare the hole capture cross section for various one-equivalent species.[2] To make such a comparison, one determines the concentration of one-equivalent species necessary to capture half the photoproduced holes [thus causing the excess current of Eq. (4) to decrease to half its value]. If a low concentration of ions is sufficient, the capture cross section must be high.

As discussed above, for photostability of the ZnO we require an additive that captures holes rapidly [as in Eq. (1)]; that is, one with a high hole capture cross section. Using the current-doubling technique to compare chemical species, it was found that the higher the redox potential of the couple M^{n}/M^{n+1}, the higher the hole capture cross section. This rule may be valid for other oxide semiconductors. The observation and analysis of the phenomenon of current doubling [Eqs. (3) and (4)] suggests, in addition, that a two-equivalent molecule R is unlikely to have a high capture cross section for carriers in both reactions (5) and (6)

$$R^{n} + p \rightarrow R^{n+2} + e \tag{5}$$

$$R^{n+2} + 2e \rightarrow R^{n} \tag{6}$$

and thus a two-equivalent species will seldom be a good recombination center. Therefore for an efficient recombination center, the species must be one-equivalent, that is, both forms (M^n and M^{n+1}) in Eqs. (1) and (2) must be stable but interconvertible.

Electron Capture Cross Section of Various Species M^{n+1}

Studies have been made of electron capture by oxidizing agents,[3] as required by Eq. (2), again using an electrochemical technique with the oxidizing agent in aqueous solution. The electron capture cross section σ is obtained from the relation Eq. (7) between the cathodic current J and the electron density n_s at the ZnO surface:

$$J = \sigma \, [M^{n+1}] \, \bar{c} \, n_s \tag{7}$$

Here σ is the capture cross section, $[M^{n+1}]$ is the effective surface density of the ions of interest, and $\bar{c}$ is the thermal electron velocity which has a value about 10^7 cm/sec. We determine the value of n_s by measuring the capacity of the depletion layer at the ZnO surface (using an impedance bridge), and utilizing a simple relationship between capacity and n_s.[1] Then the value of $\sigma[M^{n+1}]$ for various ions can be obtained from Eq. (7), and assuming reasonably constant $[M^{n+1}]$ at constant concentration of oxidizing agent, the values of σ can be compared. The limitations of the assumption of constant $[M^{n+1}]$ are discussed in detail in Ref. 3.

From these results and a theoretical analysis, a model has been developed relating the redox potential of the couple M^n/M^{n+1} to its surface energy level, and it has been found that species with an energy level just below the conduction band edge of the solid have the highest cross section both for electron capture and for hole capture.

As a result of these individual studies of hole capture and electron capture, the couple ferricyanide/ferrocyanide [Fe(III)/Fe(II)] was chosen as an appropriate additive for ZnO, and tentative rules were deduced for selection in the general case of an efficient recombination center: first, it must be one-equivalent; second, it must have an energy level near but below the conduction band of the pigment; and third, it must be deposited in both oxidation states. In addition, of course, it must be nonvolatile and reasonably stable to uv.

Hole-Electron Recombination by Iron Cyanide on ZnO

To test the efficiency of the Fe(III)/Fe(II) additive as a recombination center, chemical changes were monitored in the zinc oxide. One method used to follow such chemical changes was the dark conductance of ZnO (in the form of Li-doped single crystals). Changes in dark conductance caused by illumination in vacuum have been associated in earlier work with photolysis[4] and changes in optical properties.[5] The results of our study were encouraging.[6] With less than a monolayer of additive present, the initial rate of conductance change decreased up to five orders of magnitude. The "conductance degradation" rate was found to be lower by a factor of 10^2 or more with both Fe(II) and Fe(III) present than with either Fe(II) alone or Fe(III) alone, at the same total concentration (10^{16} ions/cm^2).

Figure 1 shows the conductance degradation rate and the approximate quantum yield for photodamage (donors produced per uv photon) as a function of time with various coverages of iron cyanide on the surface. The top curve shows the highest rate of damage with no iron cyanide present. It is seen that the quantum yield for damage rapidly decreases as the recombination centers are added.

The action of the additive as a simple recombination center was apparently masked during the initial uv exposure by a side reaction,

$$\mathrm{Fe(II)} + \mathrm{O_2(gas)} = \mathrm{Fe(III)} + \mathrm{O_2^-(ads)} \qquad (8)$$

which means that initially the "reducing agent" was sorbed O_2^- rather than Fe(II). After a short exposure to light in vacuum, however, the O_2^-(ads) was photodesorbed, and the normal action of Fe(II) and Fe(III) as a recombination center was observed. In Fig. 1 this transition occurs at about 10^3 min. After 10^3 min, analysis of the results shows consistency with Eqs. (1) and (2) for the Fe(II) and Fe(III) couple, and from Fig. 1 a concentration of 5×10^{15} ions/cm^2 reduces the rate of photolysis by a factor of 500.

Another method used[7] to follow chemical changes upon uv/vacuum exposure, on powders, was ESR. Using the $g = 1.96$ line, as a measure of conduction electrons, to follow the photodegradation of the ZnO, we found partial stabilization with as little as 10^{-5} monolayers of additive. The results are shown in Fig. 2. The effect of composition of the additive was also investigated. The resonance line produced by uv irradiation was composed of two parts with g values of 1.9600 and 1.9564. The 1.9600 line is unaffected by oxygen exposure as is the visible damage (optical degradation which appears in the visible region of the ZnO absorption spectrum), and the 1.9564 line is affected by oxygen as is the infrared damage. That is, oxygen exposure has little effect on either the 1.9600 line or the visible damage, but almost removes both the 1.9564 line and the infrared damage.

The qualitative aspects of the conductance degradation and ESR techniques are in agreement. The difference in the additive coverage required may be related to the difference in behavior of undoped powder samples and Li-doped single crystals. The fact that the additive is apparently so much more effective on powder ZnO pigment than on the Li-doped monocrystals is tentatively assumed to be due to the damaged structure (high dislocation density) of the single crystals after lithium doping. Such damage causes a very high rate of degradation, as confirmed by ESR in separate experiments, so a much larger concentration of recombination centers is presumably required to provide stability.

Preliminary results on the use of Fe(III)/Fe(II) additive to prevent optical degradation of ZnO indicate that with 5×10^{14} centers/cm^2 of iron cyanide, the sample shows no changes in absorption spectrum after 350 sun hours.

Conclusions

It is apparent that the use of recombination center additives as protection against photolysis of coatings is a very promising concept. We have shown, with single crystal studies, that the additives behave in accordance with quantitative expectations. We have shown, with ESR studies, that the additives are also effective on powder, and not limited to single crystals. Experiments on optical degradation of powders support this conclusion. However, it is apparent that additional work is required to develop a practical system. The next step is to determine the effectiveness of the additive when a vehicle is present. In particular, systems must be chosen to minimize vehicle-additive interactions. Techniques of uniform additive deposition must also be explored in order to develop effective coatings.

References

[1]Morrison, S. R. and Freund, T., "Chemical Role of Holes and Electrons in ZnO Photocatalysis," The Journal of Chemical Physics, Vol. 47, Aug. 1967, pp. 1543-1551.

[2]Gomes, W. P., Freund, T., and Morrison, S. R., "Chemical Reactions Involving Holes at the Zinc Oxide Single Crystal Anode," Journal of the Electrochemical Society, Vol. 115, Aug. 1968, pp. 818-823.

[3]Morrison, S. R., "Electron Capture by Ions at the ZnO/ Solution Interface," Surface Science, Vol. 15, 1969, pp. 363-379.

[4]Collins, R. J., and Thomas, D. G., "Photoconduction and Surface Effects with Zinc Oxide Crystals," Physical Review, Vol. 112, Oct. 1958, pp. 388-395.

[5]Lal, R. B., and Arnett, G. M., "Effect of Ultra-Violet Radiation on the Electrical Conductivity of Zinc Oxide Single Crystals," Nature, Vol. 208, Dec. 1965, p. 1305.

[6]Morrison, S. R., "Iron Cyanide as a Surface State to Prevent ZnO Photolysis in Vacuum," Journal of Vacuum Science and Technology, Vol. 7, Jan. 1970, pp. 84-89.

[7]Sancier, K. M., "ESR Investigation of Photodamage to Zinc Oxide Powders," Surface Science, Vol. 21, No. 1, June 1970, pp. 1-11.

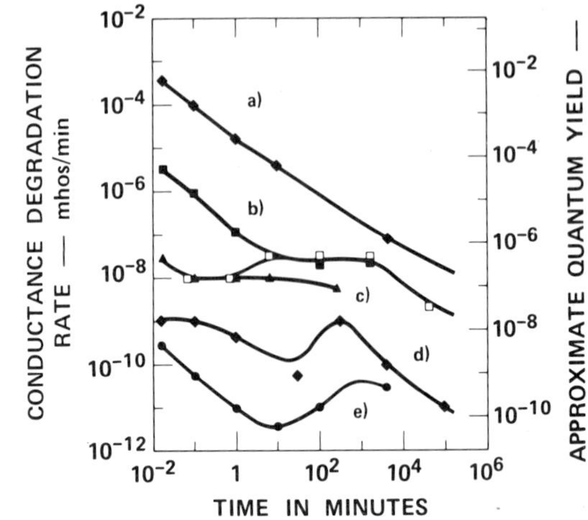

Fig. 1 Conductance degradation of ZnO due to uv illumination in vacuum. The sample was pretreated at room temperature in 0.5 torr O_2. Curve a), no iron cyanide. Curve b), $5x10^{14}$ ions/cm^2 each of ferrocyanide and ferricyanide. The two initial branches of this curve correspond to two different pretreatment times (lower curve 65 hr, upper curve 16 hr). The iron cyanide additions for the rest of the curves are: curve c), $10^{15}/cm^2$; d), $5x10^{15}/cm^2$; e), $10^{16}/cm^2$.

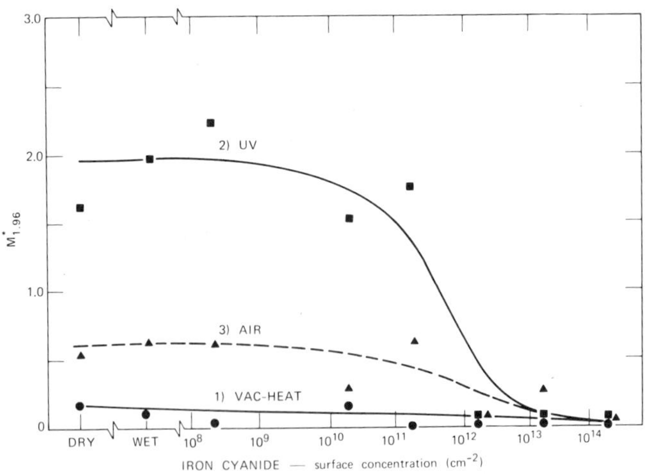

Fig. 2 Effect of iron cyanide concentration at ferri/ferro ratio = 1 on the spin density of the g = 1.96 line of ZnO. Curve 1): after heating at $525^{0}K$ in air 1 hr and then in vacuum 10 min.; curve 2): after uv irradiation for 1 hr; curve 3: after short exposure to air at room temperature.

A THEORETICAL AND EXPERIMENTAL STUDY OF LIGHT SCATTERING IN THERMAL CONTROL MATERIALS

J. E. Gilligan* and J. Brzuskiewicz+

IIT Research Institute, Chicago, Ill.

Abstract

A comprehensive model relating spectral reflectance to the fundamental properties of materials and to environment-induced changes in them was formulated in general terms. Spectral reflectance of highly characterized (BET, optical and electron microscopy, etc.) powders was measured. The structure of these paint films was investigated using scanning electron microscopy. A strong correlation between spectral reflectance and film structure (particle size, void size, agglomerate size, etc.) is evident, but not with particle size, as such. The sizes and shapes of particles, aggregates of particles, and of the void spaces between them, when their dimensions are roughly equal to those of incident wavelengths, have a maximum effect. Scattering by pigmentary particles intensifies the net absorption effect; voids tend to mitigate it.

Presented as Paper 70-831 at the AIAA 5th Thermophysics Conference, Los Angeles, Calif., June 29-July 1, 1970. This work was performed for NASA Langley Research Center under contract NAS1-8166. W. S. Slemp acted as the NASA technical program manager. The authors are indebted to G. A. Zerlaut for his technical advice and consultation. Others also deserving our appreciation include S. Shelfo, sample preparation; M. Jackson, fine particle analyses and consultation; G. Yamate, electron microscope and particle geometry analyses; O. Johari and I. Corvin, Scanning Electron Microscope; B. Boldt, DTA/TGA analysis; and G. Kimura and L. Gillilan, irradiation and gas adsorbate tests and measurements.

*Group Leader, Thermal Control, Polymer Chemistry Research.

+Research Chemist, Polymer Chemistry Research.

I. Introduction

Historically, the major task in developing low solar absorptance, pigmented, thermal control coatings ("white paints" for spacecraft temperature control) has been in making them resistant to the effects of the hostile elements of the space environment, notably ultraviolet (uv) radiation and charged particles. Increasingly, in the last few years the approach to developing stable coatings has involved fundamental studies of mechanisms of degradation. An approach that we have used,[1,2] for example, has been to elucidate mechanisms of ultraviolet degradation in specific materials and then to devise protective schemes. This has been accomplished successfully with both zinc oxide and zinc orthotitanate, both large band gap semiconductor pigments. These pigments have an ultimate α_s limit of the order 0.13 - 0.15.

In this paper we have considered the broadest aspects of reflectance and reflectance stability. An important area of investigation was the role of pigment particle size in determining spectral reflectance and in its effect upon reflectance stability. Of the many parameters controlling or affecting reflectance, the basic processes of light scattering in dense media were the least understood - both basically and in terms of their role in real paint systems. The initial efforts consequently were devoted to the development of mathematical models relating reflectance to the extrinsic (environmental) and intrinsic (materials-related) parameters. These models are examined and discussed briefly.

The context of the experimental work was in relation to the theoretical models: the role of light scattering in the overall model. A main objective was to determine whether the parameters that intensify spectral reflectance also intensify the absorption effect of an induced color center. By appropriate choice of parameters the reflectance of a paint system may be increased in selected spectral regions, but it is of major importance to understand the effects on other desired properties of the paint system, especially its environmental stability. Thus, it is important to understand not only the role of each major parameter but the relationships between them.

II. Theoretical Models

In a broad sense the ultimate objective of the program described in this paper was to develop a general model for predicting space-induced changes in solar absorptance of thermal radiative coatings, a model which accounts for reflectance in terms of both elementary and macroscopic properties of materials. Such a model, from a theoretical point of view, should relate gross spectral reflectance to at least those characteristic parameters of materials' properties to which it is most sensitive.

The general model, as we envision it, consists of three inter-related specific models. The first, one which determines the spectral reflectance properties of a dispersed pigment vehicle system, relates measurable macroscopic parameters to basic materials properties. The second model calculates the changes in basic optical properties induced by the space environment; in this case mechanisms specific to the individual components of the paint system and the damage kinetics (which are determined by both environment parameters and materials properties) are specified and related. This second model therefore predicts the rates of accumulation of all environment-induced defects which are significant in terms of solar absorptance, α_s. The third model estimates the changes in spectral reflectance which result from the additional absorption, that is, due to the combined effect of each of the individual types of induced defects. The general model has been fully developed in Ref. 3, and is in part based on the theory developed in Ref. 4.

III. Materials and Materials Characterization

A. Basic Objectives

The prime objective in the experimental studies was to determine the role of particle size in initial reflectance and in optical stability - with emphasis on the former. Particle size, however, is a very elusive quantity; it is a term which has many derivations and many interpretations. Depending on the material, particle size may have an inherent lower limit, one arising from the resistance of surface forces to further subdivision and a complementary tendency to reassociate. Therefore when particles are made microscopically small, they tend to reassociate - to aggregate into larger "chunks." To distinguish the optical (spectral reflectance) effects which are due to particle size (d_p) alone, those which are due to their aggregates (d_A), and those due to the voids between them (d_v) was thus the major objective.

The effects of these "size" variables on initial reflectance and particularly on reflectance stability can be ascertained provided the geometry characterizations can be made successfully. Their effects on reflectance and its stability can determined only if both the physical and chemical characterizations of the material are the same. Thus we required that the materials be as chemically identical as possible - that they differ in particle size and only in particle size.

B. Pigment Materials - Criteria and Selections

The four pigments selected for study in this program were Alumina (Al_2O_3), Silica (SiO_2), Zirconia (ZrO_2) and Lanthana (La_2O_3). The first two were selected mainly because of their extensive literature backgrounds and their potential use as thermal control pigments. Again, with the exception of La_2O_3, each also has been widely used in commercial coatings. Zirconia was chosen because of its potential stability to ultraviolet radiation. The resistance of La_2O_3 to the effects of protons and its poor ultraviolet behavior prompted its choice. Other materials were also used - mainly zinc sulfide (ZnS) - because of special absorption properties which could be used to demonstrate the effects of scattering. Semiconductor pigments (e.g., ZnO and TiO_2) were excluded from consideration, since the ultimate objective was a very low α_s/ε paint pigment. Organic materials, except test vehicles, were also excluded.

C. Materials Characterization

1. General Remarks. We believe that it is highly important to define terminology to avoid confusion. Particle size, for example, can be understood in many different ways. The terms particle, aggregate and agglomerate are very frequently used. The term particle, however, is usually qualified by one of the words: single or discrete. The meanings in this paper are nearly identical; they imply that the particle is a single physical entity and does not display any obvious signs of further subdivision. Our photomicrographic measurements refer in every case to an average length value corresponding to the longer of two planar coordinate dimensions; in the case of a sphere, it is the diameter; and in cubic materials, it is the length of a side. Shapes vary so greatly in some materials that their size measurements can not really be regarded as objective when only one size value is cited.

An agglomerate and an aggregate are essentially but not really the same thing - a conglomeration of single particles held together by mutual attraction, Van der Wall forces, etc. A distinction is made between the two in that the term aggregate applies to any multiparticle mass whose formation history is unknown. An agglomerate is an aggregate formed by aggregation of discrete particles and/or smaller aggregates under the influence of electrostatic forces, mechanical dispersion, evaporation of the media in which they existed, or similar external forces tending to bring them into intimate contact. These terms, of course, also refer to processes and they thus have action meanings, which are likewise distinguished.

Particle shape probably has the most subjective definition of any of the terms we will use. In general, the term shape refers to a three dimensional geometrical description of a particle or an aggregate. The subjectiveness enters into the definition through the interpretation of the various measurements which can be made. Optical determinations, the most direct, are two-dimensional; other techniques measure shape-related quantities but not shape itself. In this paper we have determined shape primarily on the basis of scanning and transmission electron microscope photographs.

2. Size Analysis. The main emphasis of this phase of work was to obtain information on the particle size and shape of the various coating materials. An Hitachi HS-6 transmission electron microscope was utilized in obtaining information on the submicron particles, and a Zeiss optical microscope was used for the larger particles and agglomerates. The basic technique utilized with the TEM in determining particle size and shape was to prepare a suspension by dispersing ultrasonically for 5 min a small amount of the sample material in water or alcohol. The data obtained from TEM and SEM observations are recorded in Table 1.

Surface area measurements were made on Linde A and Linde C alumina with the BET apparatus, the basic technique for which is described in Ref. 5. The results are tabulated in Table 2.

D. Sample Preparation

The majority of the samples prepared for reflectance measurements were in the form of binderless films on aluminum substrates. The advantages of this preparation technique are that the powder materials can be studied without contaminating them and without superimposing foreign characteristics on theirs, and that the resulting films have densities similar to those in highly pigmented paint systems. A disadvantage is that the PVC ratio can only be varied over a small range; thus, achievable variations in interparticle separations are very limited. But binderless films have the advantage that

Table 1 Particle size and shape of raw materials[a]

Pigments	Designation	Smallest particle, μ	Largest particle, μ	Average size, μ	Shape
ZnS (no grinding)	ZnS	0.05	0.4	0.1-0.2	Near spherical
ZrO_2	ZrO_2	0.15	0.7	0.3-0.4	Irregular spheres
La_2O_3 - 529.9 (undispersed)	La_2O_3	0.11	0.5	0.2-0.4	Irregular agglomerates
La_2O_3 - 529.9 (dispersed)	La_2O_3	0.05	0.5	0.1-0.4	Irregular agglomerates
Linde A (dispersed)	LA	0.1	0.8	0.25-0.5	Fused curves
Linde C (dispersed)	LC	0.25	1.0	0.5-0.75	Fused curves
Meller's Alumina (0.06μ)	M-0.06	0.01	0.10	0.06	Agglomerated and small
Meller's Alumina (0.3μ)	M-0.3	0.05	0.3	0.2	Agglomerated and small
Meller's Alumina (1.0μ)	M-1.0	0.2	0.6	0.4	Agglomerated and small
Meller's Alumina (3.0μ)	M-3.0	0.2	0.8	0.5	Agglomerated and small

[a]( ) = Manufacturer's sizing data.

Table 2 Surface area measurements, Linde A and Linde C alumina

Material	Area (B. E. T.)	Surface mean diameter[a]
Linde, type A Al_2O_3 (0.3μ)	9.18 m^2/g	0.164[b]
Linde, type C Al_2O_3 (1.0μ)	3.07 m^2/g	0.49[c]

[a]Diameter of monosized alumina (assuming a spherical shape, and a density of 3.99 g/cm^3) that will have the measured surface area.

[b]Manufacturer claims diameter is 0.3μ.

[c]Manufacturer claims diameter is 1.0μ.

their reflectances are much higher than pigment-vehicle dry films because the effective refractive index of the "vehicle" (voids) in a binderless film is unity.

For most materials hot spraying generally gave the best results. This technique involves spraying a liquid dispersion of the material onto a heated aluminum coupon. The resulting binderless films were characterized with respect to weight, thickness and spectral reflectance. Because the film density varied from one coupon to the next under apparently identical spraying conditions, we characterized films in terms of a "fluff factor" - the ratio of theoretical density to apparent film density. Fluff factors as high as 6 to 8 have been obtained with Linde A, for example. A "fluff factor" of 8 implies that the film is composed of 7 parts of voids to one part of material. The importance of film density (or "fluff factor") has been apparent, although the dependence of R_λ on it has not been quantitatively established. Also the effect of "fluff" has not been consistent.

IV. Experimental Results

A. Introduction

In a unit area, the greater the deposited thickness for a given weight of a material, the "fluffier" it is. With very few exceptions, the fluffier (less dense) films have the higher reflectance. The first group of test results show, for example, spectra of binderless films with various fluff factors. The term "undispersed" means that the powder material was used without any preparative treatment, except hand mixing. The term "dispersed" means that the powder material was dispersed in the liquid spraying vehicle by a sonifier treatment. In the abbreviated nomenclature we use "D" for dispersed and "U" for undispersed; for example, M-0.3-U refers to Meller's 0.3-micron Alumina, undispersed.

The second group consists of reflectance spectra of undispersed and dispersed films of various powder films and of their corresponding SEM photomicrographs (see Figs. 1-4).

The third group shows a series of reflectance spectra of films of Meller Alumina powders of various "particle sizes" in a pure methyl methacrylate binder (DuPont's Elvacite 6011), and also in a methyl silicone binder (GE RTV 602). Reflectance spectra are shown with particle size, PVC and film thickness as parameters. Transmittance spectra also appear in appropriate cases to show that an apparent loss of spectral reflectance does not necessarily imply an increase in spectral absorptance.

The fourth and final group consists of spectra of highly characterized films before and after 339 ESH of ultraviolet radiation (in vacuum) and after exposure to 625 Torr of O_2. The irradiation test series was intended to indicate the effect of particle size on the nature and extent of ultraviolet induced damage.

B. Discussion of Test Results

In Table 1 the reader may find helpful the designations which identify the materials we are discussing and the summaries of particle size data. Finally, the discussions and remarks here are based on analyses of the original spectral data in which the spectral regions 0.25-0.35μ, 0.325-0.7μ and 0.7-2.60μ were each recorded on separate Beckman DK-2A chart records.

1. Group I. Figure 5 shows the reflectance spectra of samples of ZnS. The sample prepared by hand dispersion (no ball-milling) obviously has a greater absorption effect at 0.72μ than the sample which was ball-milled for 96 hr. The reflectance values near the 0.72μ absorption band are very close, but the dip in the spectrum of the sample prepared from untreated ZnS at 1.5μ may suggest a shift of the scattering from this wavelength region to that near the 0.72μ absorption system. The SEM photomicrographs (hereafter

called pix) show that ball milling did not affect the particle size, it broke up some of the larger aggregates, and the resulting coatings were more smooth. Figures 6 and 7 are SEM pix at 10,000X of ZnS films, untreated and dispersed, respectively. The compactness of the ball-milled coating is indicated by its high film density factor (16 vs 3.6 for the untreated sample). Unfortunately, SEM pix at magnifications greater than 300X were not taken, so that the size distributions of the aggregates before and after dispersion could not be assessed. For reference and identification of all materials used, see Table 3.

2. Group II. The reflectance spectra just discussed and many others along with them together with their SEM pix have suggested that voids can also reflect. In Fig. 8 we have shown the reflectance spectra of five different millipore filters. As the SEM pix (Figs. 9, 10, 11, 12, 13 and 14) attest,

Table 3 Pigment materials designations and particle size data

Material	Designation	$\overline{d}_p$, μ: Mfg.[a]	$\overline{d}_p$, μ: IITRI[b]	d_p Range[c]
Linde A	LA	0.3	0.3	0.1 -0.8
Linde C	LC	1.0	0.62	0.25-1.0
Meller 0.06	M-0.06	0.06	0.06	0.01-0.1
Meller 0.3	M-0.3	0.3	0.2	0.05-0.3
Meller 1.0	M-1.0	1.0	0.4	0.2 -0.6
Meller 3.0	M-3.0	3.0	0.5	0.2 -0.8
		Pore sizes		
Millipore PH	MP-0.3	0.3 ± 0.02		0.5 -1.0
Millipore RA	MP-1.0	1.2 ± 0.3		2.5 -3
Millipore SS	MP-3	3.0 ± 0.2		4-6

[a]The average or nominal particle size claimed by manufacturer, microns.

[b]The average or nominal particle size determined at IITRI from SEM and TEM pix, microns.

[c]Same as "b" above except that range is smallest to largest particle, microns.

the major difference between these filters is the void space. The particle or fiber geometry is very uniform (eight SEM pix at various magnifications from 1000X to 20,000X taken of the type SS (nominally 3.0μ) Millipore filter, for example, showed extremely good uniformity). The reflectance spectra behave as expected from theory: at the longer wavelengths the smaller pore size materials reflect least - and so on. Even though it is not evident in these figures, due to scale compression, the spectral reflectance in the visible and ultraviolet regions show that the smaller pore size materials are the most reflective in this region. The combination of reflectance spectra and SEM pix has in this instance, conclusively demonstrated the importance of voids.

3. Group III. In this group of materials we studied the spectral reflectances of paint films, in which film thicknesses are more easily controlled and measured. Figure 15 displays the reflectance spectra of unpigmented methyl methacrylate (DuPont's Elvacite 6011) films. The very high reflectance values in the visible region and the decreasing reflectance with increasing thickness deserve special note. Figure 16 shows that the latter is not due to absorption. In applying this material the spray entraps air which forms voids in the wet film. The thin films cure very quickly and the voids are locked in; the thick films take a longer time to cure during which the air bubbles can escape. The point we wish to make is that the inclusion of stable air bubbles in a paint film strongly affects its reflectance. To study the effects of pigments in vehicles, therefore, precautions must be taken to prevent the formation of voids. For real paint systems where high reflectance is desirable, air turns out to be an excellent scattering center; and it is environmentally stable as well as inexpensive.

Figure 17 presents transmittance spectra for several thicknesses of unpigmented methyl methacrylate (m-m) polymer films made by a doctor blade technique. These can be compared to the spectra in Fig. 16. Interesting also is the comparison of the transmission spectra of m-m films pigmented to 30% PVC of M-3.0 with those of the films in Fig. 17, in which the PVC of voids is unknown. Figure 17 is a comparison of transmittance and reflectance spectra of unpigmented m-m films. Figure 18 shows R_λ spectra for sprayed films of M-0.06 pigment and these can be compared with those from doctored films in Fig. 19. The effect of particle size is not as expected. The reflectance peaks do not occur at wavelengths related to their particle sizes; in fact, they all seem to peak between 0.4 and 0.5μ. The steadily decreasing reflectance with increasing wavelength, which is typical even with larger particle sizes, contrasts with the spectral reflectance behavior of binderless films. Several explanations are possible. The most likely is that the ball milling effectively broke up the aggregates and thereby decreased reflectance at long wavelengths. If the ratio of the indices of refraction of Al_2O_3 to m-m decrease with increasing wavelength, a corresponding decrease in spectral reflectance is also expected. The effect may be a combination of these. The possibility of an increasing spectral absorptance in the vehicle is ruled out by transmittance spectra.

The effects of scattering on absorption in pigmented and clear m-m films have been investigated by calculating the extinction coefficients at a wavelength of 2.26μ. The latter is a natural absorption band in the methyl methacrylate. These were calculated from both the reflectance and transmittance spectra using the following formulae:

$$K_R = \ln(R_2/R_1)/(X_1 - X_2) \quad (1)$$

$$K_T = \ln(T_2/T_1)/(X_1 - X_2) \quad (2)$$

The parameters R, T, and X are reflectance, transmittance and film thickness, respectively, with the numerical subscripts referring to different films. The derivation of these formulae removes the thickness-independent terms such as Fresnel reflectance, making K_R and K_T internal values. Note that in both equations, K has a positive value when the log term is negative and when X_2 is greater than X_1. This is the usual case in transmittance; the thicker the film, the less the transmittance.

Table 4 presents these coefficients. The trends are generally consistent. Because the negative values of K_R imply that an increase in thickness produces an increase in reflectance, the positive values are highly significant. There are two general trends to look for in these data - the effect of particle size and the effect of PVC. The trends are obvious; but, unfortunately, the effects of voids are also evident. The average (or effective) particle size(s) of the voids being unknown makes an interpretation difficult. In Fig. 20, we have plotted K_R and K_T vs the size parameter, $\alpha(= 2\pi d/\lambda)$ for both doctored and sprayed films. The transmittance extinction coefficients, K_T, contrast sharply; the spray curve appears to have a minimum where the doctor curve has a maximum. The K_R curves also exhibit some interesting behavior. If the doctor films have an optically insignificant void fraction, then

Table 4 Effects of particle size on reflectance and transmittance extinction coefficients

a) K_R and K_T (in cm^{-1}) of Meller Alumina pigmented films of DuPont Elvacite 6011 at $\lambda = 2.26\mu$

	d_p	0.06μ	0.3μ	1.0μ	3.0μ
	a	0.167	0.834	2.78	8.34
	d_p^a	0.06	0.2	0.4	0.5
	a	0.167	0.556	1.112	1.39
K_T(doctor)	30%	73.3	94.5	113.5	85.2
K_T(spray)	30%	208	146	113	134
K_R(doctor)	30%	14.6	78.3	-28.2	-49.0
K_R(spray)	30%	-22	-13.4	-32.1	-36.3

b) K_R and K_T (in cm^{-1}) of Meller Alumina pigmented films of DuPont Elvacite 6011 at $\lambda = 2.26\mu$, $d_p = 3.0\mu$

PVC	K_T (doctor)	K_T (spray)	K_R (doctor)	K_R (spray)
0	46.4	23.8	...	138
0.11	46.6	38.0	...	-28.7
1.0	56.4	76.5	...	-87.1
10.0	101/112.2[b]	94.1/104.5	...	-41.5/-46.1
30.0	85.2/121.5	134/191.5	-49/-70	-36.3/-51.8

[a] $\alpha = 2\pi d/\lambda$; α_1 is calculated using nominal particle size, d_p, α_2, using TEM estimates of discrete particle size d_p^e.
[b] Value corrected by factor $(1-PVC/100)^{-1}$ is on the right of slash.

the effect of voids on reflectance can be very substantial, at $\lambda = 0.5\mu$ void scattering makes the difference between $K_R = 70\ cm^{-1}$ and $K_R = -15\ cm^{-1}$.

The role of voids must be regarded as extremely important. In Table 4, note that the K_T values for unpigmented films are roughly a factor of two different. The ratio K_T (spray)/K_T (doctor), however, increases strongly with PVC indicating that the effective absorption cross-section increases with particle scattering. Although attention was directed to the K_T values in Table 4, the ratio K_T (spray)/K_T (doctor) is important; as can be seen in Table 4, it always exceeds unity, except for the 1.0μ particles. The ratio K_R (spray)/K_R (doctor), deserves special note, because it indicates the magnitude of intensification that scattering has on basic absorption processes.

A set of data similar to those in Table 4 were obtained for $\lambda = 1.05\mu$. This wavelength was chosen because there is very little absorption in this region. The unpigmented films showed that air voids actually increase transmittance and decrease reflectance. This situation is reversed, however, when pigment is added. The values of transmittance extinction coefficients exceed those of reflectance extinction coefficients at this wavelength also. With absorption absent, the ratio of K_T (spray)/K_T (doctor) is at least two orders of magnitude higher. It changes sign and drops sharply, with the addition of pigment, however, to values nearly the same as those for $\lambda = 2.26\mu$, a wavelength of strong absorption. In comparison of the data at $\lambda = 1.05\mu$ with those at $\lambda = 2.26\mu$, we assumed that all of the absorption is in the vehicle and none is due to the Al_2O_3 pigment. As long as the absorption in the m-m exceeds that in the Al_2O_3, the comparison of data at $\lambda = 1.05\mu$ with those at $\lambda = 2.26\mu$ will be at least qualitatively correct. The ratios of K_R and K_T should again be compared. Void-scattering reduces the absorption effect; the scattering due to pigments, increases it. These are very important conclusions, which must be verified by further experiments. They follow from analyses of the data in Table 4 and Fig. 20.

4. Group IV. Very careful and exact measurements were made on a final series of binderless coatings. Samples to be studied in the SEM were prepared simultaneously and as identically as possible with those for ultraviolet irradiation. The reflectances were first measured on the Beckman DK-2A, after which they were installed in the IRIF-II, remeasured, irradiated to an ultraviolet exposure of 339 ESH, and finally, exposed to 625 mm Hg of O_2. The intent in this procedure was to associate spectral reflectance changes with particle size; from our previous work, however, it was obvious that parameters other than particle size would be involved.

From the analyses of these latter data, we have found that the question of what effect particle size has on damage spectra is essentially the same as that of its effect on initial spectral reflectance. The film structure, the physical and geometrical relationships existing between particles, voids and aggregates, almost entirely determines the spectral reflectance characteristics. In our ultraviolet exposure tests we found that dispersion apparently increases the ultraviolet damage effect. This, we think, is because the void structure between aggregates is reduced and, as we observed earlier, voids tend to mitigate the absorption effect, and consequently the lesser the void density, the greater is the net absorption effect.

V. Conclusions

The primary aim of the program was to correlate spectral reflectance with particle size. No such correlation, however, seems possible in dense powder films in which particle size has a dubious meaning. Particles rarely exist discretely, that is, unattached to or unassociated with any other particle or group of particles. Particle shape influences the surface electro-optical characteristics and thus the tendency to aggregate into much larger masses which very often have their own characteristic sizes and shapes. The basic light scattering behavior of a system of particles therefore depends on the action of individual, ("discrete") particles, of aggregated assemblies of them, upon their respective physical size and shape distributions, and also upon their physical arrangement and how this is affected by the distribution of particle geometries. The expectation, therefore, that reflectance should be correlatable with particle size alone is not realistic, because agglomerates are not discrete particles, because particle size and shape strongly influence the geometry and statistics of the void spaces between particles and agglomerates, and because particle-particle, particle-agglomerate, and agglomerate-agglomerate separations are equally important parameters.

We have observed that particle size and shape and their distributions affect film packing density which in turn affects void sizes and their distributions, all of which affect spectral reflectance. Still another observation is that agglomerate sizes may depend on discrete particle size(s). The overall effect of either of these may, in addition, be modified by absorption and refractive indices. These are highly significant results which require further study. Light scattering in dense media does depend in part on particle size but not in any manner predicted by theory.

Light scattering in dense dielectric powders cannot be treated accurately either by Mie Theory or by Monte Carlo applications of Mie Theory, because in this case neighboring particles bear more than a simple geometric relationship to one another. Compare, for example the case of two identical spherical particles both of diameter, d_p, and separated by a distance, say 10 d_p between centers, with the case where these particles are touching or nearly touching. To an incident beam of light, is the latter system of two particles equivalent to the first? We suspect not, on the grounds that in the latter case, the surface forces of the particles are interacting and the electromagnetic fields of each are modified by the presence of the other.

Scattering by pigmentary particles intensifies the net absorption effect. Hence, the wavelength of maximum particle scattering should not be allowed to be in the spectral region of induced damage. As a general rule, assuming negligible intrinsic absorption in the visible and infrared, materials which degrade principally in the infrared should be optimized for high visible reflectance; those which degrade in the visible and near ultraviolet should be optimized for high reflectance in the long wavelength visible and near infrared region. Ultraviolet reflectance should not be maximized in dielectric materials (e.g., Al_2O_3, SiO_2) because of very high intrinsic absorption in the far ultraviolet and significantly high impurity and induced absorption in the near ultraviolet.

The contribution of voids to the total spectral reflectance is a major one, because voids increase spectral reflectance, yet tend to mitigate the absorption effect of intrinsic absorbers. Voids may occur either between pigment particles or in the binder (air occlusions, for example) and techniques can be developed to maximize their effect. Particle scattering also increases reflectance, but tends also to increase the absorption effect of intrinsic or induced absorbers.

References

[1] Zerlaut, G.A. and Ashford, N., "Development of Space-Stable Thermal Control Coatings," Report U6002-73 (Triannual Report), NASA Contract NAS8-5379, Jan. 31, 1969, IIT Research Institute.

[2] Gilligan, J.E. and Zerlaut, G.A., "The Role of Gaseous Adsorption in Thermal Coatings Degradation", AIAA Paper 69-1025, Los Angeles, Calif., 1969.

[3] Gilligan, J.E. and Brzuskiewicz, J., "Development of Space Stable, Low Solar Absorptance, Pigmented Thermal Control

Coatings," Report C6166-12 (Final Report) , NASA CR-66197, November 1969, IIT Research Institute.

[4]Dexter, D. L., "Theory of the Optical Properties of Imperfections in Nonmetals," Solid State Physics, Vol. 6, edited by F. Seitz and D. Turnbull, Academic Press, New York, 1958, pp. 353-411.

[5]Brunauer, S., Emmett, P. H., and Teller, E., "The Adsorption of Gases in Multimolecular Layers," Journal of the American Chemical Society, Vol. 60, 1938, pp. 309-319.

Bibliography

Giese, R. H., "Light Scattering by Small Particles and Models of Interplanetary Matter Drived from the Zodical Light," Space Sciences Review, Vol. 1, 1963, pp. 589-611.

Sinclair, D., "Light Scattering by Spherical Particles," Journel of the Optical Society of America, Vol. 37, No. 6, June 1947, pp. 475-480.

Adams, J. B. and Filice, A. L., "Spectral Reflectance 0.4 to 2.0 Microns of Silicate Rock Powders," Journal of Geophysical Research, Vol. 73, No. 22, 1967, pp. 5705-5715.

Maples, D and Wiebelt, J., "Experimental and Theoretical Investigation of Electromagnetic Scattering from a Known Rough Surface," AIAA Paper 68-29, New York, 1968.

Schatz, E. G., "Spectral Reflectance of Sintered Oxides," Journal of the American Ceramic Society, Vol. 51, No. 5, May 1968, pp. 287-289.

DeVore, J. R. and Pfund, A. H., "Optical Scattering by Dielectric Powders of Uniform Particle Size," Journal of the Optical Society of America, Vol. 37, No. 10, Oct. 1947, pp. 826-832.

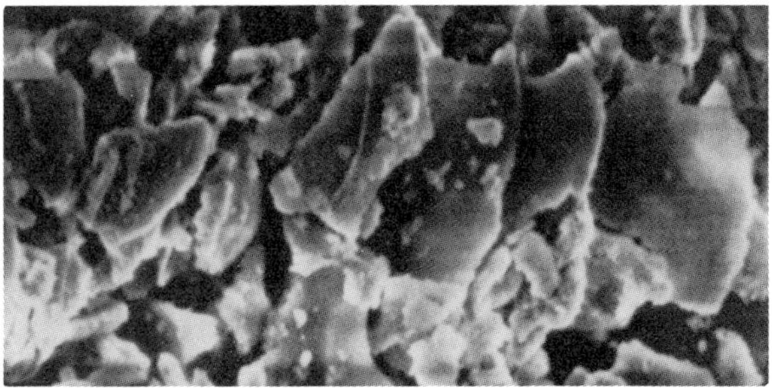

Fig. 1 SEM of M-1.0-U binderless (300X).

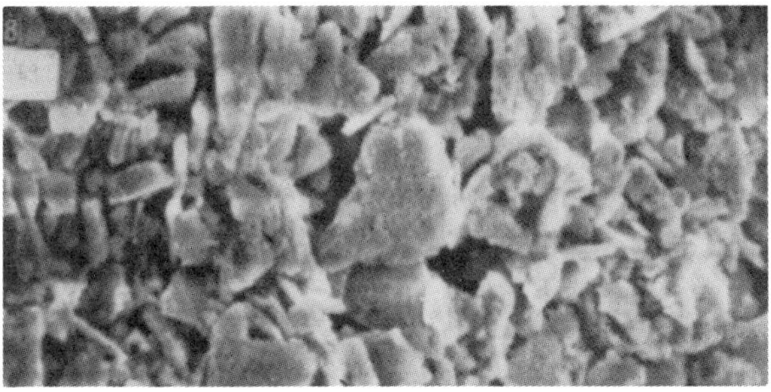

Fig. 2 SEM of M-1.0-D binderless (300X).

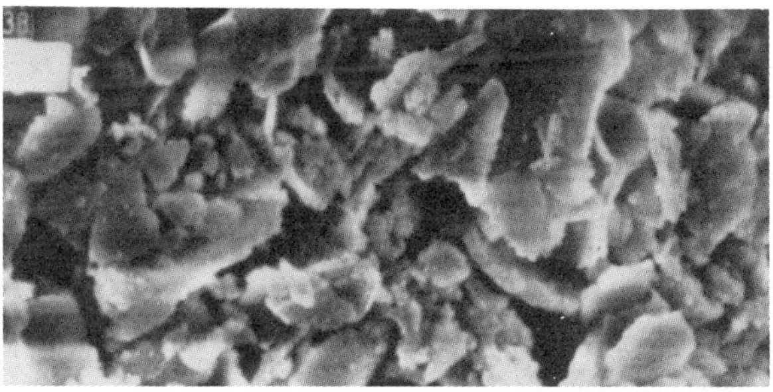

Fig. 3 SEM of Linde A-U binderless (300X).

Fig. 4 SEM of Linde A-D binderless (300X).

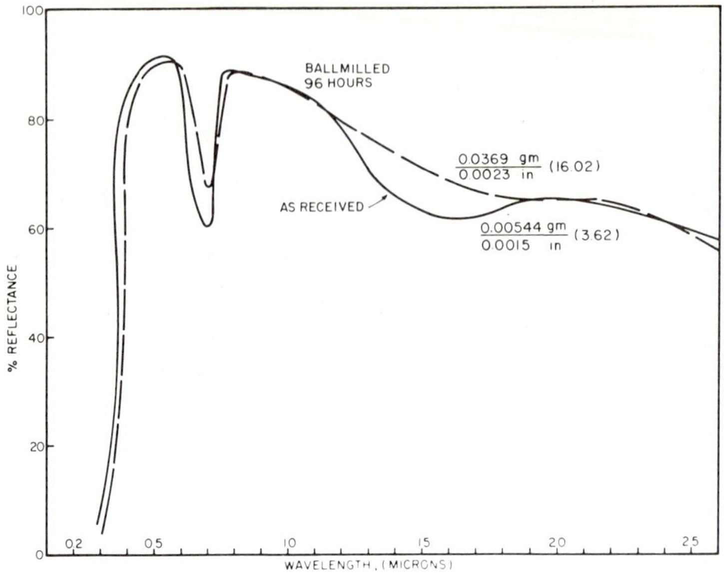

Fig. 5 Reflectance spectra of zinc sulfide.

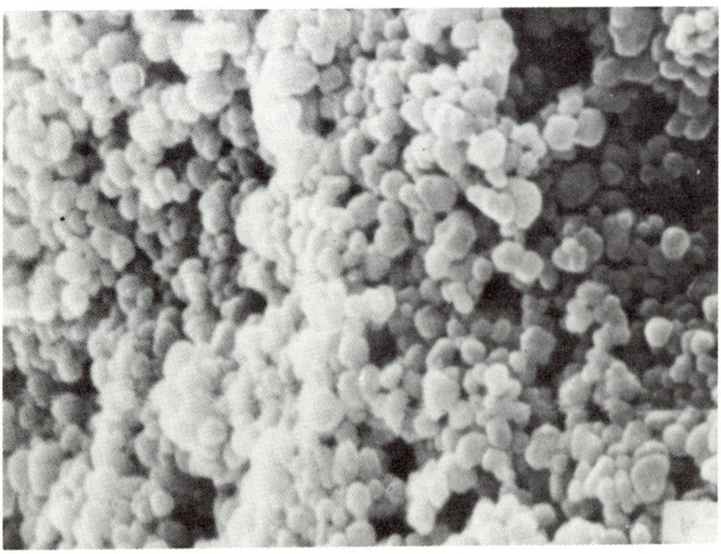

Fig. 6 SEM of ZnS-U (10,000X).

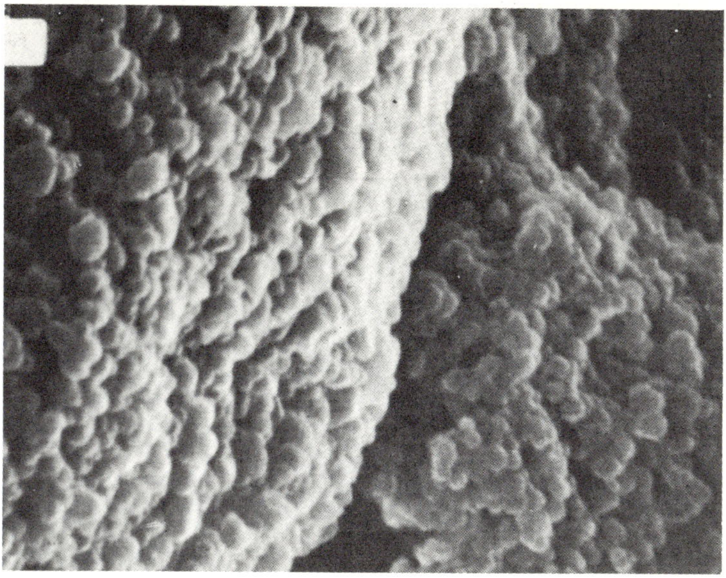

Fig. 7 SEM of ZnS-D (10,000X).

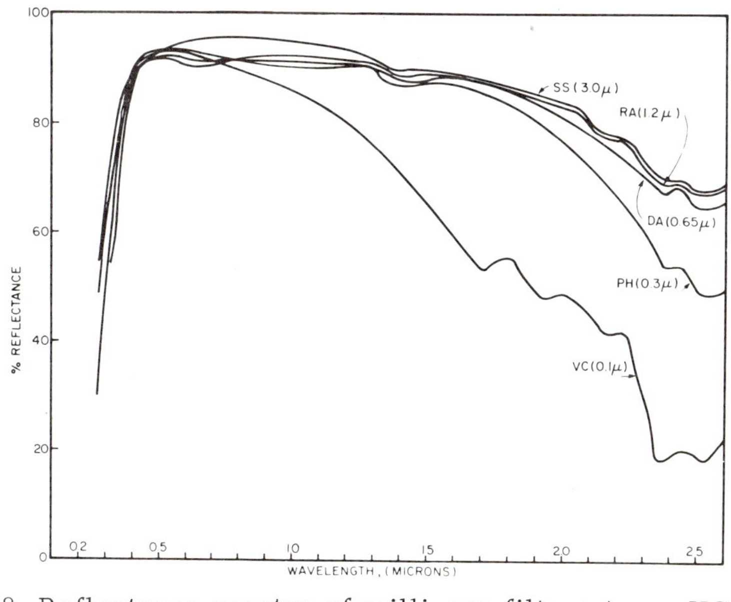

Fig. 8 Reflectance spectra of millipore filters types VC, PH, DA, RA and SS.

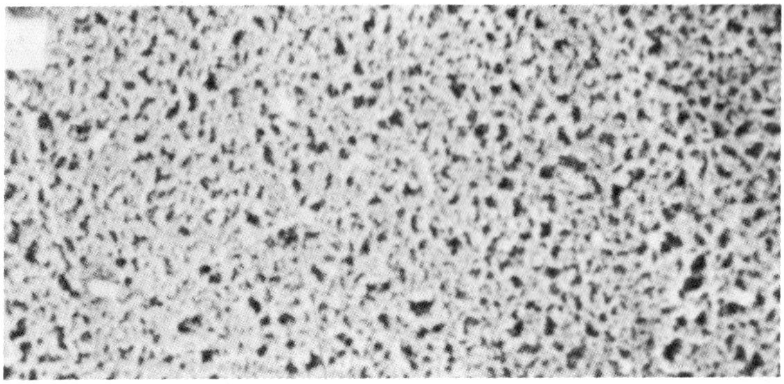

Fig. 9 SEM of 0.3μ millipore filter type PH (1000X).

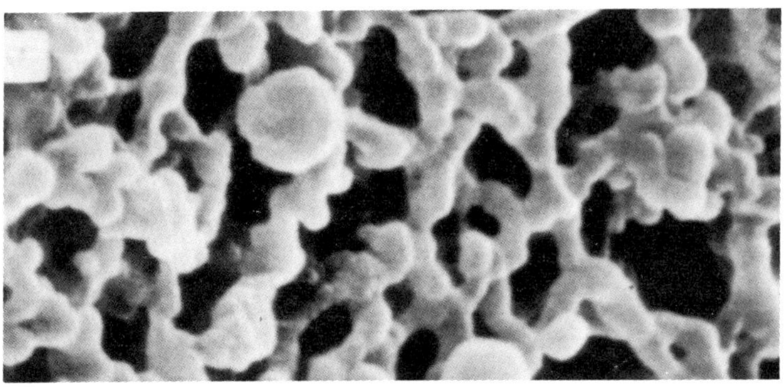

Fig. 10 SEM of 0.3μ millipore filter type PH (10,000X).

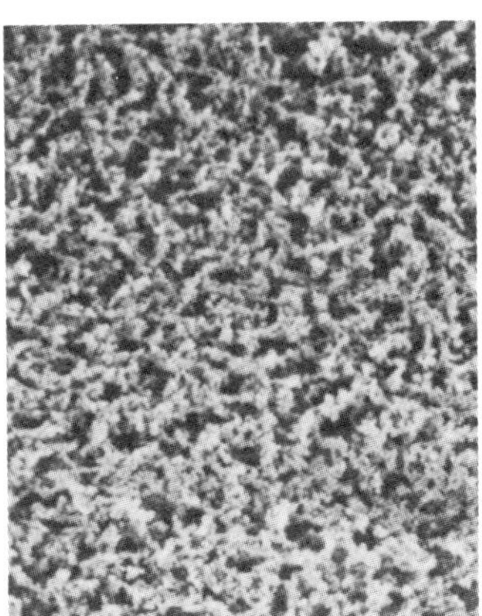

Fig. 11 SEM of 1.2μ millipore filter type RA (1000X).

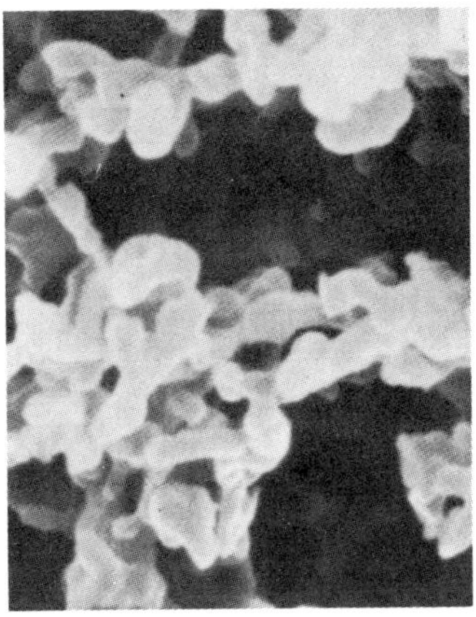

Fig. 12 SEM of 1.2μ millipore filter type RA (10,000X).

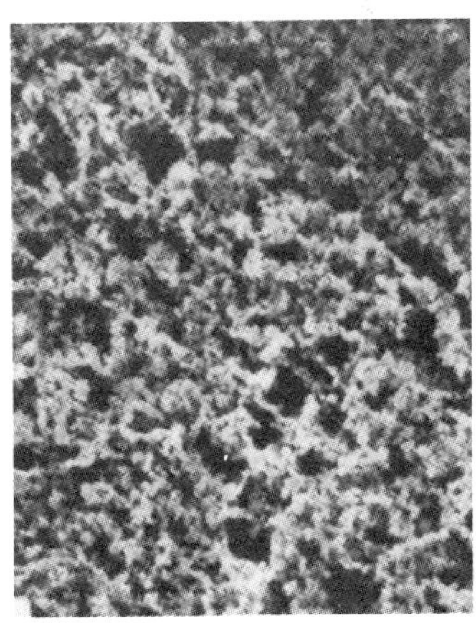

Fig. 13 SEM of 3.0μ millipore filter type SS (1000X).

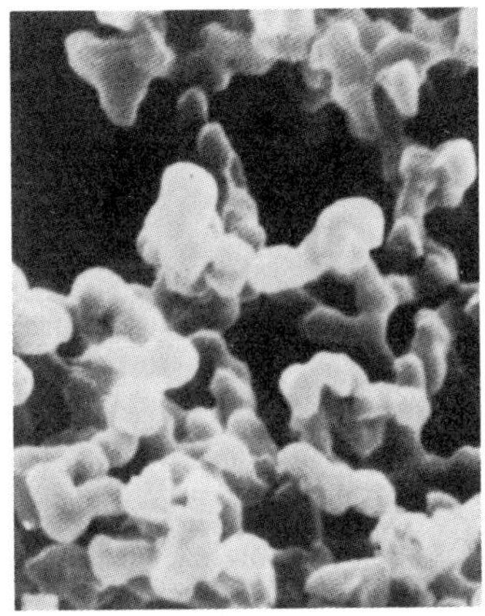

Fig. 14 SEM of 3.0μ millipore filter type SS (10,000X).

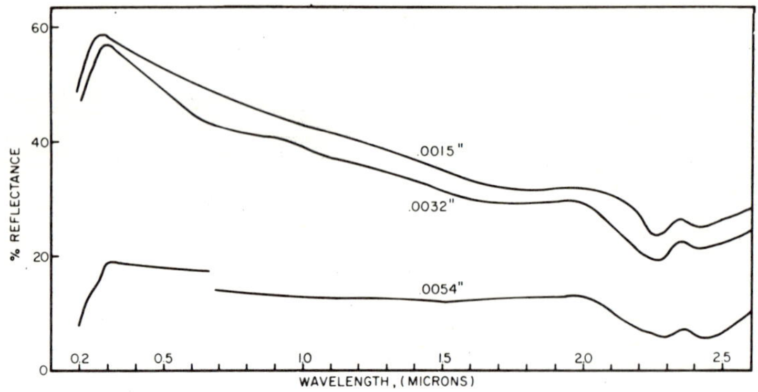

Fig. 15 Sprayed films of DuPont Elvacite 6011 reflectance vs thickness.

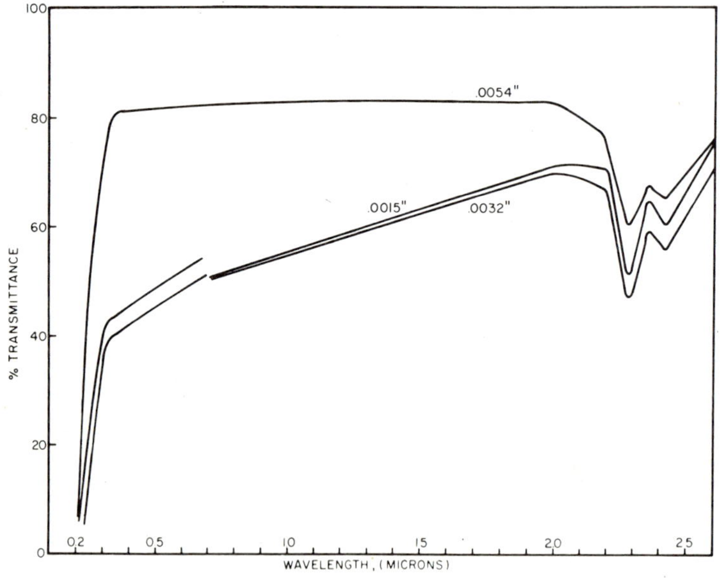

Fig. 16 Sprayed films of DuPont Elvacite 6011 transmittance vs thickness.

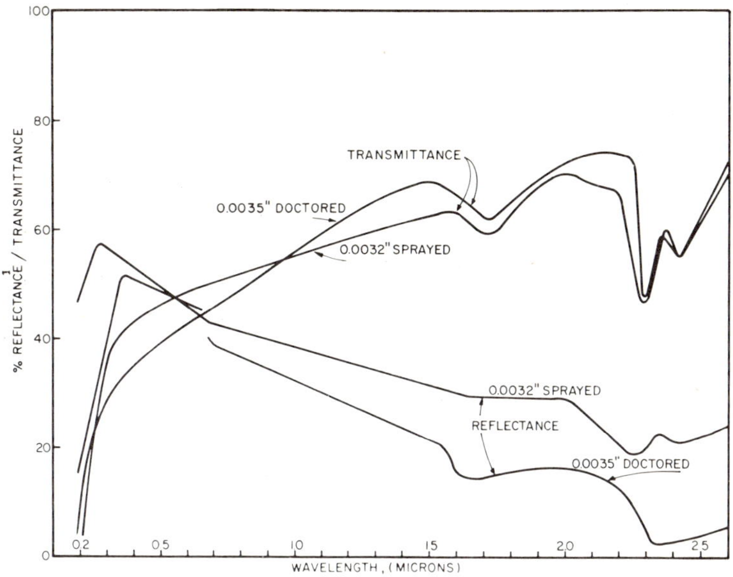

Fig. 17 Reflectance and transmittance spectra of sprayed and doctored films of unpigmented DuPont Elvacite 6011.

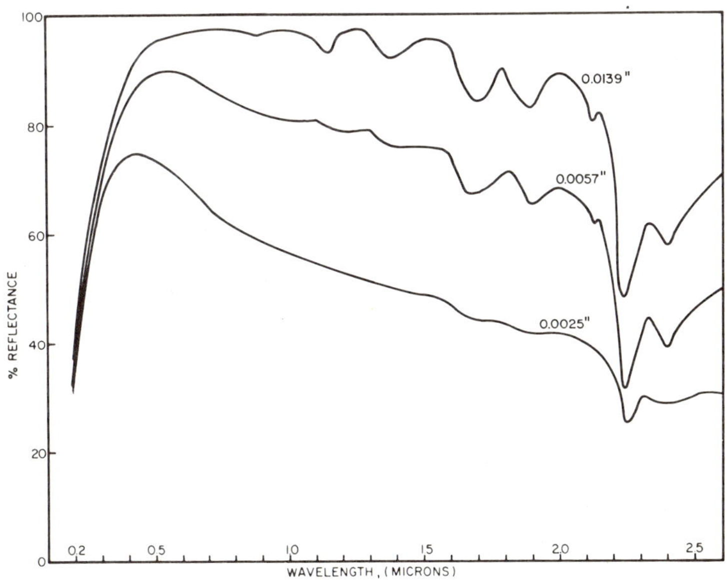

Fig. 18 Reflectance spectra of sprayed films of Meller 0.06μ Al_2O_3 at 30% PVC in DuPont Elvacite 6011 vs thickness.

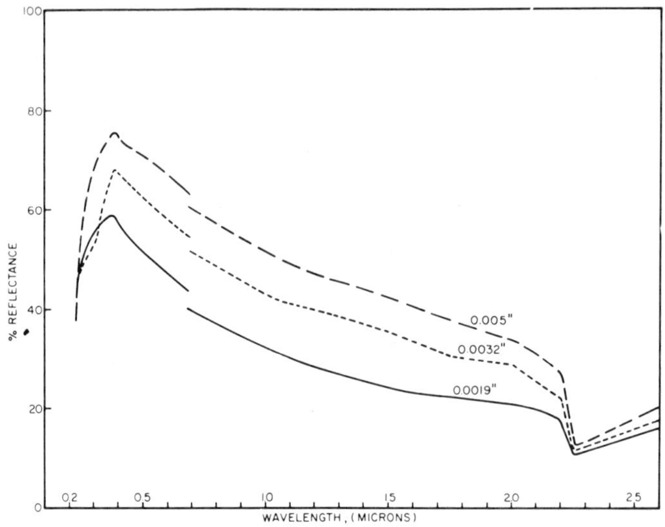

Fig. 19 Reflectance spectra of doctored films of Meller 0.06μ Al_2O_3 at 30% PVC in DuPont Elvacite 6011 vs thickness.

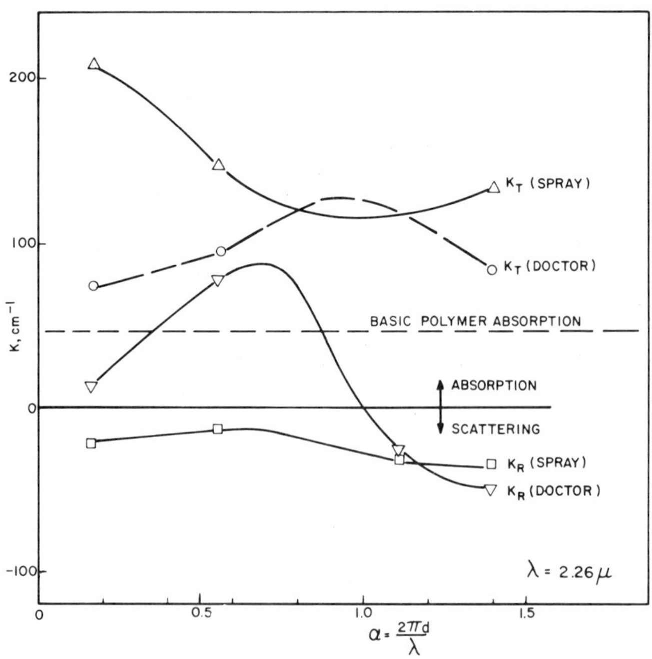

Fig. 20 Extinction coefficients vs size parameter for Meller Al_2O_3 at 30% PVC in DuPont Elvacite 6011 (methylmethacrylate) at λ = 2.26μ.

AN INVESTIGATION OF THE VALIDITY OF KIRCHHOFF'S LAW FOR FREELY RADIATING METALLIC SURFACES

T. C. Grimm*

McDonnell Aircraft Company, St. Louis, Mo.

Abstract

An emissometer was constructed for accurately (within ± 2%) measuring the total hemispherical emittance of surfaces in the temperature range of 75-200°F. The experimental technique consisted of measuring the total hemispherical emittance of metallic test surfaces under the following conditions: 1) surrounding temperature relatively close to sample temperature (ice water) and 2) sample nearly freely radiating (liquid nitrogen). For Kirchhoff's law to have been valid, emittances measured in the two environments would have to have been equal. Based upon the experimental results for aluminum pigmented silicone paint; polished copper, platinum, and tantalum; and electroplated and vapor-plated gold on polished copper, it can be concluded that Kirchhoff's law is valid within the accuracy of the experimental technique (i.e., ± 2%) for freely radiating metallic surfaces. Thus, the power radiated by a metallic surface is the same when it is freely radiating as it would be under equilibrium conditions. These experimental results are supported by calculations which show that the validity of Kirchhoff's law depends only upon the material states having their equilibrium distribution.

Nomenclature

A_1	=	area of surface 1
A_2	=	area of surface 2
B_ν	=	Planck function, specific intensity of a black body
c	=	velocity of light in vacuo

Presented as Paper 70-858 at the AIAA 5th Thermophysics Conference, Los Angeles, Calif., June 29 - July 1, 1970. This research was performed at Washington University under the direction of E. Weger in partial fulfillment of the requirements for the degree of Doctor of Science.

*Group Engineer, Materials Laboratories.

F_{21}	=	shape factor - surface 2 to surface 1
h	=	Planck's constant
I_ν	=	specific intensity
J_ν	=	source function, $j_\nu/\varkappa_\nu$
j_ν	=	emission coefficient
k	=	Boltzmann constant
l_j	=	direction cosine for the jth direction, cos θ
M^2	=	some appropriately averaged square of a sum of matrix elements
n	=	index of refraction of the medium
n	=	occupation number for photons
Q_{12}	=	net heat flow rate from surface 1 to surface 2
q_{12}	=	net heat flux from surface 1 to surface 2
R_ϵ	=	power radiated by a gray metallic surface per unit area
$R_{\nu,a}$	=	rate of absorption of radiant energy ν per unit area per unit frequency interval by a metallic surface
$R_{\nu,\epsilon}$	=	rate of emission of radiant energy ν per unit area per unit frequency interval by a metallic surface
T	=	temperature
T_1	=	**temperature of surface 1**
T_2	=	temperature of surface 2
T_r	=	temperature of radiation (surroundings)
T_s	=	temperature of surface
$\tilde{T}$	=	$\sqrt{T_1 T_2}$
x_j	=	distance in the x_j direction
α	=	total hemispherical absorptance (or absorptivity)
α_1	=	total hemispherical absorptance of surface 1
α_2	=	total hemispherical absorptance of surface 2
α_ν	=	spectral hemispherical absorptance (or absorptivity)
ϵ	=	total hemispherical emittance (or emissivity)
ϵ_1	=	total hemispherical emittance of surface 1
ϵ_1'	=	total hemispherical emittance of surface 1 based on gray assumption
ϵ_{eff}	=	effective total hemispherical emissivity
ϵ_ν	=	spectral hemispherical emittance (or emissivity)
η	=	ratio of ϵ_ν to α_ν, ϵ_ν/α_ν
$\varkappa_\nu$	=	absorption coefficient
ν	=	frequency of the radiation

ρ	=	density of the medium
ρ_ν	=	density of states factor
σ	=	Stefan-Boltzmann constant

Introduction

In general, radiant heat-transfer calculations are based upon the assumption that the fundamental laws of radiation heat transfer are valid at conditions far removed from thermal equilibrium, even though the classical derivations of these laws require the assumption of thermal equilibrium. Weinstein[1] has advanced an argument (based on calculations for a uniform, nondegenerate, dilute gas) that Kirchhoff's law is valid for freely radiating bodies and, consequently, that the power radiated by a body is the same whether or not the radiating body is in thermal equilibrium with its environment. Ashby and Schocken,[2] on the other hand, contend that although Weinstein's result may be true for gases and transparent crystals where the body may be stimulated by radiation which has been emitted from deep within the body itself, it is not true for metals because electromagnetic waves of thermal wavelengths are not propagation waves within the metal. Calculations made by Ashby and Schocken indicate that for a very hot body in a cold environment, the power radiated would be only 93% of what it would be at thermal equilibrium, whereas for a cold metallic body in a hot environment the power radiated would be greater than what it would be if the body and its surroundings were in thermal equilibrium.

The objective of this investigation was to determine the validity, or nonvalidity, of Kirchhoff's law for freely radiating metallic surfaces. To achieve this objective, an emissometer was constructed that was capable of calorimetrically measuring the total hemispherical emittance of optically smooth metals within an accuracy of $\pm$ 2%. To attain this accuracy, it was necessary to make corrections (based upon actual measurements) for heat losses from the test sample due to solid and gaseous conduction.

Theoretical Considerations of the Validity of Kirchhoff's Law at Nonequilibrium

Arguments for the Validity of Kirchhoff's Law at Nonequilibrium

To prove his hypothesis - that Kirchhoff's law is valid for a freely radiating body whose distribution over the material states of the system is the equilibrium distribution - Weinstein begins with the steady-state equation of radiative transfer, which is

$$l_j \,(dI_\nu/dx_j) = \rho\,(j_\nu - \varkappa_\nu I_\nu) \tag{1}$$

He then considers a uniform slab of a nondegenerate, dilute gas and, by considering induced emission to be negative absorption, arrives at the Kirchhoff-Planck law

$$J_\nu = I_\nu = n^2 B_\nu \tag{2}$$

a relationship which holds for complete thermodynamic equilibrium as well. He finally obtains Kirchhoff's law

$$\epsilon_\nu = \alpha_\nu \tag{3}$$

by solving the equation of transfer, Eq. (1), making use of the Kirchhoff-Planck law and the boundary conditions for a freely radiating body.

Weinstein's basic hypothesis is similar to a hypothesis made by Einstein[3] in a paper published in 1917. In this paper, Einstein asserts that molecules with states distributed according to thermal equilibrium are in dynamic equilibrium with Planckian (i.e., black body) radiation. It is important to note, however, that Einstein considers dilute gases just as Weinstein does.

Vincenti and Kruger[4] discuss the application of the equation of radiative transfer to gas dynamics. In their text, these authors introduce the assumption of quasi-equilibrium, also called the assumption of local thermodynamic equilibrium. This assumption requires that the populations of the various atomic and molecular states that participate in the emission and absorption of radiation be given by their equilibrium values, even though the radiant energy density differs from its equilibrium value. Making this assumption, ϵ_ν and α_ν for a given value of ν depend only upon the nature of the surface and the surface temperature but not upon the radiation field to which the surface is exposed. Thus, whether or not the incident intensity is equal to the equilibrium value B_ν, Kirchhoff's law, i.e., Eq. (3), is valid.

Considering solids, Sparrow and Cess[5] state that although Kirchhoff's law was derived under the condition that a given surface is in thermodynamic equilibrium with its surroundings, the law is a general relationship that applies to non-equilibrium conditions as well. These authors support this assertion by stating that α_ν and ϵ_ν are surface properties and they therefore depend solely upon the nature of the surface and its temperature and not upon the conditions of the surroundings.

In a private communication[6] to the author, Cess further explained that the applicability of Kirchhoff's law to a material surface depends solely upon the existence of local thermodynamic equilibrium within the material. He also indicated, however, that it is possible for a large incident radiation field to destroy local thermodynamic equilibrium in the vicinity of a material surface.

Arguments Against the Validity of Kirchhoff's Law at Nonequilibrium

Oster[7] states that Kirchhoff's law, i.e., absorption and emission are equal to each other for every frequency, is true only in thermodynamic equilibrium. At thermodynamic equilibrium, the distribution functions of all particles (in a plasma) are Maxwellian with the same temperature, and the radiation field is isotropic and Planckian. According to Oster, the essential criterion of thermodynamic equilibrium is the equality of the rates of each atomic process and its reverse, which is the principle of detailed balancing as originally applied by Einstein to absorption and emission by bound states.

In their paper, Ashby and Schocken state that Kirchhoff's law, whether applied to total radiation or spectral radiation, is strictly valid only for the case of thermal equilibrium. These authors argue that, according to Einstein, thermal equilibrium between a body and black body radiation is maintained partially by spontaneous emission and partially by stimulated emission. The stimulated emission is proportional to the intensity of the incident radiation; therefore, if this radiation is not black body radiation of the same temperature as the solid, the stimulated emission may be more or less than that predicted by Kirchhoff's law. This is particularly true for metals since electromagnetic waves of thermal wavelengths are not propagating waves within the metal and, therefore, none of the photons causing the stimulation can be considered to come from the metal itself. Hence, the stimulated emission is characteristic only of the environment.

Ashby and Schocken begin their analytical treatment by developing a set of basis functions for the quantized vector potential of the electromagnetic field which are exponentially attenuated within the metal, so that only electrons near the surface will participate in emission and absorption. On this basis, a perturbation theory formalism is developed with which the emissivity may be calculated as a function of wavelength, angle of emission, polarization, and temperature whenever the electronic wave functions are known. By means of this perturbation theory, Ashby and Schocken arrive at the following modified form of Kirchhoff's law:

$$\epsilon_\nu = \eta\alpha_\nu \tag{4}$$

where

$$\eta = \{[n_\nu(T_r) + 1]\, e^{-h\nu/kT_s}\}/n_\nu(T_s) \tag{5}$$

If the temperature of the metal T_s is equal to the temperature of the radiation T_r, then $\eta = 1$ and, by Eq. (4), $\epsilon_\nu = \alpha_\nu$, which is Kirchhoff's law.

At this point, Ashby and Schocken consider a metallic surface of constant absorptivity α (i.e., a gray metallic surface) and, assuming that the incoming radiation is black body radiation, arrive at the following relationship for the rate of emission per unit area of the gray metallic surface:

$$R_\epsilon = \alpha\sigma T_s^4 \frac{90}{\pi^4} \sum_{n=0}^{\infty} 1/[1 + (nT_s/T_r)]^4 \tag{6}$$

When the temperature of the metal and the radiation field are equal, i.e., at thermal equilibrium,

$$\sum_{n=0}^{\infty} \frac{1}{(1+n)^4} = \frac{\pi^4}{90} \tag{7}$$

and

$$R_\epsilon = \alpha\sigma T_s^4 = \epsilon\sigma T_s^4 \tag{8}$$

For nonequilibrium, Ashby and Schocken define an effective emissivity ϵ_{eff} as

$$\epsilon_{eff} = R_\epsilon/\sigma T_s^4 \tag{9}$$

and thus

$$\epsilon_{eff}/\alpha = R_\epsilon/\alpha\sigma T_s^4 \tag{10}$$

This ratio, which shows the effect of the radiation environment on the emissivity of the metal, is plotted as Fig. 1 in Ref. 2. In accordance with Eq. (9), the power radiated by a metal per unit area at a temperature T_s is, in general, given by

$$R_\epsilon = \epsilon_{eff} \sigma T_s^4 \tag{11}$$

and only at thermal equilibrium does the Stefan-Boltzman equation [Eq. (8)] apply. Since the previous analysis was carried out for a gray surface, both the spectral and total values of emittance and absorptance should be equal even though nonequilibrium prevails.

It is very instructive to see what result Ashby and Schocken would have obtained had they assumed, as Weinstein did, that stimulated emission is to be treated as negative absorption. Using this assumption, the equations of Ashby and Schocken for the rate of absorption and the rate of emission of radiant energy ν per unit area per unit frequency of the metallic surface become

$$R_{\nu,\alpha} = n_\nu(T_r)[\,|\overline{M}|^2 h\nu\rho_\nu - |\overline{M}|^2 h\nu\rho_\nu e^{-h\nu/kT_s}\,] \tag{12}$$

and

$$R_{\nu,\epsilon} = |\overline{M}|^2 h\nu\rho_\nu e^{-h\nu/kT_s} \text{ (spontaneous emission only)} \tag{13}$$

Thus,

$$\alpha_\nu = \frac{R_{\nu,\alpha}}{ch\nu n_\nu(T_r)\rho_\nu} = \frac{M^2(1 - e^{-h\nu/kT_s})}{c} \tag{14}$$

$$\epsilon_\nu = \frac{R_{\nu,\epsilon}}{ch\nu n_\nu(T_s)\rho_\nu} = \frac{M^2 e^{-h\nu/kT_s}}{cn_\nu(T_s)} \tag{15}$$

and

$$\epsilon_\nu = \eta\alpha_\nu \tag{4}$$

where

$$\eta = \frac{e^{-h\nu/kT_s}}{n_\nu(T_s)} \cdot \frac{1}{1-e^{-h\nu/kT_s}} = \frac{1}{n_\nu(T_s)(e^{h\nu/kT_s}-1)} \tag{16}$$

Therefore, if it is assumed that the material states have their equilibrium distribution (the quasi-equilibrium hypothesis), then for a closed, nondegenerate system

$$n_\nu(T_s) = 1/(e^{h\nu/kT_s} - 1) \tag{17}$$

and $\eta = 1$ so that $\alpha_\nu = \epsilon_\nu$, whether or not the metal is in thermal equilibrium with the radiation field.

This calculation was made by the author during the course of his thesis[8] work (1964-69). A similar calculation giving an identical result is included in a recent (not dated) report prepared by Ashby et al[9]. This result is in complete agreement with the result obtained by Weinstein. In his paper, Weinstein stressed that he would not have obtained the result he did if he had not considered induced emission to be negative absorption.

Experimental Approach

The experimental technique consisted of very accurately (within ± 2%) measuring the emittance of a small, flat test sample in a large environment whose temperature was colder than the test sample temperature. Two different environment temperatures were used. In the one case (ice water), the temperature of the environment was relatively close to the temperature of the test sample. In the other case (liquid nitrogen), the temperature of the environment was low enough so that the sample was nearly freely radiating. If, as assumed, the only change in test conditions for measurements made in the two environments was the temperature of the surroundings, then, for Kirchhoff's law to have been valid, the emittances for the same sample temperature would have to have been equal regardless of the temperature of the surroundings. If, however, the emittances measured in the two environments were not equal, and, in particular, if the emittances measured in the higher temperature surroundings were greater than the emittances measured in the lower temperature surroundings, then Kirchhoff's law would not have been valid.

Basic Equation

If a nonconcave surface A_1 at temperature T_1 is completely surrounded by a surface A_2 at temperature T_2, the net radiant heat exchange between the two surfaces is

$$Q_{12} = \frac{\sigma A_1 [(\epsilon_1/\alpha_1)T_1^4 - (\epsilon_2/\alpha_2)T_2^4]}{1/\alpha_1 + F_{21}\ [(1/\alpha_2)-1]} \tag{18}$$

When surface 2 is perfectly diffuse, the reciprocity law can be used to obtain $F_{21} = A_1/A_2$. If, in addition, surface 1 is very small relative to surface 2 so that $A_1/A_2 \approx 0$, then $F_{21} \approx 0$ and

$$q_{12} = \alpha_1 \sigma[(\epsilon_1/\alpha_1)T_1^4 - (\epsilon_2/\alpha_2)T_2^4] \tag{19}$$

Gray Case

If it is assumed that both surfaces are gray, then $\epsilon_1 = \alpha_1$ and $\epsilon_2 = \alpha_2$ and Eq. (19) simplifies to

$$q_{12} = \epsilon_1 \sigma(T_1^4 - T_2^4) \tag{20}$$

Nongray Case

In general, metallic surfaces are not gray but exhibit total emittances that increase linearly with absolute temperature. As pointed out by Agababov,[10] the emittances of polished metals calculated by means of Eq. (20) (gray analysis) can be significantly different than the true emittances given by Eq. (19). Following Agababov, the ratio of the calculated emittance using gray analysis ϵ_1' to the true emittance ϵ_1 is found by first multiplying and dividing the right-hand side of Eq. (19) by $(T_1^4 - T_2^4)/T_1^4$, yielding

$$q_{12} = \epsilon_1' \sigma(T_1^4 - T_2^4) \tag{21}$$

where

$$\epsilon_1' = \alpha_1 \frac{(\epsilon_1/\alpha_1) - (T_2/T_1)^4(\epsilon_2/\alpha_2)}{1 - (T_2/T_1)^4} \tag{22}$$

Finally, rearranging Eq. (22) gives

$$\epsilon_1'/\epsilon_1 = [1 - (T_2/T_1)^4 \ (\epsilon_2/\alpha_2)\ (\alpha_1/\epsilon_1)]/[1 - (T_2/T_1)^4] \tag{23}$$

Inspection of Eq. (23) reveals that for a closed system not in thermodynamic equilibrium, ϵ_1' and ϵ_1 are equal only if both surfaces are gray. Also, as $(T_2/T_1)^4$ approaches zero, ϵ_1' tends to approach ϵ_1. Since, in general, real surfaces are not gray, most emittances measured in ice water surroundings using the gray assumption had to be converted to true emittances by means of Eq. (23) (assuming $\epsilon_2/\alpha_2 = 1.0$) and the Eckert[11] relationship, which is

$$\alpha_1 = \epsilon_1(\tilde{T}) \text{ where } \tilde{T} = \sqrt{T_1 T_2} \tag{24}$$

Any emittance measured in liquid nitrogen surroundings, however, could be used directly because of the smallness of (T_2/T_1). Sample calculations showed that, even for a sample temperature as low as 75°F, the true emittance measured in a liquid nitrogen environment could be no more than approximately

0.3% less than the emittance calculated using gray analysis. Hence, liquid nitrogen data could be used as true data when correcting ice water data for nongray effects.

Ice water emittances were first calculated by gray analysis and then corrected for nongrayness to show quantitatively the errors which could occur in an ice water environment as a result of the gray assumption. The more direct approach would have been to calculate all emittances by means of Eq. (19).

Experimental Apparatus

The experimental apparatus, described in detail in Ref. 8, is an emissometer capable of measuring the total hemispherical emittance of materials as a function of temperature over the temperature range of 75 - 200°F.

Test Chamber

A cross section of the test chamber is shown in Fig. 1. The blackened hemisphere (surface 2), which surrounds the radiating surface of the test sample (surface 1), consisted of a Type 304 stainless steel, 12-in.-diam hemisphere with the inner surface sand-blasted and sprayed with 3M Company 101-C10 black velvet coating. The sand-blasted, flat black painted surface provided diffuse, gray surroundings for the radiating surface of the test sample and had a very low reflectance which limited the amount of background radiation reaching the test sample surface to a very low level. The base plate and 6 5/8-in.-o.d. tubing were also painted with 3M black velvet coating.

The surface of the blackened hemisphere was maintained at various temperatures by means of the following constant temperature baths: hot water, ice water, or liquid nitrogen. When a hot water bath was used for measuring the solid conduction losses from the test sample, the bath was heated by a 400-w quartz immersion heater and controlled to within ± 0.1°F by means of a mercury thermoregulator. When a liquid nitrogen bath was used, the level of the liquid nitrogen was automatically controlled by means of a Cryogenics, Inc. Cryotrol Model 100L.

The mean temperature of the blackened hemisphere was determined by means of three Teflon-insulated 24 B&S gage copper-constantan thermocouples tackwelded to the hemisphere at the positions shown in Fig. 1. Based on the three thermocouple measurements, a mean surroundings temperature was calculated assuming that the test surface was a diffuse absorber.

Vacuum System

The vacuum pumping system consisted of a Cenco Hyvac 14 vacuum pump in series with a Veeco Model EP-25W, 2-in., water-cooled diffusion pump. With this pumping system, it was possible to attain ultimate pressures in the test chamber of approximately 0.1 and 0.03 μ with ice water and liquid nitrogen surroundings, respectively. The total pressure was monitored by means of a CVC Type PHG-T-01 combination Philips-thermocouple gage.

Test Sample Substrates

The test sample substrates were 1.125 x 0.640-in. x 16-oz (0.0216-in.-thick) pieces of ABCO alloy No. 100 copper sheet. This alloy was 99.9+% pure copper and, consequently, had a high thermal conductivity which resulted in a uniform temperature throughout the substrate.

The temperature of each test sample substrate was measured by means of a Teflon-insulated 30 B&S gage copper-constantan thermocouple tackwelded to the center of the substrate, the wires being recessed in a groove. The thermocouple wire was premium grade and two thermocouples, one made with wire from each end of the two 500-ft. spools, were calibrated and found to give uniform readings within the equivalent of 0.02°F.

The copper substrate was heated by means of a Budd RP-7 metalfilm precision resistor, which consisted of a constantan element on an epoxy backing material, bonded to the back of the substrate with Eastman 910 adhesive. These resistors had a resistance of 600 $\pm$ 1 ohms over their entire operating temperature range of 75 - 200°F and, therefore, made excellent, extremely thin (0.01-in.-thick) heaters of constant resistance for relatively low-power requirements (500 mw or less). After a resistor had been bonded to the copper substrate, Teflon-coated 34 B&S gage stranded copper lead wires were soft soldered to the resistor and the bare constantan element was coated with a clear electrical varnish.

A photograph of an instrumented copper substrate is shown in Fig. 2. To protect the instrumentation from damage, it was backed with a piece of 3M Company Y9050 aluminized tape. (This tape was applied after the photograph was taken.)

Power to the heater was provided by a Harrison Laboratories Model 800 A-2 low-voltage d.c. power supply. Although this power supply had a coarse and fine voltage adjustment, a Beckman Model A, linear, 10-turn potentiometer with a dial indicator was placed in series with the power supply for extra fine voltage adjustment. The current through the heater circuit was determined very accurately by measuring the voltage drop across a Leeds and Northrup Catalogue No. 4020-B 1-ohm d.c. standard resistor with a Leeds and Northrup Type K-3 potentiometer.

Guard

The function of the guard (see Fig. 3) was to provide a support for the instrumented test sample which would allow the top surface of the test sample to exchange energy with the blackened hemisphere while limiting to a small, measurable amount the solid conduction heat losses from the edges and back of the sample and from the test sample thermocouple and heater leads. The guard shell consisted of a copper box (3-1/2 x 3 x 1-1/4 in.) fabricated from a single piece of 32-oz (0.0431-in.-thick) ABCO alloy 100 and having a Watlow 100-w brass sheathed strip heater (3 x 2-1/2 in.) soft soldered to its outer bottom surface. A block of Type FR Styrofoam was bonded to the inside of the box with GE RTV-102 adhesive/sealant; the top surface of the Styrofoam had a place hollowed out for mounting the instrumented test sample. To minimize radiation heat losses from the top surface of the guard, two layers of 1-mil-thick Mylar sheet, aluminized on both faces, were bonded to the top surface of the Styrofoam using RTV-102.

The guard temperature was measured and controlled by means of a Teflon-insulated 30 B&S gage copper-constantan thermocouple and a Yellow Springs Instrument Co. (YSI) Catalogue No. 409 thermistor probe, respectively. The thermocouple was tackwelded to the center of the inner surface of one long side of the copper box and the thermistor probe was bonded to the center of the opposite inner surface with Eastman 910 adhesive. To prevent excessive heat losses from these leads and, hence, erroneously low readings, the leads were maintained in thermal contact with the guard for a distance of several inches by being held securely to the outer surface of the guard with Y9050 aluminized tape.

The temperature of the guard was controlled by means of a YSI Model 71 temperature controller. To prevent the temperature of the guard from overshooting the set point, the heat input was reduced as necessary by adjusting the voltage to the heater by means of a variable transformer.

Instrumentation Used for Temperature Measurements

Millivolt outputs of the five thermocouples were measured by means of a Leeds and Northrup Type K-3 Universal potentiometer, Model 7553, using an ice bath to maintain a 32°F reference junction. To prevent any spurious emfs from developing as a result of temperature gradients at terminals, the only dissimilar metal junctions occurred at the points of measurement and at the reference junction. This necessitated using a copper-constantan plug and copper-constantan jacks to switch from one thermocouple to another. A Leeds and Northrup Catalogue No. 3294 pinch type reversing switch was used to switch from thermocouple readings to voltage drop readings across the 1-ohm standard resistor.

Test Sample Surfaces

Sample 1 consisted of Parsons optical black lacquer (topcoat) applied with a camel hair brush to a thickness of 0.005 in. on copper which had been roughened with No. 1 emery paper.

Sample 2 consisted of an aluminum pigmented silicone base paint sprayed to a thickness of 0.004 in. on copper which had been roughened with No. 1 emery paper. The paint formulation used was the following: TT-468-A aluminum - 60, glass resin No. 650 - 40, and butanol - 60. This formulation was ball milled for 3 hr and had to be thinned with additional butanol for spraying. The green coating was heat cured for 8 hr at 180°F.

Sample 3 was a highly polished ABCO alloy No. 100 (99.9+ copper) substrate. Immediately after polishing, the sample was stored in a desiccator. Despite this, the test surface was slightly tarnished when its total hemispherical emittance was measured. The root mean square roughness height of a similar surface was measured with a Lufkin Model 941 Mezursurf console and was found to vary between 2 and 5 μ in.

Sample 4 was prepared by bonding 5-mil platinum foil (99.9+ pure) to copper (as received) with Eccobond solder 56C, a high thermal conductivity cement. After bonding, the platinum surface was highly polished.

Sample 5, highly polished tantalum (99.9+ pure), was prepared in the same manner as Sample 4.

Sample 6 consisted of 0.2-mil-thick electroplated gold applied to a highly polished copper substrate. The electroplating was performed per Mil Spec MIL-G-45204A Type II using

Sel Rex Corporation Autronex C gold plating solution. The purity of the deposit was 99.0+.

Sample 7 consisted of 0.3-mil-thick vacuum vapor-plated gold applied to a highly polished copper substrate. The purity of the gold deposit was 99.9+.

All polishing was done using standard metallurgical procedures. The final polishing was done using Linde Type B-5125 alumina abrasive. After polishing, test samples were rinsed in tap water, rinsed in acetone, and air dried.

The test surface areas were measured with a micrometer at room temperature (approximately 75°F). The corresponding areas at higher temperatures were found by means of the appropriate coefficient of linear thermal expansion obtained from the Handbook of Chemistry and Physics.[12]

After the total hemispherical emittance of Sample 6 was measured and before Sample 7 was mounted in the emissometer, a Gier Dunkle Model DB100 infrared reflectometer became available for rapid measurement of total normal reflectance of test surfaces at room temperature. This instrument utilizes a comparative technique based on a high-reflectance gold standard and a low-reflectance black standard. The total normal emittances of Samples 3 through 7 as measured with the Gier Dunkle reflectometer were 0.019, 0.037, 0.042, 0.037, and 0.021, respectively. These reflectance data agree closely with the handbook[12] data for Samples 3, 4, and 7. Considering that most of the reflectance data were taken after the total hemispherical emittance measurements were made, this agreement indicates that the quality of preparation of the metallic surfaces for Samples 3, 4, and 7 was high and that the total hemispherical emittance testing did not significantly alter the surface characteristics.

Experimental Procedures

Test Sample Mounting

The resistance of each test sample heater was measured with a Millivac Instruments RM-279C linear-precision-ohmmeter having an accuracy of better than ± 1% of full scale. After it was verified that the heater resistance was 600 ohms, as advertised by the manufacturer, the instrumented test sample was mounted in the guard (see Fig. 3) with its test surface just slightly (0.01 to 0.02 in.) below the top surface of the guard.

The test sample thermocouple and power leads were passed through a 3/16-in. diam hole in one end of the copper box, wrapped once around the outer surface of the box, and passed through a cylindrical radiation shield and out the test chamber by way of the 1/4-in. o.d. tube section in the bottom plate. The leads were held in good thermal contact with the outer surface of the guard by means of Y9050 aluminized tape. With this arrangement, heat losses (or gains) from (or to) the test sample by conduction along the four leads could be held to a small, measurable amount which depended only upon the temperature difference between the test sample and the guard.

Ice Water Surroundings

These experiments were conducted to measure the total hemispherical emittance of the test samples in an environment with a temperature relatively close to the temperature of the radiating surface. A typical series of test runs with ice water surroundings was performed as follows:

a) An ice water mixture was added to the bath and the temperature of the test chamber was allowed to attain steady state.

b) Simultaneously with step a, the test chamber was evacuated until its pressure was stabilized in the 0.01 to 0.1 μ range.

c) The guard temperature controller was set at a nominal temperature of 75°F and power was applied to the guard and the test sample.

d) Once the guard temperature had stabilized, the power to the test sample was adjusted until the mean temperature of the guard and the temperature of the test sample were nearly identical.

e) After the guard and test sample temperatures had reached steady state, the following measurements were taken: the test chamber pressure (as indicated by the Philips gage), the emf outputs of the five thermocouples (guard, test sample, and three on hemisphere), the voltage drop across the 1-ohm standard resistor, and the temperature of the resistor.

f) Once the measurements for a nominal test sample temperature of 75°F had been satisfactorily completed, steps c, d, and e were repeated for nominal temperatures of 100, 150, and 190°F. (All samples were not measured at all four temperatures.)

The total hemispherical emittance of all test samples was first measured in ice water surroundings before being measured in liquid nitrogen surroundings. Consequently, to determine what effect, if any, the test environment had on the sample surface characteristics, an additional one or two ice water runs were made on several of the polished metal samples after the liquid nitrogen runs had been completed.

Liquid Nitrogen Surroundings

These experiments were conducted to measure the total hemispherical emittance of the test samples in an environment whose temperature was low enough so that the sample was nearly freely radiating. Test runs were made in the same manner as those made with ice water surroundings. After the total hemispherical emittance measurements had been taken, the resistance of each test sample heater was remeasured with the linear-precision-ohmmeter and the test surface of each sample was visually examined. Within the accuracy of the ohmmeter ($\pm$ 1% of full scale), the resistance of each of the seven heaters was 600 ohms, as was the case before testing. Finally, other than the crazing of Sample 1, Parsons black, no changes were noted in the conditions of the test surfaces before and after testing.

Solid Conduction Effects

Whenever a difference between sample and guard temperatures occurred, heat was either lost or gained by the sample as a result of solid conduction through the four lead wires and through the Styrofoam block. These heat effects were experimentally determined for each test sample using the technique described in Ref. 8.

Gas Conduction Losses

In the pressure range in which the present experiments were conducted (on the order of 0.1 μ), convection heat transfer did not exist as there was no bulk flow of the gas in the test chamber. Gas conduction, however, can be significant (nearly 10% of the total heat lost by the test sample) in this pressure range, especially when the test sample is a polished metal having a low emittance. The amount of gas conduction between the test sample and the blackened hemisphere was determined experimentally for each sample for both ice water and liquid nitrogen surroundings.

Background Radiation

Using measured temperatures and knowing the appropriate dimensions and thermo-optical properties, it was possible to estimate the amount of background radiation received by the test sample in both ice water and liquid nitrogen surroundings. The results of the background radiation calculations show that the amount of background radiation absorbed by the test sample is quite small, a maximum of 0.5% of the power radiated by the test sample for ice water surroundings and a test sample temperature of 75°F. These calculations were also used as corrections to the measured power losses of the test samples in addition to the solid conduction and gas conduction corrections described previously.

Results and Discussion of Results

The total hemispherical emittance data for the metallic test samples in ice water and liquid nitrogen surroundings are presented in Table 1 and Figs. 4 through 8.

Table 1 Comparison of total hemispherical emittances of metals measured in ice water surroundings with those measured in liquid nitrogen surroundings [a]

Sample No.	Sample Temp (°F)	ϵ (Ice H_2O, Nongray [b])	ϵ (LN_2)	$\left[\frac{\epsilon(\text{Ice } H_2O) - \epsilon(LN_2)}{\epsilon(LN_2)}\right]100$
2 (Al Paint)	75	0.374	0.371	+ 0.81
	100	0.377	0.375	+ 0.53
	150	0.381	0.382	− 0.26
	190	0.387	0.387	0.00
3 (Cu)	100	0.103	0.106	− 2.83
	150	0.104	0.107	− 2.80
	190	0.105	0.108	− 2.78
4 (Pt)	75	0.086	0.092	− 6.52
	100	0.091	0.095	− 4.21
	150	0.099	0.102	− 2.94
	190	0.105	0.108	− 2.78
5 (Ta)	75	0.119	0.120	− 0.83
	100	0.120	0.122	− 1.64
	150	0.128	9.126	+ 1.59
	190	0.133	0.130	+ 2.31
6 Au-Electro plated	75	0.140	0.143	− 2.10
	100	0.142	0.144	− 1.39
	150	0.146	0.145	+ 0.69
	190	0.148	0.147	+ 0.68
7 Au-Vapor-Plated	100	0.078	0.080	− 2.50
	150	0.078	0.080	− 2.50

a Average Deviation = −1.40%.

b ϵ (Ice H_2O, nongray) values were obtained from ϵ (ice H_2O, gray) data by means of Eqs. (23) and (24).

Parsons Optical Black Lacquer

The present data for Parsons black [ϵ (avg.) = 0.984] agree closely with the data published by the Eppley Foundation for Research,[13] which gives an infrared absorptance between 0.99 and 0.98 for a wavelength range of 1 - 18 μ, and with the data of Faugère and Delpont,[14] who report a total hemispherical emittance of 0.98 at 300°K (80°F).

Aluminum Pigmented Silicone Base Paint

The data on aluminum pigmented silicone base paint (Fig. 4) illustrate the need for making a nongray correction to the data taken with ice water surroundings. Without the nongray correction, it appears that the ice water surroundings produce a higher emittance. With this correction, however, it is seen that both sets of data very nearly coincide.

Optically Smooth Metals

The data on polished copper, tantalum, and platinum, and on electroplated and vapor-plated gold on polished copper (Figs. 5-8) lead to the conclusion that Kirchhoff's law is valid for freely radiating metallic surfaces. If the ice water surroundings had stimulated the test surface to emit more radiant energy from the test surfaces than the liquid nitrogen surroundings, then the emittances measured in ice water surroundings would have been larger than those measured in liquid nitrogen surroundings. In particular, had the theory of Ashby and Schocken been correct, the emittances measured in ice water surroundings would have averaged 5.29% larger (Table 2). No such effect was found. In fact, the emittances measured in ice water surroundings averaged 1.40% less (Table 1), which was within the experimental error. Thus, within the experimental error, $\alpha = \epsilon_{eff}$, and Kirchhoff's law is valid.

Although an average difference of 1.40% is within the experimental error, there are three possible effects, all slight, but which, in combination, could account for this difference. These effects are the following:

a) Most of the ice water data were measured first and, hence, any contamination which would have progressed with time would have caused an increase in the liquid nitrogen emittance. For example, ice water emittance data on platinum and gold were slightly higher when measured after the liquid nitrogen data had been measured (Figs. 6 and 8).

b) Although sample nongray effects are very small for liquid nitrogen surroundings (on the order of 0.3%), they do exist and, therefore, the true liquid nitrogen emittance is somewhat smaller than that presented.

c) Polished metals emit and absorb more strongly at the larger angles to the normal; hence, the mean temperature of the surroundings based on the diffuse assumption will be somewhat low, giving too large a $T_1^4 - T_2^4$. This effect will be most significant for ice water surroundings (approximately 0.1%) and will result in calculated values which are slightly too small.

Table 2 Comparison of theoretical total hemispherical emissivities of metals for ice water surroundings with those for liquid nitrogen surroundings – Ashby and Schocken theory

Sample Temp (°F)	ϵ (Ice–H_2O) / a	ϵ (LN_2) / a	$\left[\frac{\epsilon(\text{Ice-}H_2O) - \epsilon(LN_2)}{\epsilon(LN_2)}\right]$ 100
75	0.987	0.927	+ 6.47%
100	0.980	0.926	+ 5.83%
150	0.970	0.926	+ 4.75%
190	0.963	0.925	+ 4.11%

a Average deviation = + 5.29%.

In general, the total hemispherical emittance values reported in this paper are higher than those reported by other investigators. A notable exception is the total hemispherical emittance data of polished platinum reported by Butler and Inn,[15] which are 0.02-0.03 emittance units higher than the current values. No attempt to explain this situation will be made in this paper; however, it is the author's opinion that the difference between hemispherical and normal emittance can be much greater than the difference predicted by classical electromagnetic theory[+] and that much of the previous data was prejudiced by the results of classical electromagnetic theory. Based on their total hemispherical emittance measurements of polished nickel, tantalum, and molybdenum, Butler, Jenkins, Rudkin, and Laughridge[16] were also of the opinion that hemispherical emittance could not be calculated from the normal.

[+]Classical electromagnetic theory predicts that the total hemispherical emittance of an optically smooth surface will at most be only a factor of approximately 1.33 greater than the total normal emittance.

The large difference (approximately 80%) in the measured total hemispherical emittances of the electroplated and vapor-plated gold, as shown in Fig. 8, is quite surprising. However, the total normal emittance data for the two golds also shows a difference of (0.037-0.021/0.021) 100 ≈ 80%, which verifies the large difference in total hemispherical measurements. The reason for this difference in emittance is most likely a combination of purity (electroplated 99.0+ and vapor-plated 99.9+) and the greater porosity of the electroplated deposit which was observed by means of a scanning electron microscope. According to Fulk and Reynolds,[17] there is a wide spread in the properties of electroplated surfaces, and purity and density are more important than the highly polished look, especially beyond a wavelength of 4 μ.

Surface Contamination

The radiation process in metals normally takes place within a depth of only a few thousand angstroms; hence, the characteristics of the surface can significantly affect the thermo-optical properties of a metal. Contamination and damage to an optically smooth metal surface can occur in a wide variety of forms which include the following: fingerprints, scratches, water films, oil films, oxidation, adsorbed gases, and disorders in the lattice structure.

Although a reasonable amount of care was taken not to contaminate or damage the test surfaces between the time they were prepared and when the measurements were taken, it was impossible to eliminate completely all forms of contamination. According to Caren,[18] total hemispherical emittance values at low temperatures are often too high as a result of contamination on the surface in the form of water and oil films. Water films were not a problem in the current experiments as all measurements were taken in vacuo at pressures substantially below 22.3 torr, the vapor pressure of water at the lowest test sample temperature of 75°F. Oil contamination, however, was quite probable, even if only to a very slight extent.

Another possible source of contamination during test runs was outgassing products of materials in the test chamber, such as the adhesive on the aluminized tape. At any rate, visual inspection of the test surfaces after testing and experimental results indicated that surface contamination by outgassed products and diffusion pump oil had little or no measurable effect on the total hemispherical emittance. Figs. 6 and 8, however, reveal that the ice water emittance data were slightly higher for Samples 4 and 7 when measured after the liquid nitrogen data runs than the previous measurements made before the liquid nitrogen runs. Some contamination, probably oil, may have occurred so far as these two samples were concerned.

Oxidation of the test surface was a serious problem with copper and to some extent with tantalum, but it was not a problem with gold, and, in particular, with platinum. Visual examination of the copper sample showed it to be slightly tarnished before and after its emittance was measured.

If the results of Parker and Abbott[19] are correct, then adsorbed gases could have a quite significant effect on the measured emittance of the tantalum sample. According to these authors, tantalum exhibits a gettering action for gases in a vacuum system at temperatures below 1500°K with a corresponding increase of approximately 35% in the total hemispherical emittance. Finally, contamination in the form of disorders in the lattice structure can arise in metal surfaces as a result of mechanical polishing.[17, 20, 21] Test surfaces of Samples 3, 4, and 5 were mechanically polished and, therefore, possessed this type of contamination which could have been largely avoided had the specimens been electropolished.

When all possible types of contamination are considered, Sample 7, vapor-plated gold, had the lowest probability of being significantly contaminated. Even so, the measured total hemispherical emittance of this sample (approximately 0.08) was considerably larger than the measured total normal emittance (approximately 0.02) and therefore much larger than the handbook value for gold.

To what extent contamination was present in each test sample is not known; however, any contamination that did exist would result in total hemispherical emittance values higher than those for the pure, uncontaminated metals. If the degree of contamination did not increase during the time that emittance measurements were being made, however, contamination would not have affected the comparison of the total hemispherical emittances measured in the two environments.

Conclusions

Based upon the experimental results of this investigation, it can be concluded that Kirchhoff's law is valid within $\pm$ 2% for freely radiating metallic surfaces. Thus, the power radiated by a metallic surface is the same when freely radiating as it would be under equilibrium conditions. These experimental results are supported by calculations which show that if the material states have their equilibrium distribution, then Kirchhoff's law is valid whether or not the metal is in thermal equilibrium with the radiation field, provided that induced emission is considered to be negative absorption. In other words, Kirchhoff's law should be interpreted to be

$$\alpha(\text{net}) = \alpha(\text{true}) - \epsilon(\text{stimulated}) = \epsilon(\text{spontaneous})$$

where stimulated emission is contained in the definition of absorption rather than

$$\alpha(\text{true}) = \epsilon(\text{total}) = \epsilon(\text{spontaneous}) + \epsilon(\text{stimulated})$$

Therefore, even though the radiating body and the radiation field are not in thermal equilibrium, if the material states have their equilibrium distribution, the spontaneous emission (which does not directly depend upon the incident radiation) will be the same as it would be at thermal equilibrium, and Kirchhoff's law will be valid. If the amount of induced emission is negligible in comparison with the amount of spontaneous emission, then the way in which Kirchhoff's law is interpreted is of no significant consequence.

References

[1]Weinstein, M. A., "On the Validity of Kirchhoff's Law for a Freely Radiating Body," American Journal of Physics, Vol. 28, No. 2, Feb., 1960, pp.123-125.

[2]Ashby, N. and Schocken, K., "The Theory of Emissivity of Metals," Symposium on Thermal Radiation of Solids, NASA SP-55, March, 1964, pp.63-71.

[3]Einstein, A., "Zur Quantentheorie der Strahlung," Physik Zeitschrift, Vol. 18, 1917, pp.121-128.

[4]Vincenti, W. G. and Kruger, C. H., Jr., Introduction to Physical Gas Dynamics, Wiley, New York, 1965.

[5]Sparrow, E. M. and Cess, R. D., Radiation Heat Transfer, Brooks/Cole, Belmont, Calif., 1966.

[6]Cess, R. D., letter, State University of New York at Stony Brook, Sept. 19, 1967.

[7]Oster, L., "Radiation Laws under Non-Equilibrium Conditions," U. S. Air Force, ARL 62-320, April, 1962.

[8]Grimm, T. C., "An Investigation of the Validity of Kirchhoff's Law for Freely Radiating Metallic Surfaces," D. Sc. Thesis, Washington University, St. Louis, Mo., Jan., 1970.

[9]Ashby, N. et al, "Theoretical Study of the Radiative Emissivity of Materials," Contract No. NAS8-20365, Prepared for NASA MSFC by P. E. C. Research Associates, Inc.

[10]Agababov, S. G., "Some Aspects of Calorimetric Determination of Degree of Blackness," translated from Teplofizika Vysokikh Temperatur, Vol. 3, No. 6, Nov./Dec., 1965, pp. 933-935.

[11]Eckert, E., "Messung der Reflexion von Wärmestrahlen an Technischen Oberflachen," Forschung auf dem Gebiete des Ingenieurwesens, Vol. 7, 1936, pp. 265-270.

[12]Handbook of Chemistry and Physics, 44th ed., The Chemical Rubber Publishing Co., Cleveland, 1963.

[13]Harris, L., "The Optical Properties of Metal Blacks and Carbon Blacks," Monograph Series No. 1, MIT and The Eppley Foundation for Research, Dec., 1967.

[14]Faugère, J. R. and Delpont, J. P., "Mesure Thermique de l'Emittance Totale Hémispherérique d'un Revetement," Centre Natl D'Etudes Spatiales, NT-4, April, 1966.

[15]Butler, C. P., and Inn. E. C. Y., "The Total Hemispherical Emissivity of Metals," U. S. Naval Research Development Lab., TR-327, May 28, 1959.

[16]Butler, C. P. et al., "High Temperature Surface Parameters for Solar Power, Coatings for the Aerospace Environment," Wright Air Development Division, TR-60-773, Nov. 9-10, 1960, pp. 139-162.

[17]Fulk, M. M. and Reynolds, M. M., "Emissivities of Metallic Surfaces at 76° K," Journal of Applied Physics, Vol. 28, No. 12, Dec., 1957, pp. 1464-1467.

[18]Caren, R. P., "Low Temperature Emittance Determinations," AIAA Progress in Astronautics and Aeronautics: Thermophysics and Temperature Control of Spacecraft and Entry Bodies, Vol. 18, edited by G. B. Heller, Academic Press, New York, 1966, pp. 61-73.

[19]Parker, W. J. and Abbott, G. L., "Theoretical and Experimental Studies of the Total Emittance of Metals," Symposium on Thermal Radiation of Solids, NASA SP-55, March, 1964, pp. 11-28.

[20]Blackman, M., Egerton, A., and Truter, E. V., "Heat Transfer by Radiation to Surfaces at Low Temperatures," Proceedings of the Royal Society of London, Vol. A 194, 1948, pp. 147-169.

[21]Bennett, H. E., "Influence of Surface Roughness, Surface Damage, and Oxide Films on Emittance," Symposium on Thermal Radiation of Solids, NASA SP-55, March, 1964, pp. 145-152.

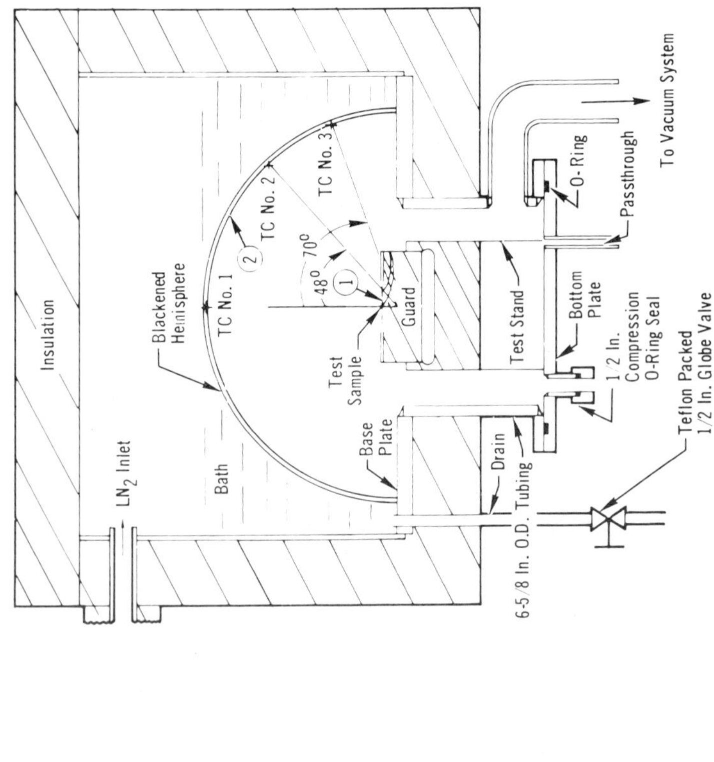

Fig. 1 Cross section of test chamber.

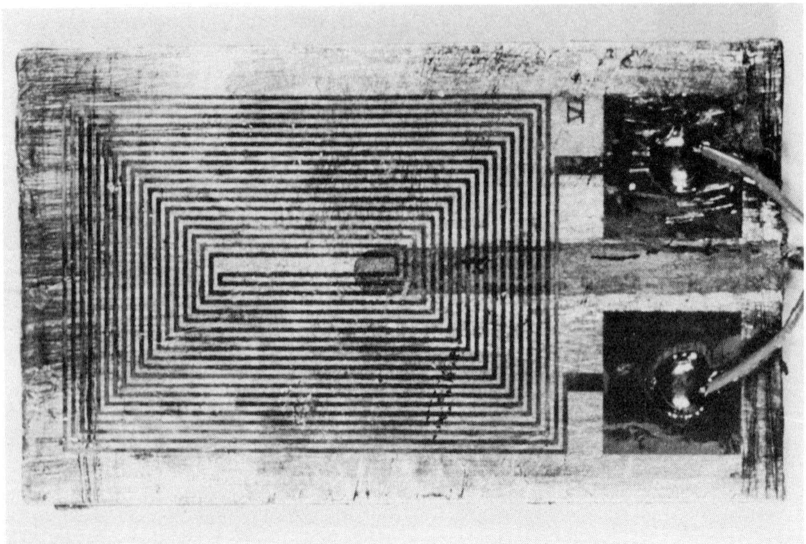

Fig. 2 Instrumentation on back of copper substrate.

Fig. 3 Test sample mounted in guard.

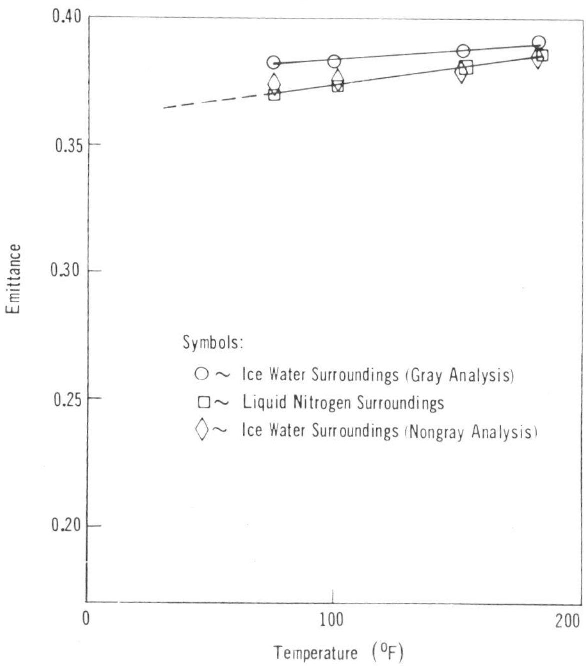

Fig. 4 Total hemispherical emittance of aluminum pigmented silicone base paint vs temperature.

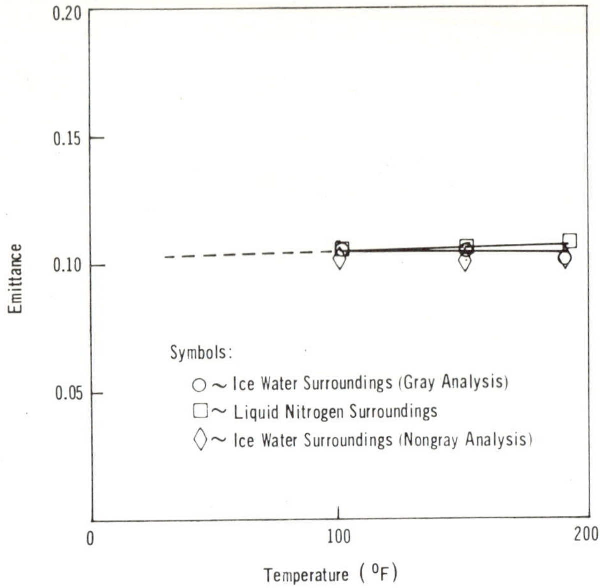

Fig. 5 Total hemispherical emittance of polished copper vs temperature.

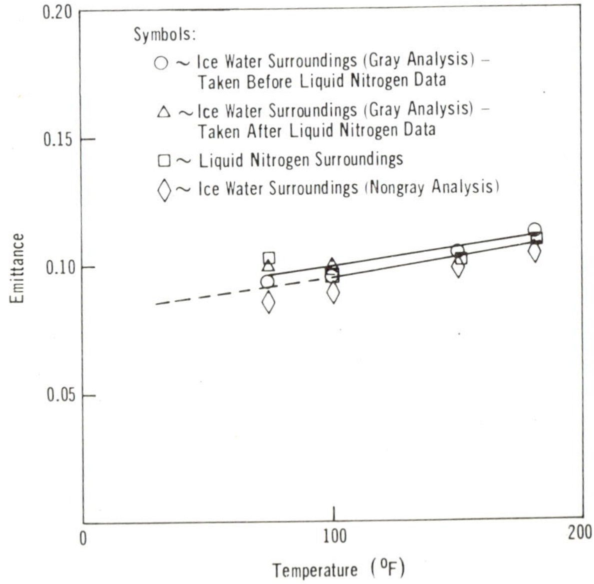

Fig. 6 Total hemispherical emittance of polished platinum vs temperature.

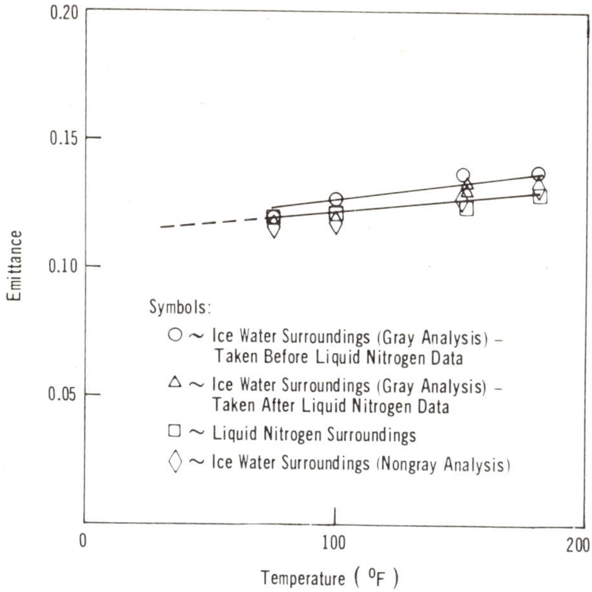

Fig. 7 Total hemispherical emittance of polished tantalum vs temperature.

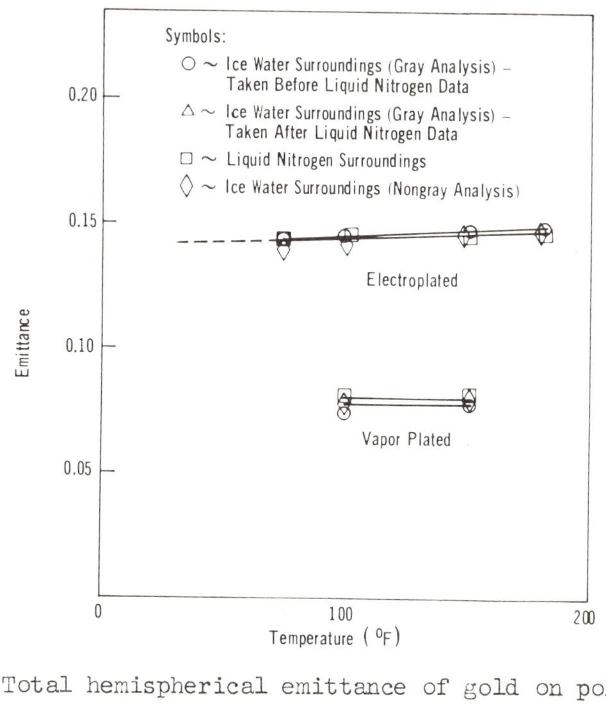

Fig. 8 Total hemispherical emittance of gold on polished copper vs temperature.

II. SURFACE RADIATION PROPERTIES

II.2 Analytical Aspects

INVESTIGATION OF THE EFFECTS OF SURFACE ROUGHNESS UPON REFLECTANCE

D. C. Look Jr.[*], University of Missouri-Rolla, Rolla, Mo.

and

T. J. Love[+], University of Oklahoma, Norman, Okla.

Abstract

The results of an analytical study, in which Monte Carlo techniques were used, illustrate the effects that statistically constructed, rounded, V-groove characteristics have on directional reflections in the plane of incidence, and when compared graphically to the experimental data of other investigators, indicate that the approach yields a close approximation for variations of the angle of incidence and the wavelengths of incident radiation. The model includes RMS peak-to-valley and peak-to-peak dimensions, a peak roundness parameter, the hypothesis of an energy bundle localization circle, multiple reflections, and shadowing.

Nomenclature

a	= peak-to-peak dimension
$\bar{a}$	= RMS peak-to-peak dimension
a'	= autocovariance length
b_1	= radius of curvature of rounded portion
F_N	= normalized detecting area (regular reflected response function)
h	= penetration depth
H_1	= center of curvature of left side of rounded portion of V-groove
$N(\psi;\theta,\phi)$	= number of energy bundles received about the direction (θ,ϕ) when ψ is the angle of incidence
P	= roundness parameter

Presented as Paper 70-820 at the AIAA 5th Thermophysics Conference, Los Angeles, Calif., June 29 - July 1, 1970.

*Assistant Professor of Mechanical Engineering.

+Professor and Director, School of Aerospace and Mechanical Engineering.

P(LIM) = limiting value of P for basic surface configuration
P(left) = probability that first incidence is on the left side of rounded V-groove
X_o = V-groove vertex position (valley)
θ = angle of reflection measured from the mean surface normal
θ_1 = $\tan^{-1}$ (slope of left side of V-groove)
λ = wavelength of incident energy bundle
μ = micron (10^{-6}m)
σ = peak-to-valley dimension
$\bar{\sigma}$ = RMS peak-to-valley dimension
σ_o = RMS optical roughness
ϕ = azimuthal angle of reflected energy measured from the plane of incidence
ψ = angle of incidence measured from the mean surface normal

I. Introduction

It is known from experimental and theoretical approximation that the roughness of surfaces of engineering interest drastically modifies the spatial distribution of reflected energy. Thus, these surfaces exhibit a superposition of regular and nonregular reflection characteristics to varying degrees depending upon such variables as wavelength, surface geometry, and small scale surface irregularities. Investigations of these surfaces have produced theories for the cases of $\sigma/\lambda >> 1$ and $\sigma/\lambda < 1$, but in cases where $\sigma/\lambda \sim 1$, there is extreme difficulty in handling the analyses. Experimental investigations of the interrelation of the various physical variables involved in the accurate prediction of reflected energy distributions have not been completely successful.

The primary characteristic dimension of a surface is the RMS surface roughness or the height of the irregularities. Obviously this dimension will not describe a rough surface adequately, because it tells nothing about the distances between the hills or the valleys of the surface. Thus a peak density or asperity function becomes important. In the literature, the second dimension, if discussed, is the autocorrelation or autocovariance length. This length is related to the RMS surface roughness and the RMS slope of the surface.[1]

To investigate the effects of certain predescribed surface parameters on the bi-directional reflectance, an analytical study using an empirical sampling technique was initiated. The results of this study illustrate the effects of statistically constructed, rounded, V-groove surfaces upon the directional reflectance in the plane of incidence. By comparing the output of this model with the experimental results published by other investigators, it can be stated that this approach yields bi-directional reflectances which are a close approximation to the empirical results for variations of the angle of incidence and the wavelength of the incident radiation. In addition, the definition of the two parameters of height and length and their associated distributions, plus knowledge of the electrical properties of the material for the wavelength considered and the conditions of irradiation and observation, is not all that is needed to determine the reflected energy distribution. This investigation has indicated that a dimension characteristic of that part of the surface which is essentially flat must be stated.

II. Description of the Model

It is assumed that the vast majority of the energy received in the plane of incidence is the result of interreflection within the rounded V-grooves, the sides of which are generally perpendicular to the plane of incidence.

The Surface

Surfaces which yield nonregular reflection are "rough" and the exact character of the "roughness" cannot be stated in a general form. During this investigation an attempt was made to bring the model into agreement with other investigators[2,3] by assuming that the general character of "roughness" is that of a V-groove. It was also assumed that the tops of the grooves are rounded and that the rounded as well as the V-groove portions reflect in a specular fashion.

Experimentally, the parameters, which are usually used to describe a surface, are an RMS peak-to-valley dimension, $\bar{\sigma}$, and an autocovariance function, a'. Because of the lack of experimentally verified information concerning a', the authors prefer to consider a parameter as indicative of the peak density. This parameter is an RMS peak-to-peak dimension, $\bar{a}$. Whether or not this dimension is related to the autocovariance length is of no concern.

To make the model as real as possible, the surface heights, σ, were distributed in a Rayleigh distribution with $\bar{\sigma}$ as the defining parameter. The peak-to-peak dimension, a, was also distributed according to a Rayleigh distribution but with $\bar{a}$ as the defining parameter. The apex of the V-groove, X_o, was assumed to have been selected in a uniformly random fashion

(Fig. 1). For each energy bundle, σ, a, and X_o were selected from the approximate random distributions, thus defining the surface. The energy bundle was followed in this geometry up to four reflections unless it escaped to the hemisphere above, was refracted at a surface, or was trapped by being reflected four times.

A roundness parameter, P, was used to determine the character of the circular caps of the rounded V-groove, because P is a fraction of σ and denotes the actual height of the irregularity. The transition from the circular cap to the V-groove is continuous. The surface shape is very much dependent on the roundness parameter P. When P=1, the surface is triangular. As P decreases, the radius of the round portion, required for continuity, increases until for a particular value of P, given σ and a, the transition point is superimposed on X_o. The value of P which produces this geometrical configuration for the particular case of X_o =a was defined as P(LIM). From a geometrical consideration, an expression for P(LIM) was derived. The result was

$$P(LIM) = \frac{a}{\sigma}\left[\sqrt{1 + \left(\frac{a}{\sigma}\right)^2} - \frac{a}{\sigma}\right]$$

It should be noticed that P(LIM) is never greater than 0.5 and decreases monotonically to zero as σ/a approaches infinity. Thus for given values of σ and a, various cases must be considered if P<0.5. Figure 2 illustrates three of the six unusual cases that may be encountered when P<P(LIM). These were included in the analysis. The other three cases are related similarly but are on the left side of X_o.

Hypothesis of a Localization Circle

Quantum mechanics indicate that it is not possible for an energy bundle to be localized in a region less than the wavelength of the bundle before a collision or interaction occurs.[4] In an attempt to analytically reproduce the experimental results of other investigators and to utilize the aforementioned statement (i.e., to couple the surface parameters to the wavelength of the incident radiation), the quantum mechanical wavepacket, in which the energy bundle is localized, was assumed to be a sphere, or a circle in the two-dimensional model that was used in the investigation. This "billiard-ball" model was used to indicate by means of a geometric calculation the initial, maximum depth of penetration of the energy bundle into the rounded V-groove. The point of first incidence cannot be deeper than this maximum depth. The calculation assumed a "localization circle" of radius λ and dependent on the ratios σ/a and σ/λ.

The interpretation of this depth of penetration is as follows: The intersection of the height with the sides of the V-groove for the $\psi \leq \pi/2 - \theta_1$ case and the right side of the V-groove for $\psi > \pi/2 - \theta_1$ case defines a forbidden zone in which the energy bundle cannot strike the V-groove for first incidence. This geometry is illustrated in Fig. 3. Notice that an account for shadowing is made. That is for the $\psi > \pi/2 - \theta_1$ case, the shadow effectively widens the forbidden zone. The assumption of the "localization circle" is based on the experimental evidence of many investigators who state that if the relationship of σ and λ is such that only nonregular reflection is obtained, increasing the wavelength will eventually produce a large regular reflection component. Thus for the model used in this investigation, as λ increases, the first incidence point of the energy bundle comes close to the tops of the V-groove. The tops of the V-groove are rounded and are exactly flat at the peaks, thus, the model follows the experimental evidence in a gross fashion.

Determination of Point of First Incidence

A decision must be made as to which side of the rounded V-groove the energy bundle strikes first. This decision must depend on σ, a, λ, and the depth of penetration.

The hypothesis for a depth of penetration introduced the idea of a forbidden zone in which the energy bundle cannot strike the surface for first incidence. Thus the determination of the side of the first incidence depends on the area available for incidence, or in the case of this two-dimensional model, the line available. For example, in Fig. 1 the probability of an energy bundle to strike the left side of the rounded V-groove is defined as

$$P(\text{left}) = \overline{AB}/(\overline{AB} + \overline{DC})$$

It is assumed that this equation is valid for all angles of incidence.

To determine the first point of incidence on either the right or left side of the rounded V-groove, it is assumed that the position probability is uniform across the beam between lines along the incident direction that intersect the surface at the penetration depth and the peak on the right side and intersect the surface at the peak and either the penetration depth or the shadow boundary on the left side. If the angle of incidence and surface geometry are such that a shadow does occur, the shadow boundary will define the beginning of the forbidden zone.

Procedure

The construction of a new rounded V-groove was required for each energy bundle, so the height, σ, width, a, and V-groove vertex point, X_0, were different for each sample. Therefore, the lobing effects of periodically rough surfaces were not observed.[5]

The analyses were programed for an IBM 360-40 system under the code name MCR. The primary requirements for a run were: angle of incidence, ψ, RMS height, $\bar{\sigma}$, RMS width, $\bar{a}$, wavelength, λ, roundness parameter, P, and the components of the index of refraction. With this information, surfaces were constructed in random fashion. A penetration depth was computed as well as the first incidence probability. Comparison of the probability to a uniformly distributed random number decided whether the first incidence was on the left or right side of the rounded V-groove. After determining the first incidence point, the Fresnel reflection coefficient for unpolarized light was computed for local conditions. As before, a random number comparison with this coefficient determined whether the energy bundle was refracted to reflected. If refracted, that bundle was considered to have been absorbed and a new surface was constructed. If reflected, it was regarded as obeying the laws of geometrical optics for specular reflection. The determination was then made as to whether the bundle escaped the rounded V-groove or struck the other side. If it was determined that the bundle escaped the geometry, its angle of escape was placed into one of 36 five-degree intervals and counted, and a new surface was constructed. If the bundle did not escape the geometry, a straight line trajectory was used to determine the point of incidence on the other side of the rounded V-groove. The depth of penetration restriction was removed, thus, the energy bundle could strike any line-of-sight position on the rounded V-groove after first incidence. When the point of second incidence was determined, the Fresnel coefficient was computed, and the reflected-refracted decision was made with the same alternatives. This procedure was continued up to four times, and if it was determined that the energy bundle did not escape by the fourth reflection, it was defined as absorbed and a new surface was constructed.

An analysis consisted of, at most, 10,000 energy bundles with an execution time of 4 to 5 minutes per 1000 energy bundles. The output of MCR is the ratio of the number in a five-degree interval and the maximum number in any interval.

III. Finite Aperture Effects

Before attempting to compare the model with the experimental data, consideration must be given to the set-up of the apparatus and to the question of what the resultant data from the apparatus truly represent. One point must be discussed: the detected distribution distortion produced by finite aperture sizes. Investigation of this point is important because the model presented in this article does not explicitly account for a regularly reflected component and no portion of the surface is effectively flat. MCR accounts for the radiation reflected from the rough portion of the surface, and the distribution distortion function, discussed next, is used to fit the regular component.

Because finite angles of acceptance are necessary in experimental apparatus, a certain degree of distortion will be exhibited.[6,7] This is particularly evident when the reflected energy distribution displays a sharp increase such as is encountered with a large regular component. The angle of acceptance, as used in this discussion, is the generating angle of the solid angle cone of radiation, incident or received. A simple model can be constructed in an attempt to account for this distortion. Assuming that the area viewed by the detector is circular and of radius c and that the sensitivity is uniform across this area and that the reflected beam is uniform and circular and of radius b, the output of the detector will be directly proportional to the area common to both the beam and the detecting area. Figure 4 illustrates the geometry. If u = y/c, then

$$\eta = \tan^{-1}\left[\frac{\sqrt{c^2-y^2}}{y}\right] = \tan^{-1}\left[\frac{\sqrt{1-u^2}}{u}\right]$$

$$\xi = \tan^{-1}\left[\frac{\sqrt{c^2-y^2}}{x}\right] = \tan^{-1}\left[\frac{c\sqrt{1-u^2}}{x}\right]$$

Thus, the common area is

$$S = (\eta c^2 - c^2 u\sqrt{1-u^2}) + (\xi b^2 - cx\sqrt{1-u^2})$$

Normalizing S to the detecting area yields

$$F_N = \frac{S}{\pi c^2} = \frac{1}{\pi}\left[\eta + \xi\left(\frac{b}{c}\right)^2\right] - \frac{\sqrt{1-u^2}}{\pi}\left[u + \frac{x}{c}\right]$$

$$= \frac{1}{\pi}\left[\cos^{-1} u + \left(\frac{b}{c}\right)^2 \cos^{-1}\left(\frac{x}{b}\right)\right] - \frac{\sqrt{1-u^2}}{\pi}\left[u + \frac{x}{c}\right]$$

where

$$x = c\sqrt{\left(\frac{b}{c}\right)^2 - 1 + u^2}$$

IV. Comparison of Model to Experimental Results

All theories and models, for whatever physical phenomena they are supposed to describe, must conform to reality. Thus by fitting the output of the MCR digital computer program to experimental data, statements can be made concerning the parameters that are vital to a thorough understanding of the effect of rough surfaces on reflected distribution based upon this model. Heretofore, the only surface character actually presented in a reflectance measurement was the RMS roughness, $\bar{\sigma}$. This dimension has been used in the model. In addition, an RMS peak-to-peak dimension, $\bar{a}$, has been used in lieu of the autoconvariance length introduced in the theories available. Also a peak roundness parameter, P, has been introduced. The usefulness of these parameters in describing the reflected energy distribution must be examined.

The first data to be considered were presented by Birkebak[6] who used an over-detect method in which the angles of acceptance were approximately two degrees. His test samples were aluminum coated ground glass and roughened nickel, and his data are presented as a ratio of two biangular measurements. Birkebak's only usable surface character is the RMS optical roughness, σ_o. Thus, the authors had to establish the magnitudes of $\bar{\sigma}/\bar{a}$ and P by trial and error. Figures 5 to 7 illustrate the comparisons for various data. The data for Fig. 5 were taken on a nickel sample with $\sigma_o/\lambda = 0.60$. The MCR output for $\bar{\sigma}/\bar{a} = 0.5$, $\bar{\sigma}/\lambda = 0.6$, and $P = 0.5$ fits quite well. Figure 6 is an illustration of the direct superposition of the regular and nonregular components. For the angles of acceptance given by Birkebak, the regular reflection response function was fitted to the spiked portion. This implies that a portion of the surface is effectively flat. The output of MCR was fitted to the remainder of the curve. This curve was plotted for a test sample of nickel with $\sigma_o/\lambda = 0.32$ and MCR output for $\bar{\sigma}/\lambda = 0.3$, $\bar{\sigma}/\bar{a} = 0.5$, and $P = 0.5$. Figure 7 shows the final data presented by Birkebak for which the output of MCR and the regular response were fitted. Again, the fit is quite good.

The second set of data considered for comparison was that presented by Torrance and Sparrow.[8] The material used in the experiments of these investigators was a nonconducting, fused, polycrystalline, magnesium-oxide ceramic. The acceptance angles of approximately two degrees in an under-detect experimental set-up were used.

The analysis for nonconducting material is somewhat more complicated than for a conductor. The reason for this is that there is a contribution to the received energy due to a volume effect. This effect is available when considering reflection from metals, but the magnitude of the component is quite small due to the large absorption coefficients associated with metals. In the case of dielectrics, this contribution may be sizeable. In agreement with other investigators,[3,9] it is assumed that this volume contribution is diffuse for the analysis used in this work.

As in the previous case, the data are presented as a ratio of biangular reflectances. A graphic comparison can be made from Fig. 8 and 9. These data were taken on the same sample at two angles of incidence and two different wavelengths. The variables used in the MCR digital computer program which resulted in the best fit appear in the figures. In Fig. 8A, the volume effect is represented by the constant value 0.75. The difference between 8A and 8B is only the angle of incidence. In Fig. 9, in addition to the volume effect and the rough surface effect, a sizeable regular component can be noted. This component was fitted exactly as was done in the preceding examples. The difference between Fig. 8 and 9 is that the wavelength was increased from 0.5μ to 1.5μ and the difference between 9A and 9B is the angle of incidence. It can be noted that the values used in MCR were chosen in accordance with the data presented by Torrance and Sparrow and that the fit is quite good.

Some of the data presented by Herold and Edwards[10] were considered next. These investigators used an over-detect method with very large angles of acceptance; incident 3.4°, emergence 4.7°, shadowed to 4.1°. With acceptance angles of this size and in particular with the receiving angle larger than the incident angle, the resulting detected distribution is not truly representative of the actual distribution. Data were taken on six samples of which only those for sand blasted aluminum are present for comparison. Herold and Edwards correlated the "specular" component with the theory of Davies[11] by using an RMS optical roughness of 0.8μ. The incident wavelength was 5.0μ yielding a σ_o/λ of 0.16. A graphical comparison can be made with Fig. 10. Notice that the regular component was fitted as before. As in the preceding examples, once a close fit is made for one set of parameters, only the parameter that is experimentally varied, in this case the angle of incidence, is altered. As can be seen in Fig. 10, the fit is not as good as in the preceding examples, but it still indicates an appropriate general trend.

As a final example, the data presented by Francis[7] were considered. He used an under-detect method with extremely large angles of acceptance of 6.4° incident and 5.3° receiving. The sample for consideration was sandblasted stainless steel (type 302). Some results of the comparison are presented in Figs. 11 and 12.‡ The RMS optical roughness and the wavelength appear on the figures. The parameters $\bar{\sigma}/\bar{a}$ and P were determined by trial and error for the 30° angle of incident case illustrated in Fig. 11. These numbers were then used throughout the remainder of the calculations. The regular components were of course included.

A word must be said concerning the error involved in this type of analysis. As a representative example, a 90% confidence level was chosen and the data used in Fig. 5 were considered. The results appear in Table 1. Thus, there is 90% confidence that the position of the points on the solid curve are true to within the stated error bands. These error bands are at least as good as the experimental error.

Table 1 Example error band of MCR for 90% confidence level and 10,000 samples

θ	Hits	% Error
10	670	±3.0
20	650	±3.2
30	580	±3.4
40	410	±4.0
50	270	±4.6
60	150	±7.0

‡In Figs. 11-12:

Data		MCR
•	ψ=10°	— - —
■	ψ=30°	———
▼	ψ=60°	— —

V. Summary

It is hoped that this research will lead to a better understanding of the effects of rough surface characteristics upon reflected distributions through its introduction of a penetration depth for first incidence. The localization circle, which has been used to determine the penetration depth and the radius of which is the wavelength of the incident radiation, is a unique innovation but basic to the research.

The results of the investigation have shown that the MCR outputs fit the data quite well except for one case. From this close fit, it can be inferred that not only must a peak-to-valley dimension be measured but a peak-to-peak dimension as well. Additionally, another insurface parameter is essential for an accurate description of the total reflected distribution. This dimension is in the plane of the surface and is representative of the portion of the surface that contributes to the regularly reflected component. The last two parameters have not received attention experimentally.

The graphical illustration of the examples indicate that the model developed in this investigation can represent experimental data. In fact, if the σ_o/λ ratio is known and the $\bar{\sigma}/\bar{a}$ ratio and P can be determined from one set of data, the wavelength and angle of incidence of the radiation can be varied, and MCR will yield a close approximation to the contribution of the reflection from the rough portion of the surface.

VI. References

[1]Bennett, H. E. and Porteus, J. O., "Relation Between Surface Roughness and Specular Reflectance at Normal Incidence," Journal of the Optical Society of America, Vol. 51, No. 2, Feb. 1961, pp. 123-129.

[2]Nelson, H. F. and Goulard, R., "Reflection from a Periodic Dielectric Surface," Journal of the Optical Society of America, Vol. 57, No. 6, June 1967, pp. 769-771.

[3]Torrance, K. E. and Sparrow, E. M., "Theory for Off-Specular Reflection from Roughened Surfaces," Journal of the Optical Society of America, Vol. 57, No. 9, Sept. 1967, pp. 1105-1114.

[4]Bohm, D. Quantum Theory, Prentice Hall, Englewood Cliffs, N.J., 1951.

[5]Beckman, P. and Spizzichino, A., The Scattering of Electromagnetic Waves From Rough Surfaces, MacMillan, New York, 1963.

[6]Birkebak, R. C. and Eckert, E. R. G., "Effect of Roughness of Metal Surfaces on Angular Distribution on Monochromatic Reflected Radiation," Journal of Heat Transfer, Vol. 87C, No. 1, Feb. 1965, pp. 85-93.

[7]Francis, R. E., "The Experimental Determination of Reflectance Functions for Type 302 Stainless Steel," Master's Thesis, 1967, Aerospace and Mechanical Engineering Department, University of Oklahoma, Norman, Okla.

[8]Torrance, K. E. and Sparrow, E. M., "Biangular Reflectance of an Electric Nonconductor as a Function of Wavelength and Surface Roughness," Journal of Heat Transfer, Vol. 87C, No. 2, May 1965, pp. 283-292.

[9]Pokrowski, G. I., "Zur Theorie der Diffusen Lichtrefexion," Zeitschrift fur Physik, Part I: Vol. 30, 1924, pp. 66-72; Part II: Vol. 35, 1925, pp. 34-37; Part III: Vol. 35, 1925, pp. 390-393; Part IV: Vol. 36, 1926, pp. 472-476.

[10]Herold, L. M. and Edwards, D. K., "Bidirectional Reflectance Characteristics of Rough, Sintered-Metal," AIAA Paper 24-26, New York, 1960.

[11]Davies, H., "Reflection of Electromagnetic Waves From a Rough Surface," Proceedings of the Institute of Electric Engineers, Vol. 101, 1954, pp. 209-214.

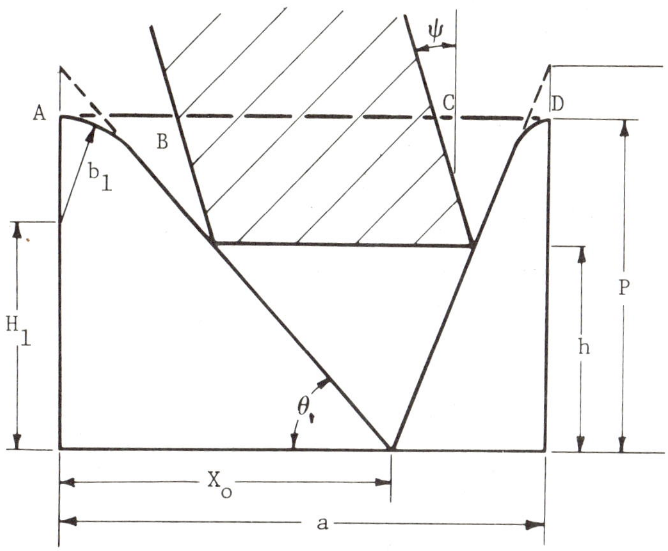

Fig. 1 Basic V-groove geometry.

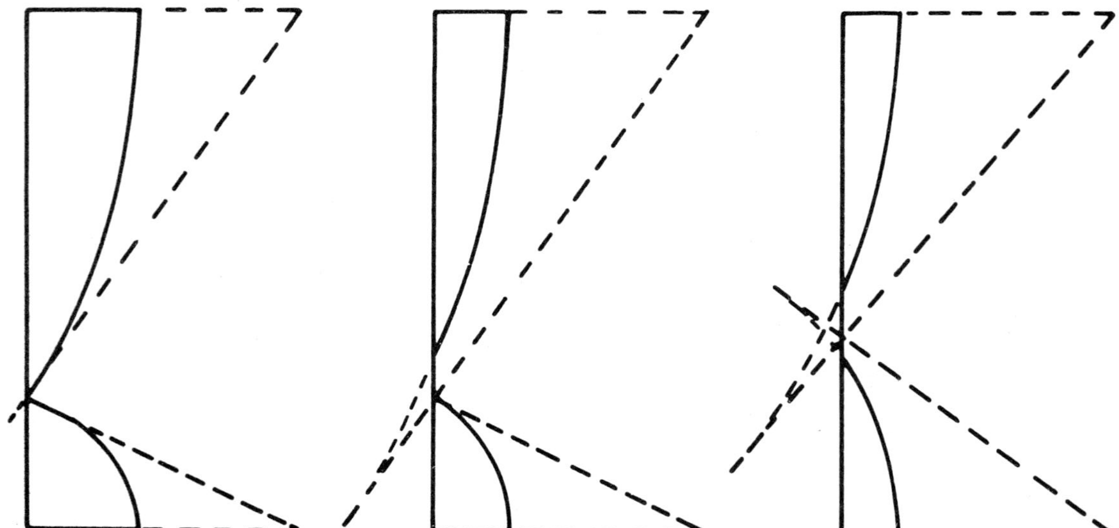

Fig. 2 Illustration of various unusual geometries encountered when P<P(LIM).

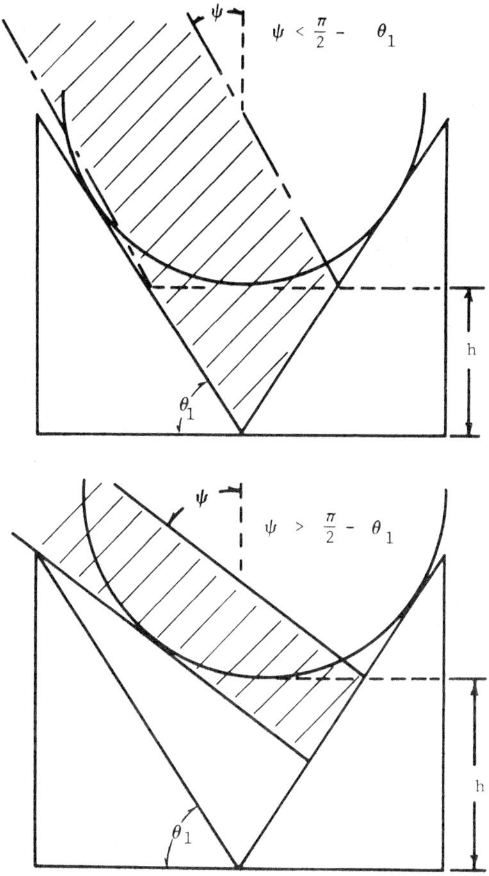

Fig. 3 Illustration of forbidden regions; /// represents the forbidden regions.

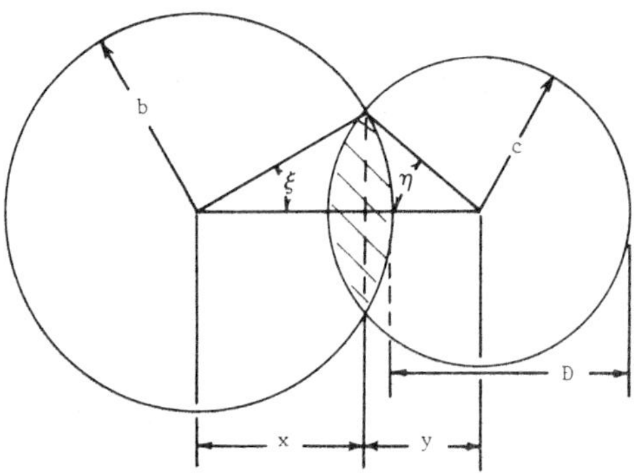

Fig. 4 Geometry used in finite receiving aperture distortion calculation.

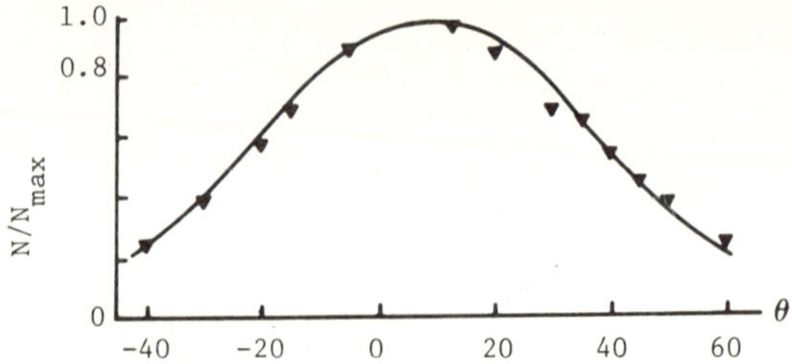

Fig. 5 Graphical comparison. ▼ Nickel: $\sigma_o/\lambda = 0.6$; - MCR: $P = 0.5$, $\bar{\sigma}/\lambda = 0.6$, $\bar{\sigma}/\bar{a} = 0.5$.

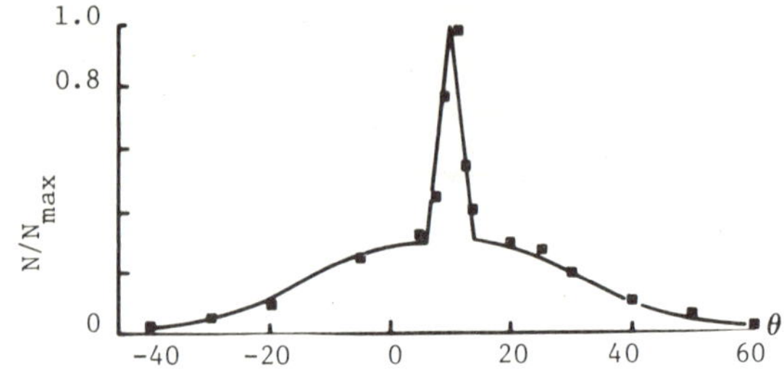

Fig. 6 Graphical comparison. ■ Nickel: $\sigma_o/\lambda = 0.32$; - MCR: $P = 0.5$, $\bar{\sigma}/\lambda = 0.3$, $\bar{\sigma}/\bar{a} = 0.5$.

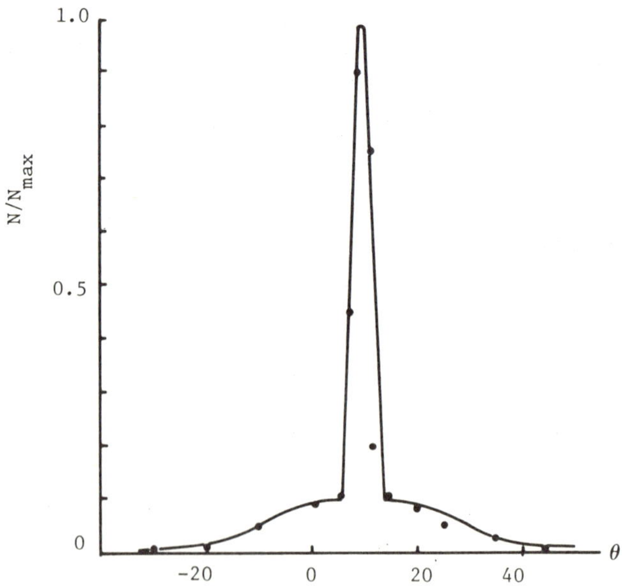

Fig. 7 Graphical comparison. ● Al-Coated Grd. Glass: $0.11 < \sigma_o/\lambda < 0.19$; - MCR: $P = 0.5$, $\bar{\sigma}/\lambda = 0.15$, $\bar{\sigma}/\bar{a} = 0.5$.

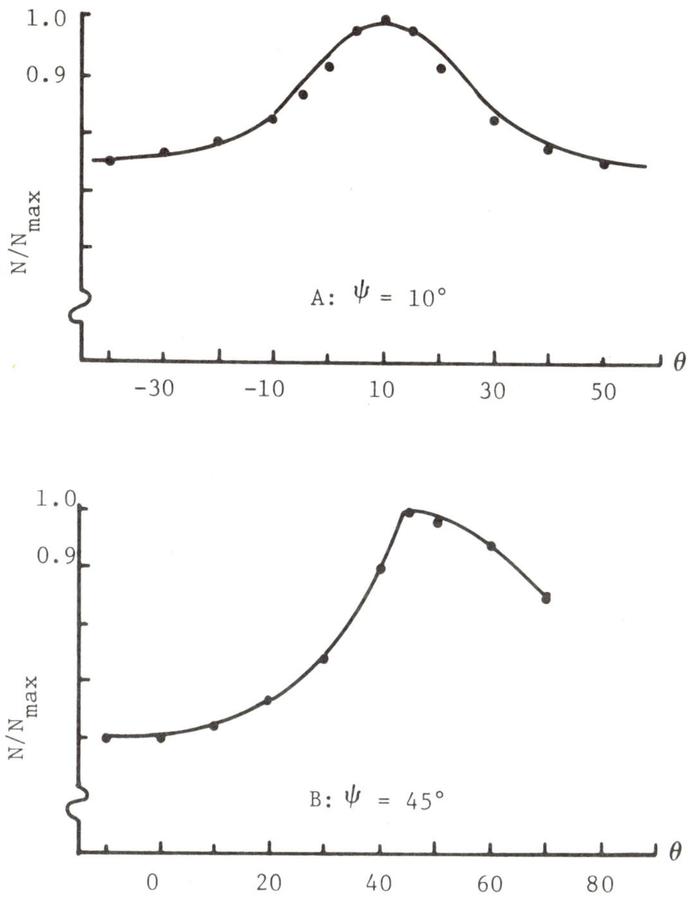

Fig. 8 Graphical comparison. • Ceramic: $\sigma_o/\lambda = 1.02$, $\lambda = 0.5\mu$; - MCR: $P = 0.3$, $\bar{\sigma}/\lambda = 1.0$, $\bar{\sigma}/\bar{a} = 5.0$.

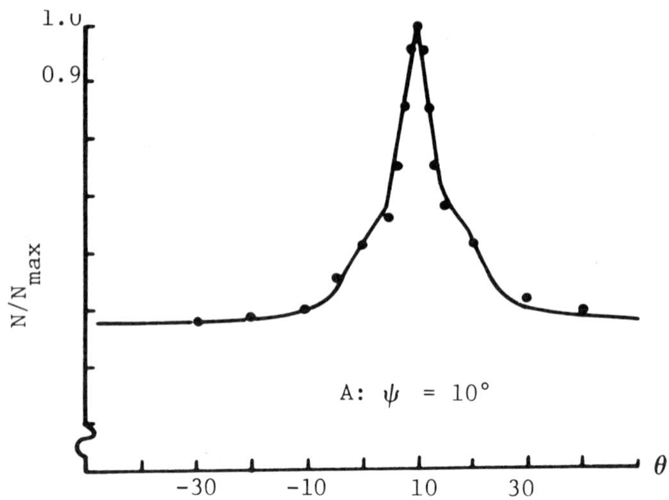

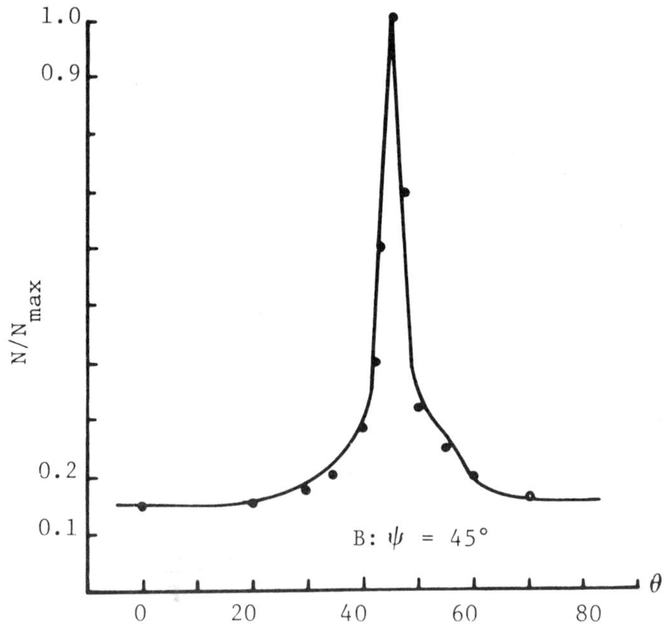

Fig. 9 Graphical comparison. • Ceramic: $\sigma_o/\lambda = 0.34$, $\lambda = 1.5\mu$; – MCR: $P = 0.3$, $\bar{\sigma}/\lambda = 0.34$, $\bar{\sigma}/\bar{a} = 5.0$.

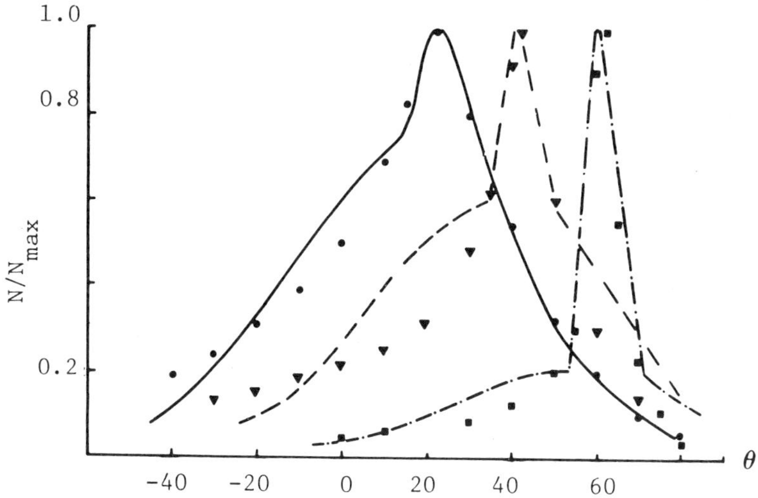

Fig. 10 Graphical comparison. SDBT. Al: $\sigma_o/\lambda = 0.16$, $\lambda = 5.0\mu$; MCR: $P = 0.95$, $\bar{\sigma}/\lambda = 0.16$, $\bar{\sigma}/\bar{a} = 8.0$.

Data		MCR
●	ψ=20°	——————
▼	ψ=40°	— — —
■	ψ=60°	—·—·—

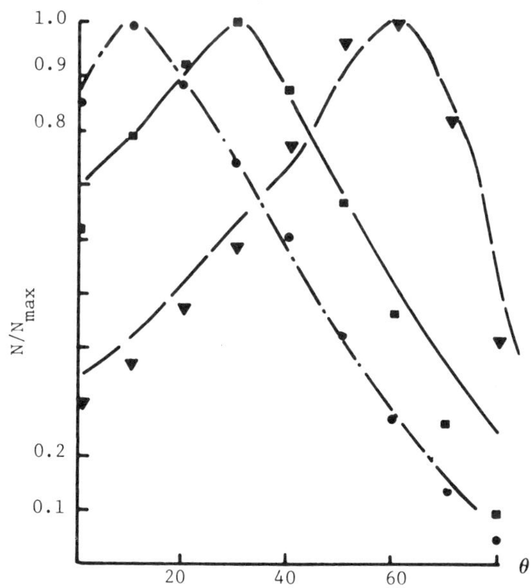

Fig. 11 Graphical comparison. SDBT. Steel: $\sigma_o/\lambda = 0.94$, $\lambda =$ 1.5μ; MCR: $P = 0.62$, $\bar{\sigma}/\lambda = 0.94$, $\bar{\sigma}/\bar{a} = 0.4$.

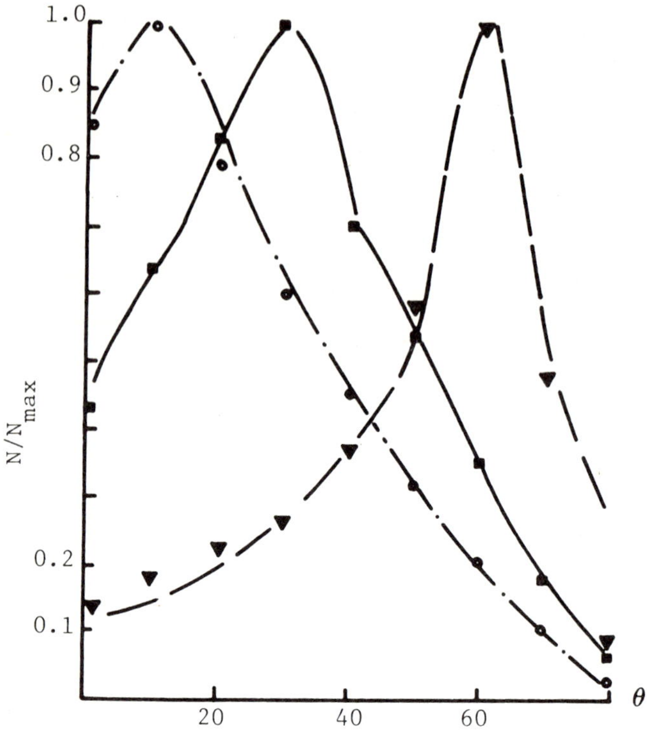

Fig. 12 Graphical comparison. SDBT. Steel: $\sigma_o/\lambda = 0.282$, $\lambda = 5.0\mu$; MCR: $P = 0.62$, $\bar{\sigma}/\lambda = 0.282$, $\bar{\sigma}/\bar{a} = 0.4$.

IMPERFECT REFLECTIONS IN THERMAL RADIATION TRANSFER

D. K. Edwards* and I. V. Bertak†

University of California, Los Angeles, Calif.

Abstract

The difficult question of how to obtain a measure of diffuseness or specularity to characterize a rough surface for radiant energy transfer calculations was examined. Two discoveries were made: 1) When a specularity is assigned to make the specular-diffuse model agree with detailed calculations for a short slot passage, predictions based upon the specular diffuse model for other passage geometries agree well with detailed calculations. 2) The Torrance-Sparrow model of a rough surface is as well suited as the specular-diffuse model for Monte Carlo numerical calculations. Both experimental data and the Torrance-Sparrow rough surface model were used to fix the bi-directional reflectances of the imperfectly diffuse passage walls investigated. System transmittances and reflectances of both square and slot passages with length to side ratios of one to eight were calculated. In addition, an analytical moment formulation was used for approximate solution and was found to support discovery 1 above. It was concluded that a slot-transmittance measurement offers a quick and convenient method to fix a measure of diffuseness or specularity, whether the specularity is positive or negative.

I. Introduction

Applied workers who need to predict thermal radiation transfer in a spacecraft or other system have been concerned with the problem of accounting for reflections which are neither perfectly diffuse nor perfectly specular. Measurements of

---

Presented as Paper 70-860 at the AIAA 5th Thermophysics Conference, Los Angeles, Calif., June 29-July 1, 1970.

*Professor of Engineering, Energy and Kinetics Department, School of Engineering and Applied Science.

†Research Assistant, School of Engineering and Applied Science.

spectral bidirectional reflectance[1-4] and total bidirectional reflectance[5,6] show that many surfaces occurring in technology and employed by the designer are neither perfectly diffuse nor perfectly specular. Theoretical diffraction models[7,8] investigated previously[9] do not indicate a marked departure from specular behavior on the part of slightly rough surfaces, as far as power transfer is concerned,[10] but models based on geometrical optics, for example, the Torrance-Sparrow[11-13] model, agree with experiments in showing a marked departure from either perfectly specular or perfectly diffuse reflection on the part of very rough surfaces.

A specular-diffuse model has been widely described in the literature.[14-16] In this model, a portion of the reflected flux is imagined to be reflected perfectly diffusely and the remainder specularly. With such a model, radiation heat transfer can be calculated using the mirror-image concept[17,18] and the standard discrete surface network[19,20] or radiosity[21] formulations. Alternatively, a Monte-Carlo procedure can be employed.[22-24]

In view of the wide discussion of the specular diffuse model, there has been notably lacking any guideline for the analyst in assigning a specularity X_S, that is, a number giving the fraction of the reflectance to be regarded as specular, or a "diffuseness" X_D, where X_D and X_S sum to unity. Without a procedure for assigning such a factor the specular-diffuse model is essentially incomplete. Even given a complete set of accurate spectral bidirectional reflectance data, the analyst has had no method for fixing a "specularity." More to the point, however, is the fact that obtaining such data is lengthy and expensive. For parametric design studies, the delay and expense may not be warranted.

It is the purpose of this paper to examine radiant transfer through passages (since it is in such cases that the effect of specularity manifests itself particularly) with a goal of establishing a practical measure of diffuseness or specularity. It is desired to find a measure which can be readily determined by a simple experiment and which will permit reasonably accurate calculations of radiation transfer in enclosures.

II. Calculation of Transmission Through Passages

Geometries Considered

Attention is directed to two geometries, the square passage with sides d x d and length L, Fig. 1, and the infinite slot formed by two plates a distance d apart, L long and infinite in width, Fig. 2. The ends are black, one with $\sigma T^4 = 1$ and the other with $\sigma T^4 = 0$. The passage walls are imperfectly diffuse with $\sigma T^4 = 0$. Note that the equation of transfer is linear in σT^4, and solutions can be superposed (e.g., Ref. 25).

Bidirectional Reflectance

Let an element of the source area be dA. The power radiated in solid angle $d\omega$ about direction θ,ϕ is given by

$$dQ = I \cos\theta \, d\omega \, dA \quad (1)$$

where I is the intensity which for a black body source is

$$I = \sigma T^4/\pi \quad (2)$$

and $d\omega$ may be written in terms of $d\theta$ and $d\phi$

$$d\omega = \sin\theta \, d\theta \, d\phi \quad (3)$$

The direction θ,ϕ shown in Fig. 3 is given by polar angle θ measured from the surface normal and azimuthal angle ϕ measured to the projection of the ray in the plane of the surface from an arbitrary reference line.

The power radiated in the θ,ϕ direction may pass directly to the receiver, or it may strike the passage wall at some direction θ_1,ϕ_1. In this case, a portion $\varepsilon(\theta_1,\phi_1)$ is absorbed and the remainder reflected. In order to focus on effects of imperfectly diffuse reflection, other phenomena such as spectral selectivity and polarization (e.g., Refs. 25 and 26) are not considered in what follows, and the walls are taken to be opaque.

If the walls are imperfectly diffuse, that is, neither perfectly diffuse nor perfectly specular, the reflected radiation will be distributed nonuniformly in direction θ_2,ϕ_2 measured at the irradiated spot on the passage wall. It can be appreciated intuitively that the nature of the reflection will have a profound influence on the transmission from the source to the sink when ε is rather small and the passage long. For example, if the passage were specular, upon reflection a fraction $1 - \varepsilon$ of the radiation would continue down toward the receiver. But, if the passage were perfectly diffuse, half of the fraction $1 - \varepsilon$ would be turned back toward the source, and, if the passage backscattered appreciably, still a larger fraction would be turned back.

The intensity of the imperfectly diffused radiation may be described by bidirectional reflectance

$$\rho_{Bi}(\theta_2,\phi_2,\theta_1,\phi_1) = \pi dI^+(\theta_2,\phi_2)/I^-(\theta_1,\phi_1)\cos\theta_1 \, d\omega \tag{4}$$

where $d\omega$ in this case is the solid angle subtended by the source element dA, a distance r away

$$d\omega = dA \cos\theta/r^2 \tag{5}$$

It is necessary to know ρ_{Bi} from experiment or have a model capable of specifying it in order to treat radiant heat transfer, taking account of imperfect reflections.

Calculation Methods

As a mathematical exercise, one can readily set up an integral equation for intensity using Eq. (4) and making the observation that the intensity incident on one spot was emergent from another. For an arbitrary ρ_{Bi}, the governing equation appears analytically intractable.

A method of solving approximately by subdividing the solid angle above the surface into a number of zones or fluxes will approach becoming exact only when the number of zones is made prohibitively large. Bevans and Edwards' "third approximation"[27] appears cumbersome for the same reason.

Ray tracing may be envisioned. The source could be subdivided, say into equal areas. A hemisphere over the centroid of each of the subdivisions would be subdivided and a ray sent out in each direction with power given by Eq. (1) with intensity

$$I(\theta,\phi) = \varepsilon(\theta,\phi)\,\sigma T^4/\pi \tag{6}$$

The closest intersection S' of the ray with the enclosure surface is found, and the angles of incidence θ_1',ϕ_1'. An amount of energy $\varepsilon(\theta_1',\phi_1')dQ$ would be absorbed by that surface and the remainder reflected. But the reflected radiation is distributed over the hemisphere centered over position S' as indicated by Eq. (4). One could subdivide this hemisphere and start to trace each of several rays, but this multiple branching upon a second, a third, and each succeeding reflection leads to an absurd algorithm. Monte Carlo[22-24] techniques, however, substitute for this branching a random choice of a single direction at each step in the ray tracing procedure. A numerical computation is then easily executable.

Monte Carlo Alternatives

The essence of a Monte Carlo technique thus is to avoid branching by a random choice of one direction out of the infinity possible at each reflection. Random choices may also be made to decide whether or not to terminate the ray tracing; for example, to decide whether a ray is to be regarded as completely reflected, but this aspect of Monte Carlo is not essential.

Two alternatives[24] present themselves in choosing directions: 1) Choose two random numbers P_1 and P_2 which fix θ and ϕ through two independent functional dependencies[24] or 2) choose two random numbers in succession[28] to fix say ϕ first and then θ, the second function depending upon ϕ. The first scheme is much simpler, but the second might be numerically more efficient. Consider, for example, the first choice for a ray emitted by a nonblack source. After first choosing a pair of random numbers to fix a source point, such that the points would cover the source with a uniform density as the number goes to infinity, one might fix θ and ϕ independently by choosing P_1 and P_2 in the range 0 to 1.

$$\theta = \sin^{-1}(P_1^{1/2}), \ \phi = 2\pi P_2 \tag{7}$$

The ray would then be weighted by $\varepsilon(\theta,\phi)$. Alternatively one might take

$$\phi = 2\pi P_1 \ , \ \theta = P_\phi^{-1}(P_2) \tag{8}$$

where P^{-1} means the inverse (like $\sin^{-1}$) and

$$P_\phi = \frac{\int_0^\theta \varepsilon(\theta,\phi) \sin\theta \cos\theta \, d\theta}{\int_0^{\pi/2} \varepsilon(\theta,\phi) \sin\theta \cos\theta \, d\theta} \tag{9}$$

The trouble with the first scheme [Eq. (7)] is that a number of uninteresting (low ε) rays may be chosen for tracing so that a long computation time may result for a given accuracy. The trouble with the second scheme, Eqs. (8) and (9) is that an array of tables must be computed and stored. While this is no hardship for emittance, a function of only θ and ϕ, it is an inconvenience for the case of ρ_{Bi} $(\theta_1,\phi_1,\theta_2,\phi_2)$, even if the surface is isotropic, that is, even if

$$\rho_{Bi}(\theta_2,\phi_2,\theta_1,\phi_1) = \rho_{Bi}(\theta_2,\theta_1,\phi_2-\phi_1) \tag{10}$$

In making calculations using experimental data, the present authors favor scheme 1, although not necessarily the exact combination of weighting functions indicated in Eq. (7). An important finding in the present work is, however, that a third alternative can be employed when the experimental data can be fit with the Torrance-Sparrow model, or, when in the absence of experimental data, the Torrance-Sparrow model is used to make estimates. The details of this third alternative are given in Sec. III.

An Approximate Two-Flux Formulation

Some insight into the effect of imperfectly diffuse reflection on radiation transfer can be gained from an approximate analysis of transmission through a long passage. The approximate analysis is made somewhat in the manner of Schuster[29] who treated transmission through fog. Let the radiant intensity actually existing within the passage be approximated by an isotropic flux

$$q^+ = \pi I^+$$

directed up the passage in the positive x direction and a second isotropic flux

$$q^- = \pi I^-$$

directed back down the passage in the negative x direction. We formulate in terms of the bidirectional reflectance what changes occur in q^+ and q^- upon reflection from a strip of the passage wall dx long.

Let the cross-sectional area of the passage be A and the perimeter P. The power striking the side wall area Pdx from the q^+ flux at location x would be, under the assumption of a uniform perfectly diffuse intensity,

$$AF_{cs}q^+ = PdxF_{sc}q^+ = \frac{1}{2} P\, dxq^+ \tag{11}$$

since the shape factor from the strip s to the cross-section c is 1/2. A fraction of this power is reflected into the forward direction, a fraction into the backward direction, and the remainder absorbed. Let ρ_F be the fraction reflected into the forward half hemisphere located over a point at the center of the strip. From Eqs. (1) and (4),

$$\rho_F = \int_{\phi_2=-\pi/2}^{\phi_2=+\pi/2}\int_{\theta_2=0}^{\theta_2=\pi/2}\int_{\phi_1=\pi/2}^{\phi_1=3\pi/2}\int_{\theta_1=0}^{\theta_1=\pi/2}$$

$$\frac{2}{\pi^2}\,\rho_{Bi}(\theta_2,\phi_2,\theta_1,\phi_1)\cos\theta_1\ \sin\theta_1\ d\theta_1\ d\phi_1$$

$$\cos\theta_2\ \sin\theta_2\ d\theta_2\ d\phi_2 \tag{12}$$

Similarly let ρ_B be the fraction reflected into the back half hemisphere.

$$\rho_B = \int_{\pi/2}^{3\pi/2}\int_0^{\pi/2}\int_{\pi/2}^{3\pi/2}\int_0^{\pi/2} \frac{2}{\pi^2}\,\rho_{Bi}$$

$$\cos\theta_1\ \sin\theta_1\ d\theta_1\ d\phi_1\ \ \cos\theta_2\ \sin\theta_2\ \ d\theta_2 d\phi_2 \tag{13}$$

Similar considerations apply for the backward flux, and the same two forward and backward reflectances arise, because of reciprocity

$$\rho_{Bi}(\theta_2,\phi_2,\theta_1,\phi_1) = \rho_{Bi}(\theta_1,\phi_1,\theta_2,\phi_2) \tag{14}$$

Equation (11) indicates that the power Aq^+ is decreased by (1/2) $Pdxq^+$ but a fraction ρ_F of this power is reflected into the forward direction. A fraction ρ_B of the power (1/2) $Pdxq^-$ also is reflected by the dx strip into the forward direction.

$$d(Aq^+) = -(1-\rho_F)\,\frac{1}{2}\,Pdxq^+ + \rho_B\,\frac{1}{2}\,Pdxq^-$$

$$\frac{2A}{P}\,\frac{dq^+}{dx} = -(1-\rho_F)\ q^+ + \rho_B\ q^- \tag{15}$$

Let an optical depth t be defined

$$t = Px/(2A)$$

Then Eq. (15) becomes

$$dq^+/dt = -\ (1-\rho_F)\ q^+ + \rho_B\ q^- \tag{16}$$

A similar expression for q^- may be found by interchanging +t with -t and q^+ with q^-.

$$-dq^-/dt = -(1-\rho_F)\ q^- + \rho_B\ q^+ \tag{17}$$

Equations (16) and (17) constitute two simultaneous ordinary linear differential equations subject to boundary conditions

$$t = 0\ ,\ q^+ = 1,\quad t = t_L,\ q^- = 0 \tag{18}$$

The solutions may be readily found, and when they are, the desired system parameters are transmittance τ and reflectance ρ

$$\tau \equiv q^+(t_L),\qquad \rho \equiv q^-(0) \tag{19}$$

The solutions yield

$$\tau = m/[(1-\rho_F)\sinh mt_L + m \cosh mt_L] \tag{20}$$

$$\rho = \rho_B \sinh mt_L/[(1-\rho_F)\sinh mt_L + m \cosh mt_L] \tag{21}$$

$$m = [(1-\rho_F + \rho_B)(1-\rho_F - \rho_B)]^{1/2} \tag{22}$$

$$t_L = \frac{PL}{2A} \quad \begin{array}{l} = 2L/d \text{ for a square} \\ = L/d \text{ for a slot} \end{array} \tag{23}$$

If the passage walls are perfectly diffuse, there would follow from Eqs. (12) and (13)

$$\rho_F = \rho_B = \rho/2 \tag{24}$$

If they were perfectly specular,

$$\rho_F = \rho \qquad \rho_B = 0 \tag{25}$$

In the specular-diffuse approximation

$$\rho_F = \rho_S + (\rho_D/2)\quad \rho_B = \rho_D/2 \tag{26}$$

The approximate solution thus is interesting in several respects:

1) It suggests that passage reflectance and transmittance depend in the main on moments of the bidirectional reflectance, such as Eqs. (12) and (13), rather than the detailed behavior of the bidirectional reflectance.

2) It suggests that any convenient surface reflection model, such as the specular-diffuse model or the Torrance-Sparrow model, can be used to represent imperfectly diffuse reflection well enough for calculation of passage transmittance and reflectance, provided the model can accomodate both forward and backward scattering.

3) It suggests that the passage transmittance and reflectance characteristics for one geometry are approximately equal to those for another geometry provided the PL/A values and the passage wall properties match.

III. Numerical Procedures

Cases Investigated

In order to test the suggestions inferred from the approximate analytical solution, exact numerical results were obtained for square and infinite slot passages with the following side walls: 1) Walls with bidirectional reflectance indicated by Fig. 9 of Ref. 4, which gave actual experimental data for a rough aluminized sintered bronze. 2) Walls with randomly rough slopes and azimuthal orientations prescribed by the Torrance-Sparrow model. 3) Walls with specular-diffuse reflection characteristics.

Calculations Based Upon Experimental Data

The data itself consisted of seven sets of azimuths, $0^\circ - 180^\circ$ by 30° increments, at nine values of θ_2, $0^\circ - 80^\circ$ by 10°, and 4 values of θ_1, $0^\circ - 60^\circ$ by 20°. These were taken directly from Table 2 and Fig. 9 of Ref. 4; Eqs. (9-19) in that reference were not employed. By interpolation, extrapolation, and use of reciprocity [Eq. (14) in Sec. II] seven ten by ten arrays were obtained, each of which was averaged across the diagonal by virtue of Eq. (14). A look-up-interpolate routine was then used to obtain ρ_{Bi} for any set of θ_1, θ_2, and ϕ.

The directional reflectance was calculated from ρ_{Bi} by numerical integration

$$\rho(\theta) = \frac{1}{\pi} \int_0^{2\pi} \int_0^{\pi/2} \rho_{Bi} \sin\theta_2 \cos\theta_2 \, d\theta_2 \, d(\phi_2-\phi_1) \qquad (27)$$

Values were stored in a table, and a look-up-interpolate routine was used to find a value.

The Monte Carlo procedure employed was straightforward. A point on the source was chosen randomly, as were a value of

$\sin^2\theta$ and a value of ϕ in the range 0 to $\pi/2$ (the range could be restricted because of symmetry for the passages investigated). Each ray was assigned a weight by unity. The intersection of the ray with the passage walls was found. If the ray struck a side wall, a portion of the ray $[1-\rho(\theta_1)]$ was allotted for that side and new random numbers were chosen for $\sin^2\theta_2$ and ϕ_2. The value of ρ_{Bi} was found, the weight of the ray was multiplied by ρ_{Bi}, and the process continued. If the ray struck an end wall, the remaining weight was alloted to that end, and the entire procedure repeated. Mean values of system transmittance, reflectance, and absorptance and root-mean-square deviations were then calculated.

Calculations Based Upon the Torrance-Sparrow Model

The surface was imagined to be composed of rough elements in the form of symmetrical vee grooves whose upper edges were in the same plane. The probability of a facet having its normal fall in solid angle $d\omega$ and angle α from the plane was taken to be

$$be^{-C^2\alpha^2}, \quad \alpha \leq \pi/2$$

All azimuths of the v grooves relative to a fixed axis were therefore equally likely. The probability of a beam striking a given facet-pair is the fraction of the surface area taken up by that pair. The probability of a beam striking a facet-pair with an angle less than α is therefore

$$P(\alpha) = \int_0^{\alpha} e^{-C^2\alpha^2} \sin\alpha \cos\alpha \, d\alpha / \int_0^{\pi/2} e^{-C^2\alpha^2} \sin\alpha \cos\alpha \, d\alpha \tag{28}$$

This quantity was calculated by numerical integration for a given C and stored in such a way that, given P, α could be found by a look-up-interpolate routine (Fig. 4).

To complete the Torrance-Sparrow model, it is necessary to specify a facet reflectance. For simplicity, the facets were taken to be perfectly specular and perfectly reflecting; however, all rays reflected from one side of a v groove to the other were taken to be completely absorbed. This latter convention is completely arbitrary and unnecessary; our computer routine does have the capability of tracing multiple reflections in the v groove, but by not allowing such interreflections, computation time was minimized without compromising the goals of the investigation. Reciprocity was not violated, and the surface model was complete so that bidirectional and directional reflectances could be calculated.

The following Monte Carlo algorithm was employed for exact numerical calculation of passage transmittance and reflectance: A source location and ray directions were chosen randomly as before, and the intersection on the passage wall was found. Then random numbers established α via $P(\alpha)$ given by Eq. (28) and the azimuth of the v groove. Projection of the incident ray onto the plane perpendicular to the v groove was then obtained so that the shadowing and masking factors could be calulated according to Ref. 13. These factors weighted by the facet area projected normal to the beam established the fraction reflected, G. A portion 1-G of the weight of the ray was allotted to side wall absorption. The unmasked and unshadowed portions of the v groove facets were weighted by projected area, and a random number was chosen to establish whether the left facet or right facet could be regarded as establishing the direction of the reflected ray. As before, if a ray struck an end, the remaining weight of the ray was allotted to that surface, and the procedure repeated. The means and root-mean-square deviations were then computed for each surface.

Calculations Based upon the Specular Diffuse Model

In this case, the procedure was nearly the same as that employed for the calculations using experimental data, except the diffuse bidirectional reflectance was constant. A random number was selected to decide whether a given ray striking a side wall was specularly or diffusely reflected. If the number was less than X_S, all the energy remaining after reflection was assumed to be specularly reflected. The fraction absorbed during the reflection was taken to be $1 - (\rho_D+\rho_S)$ where ρ_D is the diffuse reflectance and ρ_S the specular reflectance.

$$X_S = \rho_S/(\rho_D + \rho_S) \tag{29}$$

IV. Transmission and Reflection Results

Table 1 shows the directional reflectance, forward reflectance, backward reflectance, and hemispherical reflectance of the four surfaces considered, the experimental one and the Torrance-Sparrow surfaces for $C^{-1} = 10^\circ$, 20°, and 40°. The surfaces are all fairly highly reflecting and range from nearly perfectly specular to quite diffuse, although not perfectly so, as judged from the forward and backward reflectances. Table 2 shows the passage-system transmittance and reflectances calculated for passages with these surfaces. Tables 3 and 4 show results for the specular-diffuse model.

Table 1 Directional, forward, backward and hemispherical reflectances of surface systsms investigated

Surface bidirectional reflectance	Directional reflectance $\psi=15^\circ$	$\psi=30^\circ$	$\psi=45^\circ$	$\psi=60^\circ$	$\psi=75^\circ$	Forward reflectance ρ_F	Backward reflectance ρ_F	Hemispherical reflectance ρ
T.S.[a] $C^{-1}=10^\circ$	1.00	1.00	1.00	0.987	0.844	0.896	0.076	0.972
T.S. $C^{-1}=20^\circ$	0.978	0.962	0.941	0.922	0.949	0.752	0.181	0.933
T.S. $C^{-1}=40^\circ$	0.712	0.730	0.771	0.838	0.929	0.523	0.262	0.785
H.E.[b] F-40	0.791	0.821	0.832	0.857	0.943	0.475	0.362	0.837

[a] T.S. refers to Torrance and Sparrow[13] (theoretical model).

[b] H.E. refers to Herold and Edwards[4] (experimental data).

Table 2 System transmittance and reflectance for passages with the wall characteristics investigated

Surface bidirectional reflectance	Exact Monte Carlo results: System transmittance L/d=1	L/d=2	L/d=4	L/d=8	System reflectance L/d=1	L/d=2	L/d=4	L/d=8
Infinite slot passage: length L, spacing d								
T.S.[a] $C^{-1}=10^{o}$	0.950	0.908	0.846	0.744	0.034	0.063	0.097	0.148
T.S. $C^{-1}=20^{o}$	0.856	0.764	0.622	0.440	0.087	0.124	0.188	0.226
T.S. $C^{-1}=40^{o}$	0.703	0.549	0.384	0.205	0.125	0.168	0.188	0.205
H.E.[b] F-40	0.664	0.493	0.304	0.144	0.202	0.274	0.322	0.329
Square passage: length L, side d								
T.S. $C^{-1}=10^{o}$	0.895	0.833	0.716	0.556	0.073	0.106	0.165	0.226
T.S. $C^{-1}=20^{o}$	0.744	0.587	0.391	0.193	0.139	0.192	0.245	0.268
T.S. $C^{-1}=40^{o}$	0.522	0.316	0.138	0.042	0.181	0.210	0.212	0.223
H.E. F-40	0.466	0.253	0.098	0.016	0.287	0.349	0.350	0.350

[a]T.S. refers to Torrance and Sparrow[13] (theoretical model).

[b]H.E. refers to Herold and Edwards[4] (experimental data).

Table 3 System transmittance and reflectance for slot passages according to the specular-diffuse model.

Specularity	Hemispherical reflectance of wall	System transmittance L/d=1	L/d=2	L/d=4	L/d=8	System reflectance L/d=1	L/d=2	L/d=4	L/d=8
0	1.00	0.673	0.547	0.402	0.269	0.327	0.453	0.598	0.731
	0.95	0.662	0.503	0.338	0.181	0.292	0.404	0.493	0.552
	0.90	0.647	0.463	0.286	0.129	0.262	0.359	0.422	0.455
	0.80	0.593	0.408	0.221	0.102	0.226	0.300	0.325	0.343
	0.70	0.575	0.369	0.189	0.078	0.182	0.232	0.252	0.256
0.25	1.00	0.748	0.616	0.473	0.362	0.253	0.383	0.526	0.639
	0.95	0.725	0.563	0.408	0.241	0.230	0.340	0.426	0.496
	0.90	0.705	0.533	0.354	0.187	0.203	0.298	0.353	0.387
	0.80	0.650	0.464	0.278	0.141	0.173	0.240	0.265	0.276
	0.70	0.615	0.414	0.241	0.117	0.140	0.181	0.200	0.203
0.50	1.00	0.825	0.711	0.594	0.468	0.178	0.295	0.410	0.533
	0.95	0.784	0.651	0.503	0.325	0.166	0.253	0.381	0.402
	0.90	0.760	0.612	0.441	0.254	0.147	0.216	0.274	0.306
	0.80	0.700	0.530	0.346	0.189	0.167	0.172	0.198	0.210
	0.70	0.660	0.474	0.296	0.153	0.098	0.130	0.140	0.150
0.75	1.00	0.909	0.822	0.732	0.617	0.092	0.178	0.265	0.384
	0.95	0.865	0.765	0.627	0.462	0.088	0.144	0.205	0.261
	0.90	0.831	0.712	0.505	0.360	0.081	0.114	0.166	0.197
	0.80	0.764	0.610	0.444	0.265	0.063	0.095	0.113	0.122
	0.70	0.713	0.544	0.367	0.213	0.050	0.071	0.076	0.080
1.00	1.00	1.000	1.000	1.000	1.000	0.0	0.0	0.0	0.0
	0.95	0.953	0.914	0.847	0.737	0.0	0.0	0.0	0.0
	0.90	0.912	0.840	0.728	0.572	0.0	0.0	0.0	0.0
	0.80	0.848	0.725	0.560	0.394	0.0	0.0	0.0	0.0
	0.70	0.773	0.622	0.405	0.297	0.0	0.0	0.0	0.0

V. Discussion

Accuracy of Monte Carlo results may be estimated readily.[23,28] In the present case, $N = 4096$ rays were used. Let the probable deviation Δp in the desired quantity divided by that quantity p be r. Then N, p, and r are related by[30]

$$N = (0.675/r^2)(1-p)/p \tag{30}$$

Suppose a value of $p = 0.200$ is desired to an accuracy of 3%, i.e., 0.200 ± 0.006. Then 3000 rays would be required. Computer calculations of rms deviations was also made by using 16 submeans of 256 rays. These direct calculations indicate an accuracy $r = \Delta p/p$ of approximately $\pm$ 0.005 in agreement with Eq. (30).

If the approximate relations, Eqs. (20) and (21) are compared with the exact numerical results, agreement is found to

Table 4 System transmittance and reflectance for square passages according to the specular diffuse model.

Specularity	Hemispherical reflectance of wall	System transmittance L/d=1	L/d=2	L/d=4	L/d=8	System reflectance L/d=1	L/d=2	L/d=4	L/d=8
0	1.00	0.535	0.358	0.231	0.141	0.465	0.642	0.769	0.859
	0.95	0.490	0.307	0.142	0.035	0.414	0.522	0.586	0.592
	0.90	0.454	0.260	0.092	0.017	0.374	0.447	0.486	0.487
	0.80	0.404	0.199	0.051	0.010	0.288	0.333	0.351	0.360
	0.70	0.351	0.155	0.039	0.007	0.234	0.259	0.267	0.262
0.25	1.00	0.600	0.448	0.302	0.192	0.392	0.552	0.698	0.808
	0.95	0.550	0.384	0.193	0.061	0.333	0.444	0.518	0.549
	0.90	0.514	0.326	0.131	0.039	0.309	0.377	0.426	0.415
	0.80	0.462	0.250	0.090	0.020	0.236	0.289	0.287	0.297
	0.70	0.400	0.193	0.056	0.015	0.194	0.213	0.219	0.216
0.50	1.00	0.699	0.562	0.415	0.276	0.301	0.438	0.585	0.724
	0.95	0.649	0.477	0.287	0.119	0.259	0.354	0.427	0.462
	0.90	0.600	0.409	0.209	0.069	0.214	0.297	0.333	0.358
	0.80	0.528	0.318	0.130	0.037	0.170	0.202	0.223	0.225
	0.70	0.453	0.255	0.092	0.026	0.134	0.152	0.159	0.162
0.75	1.00	0.830	0.717	0.579	0.438	0.181	0.283	0.421	0.562
	0.95	0.765	0.598	0.412	0.216	0.142	0.233	0.296	0.328
	0.90	0.703	0.524	0.320	0.144	0.129	0.178	0.210	0.227
	0.80	0.605	0.402	0.208	0.080	0.098	0.119	0.137	0.134
	0.70	0.526	0.321	0.138	0.049	0.065	0.084	0.090	0.091
1.00	1.00	1.000	1.000	1.000	1.000	0.0	0.0	0.0	0.0
	0.95	0.912	0.836	0.719	0.555	0.0	0.0	0.0	0.0
	0.90	0.840	0.720	0.556	0.364	0.0	0.0	0.0	0.0
	0.80	0.712	0.551	0.353	0.177	0.0	0.0	0.0	0.0
	0.70	0.608	0.412	0.232	0.095	0.0	0.0	0.0	0.0

within 10% for the short passages but is very poor for the longer ones, for which the equations underpredict τ by a factor of two or more. The reason for this discrepancy likely lies in the two flux model itself. Far down a passage the forward flux will depart from being diffuse. The approximate analysis has been described nevertheless, for the three reasons listed following Eq. (26). For example, the analytical model suggests that values of τ and ρ for a slot passage of $L/d = 4$ should pertain approximately to a square passage with the same walls of $L/d = 2$. Tables 3 and 4 show that such is the case. Although the rule breaks down somewhat for longer passages with low reflecting walls and low specularity, it does suggest that estimates for other cross sections could be based upon Tables 3 and 4 by use of Eq. (23).

An extensive literature exists on the definition and measurement of gloss. Harrison[31] gives a survey of the older literature. More recent work has been done by Hunter.[32,33] The work does not appear pertinent to the engineer interested in power transfer. However the comparison in Table 5 shows that a slot transmission measurement offers a reasonable way to fix a surface specularity for power transfer purposes.

Table 5, column 1, shows the two-flux model specularity factor derived from Eq. (26)

$$X_{S,\ \text{two flux}} = (\rho_F - \rho_B)/(\rho_F + \rho_B) \tag{31}$$

The remaining columns of the table show values of X_S derived by the following procedure: A value of system transmittance from Table 2 and the value of hemispherical wall reflectance (Table 1) are used to enter Table 3 or 4 (depending upon the geometry) to find the specularity. Tables 3 and 4 were graphed to facilitate interpolation.

The interesting feature of the table is that the values in the last four columns for a given surface are nearly the same for both geometries. This insensitivity of X_S shows that a passage transmittance measurement together with a hemispherical reflectance measurement of the wall surface itself can be used to obtain a value of X_S via Table 3 and 4 which will be approximately valid for other cross-sectional geometries and length-to-width ratios.

Although Eq. (31) does not yield values of X_S in agreement with those found as described previously, it appears to establish a lower limit. More important, however, is that it shows clearly that negative values of specularity are possible. In the specular-diffuse model, a negative specularity would denote that the specular ray would emerge in the back direction $\theta_2 = \theta_1, \phi_2 = \phi_1$ instead of in the forward direction $\theta_2 = \theta_1$, $\phi_2 = \theta_1 - \pi$. A surface containing cubical indentations would exhibit such a behavior.

Once a value of X_S is determined the analyst may use it with the specular-diffuse model or, if X_S is positive and large, say $0.5 \leq X_S \leq 1$, with the Torrance-Sparrow model. The latter appears physically more realistic and is computationally as convenient for Monte Carlo calculations as the specular-diffuse model.

VI. Conclusions

The following conclusions are reached as a result of this study:

Table 5 Comparison of "specularity" factors derived from passage transmittances

Surface bidirectional reflectance	Two-flux specularity factor	Slot transmittance specularity factor L/d=1	L/d=2	L/d=4	L/d=8
Infinite slot passage: length L, spacing d					
T.S.[a] $C^{-1}=10^{\circ}$	0.843	0.93	0.94	0.95	0.94
T.S. $C^{-1}=20^{\circ}$	0.612	0.76	0.77	0.77	0.77
T.S. $C^{-1}=40^{\circ}$	0.333	0.52	0.59	0.62	0.59
H.E.[b] F-40	0.135	0.22	0.26	0.24	0.21
Square passage: length L, side d					
T.S. $C^{-1}=10^{\circ}$	0.843	0.90	0.94	0.94	0.94
T.S. $C^{-1}=20^{\circ}$	0.612	0.73	0.77	0.77	0.76
T.S. $C^{-1}=40^{\circ}$	0.333	0.55	0.56	0.58	0.59
H.E. F-40 (Expt'l data)	0.135	0.20	0.15	0.21	0.10

[a] T.S. refers to Torrance and Sparrow[13]. [b] H.E. refers to Herold and Edwards[4].

1) The Torrance-Sparrow rough surface model is particularly suited for Monte Carlo calculations.

2) Quite rough surfaces can have appreciable values of specularity from a power transfer point of view. The Herold-Edwards sintered bronze had a value of $X_S = 0.2$ while the very rough Torrance-Sparrow surface $C^{-1} = 400$ had a value of X_S as large as 0.56.

3) A two-flux optical depth correlated transmittance and reflectance well for short passages ($L/d \leq 4$) of different cross-sections.

4) A slot transmission measurement may be used conveniently to characterize for power transfer calculations the specularity X_S of a rough surface.

References

[1]Birkebak, R.C. and Eckert, E.R.G., "Effects of Roughness of Metal Surfaces on Angular Distribution of Monochromatic Reflected Radiation," Transactions of the ASME, Ser. C: Journal of Heat Transfer, Vol. 87, 1965, pp. 85-94.

[2]Torrance, K.E. and Sparrow, E.M., "Biangular Reflectance of an Electric Nonconductor as a Function of Wavelength and Surface Roughness," Transactions of the ASME, Ser. C: Journal of Heat Transfer, Vol. 87, 1965, pp. 283-292.

[3]Torrance, K.E., Sparrow, E.M., and Birkebak, R.C., "Polarization, Directional Distribution, and OFF-Specular Peak Phenomena in Light Reflected from Roughened Surfaces," Journal of the Optical Society of America, Vol. 56, 1966, pp. 916-925.

[4]Herold, L.M. and Edwards, D.K., "Bidirectional Characteristics of Rough, Sintered-Metal, and Wire-Screen Surfaces Systems," AIAA Journal, Vol. 4, No. 10, Oct. 1966, pp. 1802-1810.

[5]Schmidt, E. and Eckert, E.R.G., "Über die Richtungsverteilung der Wärmestrahlung von Oberflächen," Forschung auf dem Gebiete des Ingenieurwesens, Vol. 6, 1935, pp. 175-183.

[6]Münch, B., "Die Richtungsverteilung bei der Reflexion von Wärmestrahlung und ihr Einfluss auf die Wärmeübertragung," Mitteilungen aus der Institut für Thermodynamik und Verbrennungsmotorenbau, No. 16, 1955, Verlag Leeman, Zurich.

[7]Davies, H., "The Reflection of Electromagnetic Waves From a Rough Surface," Proceedings of the Institution of Electrical Engineers, Vol. 101, Pt. 4, 1954, pp. 209-214.

[8]Beckmann, P. and Spizzichino, A., The Scattering of Electromagnetic Waves From Rough Surfaces, Macmillan, New York, 1963.

[9]Hering, R.G., "Radiative Heat Exchange Between Specularly Reflecting Surfaces with Direction Dependent Properties," Proceedings of the Third International Heat Transfer Conference, American Institute of Chemical Engineers, Vol. 5, 1966, pp. 200-206.

[10]Edwards, D.K., "Radiative Transfer Characteristics of Materials," Transactions of the ASME, Ser. C: Journal of Heat Transfer, Vol. 91, 1969, pp. 1-15.

[11]Middleton, W.E.K. and Mungall, A.G., "The Luminous Directional Reflectance of Snow," Journal of the Optical Society of America, Vol. 42, 1952, pp. 572-579.

[12]Christie, A.W., "The Luminous Directional Reflectance of Snow," Journal of the Optical Society of America, Vol. 43, 1953, pp. 621-622.

[13]Torrance, K.E. and Sparrow, E.M., "Theory of Off-Specular Reflection from Roughened Surfaces," Journal of the Optical Society of America, Vol. 57, 1967, pp. 1105-1114.

[14]Seban, R.A., Discussion of Reference 18, Transactions of the ASME, Ser. C: Journal of Heat Transfer, Vol. 84, pp. 299-300.

[15]Sparrow, E.M. and Lin, S.L., "Radiation Heat Transfer at a Surface Having both Specular and Diffuse Reflectance Components," International Journal of Heat and Mass Transfer, Vol. 8, 1965, pp. 769-779.

[16]Sarofim, A.F. and Hottel, H.C., "Radiative Exchange Among Non-Lambert Surfaces," Transactions of the ASME, Ser. C: Journal of Heat Transfer, Vol. 88, 1966, pp. 37-44.

[17]Eckert, E.R.G. and Sparrow, E.M., "Radiative Heat Exchange

Between Surfaces with Specular Reflection," International Journal of Heat and Mass Transfer, Vol. 3, 1961, pp. 42-54.

[18]Sparrow, E.M., Eckert, E.R.G., and Jonsson, V.K., "An Enclosure Theory for Radiative Exchange Between Specularly and Diffusely Reflecting Surfaces," Transactions of the ASME, Ser. C: Journal of Heat Transfer, Vol. 84, 1962, pp. 294-300.

[19]Oppenheim, A.K., "Radiation Analysis by the Network Method," Transactions of the ASME, Vol. 78, 1956, pp. 725-735.

[20]Oppenheim, A.K., "The Engineering Radiation Problem," Zeitschrift für Angewandte Mathematik und Mechanik, Vol. 36, 1956, pp. 81-93.

[21]Bobco, R.P., "Radiation Heat Transfer in Semigray Enclosures with Specularly and Diffusely Reflecting Surfaces," Transactions of the ASME, Ser. C: Journal of Heat Transfer, Vol. 86, 1964, pp. 123-130.

[22]Howell, J.R. and Perlmutter, M., "Monte Carlo Solution of Thermal Transfer Through Radiant Media Between Gray Walls," Transactions of the ASME, Ser. C: Journal of Heat Transfer, Vol. 86, 1964, pp. 116-122.

[23]Howell, J.R. and Siegel, R., Thermal Radiation Heat Transfer, NASA SP-164, Vol. II, 1969, pp. 177-211.

[24]Love, T.J., Radiative Heat Transfer, edited by L.W. Stockham, Merrill Publishing Co., Columbus, Ohio, 1968, pp. 273-280.

[25]Edwards, D.K and Bevans, J.T., "Effect of Polarization on Spacecraft Radiation Heat Transfer," AIAA Journal, Vol. 3, No. 7, July 1965, pp. 1323-1329.

[26]Edwards, D.K. and Tobin, R.D., "Effect of Polarization on Radiant Heat Transfer Through Long Passages," Transactions of the ASME, Ser. C: Journal of Heat Transfer, Vol. 89, 1967, pp. 132-138.

[27]Bevans, J.T. and Edwards, D.K., "Radiative Exchange in an Enclosure with Directional Wall Properties," Transactions of the ASME, Ser. C: Journal of Heat Transfer, Vol. 87, 1965, pp. 388-396.

[28] Schornhorst, J.R. and Viskanta, R., "Effect of Direction and Wavelength Dependent Surface Properties on Radiant Heat Transfer," AIAA Journal, Vol. 6, No. 8, Aug. 1968, pp. 1450-1455.

[29] Schuster, A., "Radiation Through a Foggy Atmosphere," The Astrophysical Journal, Vol. 21, 1905, pp. 1-22.

[30] Uspensky, J.V., Introduction to Mathematical Probability, McGraw-Hill, New York, 1937, p. 131.

[31] Harrison, V.G.W., Definition and Measurement of Gloss, The Printing and Allied Trades Research Association, 1945.

[32] Hunter, R.S., "Gloss Evaluation of Materials," ASTM Bulletin, No. 186, 1952, pp. 48-55.

[33] Hunter, R.S., "Those Other Aspects of Appearance," Color Engineering, Vol. 1, No. 2, 1963, pp. 8-14.

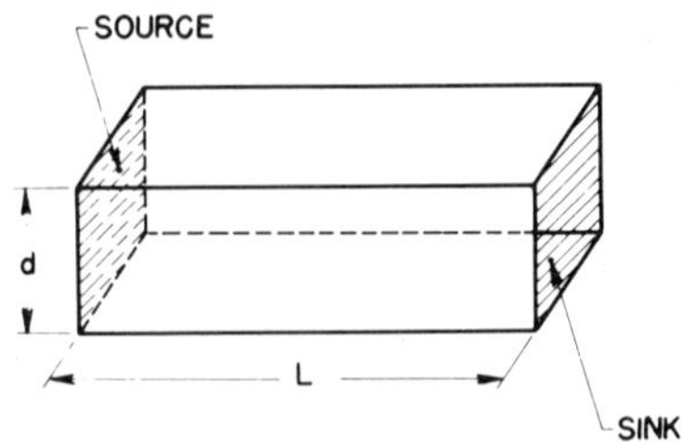

Fig. 1 The square passage.

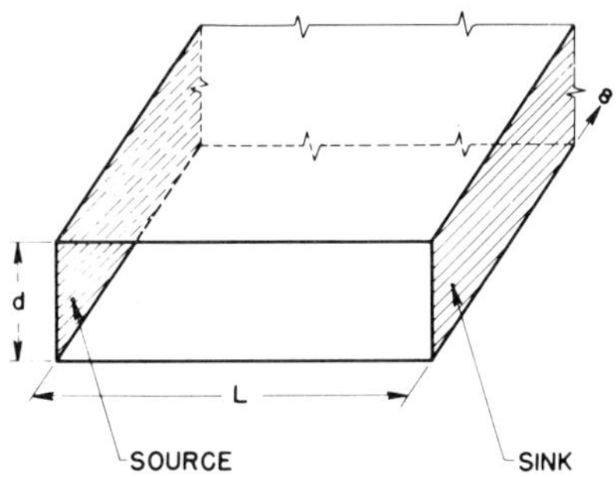

Fig. 2 The slot passage.

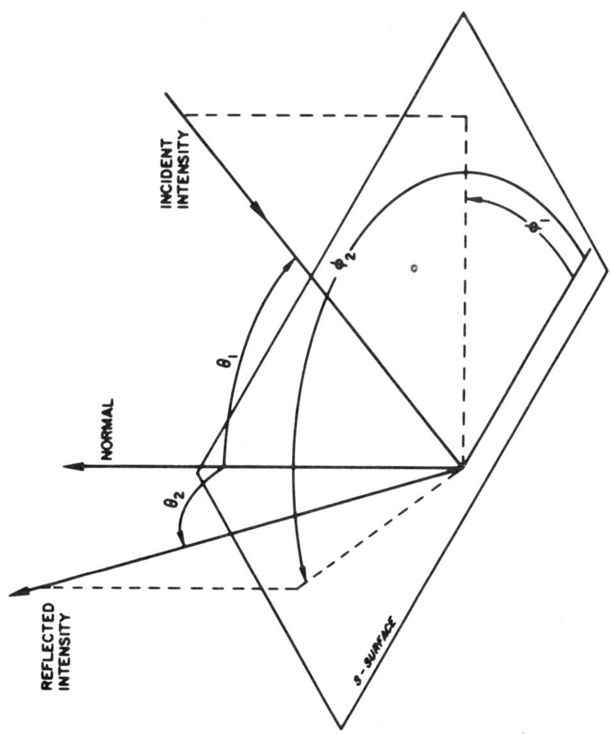

Fig. 3 Coordinates for incident and imperfectly diffuse reflected rays.

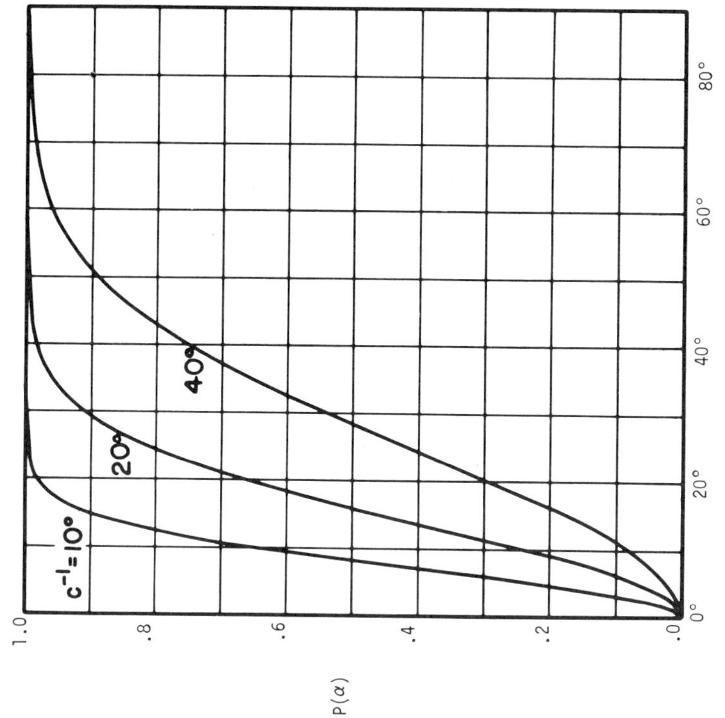

Fig. 4 Ogive to the surface facet slope distribution.

II. SURFACE RADIATION PROPERTIES

II.3 Engineering Measurements

DETERMINATION OF HEMISPHERICAL EMITTANCE BY MEASUREMENTS OF INFRARED BIHEMISPHERICAL REFLECTANCE

F. G. Sherrell*
ARO, Inc., Tullahoma, Tenn.

F. Shahrokhi†
University of Tennessee, Tullahoma, Tenn.

Abstract

A technique for the experimental determination of spectral, hemispherical emittance of surfaces in the infrared wavelength region is presented. The technique consists of using an infrared integrating sphere to measure spectral, bihemispherical reflectance. The feasibility of the technique is justified both on an analytical basis and, to the extent possible, on an experimental basis. A spectral, bihemispherical reflectometer was developed. The device produces reflectance measurements with an estimated uncertainty of $\pm 2\%$ of full scale. The reflectometer was applied to determine the spectral hemispherical emittance of fused quartz in the 5 - 16 μ range.

I. Introduction

Many problems, including space simulation and space systems testing, deal with radiative heat transfer from surfaces. Such problems require knowledge of the total hemispherical emittance ϵ of near ambient temperature surfaces for computations of radiant interchange. An ideal black surface at temperature T, having an $\epsilon = 1$, emits spectral power per unit area and wavelength as determined by Planck's law. The total emitted power per unit area of black surface is found by integrating Planck's law.

Presented as Paper 70-65 at the AIAA 8th Aerospace Sciences Meeting, New York, N. Y., January 19-21, 1970. This research was sponsored by Headquarters AEDC, Arnold Air Force Station, Tenn., under Contract F 40(600)-69-C-0001 with ARO, Inc.

*Section Supervisor, Aerospace Environmental Facility.
†Associate Professor, Space Institute.

The efficiency of a real surface as a power emitter may be specified in terms of the spectral hemispherical emittance ϵ_λ of the surface. The ϵ_λ is defined as the ratio of the spectral emissive power of wavelength λ from the surface to the spectral emissive power of the same wavelength from an ideal black surface. The total emitted power per unit area of real surface is given by Ref. 1:

$$e(T) = \int_0^\infty \epsilon_\lambda \, e_{b\lambda} \, (T) \, d\lambda \tag{1}$$

The total hemispherical emittance ϵ of a surface is defined as the ratio of the emissive power from the surface to the emissive power from an ideal black surface of the same temperature:

$$\epsilon = \int_0^\infty \epsilon_\lambda \, e_{b\lambda} \, (T) \, d\lambda / \int_0^\infty e_{b\lambda} \, (T) \, d\lambda \tag{2}$$

If the surface is a "gray" surface, the ϵ_λ is a constant and $\epsilon = \epsilon_\lambda$. In the general case the surface is nongray; i.e., ϵ_λ varies with λ and ϵ varies with surface temperature according to Eq. (2).

Regardless of whether the surface is gray or nongray, there are only three practical approaches to the determination of ϵ for near ambient, or below ambient, temperature surfaces: 1) calorimetric techniques,[2] 2) measurements of surface emissive power using cryogenically cooled thermopile detectors,[3] and 3) measurements of infrared reflectance from which, by Kirchhoff's law, ϵ can be determined.

The third possible way listed above is to measure the spectral bihemispherical reflectance ρ_λ of the surface. By Kirchhoff's law, $\epsilon_\lambda = \alpha_\lambda$,[1] and for an opaque surface $(1 - \rho_\lambda) = \alpha_\lambda$. Consequently, $\epsilon_\lambda = (1 - \rho_\lambda)$ for opaque surfaces. Therefore, by measuring ρ_λ, ϵ_λ is defined, and the total hemispherical emittance ϵ can be calculated for any desired surface temperature from Eq. (2).

If the surface is gray, as is often a close approximation, $\epsilon = \epsilon_\lambda$, and the measurement of ρ_λ becomes a particularly convenient way to determine ϵ.

However, prior to the present work, there has been no way to measure ρ_λ directly. The heated cavity reflectometer of Dunkle et al.[4] is widely used to measure the infrared, spectral, directional-hemispherical reflectance $\rho_{\lambda dh}(\theta, \varphi)$ which, with Kirchhoff's law, can be used in various ways to estimate ϵ .

In the final analysis the heated cavity reflectance data provide only estimates of ϵ, and as indicated previously, a technique for measurement of the spectral, bihemispherical, reflectance ρ_λ has not been available. Magnesium oxide (MgO) coated, integrating spheres,[5] and auxiliary spectrophotometry equipment have been widely used in commercial and noncommercial instruments for measuring spectral reflectance of surfaces in the 0.25 - 2.5μ region. Depending on the design, integrating spheres may be used to measure either directional-hemispherical reflectance or hemispherical-directional reflectance. This strongly suggests that they can also be used to measure hemispherical-hemispherical (i.e., bihemispherical) reflectance ρ_λ. As indicated previously, this is a particularly worthwhile measurement since $(1 - \rho_\lambda) = \epsilon_\lambda$ for opaque surfaces. The present work is directed toward the development of a technique for the direct measurement of ρ_λ using an infrared integrating sphere.

II. Analysis of a Technique for Measuring Bihemispherical Reflectance with an Integrating Sphere

One very useful form of Kirchhoff's law specifies that the monochromatic, directional emittance $\epsilon_\lambda(\theta,\varphi)$, of a surface is equal to the monochromatic, directional absorptance $\alpha_\lambda(\theta, \varphi)$ of the surface when the surface temperature is held constant.[6] The quantity $\alpha_\lambda(\theta,\varphi)$ is defined as the fraction of incident power, flowing from the (θ,φ) direction, which is absorbed by the surface. For an opaque surface, the conservation of energy requires that

$$\alpha_\lambda(\theta, \varphi) = 1 - \rho_{\lambda dh}(\theta, \varphi) \tag{3}$$

where $\rho_{\lambda dh}(\theta,\varphi)$ is the monochromatic, directional-hemispherical reflectance of the surface. Therefore, by Kirchhoff's law:

$$\epsilon_\lambda(\theta, \varphi) = 1 - \rho_{\lambda dh}(\theta, \varphi) \tag{4}$$

Further, the hemispherical and the directional emittances are related as follows:

$$\epsilon_\lambda = \frac{1}{\pi}\int_0^{2\pi}\int_0^{\pi/2} \epsilon_\lambda\ (\theta,\ \varphi)\cos\theta\ \sin\theta\ d\theta d\varphi \qquad (5)$$

By substituting Eq. (4) into Eq. (5) the following relation between hemispherical emittance and directional-hemispherical reflectance is obtained:

$$\epsilon_\lambda = 1 - \frac{1}{\pi}\int_0^{2\pi}\int_0^{\pi/2} \epsilon_\lambda\ (\theta,\ \varphi)\cos\theta\ \sin\theta\ d\theta d\varphi \qquad (6)$$

The integral term is the monochromatic, hemispherical-hemispherical (i.e., bihemispherical) reflectance ρ_λ[7]:

$$\rho_\lambda = \frac{1}{\pi}\int_0^{2\pi}\int_0^{\pi\ 2} \rho_{\lambda dh}(\theta,\ \varphi)\cos\theta\ \sin\theta\ d\theta d\varphi \qquad (7)$$

Equations (6) and (7) show that $\epsilon_\lambda = (1 - \rho_\lambda)$. Therefore, a measurement of ρ_λ is very desirable, since it defines ϵ_λ. To this end consider an arrangement in which a small sample of area A_s, is wall-mounted on a port looking into an integrating sphere (Fig. 1). The sphere's interior is coated with a diffusing, highly reflecting material of reflectance ρ_λ.

Let monochromatic power P_o pass into the sphere through a second port, suffer its first reflection at the sphere wall, disperse to the sphere interior and suffer a second reflection, a portion of the power will hemispherically irradiate and reflect from the sample. The reflected component will again reflect off the sphere wall and a portion of this component will strike the detector and contribute to its output. Baffle No. 2 (Fig. 1) is sized to prevent any reflected energy from going directly from sample to detector.

Therefore, with the arrangement of Fig. 1, the sample area is uniformly, hemispherically irradiated, and further, the reflected power is hemispherically collected by the sphere and a fraction of the reflected power transferred to the detector. This suggests that a portion of the detector output signal is proportional to ρ_λ as defined by Eq. (7). In fact, a straightforward, but rather lengthy, analysis will show this to be the case[8]; that is, the detector output voltage, E_o is of the form

$$E_o = E + \Delta E \qquad (8)$$

where $\Delta E = K\rho_\lambda$. The ΔE component (for $\rho_\lambda = 1$) is much smaller than E, which is generated by wall reflected power that never strikes the sample. It follows, that if ΔE is large enough, relative to E, to be measured accurately, then the integrating sphere can be applied to the determination of ρ_λ (i.e., ϵ_λ). The calibration constant, K, could be determined by measuring ΔE for a surface of known infrared ρ_λ. It is desirable at this point to evaluate the magnitude of ΔE relative to E.

The input power P_o separates into an infinite number of components by inter-reflections within the sphere. Figure 2 indicates the manner in which the input power disperses. The object at this time is to derive an expression for the power included in all sample-reflected components reaching the detector A_d. Individual subcomponents which have reflected off the sample three or more times will be excluded from the analysis. This is justified on the basis that the shap factor (A_s/A_o) for radiative transfer from the sphere wall to the sample will be of order 10^{-2}, and the power contained in components reflected from the sample will be attenuated by orders of 10^{-2N}, where N is the number of sample reflections which have occurred.

Figure 3 shows the power of each reflected component of Branch No. 1 (Fig. 2). These values follow from the straightforward application of the shape factor for energy transfer between one area and another on the surface of an ideal integrating sphere.

The magnitude of ΔE can be evaluated with the aid of Figs. 2 and 3. Summing the sample reflected components from all branches gives a detected power component of

$$\Delta P \approx \frac{\rho_s \rho_w^3 P_o \ (A/A_o)^2 \ (A_s/A_o) \ (A_d/A_o)}{(1 - \rho_w \ (A/A_o) \ )^2} \tag{9}$$

In this result, a ρ_s^2 term was neglected, since it contributed less than 1% of ΔP. Referring again to Fig. 3, the detector component calculated by summing the wall-reflected components across the top of Fig. 3 is given by

$$P = \frac{\rho_w^2 P_o \ (A/A_o) \ (A_d/A_o)}{1 - \rho_w \ (A/A_o)} \tag{10}$$

Ratioing these two components gives

$$\frac{\Delta P}{P} = \frac{\Delta E}{E} = \frac{\rho_s \rho_w \ (A/A_o) \ (A_s/A_o)}{1 - \rho_w \ (A/A_o)} \tag{11}$$

As a typical example, ρ_w = 0.95, (A/A_o) = 0.98, and (A_s/A_o) = 0.01, and from Eq. (11)

$$\Delta E/E = 0.133 \ \rho_s \tag{12}$$

This result suggests that with stable instruments and conditions ρ_s can be accurately determined by measuring ΔE. For example, a reflectance change of 0.01 (i.e., 1% on a scale of 100%) will generate a ΔE of

$$\Delta E \approx 1.33 \times 10^{-3} E \tag{13}$$

This indicates that the detector output voltage will have to be resolved to withing approximately ± 1 part per thousand in order to determine ρ_s to within ± 1 on a full scale of 100% reflectance. This resolution is feasible with modern instrumentation provided E is stable and sufficiently large.

III. Description of an Infrared Bihemispherical Reflectometer

The spectroreflectometer shown schematically in Fig. 4 and pictorially in Fig. 5 was designed and developed to implement the bihemispherical reflectance measurement technique that was analyzed in the foregoing discussion. In operation, energy from the infrared source is collected by the gold-coated, 6-in. mirror (Fig. 4) and is focused at the entrance slit of the monochromator. The energy is dispersed by a cesium bromide (CsBr) prism which is usable over the infrared region from 1 to 38 μ . After leaving the monochromator, the energy is chopped at 100 Hz and is collected by the gold-coated, 8-in. mirror and focused into the infrared integrating sphere.

The interior of the sphere wall was covered with strips of 180 grit, waterproof, silicon carbide abrasive paper. Gold was vacuum evaporated over the paper. Samples prepared in this manner exhibited a diffuse reflectance of 0.95. A small portion of the sphere input power is measured by the input power detector which is mounted on the sphere wall behind the narrow, centrally located slit between the baffles. This detector signal is used by the slit servo instrumentation to maintain the input power constant during a measurement.

The remainder of the input power is scattered over the sphere interior by an initial wall reflection from the area between the baffles. The baffles prevent any of this first reflection energy from getting to the sample port or to the hemispherical detector. The inter-reflections which occur within the sphere were fully examined in the previous analysis. The sample on the 0.9-in. diam sample port is hemispherically illuminated; and since the ratio of sample area to sphere area is $(A_s/A_o) = 0.0109$ (i.e., approximately 10^{-2}), the output of the hemispherical detector can be expected to be of the form $E + \Delta E$ where ΔE is a linear function of the bihemispherical reflectance of the sample surface as specified in Eq. (11).

Thermistor detectors were chosen for use in the reflectometer because broadband (i.e., 1 - 35μ) detectors were readily available commercially. Also, thermistor detector responsivities for 6.25 mm^2 active thermistor flakes range from 50 to 100 v/w as compared to the 5 - 15 v/w generated by thermopile infrared detectors of comparable receiver area.

The two detector signals are read out on two separate Princeton Applied Research (PAR) Model HR-8 lock-in amplifiers. These amplifiers synchronously rectify the detector signals and function as narrow band detection systesm. The amplifiers utilize the 100 Hz reference signal that is generated by the auxiliary chopper blade and the reference lamp and detector indicated in Fig. 4.

Both of the HR-8 amplifiers have differential input preamplifiers which are utilized to take the difference between two input signals. Using one amplifier, the signal from the input power detector is compared to a standard calibration signal and the difference signal is amplified and used to serve the monochromator slits for maintaining a constant, sphere-input power.

With the second amplifier, the signal from the hemispherical detector is compared to a standard signal from a PAR offset generator and is nulled when there is no sample on the sphere sample port. Nulling is accomplished by varying the amplitude and phasing of the output of the offset generator. After nulling is accomplished, the sensitivity of the amplifier can be increased to display and record the ΔE resulting from placing a sample on the sample port.

Data are collected as follows. The monochromator is adjusted to the wavelength band of interest and the hemispherical detector reading is nulled with the sample port open. The sensitivity of the amplifier is increased by a factor of 10 and final nulling is then accomplished by adjusting the offset generator for a zero recorder deflection. A standard gold mirror is placed on the sample port and the resulting $(\Delta E)_g$ is recorded. The standard is then replaced with the test sample and the resulting (ΔE) is recorded. The spectral, bihemispherical reflectance $(\rho\lambda)_s$ of the sample is then calculated:

$$(\rho_\lambda)_s = [(\rho\lambda)_g/(\Delta E)_g] \qquad (14)$$

Therefore, the measurement is relative to the spectral, bihemispherical reflectance (0.98) of the standard gold mirror. An example of the recorded data for a gold mirror and two sample surfaces is shown in Fig. 6.

IV. Summary and Analysis of Experimental Results

In the previous analysis it was found that the change ΔE in the hemispherical detector signal when a sample was placed on the sample port divided by the hemispherical detector signal E with no sample on the sample port, was closely approximated by Eq. (11).

The first performance test consisted of experimentally measuring $\Delta E/E$ as a function of bihemispherical reflectance using three specially prepared "working standard" surfaces (Fig. 7). The results indicate that the reflectometer performs according to Eq. (11) but are somewhat inconclusive because the reflectances of two of the working standard surfaces are not known accurately enough. Assuming that the gold standard (the data point at $\rho = 0.98$, Fig. 7) is correct, the deviations of the other two samples are in the expected direction. On the basis of the gold standard, the ratio $\Delta E/E$ is equal to 0.109 ρ_s. The value calculated from Eq. (11) for this reflectometer is 0.145 ρ_s. This is considered to be good agreement between theory and experiment and implies an effective wall reflectance of 93%. This also seems reasonable, since the seams in the wall coating undoubtedly reduce the effective wall reflectance slightly from the 95% reflectance of the gold-coated, 180 grit paper.

A more conclusive test of reflectometer performance was to apply the device to determine the spectral hemispherical emittance of 100 C fused quartz as shown in Fig. 8. Detector noise limited the data to the region less than 16 . The resulting (extrapolated data were used to calculate the total hemispherical emittance by rationing the areas under the curves in Fig. 9. The resulting experimental emittance of 0.80 agrees very well with the published value of 0.78[9]. On the basis of the experimental results, it was concluded that the reflectometer functions as specified by the analysis.

V. Conclusions

An integrating sphere may be applied to the direct measurement of the spectral, bihemispherical reflectance of surfaces in the infrared. This surface property is particularly useful since unity minus this quantity is the spectral hemispherical emittance of the surface. This experimental technique for determining hemispherical emittance was justified both on an analytical basis and to the extend possible with available "working standard surfaces," on an experimental basis. In its present state of development, the experimental reflectometer produces reflectance data with an estimated uncertainty of ± 0.02 ($\pm 2\%$ of full scale) at 5μ. As a result of excessive detector noise, this increases to ± 0.10 at 16 . Further development work should be directed toward improving the signal-to-noise ratio. This will extend the wavelength range of the instrument and render it a very useful analytical device.

References

[1]Sparrow, E. M. and Cess, R. D., Radiation Heat Transfer, Brooks/Cole, Belmont, Calif., 1967.

[2]Butler, C. P., and Inn, E. C. Y., "A Method for Measuring Total Hemispherical Emissivity of Metals," Surface Effects on Spacecraft Materials, First Symposium, edited by Francis J. Clauss, Wiley, New York, 1960.

[3]Schmidt, R. N. and Janssen, J. E., "Low-Temperature Total Emittance Calorimeter," Measurement of Thermal Radiation Properties of Solids, edited by C. Richmond, NASA SP-31, 1963.

[4]Dunkle, R. V., et al., "Heated Cavity Reflectometer for Angular Reflectance Measurements," Progress in International Research on Thermophysical and Transport Properties, The American Society of Mechanical Engineers, New York, 1962, pp. 541-562.

[5]McNicholas, H. J., "Absolute Methods in Reflectometry," Bureau of Standards Journal of Research, Vol. 1, U.S. Government Printing Office, Washington, D. C., 1928.

[6]Wiebelt, J. A., Engineering Radiation Heat Transfer, Holt, Rinehart and Winston, New York, 1966.

[7]Wolfe, L. and Nichodemus, E., "Radiation Theory," Handbook of Military Infrared Technology, edited by W. L. Wolfe, U.S. Government Printing Office, Washington, D. C., 1965.

[8]Sherrell, F. G., Determination of Hemispherical Emittance by Measurements of Infrared Bi-Hemispherical Reflectance, M. S. Thesis, 1969, The University of Tennessee.

[9]Gubareff, A. A., Janssen, J. E., and Torborg, R. H., Thermal Radiation Properties Survey, 2nd ed., Honeywell Research Center, Minneapolis, 1960.

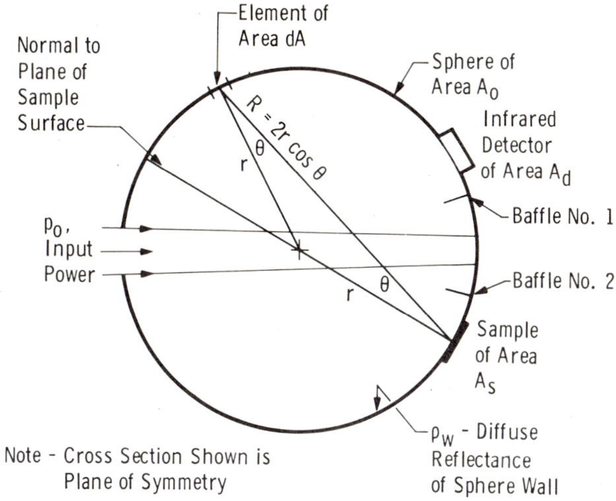

Fig. 1. Integrating sphere and associated components for bihemispherical reflectance measurements.

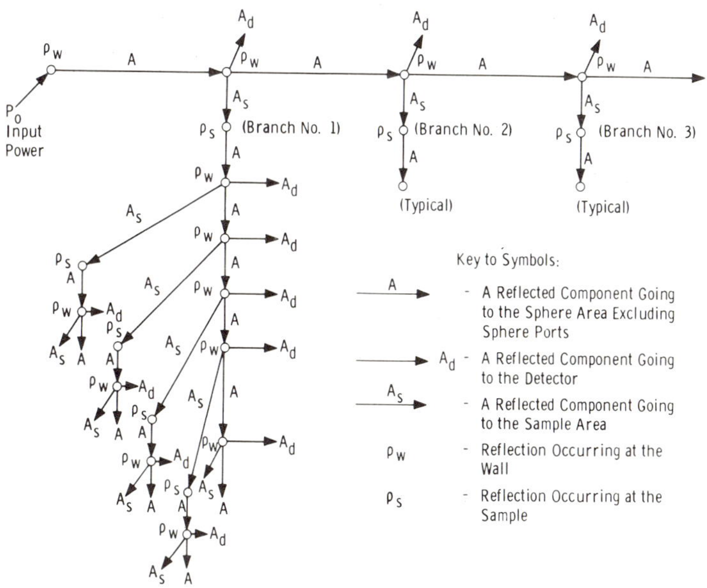

Fig. 2. Inter-reflection components.

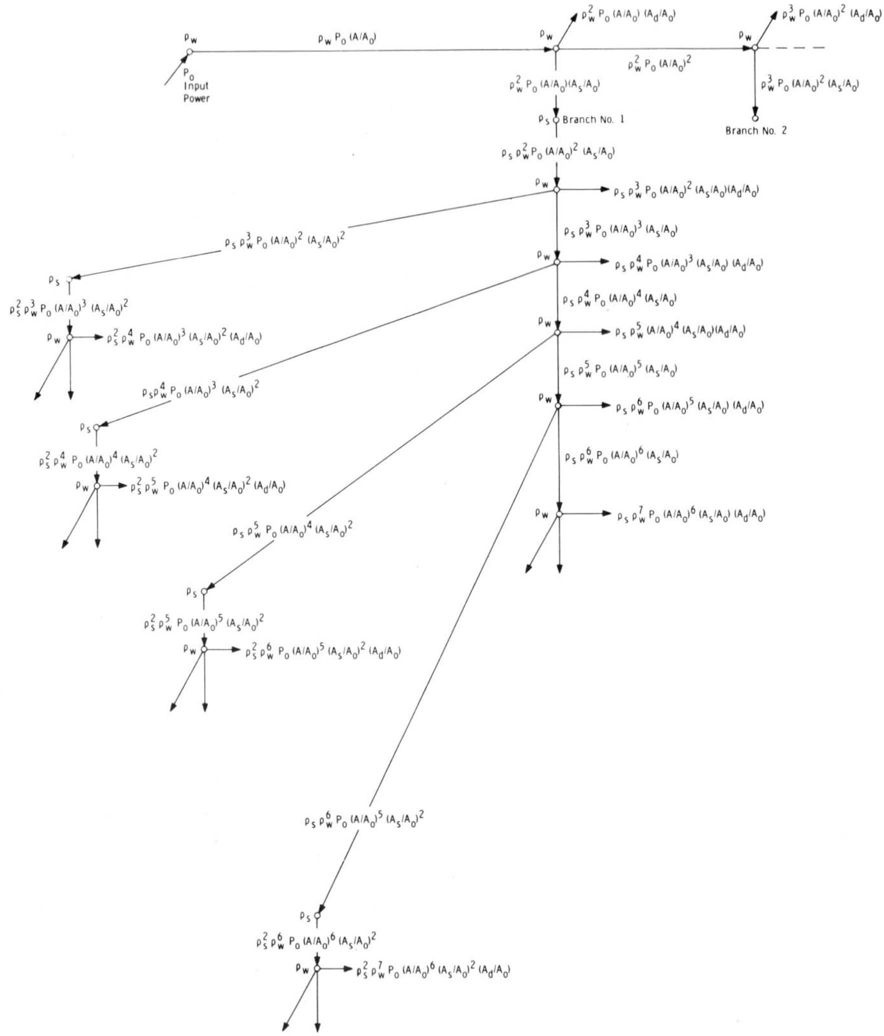

Fig. 3. Power retained in inter-reflection components.

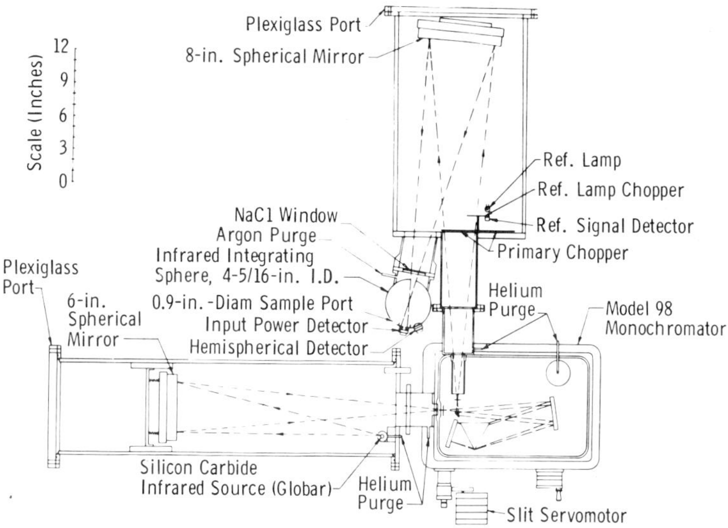

Fig. 4. Infrared bihemispherical reflectometer schematic.

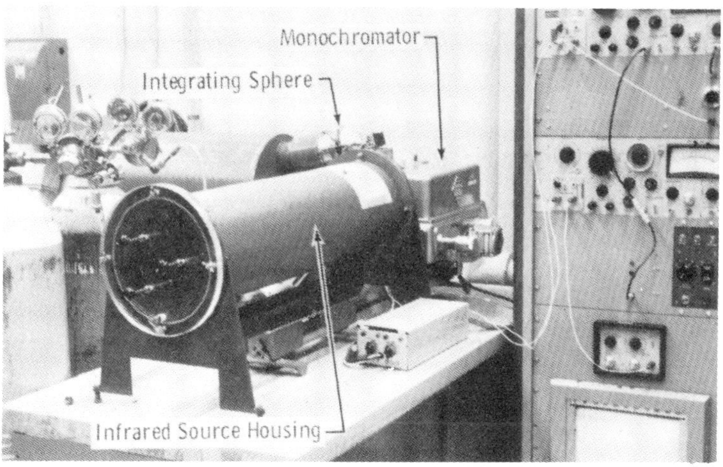

Fig. 5. Infrared bihemispherical reflectometer photograph.

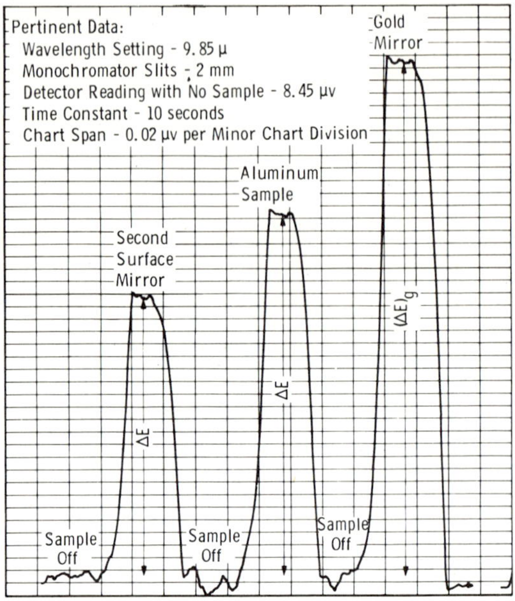

Fig. 6. Example of reflectometer data for a gold mirror, a sand blasted aluminum surface and a second surface quartz mirror.

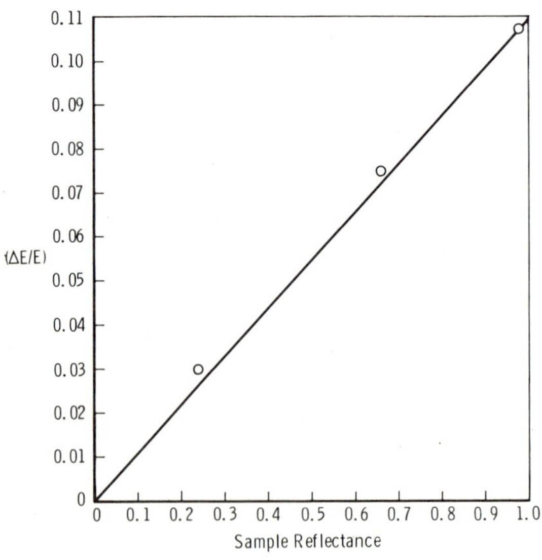

Fig. 7. Plot of fractional change in hemispherical detector signal bihemispherical reflectance at 7μ.

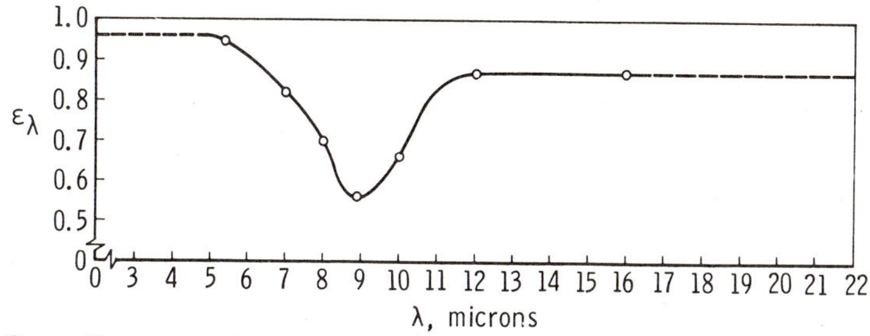

Fig. 8. Spectral hemispherical emittance of fused quartz.

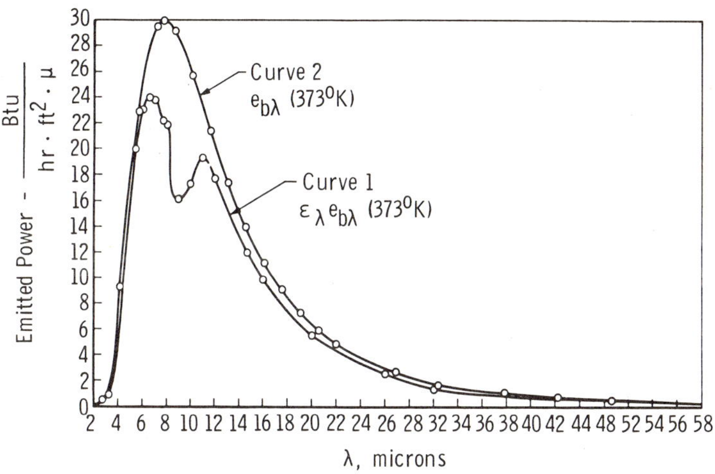

Fig. 9. Spectral power emitted from 100°C fused quartz.

HEMISPHERICAL THERMAL EMITTANCE OF COPPER AS A FUNCTION OF OXIDATION CONDITIONS

Robert L. Reid*

University of Tennessee, Knoxville, Tenn.

Carlos W. Coon†

Southern Methodist University, Dallas, Texas

Abstract

This research investigated the effect of the degree of oxidation on the emittance of copper for application to copper radiation shields or multifoil insulation systems containing copper foils. Previously, the only literature data available were extremely qualitative, and the only specification given was that the copper was lightly or heavily oxidized. The specimens in this study were oxidized in ambient air and the emittance measured in vacuum in a spherical enclosure by the calorimetric method. Emittance data for specimens prepared by surface grinding were taken for oxidation at various times at 300, 400, 500, and 600°F and for unoxidized copper. The emittances were measured in vacuum between 700 and 1400°F and varied between about 0.03 and 0.25 depending on the time and temperature of oxidation and the temperature in vacuum. An uncertainty analysis showed that the emittances were within 2.8 to 3.8% for unoxidized and heavily oxidized copper and within 4.8 to 16.0% for intermediate stages of oxidation. A parabolic oxide thickness equation was found to correlate the data satisfactorily. Some data were also obtained for surfaces finished on an automatic lapping machine and oxidized at 500°F. These emittance data were somewhat higher than the data for specimens prepared by surface grinding.

Presented as Paper 70-66 at the AIAA 8th Aerospace Sciences Meeting, New York, Jan. 19-21, 1970. This research was sponsored by the Thermal and Fluid Sciences Center of Southern Methodist University and by NASA under grant NGL 44-007-006.

*Assistant Professor, Mechanical/Aerospace Engineering Department; formerly, Research Assistant, Thermal and Fluid Sciences Center, Southern Methodist University.

†Assistant Professor, Thermal and Fluid Sciences Center.

Nomenclature

a = absorption coefficient, μ^{-1}
A = surface area of specimen, ft^2
A_w = inner surface area of sphere, ft^2
B = constant in Eq. (8), ^oR
D = oxide thickness, μ
D' = corrected oxide thickness, μ
q = net heat flow, watts
q_L = heat loss out heater leads and thermocouples, watts
q_T = total power input, watts
R = universal gas constant, 1.986 cal/gmole
s = back scattering coefficient, μ^{-1}
t = oxidation time, hr
T = oxidation temperature, ^oR
T_s = specimen temperature in vacuum, ^oR
T_w = wall temperature, ^oR
w = weight gain upon oxidation, g/cm^2
w' = weight gain corrected for decomposition, g/cm^2
α = constant in Eq. (8), $(\mu/hr)^2$
β_o = $[a/(a+2s)]^{1/2}$
ε = emittance
ε_o = emissivity of unoxidized copper
ε_w = wall emittance
ρ = $\varepsilon_o/(2-\varepsilon_o)$
σ = Stefan-Boltzmann constant, 1713×10^{-12} BTU/hr ft^2 $^oR^4$
σ_o = $[a(a+2s)]^{1/2}$

1. Introduction

Copper is frequently chosen whenever a low emissivity thermal radiation shield is needed in a vacuum environment at elevated temperatures. One such application is multifoil insulation systems where copper foils are used alone or in conjunction with aluminum or nickel foils. However, the main disadvantage of copper is its low resistance to oxidation. If the radiation shield or multifoil insulation must be heated in air before vacuum is attained, oxidation will occur and the emissivity will increase. This sometimes occurs when multifoil insulation is used around the thermoelectric elements of an isotope powered thermoelectric generator. The system must be fueled on the ground and a considerable hold time elapses during which the insulation is hot before it is exposed to the low pressures of space. Since some of the metallic foils in insulation systems have relatively poor oxidation resistance, oxide will form on the foils during the air exposure of the hot insulation system. Although in some cases it may be possible to provide an inert gas purge during ground hold and lift-off, in many cases this would not be practical since the inert gas supply system would contribute weight and an extremely pure inert gas would have to be used. It is therefore useful to be able to predict and design for this deterioration in conductivity due to oxidation.

There is the possibility of using gold to replace copper, and platinum to replace nickel, in a multifoil insulation system to increase the oxidation resistance with little or no degradation of thermal performance. However, the cost of such a system would be orders of magnitude higher than a conventional copper-nickel system. In most applications, the economics would probably not allow such a substitution. At least it would be desirable to be able to calculate the performance of an oxidized copper-nickel system in order to make a cost-performance analysis. In addition, platinum and gold foils are available only in narrow widths and short lengths which would make the installation of such a system impractical in many cases.

A first-order solution to the multifoil insulation oxidation problem has been previously presented in the literature.[1] As evident, this solution required the knowledge of emittance as a function of the degree of oxidation for the copper, aluminum, and nickel foils. This information was available in the literature for nickel[2,3] and the excellent oxidation resistance of aluminum precluded the need for the data. However, quantitative data could not be located for the effect of oxidation on the emittance of copper. The problem was solved without the data by making the bold assumption that the scattering constants were equal for copper and nickel oxides and using qualitative literature data for the emittance of heavily oxidized copper surfaces. This lack of data for relating the emittance of an oxidized copper surface to the degree of oxidation motivated the present study.

There have been no previous quantitative studies of the effect of the degree of oxidation on the emittance. However, several investigators have measured the emittance of heavily oxidized copper surfaces. Instead of the calorimetric method, a comparative method was used. In this method, the radiant flux leaving a body is intercepted by a thermopile or other radiation detector. The measured energy is compared to that received from a black body at the same temperature. The emittance then results from the division of these two quantities. The results of these measurements are given in Fig. 1. For simplicity, only representative data are given in Fig. 1. Each study will now be individually discussed.

The specimen used by Wade[4] was a 3-in.-long strip of copper with 3-in. strips of Inconel welded on each end. A current was passed through the assembly. The copper was thus heated by conduction from the Inconel strips. The specimen was oxidized in air at 1400°F and the emittance measured as a function of time during the oxidation until no further change in emittance was noted. This took about 40 min. The emittance of the sample was then measured in air at temperatures from 600 to 1400°F. The total normal emittance was calculated as mentioned above and the total hemispherical emittance was calculated from measured values of the angular dependence of the emission.

Betz et al.[5,6] measured the emittance of samples that had been oxidized at 1200°F for 30 min. The emittance was measured in both vacuum and air from 250 to 1450°F. Note from Fig. 1 that the results of the air and vacuum runs were significantly different. As will be pointed out in the next section, a transformation of oxide form in vacuum could account for this difference.

Aref[7] measured the emittance of "oxidized" copper samples at temperatures between 500 and 1500°F. No further specifications of the specimens were given.

Carroll and Pears[8] measured the emittance of samples heated in air from 600 to 1830°F. The emittance was measured as the sample heated up and oxidized. The sharp decrease in the emittance above 1600°F was probably due to the vaporization of copper or copper oxide or both.

One significant factor to be noted from the above and Fig. 1 is that the condition of the sample and the conditions of the experiments apparently have profound influence on the results. It is also evident that this previous data is for copper oxidized at higher temperatures than those of interest in the present study. The oxide thicknesses were also apparently much greater than the thicknesses to be investigated in this study.

The objective of this research was to determine the quantitative relationship between the temperature and time of oxidation of copper and the resultant dependence of emittance upon temperature in vacuum.

The primary temperature range of interest in vacuum is 900 to 1400°F. This is the temperature range of operation for a multifoil insulation system containing nickel, copper, and aluminum foils.[9] Some data were obtained below 900°F for possible other applications involving oxidized copper. The temperature range of oxidation was fixed at 400 to 600°F since it has been previously determined[1] that this would be the range of operation of the copper section of a composite insulation system when filled with air and fueled by a constant heat source radioisotope. The times of oxidation were taken as 5 hr to 30 days, since this is the length of time a system might remain fueled on the launching pad before lift-off.

The method for emittance measurement chosen was the calorimetric method. By this method, a heater is enclosed within a hollow specimen and suspended in a vacuum chamber. The heat transfer by radiation from the specimen to the enclosure is then calculated from the power measurement, making corrections for heat lost out the thermocouples and heater leads. From the heat transfer, the emittance of the specimen is then calculated. This method yields a straightforward measurement of the emittance and avoids the construction of black body references that are required for thermopile detector methods.

2. Oxidation Mechanism for Copper

The rate of oxidation of copper is a diffusion governed process after the first few angstroms have formed on the surface. This causes concentration gradients to exist in the oxide film, which leads to the formation of copper rich cuprous oxide (Cu_2O) near the metal-oxide interface and oxygen rich cupric oxide (CuO) near the oxide-air interface. It has been found[10] that at oxidation temperatures below the 600°F limit of this study the oxide is about 95% cupric oxide. At high oxidation temperatures, the oxide is mainly cuprous oxide.

In order to predict the growth rate of the oxide, the literature was examined to obtain a rate equation for the oxidation. The results of several investigators were summarized in Ref. 11 but showed severe disagreement. In order to aid in selecting the times and temperatures of oxidation, the parabolic equation from Tylecote[10] was used. The equation is

$$w^2 = 0.045 \cdot t \cdot \exp(-29540/RT) \qquad (1)$$

When copper has been oxidized and then brought into a vacuum, several phenomena may occur. One potential process is reduction by offgassing products. Primarily, this means reaction with the dissolved hydrogen that is present in almost all metals and which begins to evolve in vacuum. The hydrogen will react with both the cupric and cuprous oxides to form copper and water. Another gas that will act similarly is carbon monoxide.

These reactions are, of course, detrimental to an experimental program unless a study of a particular known gas condition was desired. However, from a practical standpoint in multifoil or radiation shield design, it might be beneficial to not outgas the materials before use. In fact, a hydrogen soak at high temperatures might be used to try to make more reducing gas available to remove the oxide. To avoid the reduction complications, all specimens were degassed by heating in vacuum at 1400 - 1450°F for 2 - 4 hr prior to testing.

Dissociation of the oxide is another means by which the oxide might disappear or transform. Dushman[12] gives the conditions under which this might occur. The first reaction to consider is

$$4CuO = 2Cu_2O + O_2 \tag{2}$$

Figure 2 from Dushman gives the oxygen partial pressure at which the reaction might occur as a function of temperature. If the given oxygen pressure is on or below the curve at a given temperature, then the cupric oxide will decompose to cuprous oxide as given above. Note that at an oxygen pressure of 10^{-5} mm to 10^{-7} mm (expected values within the vacuum chamber), this transformation will occur at about 800 - 950°F.

The other dissociation reaction that might occur would be

$$2Cu_2O = 4Cu + O_2 \tag{3}$$

The dissociation curve in Dushman reveals that the partial pressure of oxygen would have to be below about 10^{-8} mm (at 1300°F) for this reaction to take place. Pressures this low are not expected in the laboratory vacuum chamber. However, this reaction would occur in deep space. From data of atmospheric compositions and pressures as a function of altitude, it can be seen that the cuprous oxide will decompose if the mission height is greater than about 150 miles.

Since the cupric oxide will transform to cuprous oxide in the vacuum chamber, Eq. (1) must be changed to reflect the oxygen weight lost. The corrected weight gain is

$$w'^2 = 0.0119 \cdot t \cdot \exp(-29540/RT) \tag{4}$$

It is desirable for correlation purposes to change this weight gain into an oxide thickness. Assuming that the oxide has the bulk density of cuprous oxide, a conversion constant was derived in Ref. 13. The final expression is then

$$D = [2.628 \times 10^6 \cdot t \cdot \exp(-14860/T)]^{1/2} \tag{5}$$

where the value of R has also been substituted.

3. Experimental Apparatus and Procedure

The specimens were machined from a 4-in.-diam electrolytic tough pitch copper bar. Electrolytic tough pitch copper is a minimum of 99.90% pure as described in Federal Specification QQ-C-576b.

The specimens were in the shape of thin hollow disks composed of two flat milled out pieces. The complete specimens were about 4.0 in. in diameter and 0.6-in. thick. The two halves were held together at the outside edge by two copper plated flat head slotted 6-32 machine screws. On one face of the specimen, 1 1/4 in. from the center, there were two holes 3/16 in. in diameter and 5/8 in. apart for the heater posts to pass through.

A commercial disk heater rated for 600 w in air was used. It was a half-sheath model with the side opposite the posts having the ceramic refractory directly exposed. The other side and edges of the heater had a chrome steel sheath. The threads were ground off the heater posts leaving a smooth 0.1-in.-diam post. Near the end of each post, a small hole was drilled to permit the passage of 16-gage nickel wire. Short sections of mullite ceramic tubing 1/8 in. i.d. 3/16 in. o.d. were slipped over the posts to act as electrical insulators. The nickel lead wires passed through the holes and looped upward to form a hook. Hooks were also formed on the other ends of the 6-in.-long wires to hang on mating hooks of copper wire passing into the vacuum chamber.

The specimen temperature was measured with two 30 gage chromel alumel thermocouples. Holes were first punched at an angle into the copper, the bare junction inserted, and the holes then pounded closed. Care was taken to separate the wires immediately after they exited from the holes. One thermocouple was installed about 1/4 in. from the bottom on the back side (i.e. opposite the heater posts) and the other 1/4 in. from the top on the front side. A sketch of an assembled specimen appears in Fig. 3.

The method of surface preparation selected was mechanical polishing. Specimens were prepared by two mechanical polishing methods. Most of the data were obtained with surfaces prepared by surface grinding. The exact grinding procedure is given in Ref. 13. After removing the specimens from the grinder, the edges were sanded to a 600 grit on a lathe. Just before use, the specimens were wiped lightly with an oxide removing solution (BRASSO) to remove blemishes. The roughness was measured and varied between 5 and 16 μ in. for all the specimens.

Note that the edge did not receive quite the same treatment as the faces, and its roughness could not be measured. However, the edge contributes only 20-25% to the total area, and its condition is therefore of secondary importance. It did appear, however, to be quite smooth and not grossly different from the faces. The effect of this 20-25% area that may have had a slightly different initial surface condition was not considered in the error analysis to be discussed later. However, it was determined by a microscope that the surfaces roughened during oxidation so that any initial difference in surface roughness was probably obscured before the emittance was measured.

The other method of mechanical finishing used was automatic lapping on a machine manufactured by the Crane Company. The surface roughness varied between 3 and 5 μ in. The edges were sanded on a lathe and the blemishes removed with BRASSO as in the grinding process above. The majority of the specimens were prepared by surface grinding and the data analysis based on these tests. Although the lapping gave a smoother surface, as will be seen later it had a higher emittance and the grinding process was thought to be more reproducible.

The vacuum chamber consisted of an 18-in.-diam 1/4-in.-thick stainless steel sphere composed of two hemispheres joined together by an O-ring flange. The sphere was sandblasted, copper plated, and oxidized. Its emittance was estimated at 0.7. As will be pointed out later in this section, the exact value of this emittance need not be known.

Eighteen gage chromel alumel wires passed into the chamber through a Conax feedthrough. Two thermocouples went to chromel alumel connector plugs while one thermocouple was soldered to the inside chamber wall. The power was brought into the sphere by two 14-gage copper wires passed through another Conax feedthrough. The wires were bent into hooks inside the chamber to match with the hooks on the nickel wires leading to the specimen. The sphere had three adjustable legs so that it could be set on top of the vacuum pumping unit.

For oxidation, a controller was used to control the specimen temperature within about 5°. The power during emittance measurement was supplied by a constant voltage regulator. The power was measured with ammeters and a digital voltmeter.

The actual procedure followed during a given test was as follows. The sphere was first evacuated and the controller set to the desired temperature of oxidation. After this temperature had been obtained, the sphere was isolated from the pumps and a valve opened to admit atmospheric air. After the desired oxidation time, the air valve was closed and the sphere evacuated. The controller was then switched out of the circuit and the power level set with the Powerstat according to the expected emittance at a given specimen temperature in vacuum. When the millivolt recorder indicated that equilibrium had been obtained, the temperatures were measured with the potentiometer or the digital voltmeter. The current was noted from the ammeter and the pressure level from the pressure gage. The power was then increased for the next temperature point. Note that the specimen was not cooled between oxidation and emittance measurement.

The total power input to the sample, q_T, was calculated as the product of voltage and current after verifying that the heater inductance was negligible. The power losses consist of heat conducted and radiated from the heater leads and thermocouples and heat generated in the heater leads. These losses were calculated on the Univac 1108 digital computer. These calculations are described in Ref. 13. The generation in the leads was found to be negligible. The solution was also found to be insensitive to the range of wall temperatures encountered. These observations allowed the preparation of a single plot of losses, q_L, vs specimen temperature.[13] In general these losses were less than 5% of the total heat flow except for unoxidized copper where they were as much as 20% at the lower temperatures. Gas conduction was neglected since it was calculated in Ref. 13 that the system would be in the free molecule flow regime if the pressure were kept below about 3×10^{-4} torr. Small errors of at most 1% were calculated for a few tests, and this error was included in the uncertainty analysis.

The heat transferred due to radiation between two concentric spheres is given by Christiansen's equation[14]

$$q = \frac{0.293\varepsilon\sigma\, A(T_s^{\,4}-T_w^{\,4})}{1+\varepsilon\left(\frac{1}{\varepsilon_w}-1\right)\frac{A}{A_w}} \tag{6}$$

This equation is a very good approximation even if the specimen is disk shaped as long as $A_w >> A$ and/or $\varepsilon_w >> \varepsilon$. Solving Eq. (6) for ε, we have

$$\varepsilon = \frac{q/A}{(T^4-T_w^{\,4})-\left(\frac{1}{\varepsilon_w}-1\right)\frac{q}{A_w}} \tag{7}$$

The emittance can thus be calculated from Eq. (7) from the corrected power input to the specimen. Note that ε_w appears in the term that is multiplied by A/A_w and thus contributes little to the calculated heat flow unless ε_w is very small. Values of ε_w of 0.7 and 0.9 were found to affect the calculated heat flow by only 0.5%.

4. Results for Specimens Prepared by Surface Grinding

The data taken for specimens prepared on the surface grinder appear in Figs. 4-7. Although 300°F was outside the scope of the desired oxidation conditions, one test was performed at this temperature and appears in Fig. 4. The data for unoxidized copper appears in all of these figures for comparison purposes. Test 39a was a repeat of test 9 and showed a high degree of reproducibility. The error bands from the uncertainty analysis for several of the points are also shown. Note in Fig. 7 that the test with a longer oxidation time fell below the other test. At the end of this test, when the sphere was opened, the specimen appeared blistered. This blistering probably exposed bare copper and led to the lower emittance. Longer oxidation times would probably result in gross scaling and the emittance might decrease further.

The oxide thickness was calculated for each test from Eq. (5). For each test emittance values at 750, 1050, and 1350°F were then obtained from Figs. 4-7. These values were plotted versus the oxide thickness on Fig. 8. It is obvious from Fig. 8 that since a single curve cannot be drawn through all the points at a given specimen temperature in vacuum, Eq. (5) does not satisfactorily correlate the data.

There might be several reasons for the failure of Eq. (5). One reason is that perhaps the wrong equation was extracted from the literature since, as already pointed out, there is considerable disagreement among investigators as to the oxidation rate of copper. Perhaps no one has yet determined the correct relationship. Another reason might be the surface condition of the specimens before oxidation. Previous workers investigating oxidation have typically used cold rolled annealed strips 0.010 to 0.020 in. thick. The surface grinding and subsequent polishing with BRASSO in the present tests may have influenced the oxidation rate.

Since only a correlation of emittance with a number proportional to the oxide thickness is desired, it is possible to adjust the temperature dependence of Eq. (5) and change the propostional constant to insure only the right magnitude of oxide thickness. This is done by finding conditions at two different oxidation temperatures that give the same emittances in vacuum and assuming that the oxide thicknesses are then equal. Although two tests were not performed that gave exactly the same emittance, interpolation on a plot of emittance vs oxidation time gave some sets of equivalent conditions. The general form of Eq. (5) is

$$D' = [\alpha \cdot t \cdot \exp(-B/T)]^{1/2} \tag{8}$$

The expression for B for conditions giving the same thickness is

$$B = [\ln(t_1/t_2)]/[(1/T_1)-(1/T_2)] \tag{9}$$

B was determined by this process to be about 26,000.

As stated previously, α is arbitrary if only a correlation is desired but to retain the correct order of magnitude, α was fixed so that the new equation gives the same value of oxide thickness at 450°F as does Eq. (5). The constant was then determined to be 5.447×10^{11}. The final corrected equation is then

$$D' = [5.447 \times 10^{11} \cdot t \cdot \exp(-26000/T)]^{1/2} \tag{10}$$

A plot of emittance vs oxide thicknesses calculated by Eq. (10) is given in Fig. 9. As can be seen from Fig. 9, Eq. (10) gives a much better correlation than Eq. (5).

A theoretical model for emittance as a function of oxide thickness was also used to correlate the data. This is the approximate two-flux model for radiant transfer in absorbing, scattering, and emitting media. More exact models are available but these require knowledge of the refractive index of the coating and substrate as a function of wavelength and temperature in the visible and infrared wavelength ranges and from 600° to 1400°F. This information is not available for copper and copper oxide. One advantage of the two flux model is that the absorption and scattering constants can be extracted from the data to give a set of correlation equations.[1,13] The two-flux model and the more exact solutions all suffer from a critical restriction. This restriction is that the oxide thickness must be much greater than the wavelength of radiation in the media. Calculations in Ref. 13 reveal that the oxide thicknesses in this study are on the order of the wavelength of radiation in the media so that if any of the available models are accurate, it may be only fortuitous.

When the backscattering coefficient was extracted from the data, it appeared to be dependent on the thickness of the oxide, indicating a possible deficiency of the model. To be consistent with the model, the coefficients were averaged at a given temperature. Consequently, the model does not agree exactly with the emittance data. The two-flux model equation is given below with the absorption and backscattering coefficients given in Fig. 10. D' is given by Eq. (10).

$$\varepsilon = \frac{2\beta_o[(\rho-\beta_o)\exp(-\sigma_o D')+(\rho+\beta_o)\exp(\sigma_o D')]}{(1+\beta_o)(\rho+\beta_o)\exp(\sigma_o D')+(1-\beta_o)(\beta_o-\rho)\exp(-\sigma_o D')} \quad (11)$$

Figure 11 shows the experimental error bands and the calculated two-flux model values for a few points. Some of the two-flux model values are obviously considerably outside the error bands.

In addition to the theoretical reasons for the model not being valid, one additional factor may have caused the bad fit. Although Test 44 was used as the "infinite" coating and the data does level out at these conditions, there is reason to suspect that the radiation effect is not entirely due to the coating. If the coating were effectively infinite, then the tests to be discussed later, using surfaces prepared by lapping, also would have shown no values of emittance higher than about 0.25. However, these tests indicated an "infinity" at about 0.35. A possible explanation is that the coating undergoes a lattice rearrangement at these leveling out points and thereafter begins to scale. The substrate and coating therefore possibly both still contribute to the emittance when the emittance versus oxide thickness curve levels out. This also might explain why oxidized surfaces measured in air have a higher emittance than measured in vacuum as indicated in Fig. 1.

In summary, although the model could be used for approximate results, better emittances can be determined for given oxidation conditions by calculating the oxide thickness from Eq. (10) and then using Fig. 9. This oxidized copper emittance data can then be used by the designer in conjunction with the analysis in Ref. 1 to determine the effect of oxidation on the thermal conductivity of multifoil insulation systems.

5. Results for Specimens Prepared by Lapping

These results are incomplete as this method of surface preparation was abandoned when a test of the unoxidized copper revealed an emittance of 0.063 at 1022°F. It was thought that unusual conditions may have been connected with the lapping process that would be difficult to duplicate elsewhere. The major effect may be the cutting oil. For a few specimens, the wrong cutting oil was inadvertently used and the emittance measured 0.15 unoxidized. If a change in cutting oil changes the emittance from 0.15 to 0.063, then even the correct oil may be causing the emittance to be 0.063 instead of 0.03. The results are presented only for general interest. Only data for oxidation at 500°F are presented. The surface roughnesses varied from 3 to 5μ in. The emittances are given in Fig. 12 for various oxidation times near an oxidation temperature of 500°F.

6. Uncertainty Analysis

A complete uncertainty analysis can be found in Ref. 13. The over-all uncertainty in emittance consists of two parts: a measurement uncertainty and an uncertainty which includes the uncertainty in the knowledge of the oxide thickness. This last part relates to the uncertainty in the values of the oxidation temperature and time. From Ref. 13, the following observations can be made.

1) The uncertainty in the measured value of emittance ranges from 0.0013 to 0.0075 for ground specimens and from 0.0060 to 0.0310 for lapped specimens.

2) This absolute uncertainty corresponds to 2.7 to 3.7% error in the emittance for ground specimens and 2.2 to 13.5% error for lapped specimens.

3) The main contributing factor in this emittance error is the uncertainty in the sample temperature.

4) The uncertainty in oxide thickness ranges from 11.3 to 18.4% with the controlling factor being the temperature uncertainty during oxidation.

5) The over-all uncertainty in emittance is about 2.8 to 3.8% for unoxidized copper and heavily oxidized copper and controlled by the emittance measurement uncertainty.

6) For intermediate stages of oxidation, the uncertainty ranges from 4.8 to 16.0 and is generally controlled by the uncertainty in oxide thickness.

7. Summary and Conclusions

The conclusions of this study can now be iterated.

1) The emittance of copper as a function of vacuum temperature and time of oxidation is given for oxidation temperatures of 400, 500, and 600°F in Figs. 5-7. One test is also given at 300°F in Fig. 4.

2) A parabolic oxidation thickness equation, Eq. (10), can be derived to successfully correlate the data as illustrated in Fig. 9.

3) The two flux model does not provide a good correlation but the derived equation, Eq. (11), can be used if only approximate answers are desired.

4) The uncertainty in the emittance values reported ranges from 2.8 to 3.8% for unoxidized and heavily oxidized copper and within 4.8 to 16.0% for intermediate stages of oxidation.

References

[1]Reid, R. L., "Air Exposure of Multifoil Insulation Systems," ASME Paper 68-HT-50, 1968.

[2]Shelton, J. L. and Akers, W. W., "The Total Hemispherical Thermal Emittance of Nickel as a Function of Oxide Thickness in the Temperature Range 400-900°C," AIAA Progress in Astronautics and Aeronautics: Thermophysics and Temperature Control of Spacecraft and Entry Vehicles, Vol. 18 edited by G. B. Heller, Academic Press, New York, 1966, pp. 151-165.

[3]Oubre, C. L., "The Total Hemispherical Thermal Emittance of (100), (110), and (111) Single Crystals of Nickel as a Function of Oxide Thickness, Temperature Range 300-900°C," Ph.D. Thesis, 1965, Rice Univ.

[4]Wade, W. R., "Measurements of Total Hemispherical Emissivity of Various Oxidized Metals at High Temperature," TN 4206, 1958, NACA.

[5]Betz, H. T. et al., "Determination of Emissivity and Reflectivity Data on Aircraft Structural Materials," WADC TR 56-222 Pt. I, AD 110 485, 1956, Wright Air Development Center.

[6]Betz, H. T. and Olson, O. H., "Determination of Emissivity and Reflectivity Data on Structural Materials," WADC TR 56-222 Pt. II, 1957, Wright Air Development Center.

[7]Aref, M. N., "Emissivity Measurement of Industrial Surfaces Due to Thermal Radiation," ASME Paper 58-HT-18, 1958.

[8]Carroll, J. M. and Pears, C. D., "The Emittance of Calorimetric Discs," CR-55073, N64-12156, 1963, NASA.

[9]Dewitt, W. D., Gibbon, N. C., and Reid, R. L., "Multifoil Type Thermal Insulation," IEEE Transactions on Aerospace and Electronic Systems, Vol. 4, No. 5, supplement 1968, pp. 363-271.

[10]Tylecote, R. F., "The Oxidation of Copper in the Temperature Range 200-800°C," Journal of the Institute of Metals, Vol. 81, 1952-53, pp. 681-700.

[11]Ronnquist, A. and Fischmeister, H., "The Oxidation of Copper - A Review of Published Data," Journal of the Institute of Metals, Vol. 89, 1960-61, pp. 65-76.

[12]Dushman, S., Scientific Foundations of Vacuum Technique, 2nd ed., Wiley, New York, 1962, p. 742.

[13]Reid, R. L., "The Total Hemispherical Thermal Emittance of Copper as a Function of the Degree of Oxidation in the Temperature Range 700-1400°F," Ph.D. Dissertation, 1969, Southern Methodist University.

[14]Jakob, M., Heat Transfer, Vol. II, Wiley, New York, 1957, p. 5.

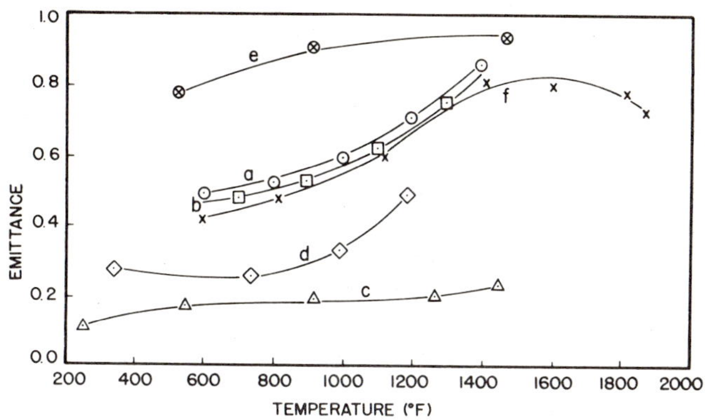

Fig. 1 Previous data for heavily oxidized copper; a) Wade,[4] Total hemispherical emittance, in air; b) Wade,[4] Total normal emittance, in air; c) Betz,[6] Total normal emittance, in vacuum; d) Betz,[5] Total normal emittance, in air; e) Aref,[7] Total normal emittance, in air; f) Carroll,[8] Total normal emittance, in air.

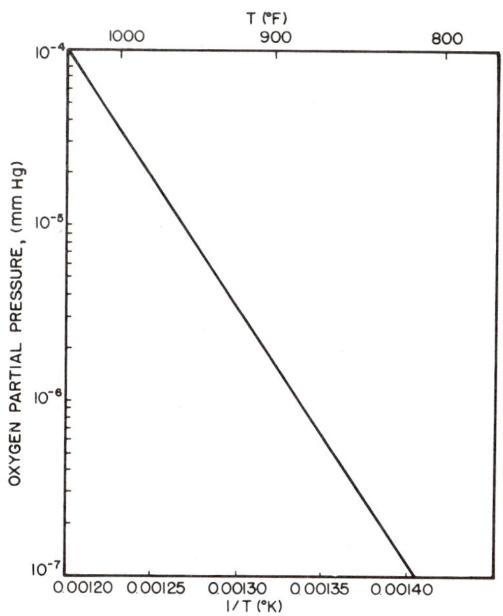

Fig. 2 Decomposition conditions for cupric to cuprous oxide.

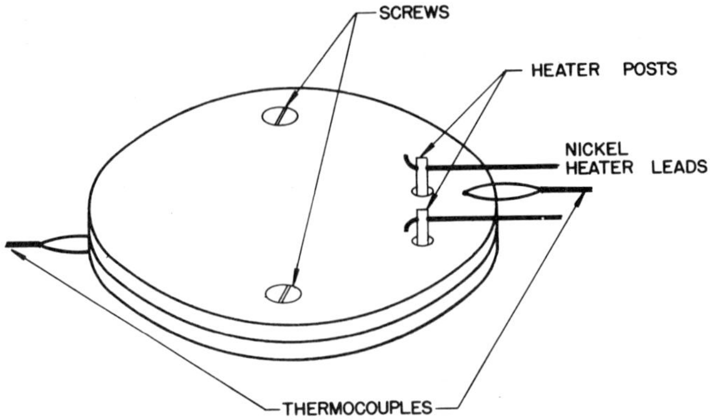

Fig. 3 Test specimen.

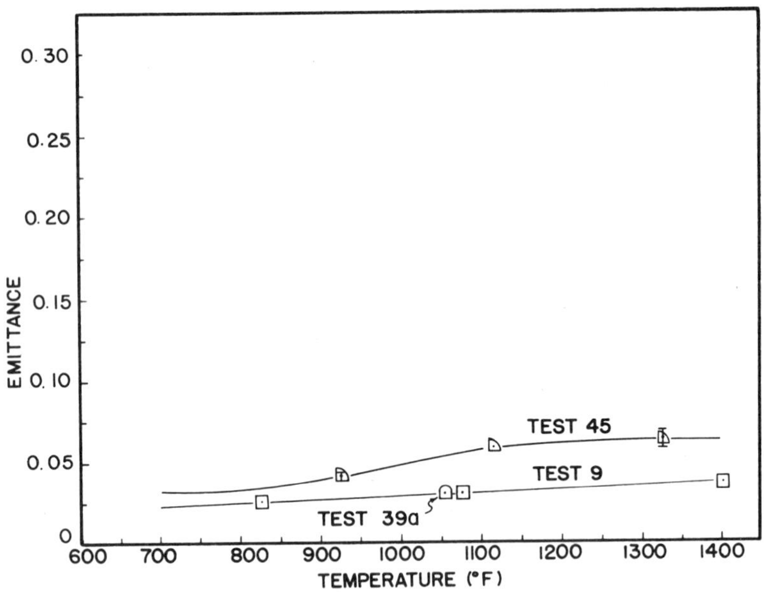

Fig. 4 Emittance for specimen prepared by surface grinding and oxidized near 300°F; Test 39a - unoxidized; Test 9 - unoxidized; Test 45 - 19.0 hr at 292.7°F.

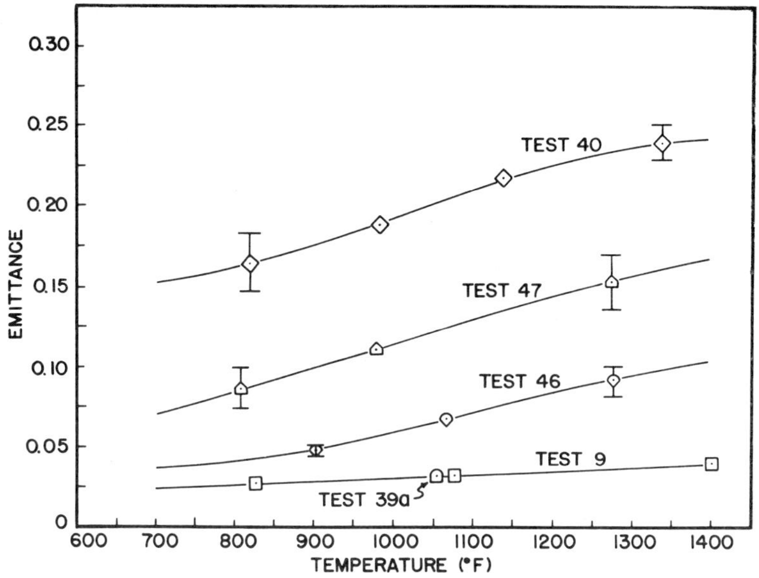

Fig. 5 Emittance for specimens prepared by surface grinding and oxidized near 400°F; Test 47 - 4.42 hr at 399.5°F; Test 46 - 45 min at 405.1°F; Test 40 - 17.5 hr at 396.8°F; Test 9 - unoxidized; Test 39a - unoxidized.

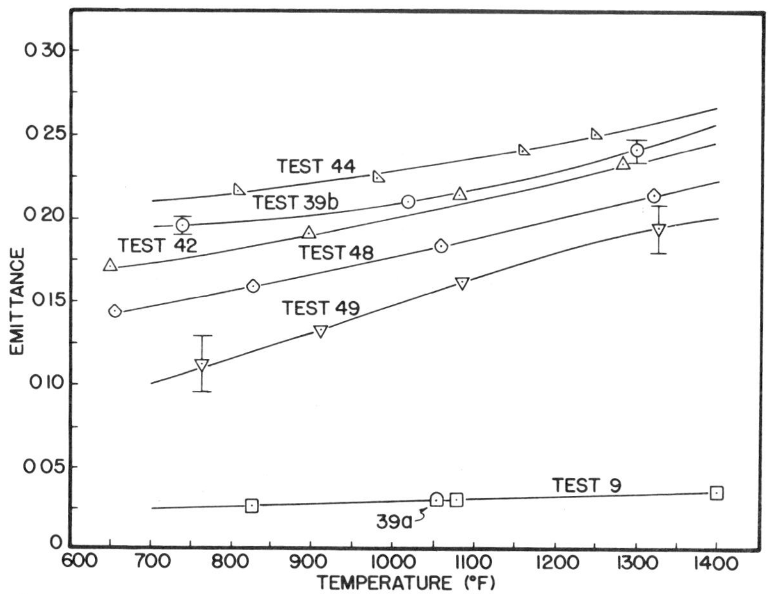

Fig. 6 Emittance for specimens prepared by surface grinding and oxidized near 500°F; Test 49 - 15 min at 511.1°F; Test 48 - 15 min at 531.7°F; Test 44 - 8.38 hr at 504.2°F; Test 42 - 43 min at 499.7°F; Test 39b - 3.03 hr at 502.5°F; Test 9 - unoxidized; Test 39a - unoxidized.

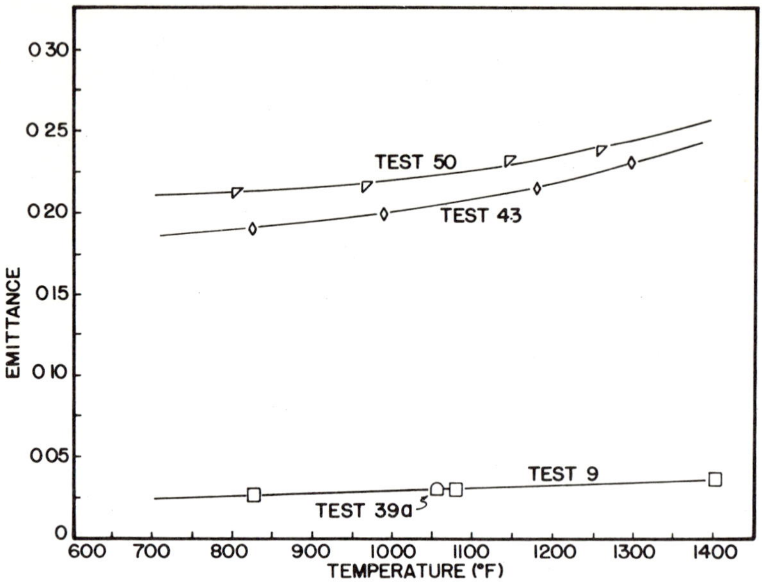

Fig. 7 Emittance for specimens prepared by surface grinding and oxidized near 600°F; Test 50 - 15 min at 605.0°F; Test 43 - 2.0 hr at 598.0°F; Test 9 - unoxidized; Test 39a - unoxidized.

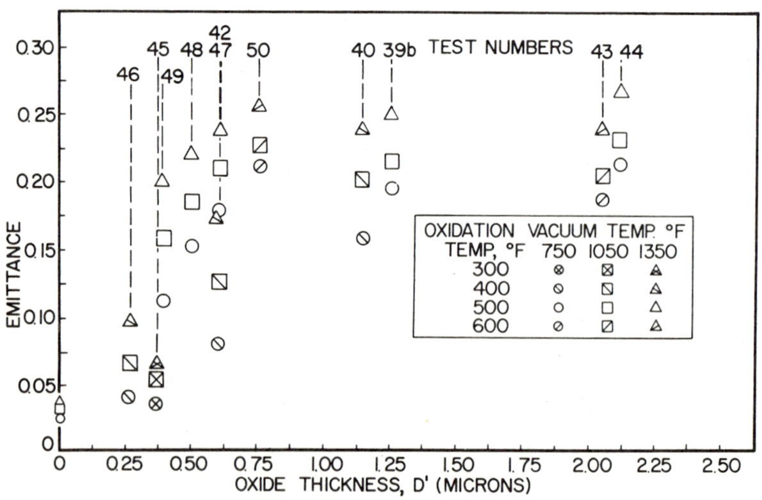

Fig. 8 Emittance as a function of oxide thickness calculated by Eq. (5).

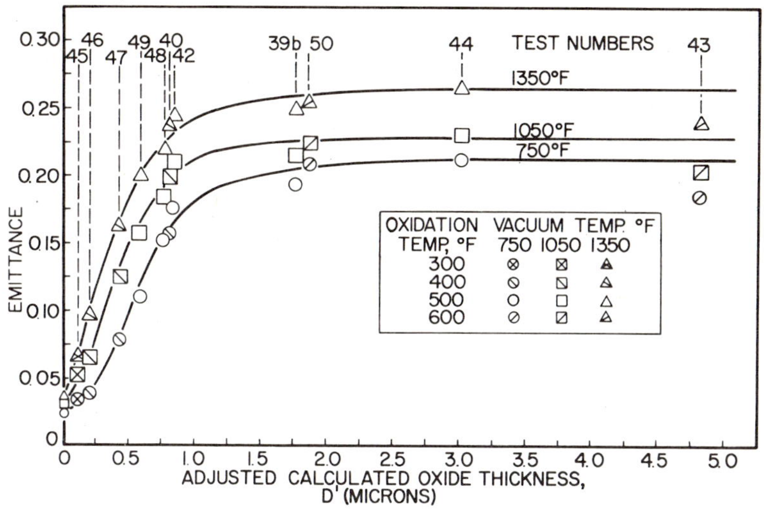

Fig. 9 Emittance as a function of adjusted oxide thickness calculated by Eq. (10).

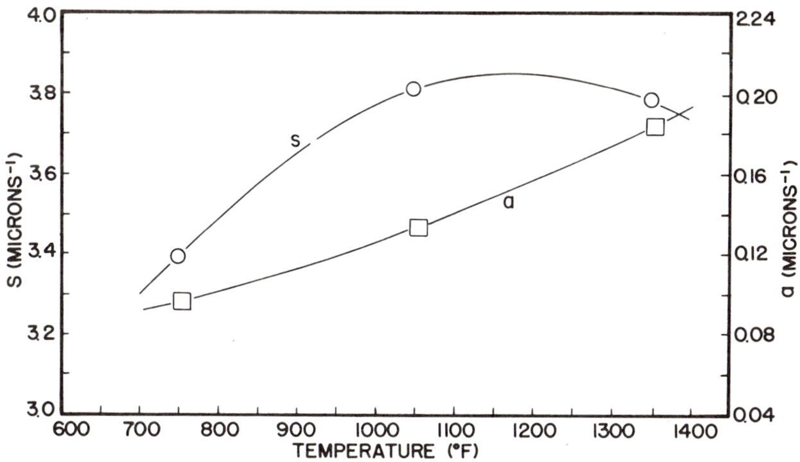

Fig. 10 Back scattering and absorption coefficients as a function of temperature as calculated from Eq. (11).

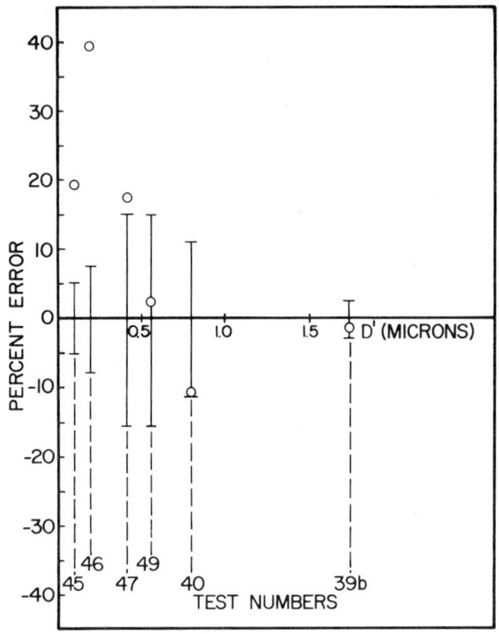

Fig. 11 Percent error of the two flux model correlation vs oxide thickness calculated from Eq. (10) and showing error bands.

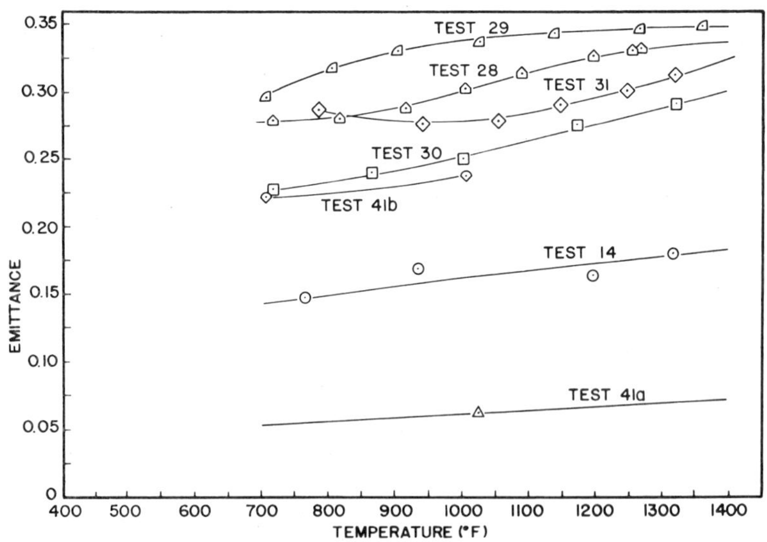

Fig. 12 Emittances for specimens prepared by lapping and oxidized near 500°F; Test 41b - 3.3 hr at 497.2°F; Test 14 - 15 min at 493°F; Test 28 - 15.6 hr at 490°F; Test 30 - 3.2 hr at 499°F; Test 29 - 13.25 hr at 500°F; Test 31 - 47.2 hr at 495°F; Test 41a - unoxidized.

INVESTIGATION OF THE EMITTANCE OF COATED REFRACTORY METALS

Karl O. Bartsch* and Walter P. Hudgins*
North American Rockwell Corporation, Los Angeles, Calif.

and

Norman M. Geyer†
Air Force Materials Laboratory,
Wright-Patterson Air Force Base, Ohio

Abstract

The effect of temperature and pressure, as functions of time, on the spectral normal and total normal emittance of several coated refractory metal alloys is investigated by the sliding specimen emittance measurement technique. The relationship between emittance and coating chemistry is explored. Two furnaces, forming part of the measuring equipment which operates at 0.01 to 760 torr, are described - one suitable for use at 2500° F and the other for use to 3000° F. Emittance data are given for Sylvania Si-20Cr-20Fe coated Cb-752 columbium alloy, TRW Cr-Ti-Si coated Cb-752, Chromalloy W-3 coated TZM molybdenum, and Sylvania Sn-Al coated Ta-10W tantalum alloy. The emittance of all coating systems is shown to vary significantly with temperature, pressure, and time.

Presented as Paper 70-68 at the AIAA 8th Aerospace Sciences Meeting, New York, N.Y., January 19-21, 1970. This work performed under Contract AF33(615)-3039 for Air Force Materials Laboratory, Wright-Patterson Air Force Base, Ohio. The authors wish to thank L. G. Kimball for his part in the design and construction of the emittance apparatus and the performance of most of the measurements, E. R. Geib for his assistance in the design of the 3000° F furnace, and H. W. Haroldson for the blackbody calculations for the 3000° F apparatus. The authors also especially thank T. Place and J. W. McCamont of Philco Ford, Aeronutronic Division, for their capable handling of the x-ray analysis.

*Members of the Technical Staff.

†Senior Project Engineer.

Introduction

Refractory metal alloys, such as columbium and tantalum, are candidates for use as hot structure for lifting re-entry and hypersonic vehicles, and as engine components. Although these alloys possess usable strength properties up to 3000° F and greater, they are rapidly attacked in an oxidizing atmosphere unless protected by a refractory coating. Coating life decreases markedly as the temperature rises above 2000° F, especially above 2500° F and at reduced pressure. Therefore, protective life can be extended by maintaining the metal-coating system at as low a temperature as possible. A high total emittance of the external surface of a hypersonic vehicle radiating to space is, therefore, desirable, and knowledge of this total emittance value is important in aerothermodynamic and heat transfer calculations. In addition, spectral emittance data may be useful in calculating radiant heat exchange in enclosures on such vehicles.

This paper illustrates that the "emittance" for a given coating system does not consist of a single total emittance number value and a single spectral emittance curve, but rather a family of numbers and curves. The emittance of a coated refractory metal is directly related to the chemistry of the coating which varies as a function of the environmental factors including temperature, time, and oxygen partial pressure.

A great deal of effort has been expended during the past 10 years upon the formulation and testing of coatings for refractory metals. The emphasis has been principally upon the attainment of coatings which will provide oxidation protection to the substrate alloys for as long a time as possible. Little consideration has been given directly to the formulation of refractory metal coatings possessing high emittance or a stable emittance. It is considered that knowledge of the relationship between spectral emittance and coating chemistry is prerequisite to the formulation of coatings providing both good oxidation stability and controlled emittance.

Numerous investigations have been conducted in the past on the emittance properties of uncoated and coated refractory alloys typically as cited.[1-4] Measurements have been made in varying environments including rather hard vacuums, inert gas, reducing atmospheres containing hydrogen, and in air at re-entry pressures. Measurements in re-entry environments are of the greatest interest to those engaged in re-entry vehicle design.

The subject of emittance as it relates to some coated refractory alloys of current interest has been investigated extensively under an Air Force contract.[5-7] The purpose of this paper is to present some of the highlights of this investigation which include 1) the adaptation of an emittance measurement method called the sliding specimen technique to refractory metal coatings in re-entry environments, 2) spectral and total normal emittance data for four different coated refractory alloys, and 3) indications of how coating chemistry and emittance are related in varying environments.

Work done by Schatz[8-9] was especially useful in correlating emittance to many of the compounds present in the refractory metal coatings investigated. In addition to the work reported herein, several other programs dealing with the emittance of refractory metal coatings of current interest are underway.[10-11]

Emittance Measurement Method

The sliding specimen technique for obtaining the total normal emittance of materials has been used by North American Rockwell for a number of years. This technique was first reported for obtaining total normal emittance at the 1962 Thermal Radiation Properties Symposium.[12] Subsequently, it was determined that this technique could be used for spectral emittance as well as for total emittance. The technique was adapted in this program for measuring both total and spectral normal emittance. The sliding specimen method is similar to that used by Hedge to measure total normal emittance.[13] In the Hedge method, the specimen is stationary and a water-cooled radiation detector unit is rapidly moved into the furnace very close to the surface of the specimen. The basic difference between methods is that, in the sliding specimen method, the specimen is moved to the radiation detector whereas, in the Hedge method, the radiation detection is moved to the specimen.

Sliding Specimen Technique

The principle of the sliding specimen technique is illustrated in Fig. 1. A cylindrical cavity, which has a large enough wall area to aperture area ratio to provide an effective blackbody, is heated to the desired test temperature. A water-cooled aperture is provided at the front of the cavity to prevent spurious radiation entering the detector optical path.

The energy emanating from the aperture at a near normal angle is collected by the Barnes radiometer for the total normal measurements.

The specimen is mounted on a suitable pushrod and positioned in the cavity through the rear of the furnace. The specimen is uniformly heated by radiation from the hot walls of the cavity to an equilibrium temperature which is the same as that of the cavity walls. When the specimen has reached the equilibrium temperature, a blackbody emittance value is obtained by reading the total energy radiated from the cavity and setting the detector output to a convenient location on the recorder. The actual distance between the detector zero line and the blackbody energy reading will then correspond to 100 percent emittance.

Figure 2 diagrammatically shows an energy trace. When the specimen is at the rear of the cavity (position A), the energy is at the blackbody energy level. As the specimen is moved rapidly forward (position B), the energy emanating from the aperture decreases until the instant the specimen reaches the front of the cavity (position C). Position C then is the point at which the energy emitted corresponds to that of the specimen at the test temperature. The specimen is held in the forward or test position for several seconds, during which time the energy level decreases at a slower rate as a result of specimen cooling. The specimen is then moved to the back of the cavity (C', B', A') to prepare for subsequent runs. The emittance is reported as the ratio of the specimen energy level at position C to the blackbody energy level at position A, expressed as a decimal value or as a percentage.

Spectral normal emittance values are obtained in much the same way as the total normal emttance values. The radiometer is eliminated, and the energy from the aperture is passed by a suitable mirror system to a monochrometer and thence to a detector of suitable sensitivity for the wavelength, illustrated in Fig. 3.

This technique, which is a single beam method, has several advantages over the more usual comparison methods of measuring spectral emittance. The main advantage lies in having the test specimen within the reference blackbody cavity. This arrangement provides an easy means of heating and controllably maintaining the specimen at the same temperature as the reference cavity. Since the cavity has a relatively large mass as compared to the specimen and specimen holder, the specimen reaches equilibrium at the cavity temperature in a short time (approximately 10 min). Such a time period allows measurement of the emittance as a function of time, and thus emittance changes that may occur due to oxidation or other chemical effects are readily observed. In the double beam comparison methods, such rapidity of measurement cannot be easily obtained, since considerable time often is required to balance the temperature of the specimen and reference cavity as well as to insure thermal stability. An additional advantage of the present method is that direct temperature measurement of the specimen is not required, and thus the normally attendant thermocouple attachment problems can be eliminated. This is especially advantageous for coated refractory specimens that contain silicon, which rapidly attacks platinum thermocouples at temperatures above 1500° F. Also, problems of inequality of beam path, introduced by mirrors and windows, are not encountered with the single beam system. The only basic problem encountered in the technique is assurance of isothermal blackbody conditions.

Emittance Furnaces

Two sliding specimen emittance furnaces were constructed, one with a temperature limit of 2500° F and the second capable of use to 3000° F. The 2500° F furnace is shown diagrammatically in Fig. 4. The cavity is located approximately 8 in. from the front of the furnace, and a deep cone water-cooled aperture is inserted with its apex at the front of the cavity. This arrangement provides a sufficient length of ceramic tube between the hot cavity and the water-cooled seal to preclude subjecting the tube to undue thermal stresses.

The design of the 3000° F test furnace is similar to that used for the 2500° F furnace (Fig. 5). The values of the cavity parameters for the 3000° F furnace were selected with the assistance of a radiosity network analysis. Solution of the network was effected by computer. The analysis indicated the final design to possess a 98% minimum blackbody efficiency. The blackbody cavity is fabricated from a beryllia tube (2 1/4 in. i.d. by 24 in. long) with a cast zirconia cavity liner. Figure 6 shows the internal dimensions and geometry of the ceramic liner. The beryllia tube is supported at both ends by platinum diaphragms which isolate the test chamber environment from the heating chamber environment. The diaphragms are attached to the beryllia tube by chemical deposition of paladium on the tube and electrodeposition of platinum to form a seal between the tube and the diaphragm. The platinum diaphragms are soldered to water-cooled copper rings which provide support for the cavity tube assembly and a seal to the center casing.

The heating element consists of tantalum wire wound directly on the beryllia tube. Guard heaters are formed by close winding at the ends of the heat zone and tapping the primary element so that the guard heaters may be driven independently. High-purity argon at a pressure close to that of the blackbody chamber is circulated through the heating element chamber.

A sodium chloride window is provided at the front of water-cooled aperture, and vacuum is provided by mechanical pumps. Air is leaked into the cavity to insure an oxidizing atmosphere.

Emittance Apparatus Performance

The performance of both the 2500° F and the 3000° F emittance apparatus was checked at 1880° F using NBS Inconel standards and an NBS standard platinum sample. Typical data are shown in Fig. 7. Agreement was within the NBS envelope of variation of all wavelengths at which measurements were made.

Room Temperature Emittance

The as-received room temperature reflectance of specimens was measured in the spectral region of 2.5 to 15μ by the Hohlraum technique. Values below 2.5μ were measured by means of a modified Beckman DK-2 spectro-reflectometer. Reflectance was converted to emittance by difference.

Coatings Tested

The emittance properties of the following coatings on the listed alloys were measured:

1. Cb-752 columbium alloy: Si-20Cr-20Fe coating (R512E), High Temperature Composites Laboratory, Sylvania Electric Products, Hicksville, New York

2. Cb-752 columbium alloy: Cr-Ti-Si coating, TRW Equipment Laboratories, TRW, Inc, Cleveland, Ohio

3. TZM molybdenum alloy: W-3 coating, Chromalloy Division of Chromalloy-American Corporation, West Nyack, New York

4. Ta-10W tantalum alloy: Sn-Al coating (R505F), High Temperature Composites Laboratory, Sylvania Electric Products, Hicksville, New York

Specimens consisted of sheet metal disks approximately 31/32 in. diam x 0.030 to 0.040 in. thick.

Study of Emittance Properties of Coatings

Spectral and total normal emittance measurements were made of the four coating systems described. Measurements were made at air pressures from 0.01 to 760 torr at temperatures from 1400° F to 2500° F. A few preliminary measurements were made of the Chromalloy W-3 coating system at 3000° F.

The spectral reflectance of the various coatings was measured at room temperature prior to high temperature exposure. Spectral reflectance was converted to room temperature spectral emittance (absorptance) which provided a basis of comparison for emittance measured at elevated temperature.

Analysis of coatings was mainly by x-ray diffraction made without disturbing the surface of the coating. Analysis was conducted typically after exposure at the same conditions reported for total normal emittance values.

Sylvania Si-20Cr-20Fe Coated Columbium Cb-752 Alloy

Figure 8 shows a typical set of spectral curves of Sylvania Si-20Cr-20Fe coated Cb-752 columbium at 2000° F and at 5 torr air pressure. Heating at 2000° F for 1/2 hr caused the emittance to increase greatly as shown. Continued heating to 3 hr at this same pressure and temperature caused a slight change from the 1/2-hr result. Figure 9 shows the effect of heating at various pressures at 2500° F for 1 hr. Varying results were obtained at different pressures. There is no general pattern, except that the emittance is quite high (0.85 to 0.90 at 1μ, and 0.70 to 0.93 at 10μ). It should be noted that the spectral curves are relatively flat after elevated temperature exposure.

Figure 10 shows the effect of temperature and pressure on the total normal emittance. The lower curve in the figure is a baseline plot of total emittance versus temperature for as-received samples prior to heating, obtained by integrating room temperature spectral emittance data (inferred from room temperature reflectance) over blackbody functions at various temperatures. This curve incorporates the effect of the Wien displacement function on emittance, and thus gives the total normal emittance which a coating would have at various temperatures if no chemical or physical changes occurred. Major departures from the baseline curve indicate effects on emittance of chemical and physical changes in the coating.

For the instance of Si-20Cr-20Fe coated Cb-752, the emittance values are far above the baseline curve; hence, it is evident that major chemical changes occurred at 1400° F and greater. This was also shown in Figs. 8 and 9. Pressure has a moderate effect at 1400° F, a major effect at 2000° F, and a moderate effect at 2500° F.

Figure 11 shows the changes in surface composition at various pressures after exposure for 3 hr at 2500° F. The total normal emittance is plotted at the top of the figure for reference. Two compounds were initially present on the surface, $(CrFe)Si_2$ and $NbSi_2$, 32 and 68%, respectively. After heating, the $(CrFe)Si_2$ disappeared. Niobium pentoxide (Nb_2O_5), silicon dioxide (SiO_2), and $NbSi_2$ were found in varying ratios depending upon pressure. It is also possible that chromium oxide was present, although not revealed by x-ray. Even though the composition is shown to vary widely, total normal emittance appeared to be insensitive to such changes once the initial composition disappeared.

Effecting correlation between chemical composition and emittance presents many problems. Figure 12 illustrates one of these problems. Curve B is emittance measured at elevated temperature; curve C the emittance (absorptance) measured at room temperature after heating; and A the room temperature emittance (absorptance) prior to heating, presented for reference. It can be seen that the emittance measured at elevated temperature is considerably higher than the room temperature values. One possible explanation is that this disparity may be caused by a layer that is translucent at elevated temperature but opaque at room temperature. The translucency would permit high emittance materials beneath the thin layer to increase the apparent spectral emittance at elevated temperature. Another explanation is that different compounds may be present at elevated temperature than are present at room temperature. It should be noted that the compositions determined by x-ray reported herein were run at room temperature.

TRW Cr-Ti-Si Coated Columbium Cb-752 Alloy

Figure 13 shows a typical set of spectral curves for TRW Cr-Ti-Si coated Cb-752 columbium at 2000° F and at 5 torr air pressure. Heating at 2000° F for 1/2 hr had only a slight effect on the emittance of this system. Heating for 3 hr, however, at this temperature and pressure caused the emittance to increase greatly as shown. Figure 14 shows the effect of heating at various pressures at 2500° F for 1 hr. The spectral curves obtained fall into two distinct groups: group B, at 0.01 and 0.1 torr; and group C, at 5, 100, and 760 torr. The spectral emittance is high (0.88 to 0.92 at 1μ, and 0.84 to 0.90 at 10μ). Curve A is the average spectral emittance derived from room temperature reflectance of as-received specimens. After sufficient time at elevated temperature, most curves are characteristically flat. Some, however, slope slightly.

Figure 15 shows the effect of temperature and pressure on total normal emittance. The lower curve shown is the baseline plot calculated from room temperature measurements as previously explained. The emittance values at 1400° F at 0.10 to 760 torr are equal to or less than the baseline values. The 0.01 torr emittance value at 2000° F is also very close to the baseline. These observations indicate that the surface has probably undergone very little reaction at these conditions. The remaining values indicate moderate to substantial reactions at temperatures of 2000° F and greater. At 2000° F, the effect of pressure is very pronounced, the greatest effect occurring at 760 torr. Pressure has little effect at 2500° F.

Figure 16 shows the changes in surface composition at various pressures after exposure for 3 hr at 2500° F. The total normal emittance is plotted as a function of pressure at the top of the curve for reference. Initially, $(Cr\text{-}Ti)Si_2$ constitutes 95 to 100% of the surface. The percentage of chromium oxide present in the coating is seen to increase as a function of increasing pressure at 2500° F. Simultaneously, there is a decreasing percentage of the complex intermetallic oxide $Cr_2Ti_2O_7$ which has been formed in a rather high concentration (60 to 65%) at the 0.01 torr condition. There appears to be a tendency for this complex oxide to be the preferred state at the lower pressure conditions, while the individual oxides predominate at the higher pressure exposure conditions. Even though the composition analysis shows a great variance in the surface constituents, the total normal emittance values are relatively insensitive to pressure and remain high.

Chromalloy W-3 Coated TZM Molybdenum Alloy

Figure 17 shows a typical set of spectral curves of Chromalloy W-3 coated TZM molybdenum at 2000° F and 5 torr air pressure. Heating at 2000° F for 1/2 hr caused the emittance to increase moderately as shown. Continued heating to 10 hr at this same temperature and pressure caused some additional increase. Figure 18 shows the effect of heating at various pressures at 2500° F for 1 hr. Exposure at 0.01 torr caused an emittance decrease over part of the spectrum, whereas exposure to higher pressures caused significant increases. Many of the spectral curves for this system characteristically slope downward, the highest emittance values being at the short wavelengths. Figure 19 shows a set of preliminary spectral curves at 2800° F and at 3000° F. At these extreme temperatures, relatively flat curves are obtained.

Figure 20 shows the effect of temperature and pressure on total normal emittance. The baseline plot is also shown. At 1400° F and 2000° F, the 5, 100, and 760 torr exposed specimens exhibit approximately the same values as the baseline, thus indicating that little chemical change occurred to affect emittance. At 2500° F, this system was more sensitive to pressure than the other coatings tested; an emittance variation of 0.5 to 0.8 was observed. The performance at 0.01 torr pressures is unusual; the 1 hr emittance at 2000° F and 2500° F are less than those of the baseline values. This is presumed to be caused by a loss of a superficial oxide layer when exposed to elevated temperature at this low pressure.

Figure 21 shows the changes in surface composition at various various pressures after exposure for 3 hr at 2500° F. The total normal emittance is plotted at the top of the chart for reference. The emittance appears to be markedly influenced by pressure at 0.01 torr. Other pressures appear to have little effect. Initially, before heating, the surface composition is essentially all $MoSi_2$. At pressures from 5 to 760 torr, mainly $MoSi_2$ and SiO_2 are present. At 0.01 and 0.1 torr, Mo_5Si_3, SiO_2, and elemental Mo mainly are present. The drop in emittance to 0.59 at 0.01 torr may be ascribed to the presence of 85% elemental molybdenum, a relatively low emittance material.

Many of the changes noted in the spectral curves for this system appear to be related to the formation and volatilization of silicon monoxide and silica. Silica may exhibit medium to high emittance from 3 to 10μ depending upon film thickness and structure. The spectral emittance increases between 3 and 10μ shown in Figs. 17 and 18 may be caused by silica formation.

Sylvania SN-Al Coated Ta-10W Tantalum Alloy

Figure 22 shows a typical set of spectral curves for Sylvania Sn-Al coated Ta-10W tantalum alloy at 5 torr air pressure. Heating at 2000° F for 1/2 hr caused the emittance to increase greatly as shown. Continued heating to 10 1/2 hr at this same temperature and pressure caused only slight additional changes. Figure 23 shows the effect of heating at various pressures at

2500° F for 1 hr. The spectral curves obtained fall into two distinct groups: group B, at 0.01 torr; and group C, at 5, 100, and 760 torr. Curve A is the envelope of values derived from room temperature reflectance of preexposed specimens. The as-received emittance values for this system display a large spread.

Figure 24 shows the effect of temperature and pressure on the total normal emittance. The lower curve shown is the baseline plot calculated from room temperature measurements as previously explained. At 1400° F and pressures of 0.01 and 760 torr, the emittance departs moderately from the baseline indicating that some chemical or physical changes occurred. It is seen that the emittance is at a maximum at 2000° F and drops to lower values at 2500° F. Pressure has a strong effect at 2000° F, but no effect at 2500° F.

Figure 25 shows the composition after exposure for 10 hr at 2000° F at varying pressures. The total normal emittance is plotted at the top for reference. Initially, the surface consists of 60 to 70% $MoAl_3$ and 20 to 30% Sn. At 2000° F and the pressure conditions shown, some tin oxide, aluminum oxide, and elemental tin may be present. The latter was found in quantity between 0.1 and 5 torr. Molybdenum, which was present initially, was not detected by x-ray in the heat-affected coatings. Elemental tin has a low emittance and probably accounts in large measure for the decreasing emittance values from 0.1 to 5 torr.

Conclusions

1) The sliding specimen emittance measurement method has been adapted to the pressure-temperature environment of refractory metal coating utilization.

2) The emittance properties of the four refractory metal coatings studied are strongly dependent upon temperature, pressure, and time of exposure.

3) The following total normal emittance values obtained after 1 hr of exposure illustrate the effects of temperature and pressure:

	1400° F	2000° F	2500° F
Sylvania Si-20Cr-20Fe (R512E) on Cb-752			
0.01 torr	...	0.76	0.76
100 torr	0.77	0.85	0.84
TRW Cr-Ti-Si on Cb-752			
0.01 torr	...	0.54	0.86
100 torr	0.46	0.63	0.86
Chromalloy W-3 on TZM			
0.01 torr	...	0.37	0.51
100 torr	0.47	0.56	0.65
Sylvania Sn-Al on Ta-10W			
0.01 torr	0.51	...	0.77
100 torr	...	0.91	0.77

4) Correlating emittance with coating composition is difficult. The chemical systems involved are complex. Although the standard x-ray diffraction analysis of a specimen surface run at room temperature yields useful information, it is possible that different compounds exist at elevated temperature than at room temperature. Additionally, it appears that thin layers of compounds on the surface which affect emittance may not appear in sufficient amount in the x-ray analysis. This is because the x-rays penetrate below the emittance active depth of the coating, and surface compounds are diluted in the reported analysis by subsurface materials.

References

[1]Fieldhouse, I. B., Lang, J. I., and Blau, H. H., "Investigation of Feasibility of Utilizing Available Heat Resistant Materials for Hypersonic Leading Edge Applications, Vol. IV. Thermal Properties of Molybdenum Alloy and Graphite," WADC Report 59-744, Oct. 1960.

[2]Coffman, J. A., Kibler, G. M., and Riethof, T. R., "Carbonization of Plastics and Refractory Materials Research," Third Quarterly Progress Report, Flight Propulsion Laboratory, General Electric AD 245 223, 30 Sept. 1960.

[3] Wurst, J. et al., "The Evaluation of High Temperature Materials, Section II, Evaluation of Coatings for Refractory Alloys," ML-TDR-54-62, Vol. II, University of Dayton Research Institute, Sept. 1964.

[4]Ault, M. et al., "Coated Refractory Metal Technology - 1965; Oxidation Resistant Coatings for Refractory Metals," Materials Advisory Board, Publication MAB-210-M, Nov. 1965.

[5]Bartsch, K. O. et al., "The Role of Emittance in Refractory Metal Coating Performance, Part I - Review and Analysis," AFML-TR-66-55, Pt I, Aug 1966, Air Force Materials Lab.

[6]Bartsch, K. O., Kimball, L. G., and Hudgins, W. P., "The Role of Emittance in Refractory Metal Coating Performance, Part II - Total and Spectral Emittance Measurements to 2500F," AFML-TR-66-55, Pt II, Aug. 1968, Air Force Materials Lab.

[7]Bartsch, K. O., Kimball, L. G., and Hudgins, W. P., "The Role of Emittance in Refractory Metal Coating Performance, Part III - Total and Spectral Emittance Measurements Above 2500F and Correlation of Emittance with Composition," AFML-TR-66-55, Pt III, Nov. 1969, Air Force Materials Lab.

[8]Schatz, E. A. and McCandless, L. C., "Research for Low and High Emittance Coatings," ASD-TR-62-483, May 1962, Air Force Materials Lab.

[9]Schatz, E. A. et al., "High Temperature, High Emittance Intermetallic Coatings," ASD-TDR 63-657, Aug. 1963, Air Force Materials Lab.

[10]Bartsch, K. O., Hudgins W. P., and Geib, E. R., "The Emittance of Refractory Metal Coatings," Air Force Contract F33615-69-C-1618, Air Force Materials Lab., to be published.

[11]Fitzgerald, B. G. and Rusert, E. L., "Evaluation of the Fused Slurry Silicide Coating Considering Component Design and Reuse," AFML-TR-70-154, Air Force Materials Lab., to be published.

[12]Konopken, S. and Klemm, R., "Evaluation of Thermal Radiation at High Temperatures," Measurement of Thermal Radiation Properties of Solids, NASA SP-31, 1963.

[13]Hedge, Jack C., "Total Normal Emittance Measurements to 2200° C in Air," AIAA Progress in Astronautics and Aeronautics, Thermophysics and Temperature Control of Spacecraft and Entry Vehicles, edited by G. B. Heller, Vol. 18, Academic Press, New York, 1966, pp. 33-46.

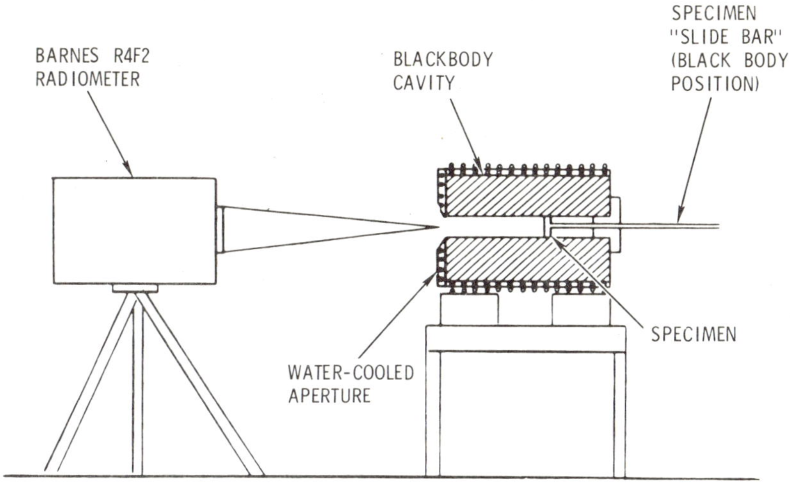

Fig. 1 Principle of sliding specimen emittance apparatus.

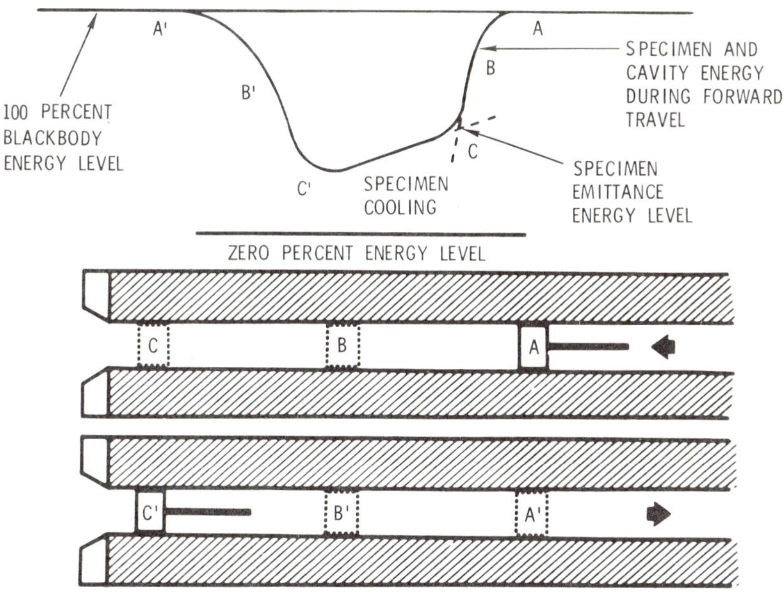

Fig. 2 Typical energy trace for sliding specimen emittance apparatus.

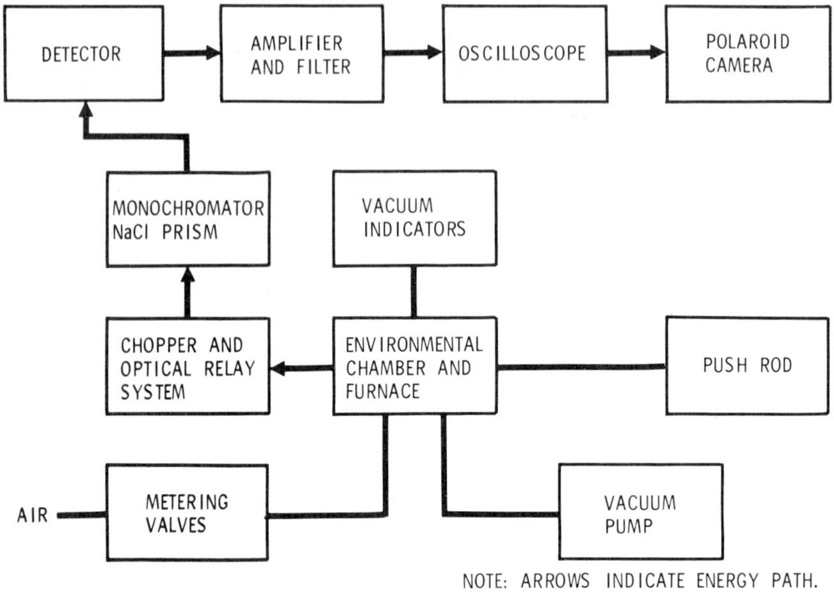

Fig. 3 Block diagram of sliding specimen spectral normal emittance system.

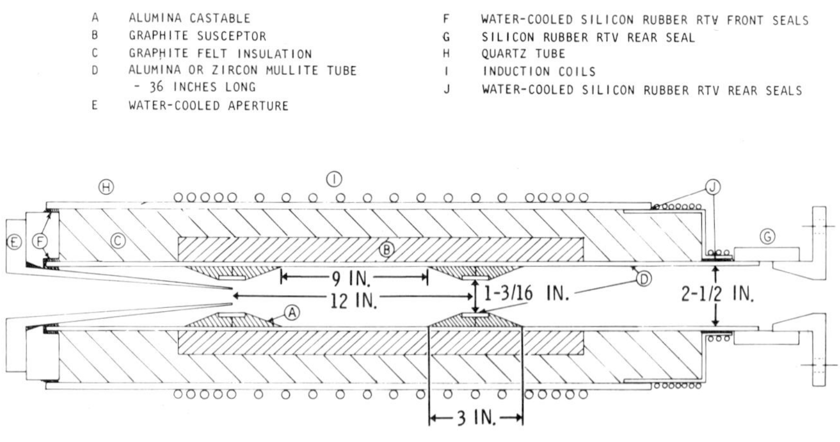

Fig. 4 Cast insert 2500° F emittance furnace.

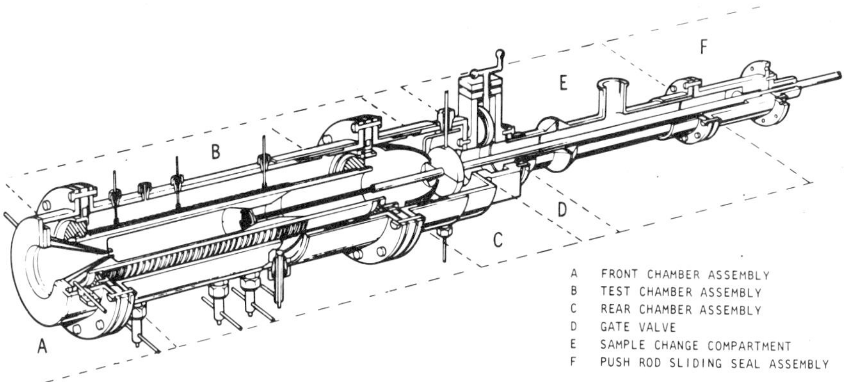

Fig. 5 3000° F emittance test furnace.

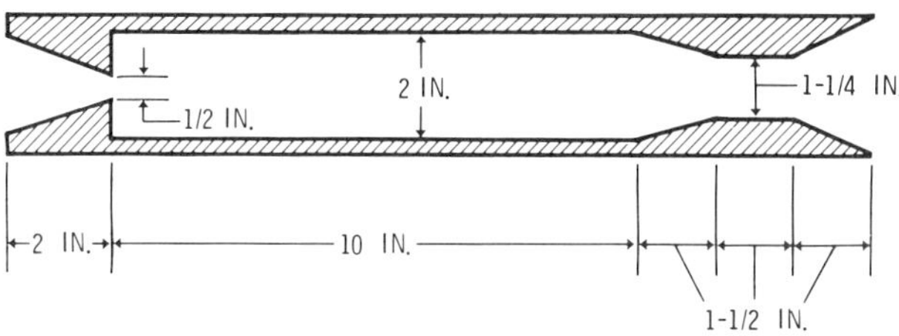

Fig. 6 Geometry of test cavity liner for 3000° F furnace.

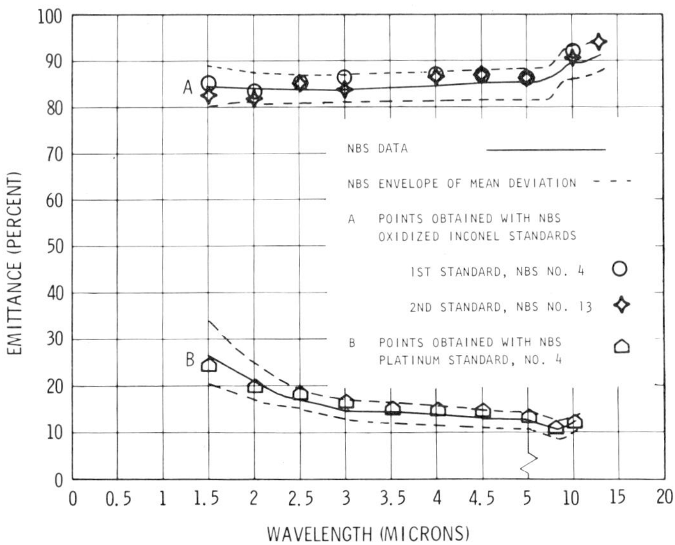

Fig. 7 Test of 3000° F furnace at 1800° F.

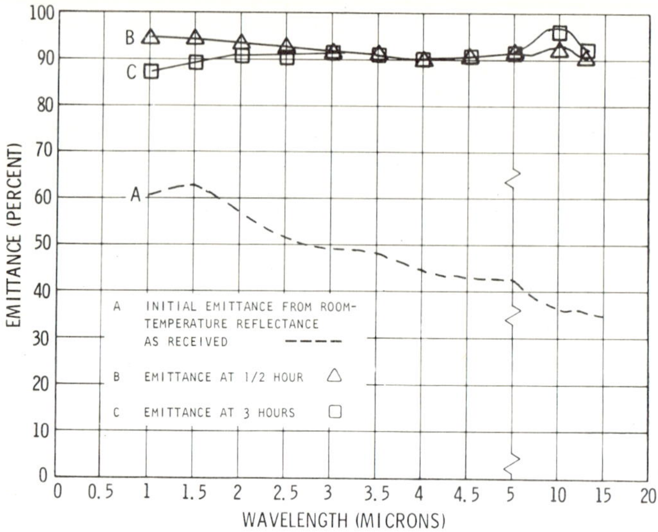

Fig. 8 Spectral emittance of Sylvania Si-20Cr-20Fe coated Cb-752 at 2000° F and 5 torr.

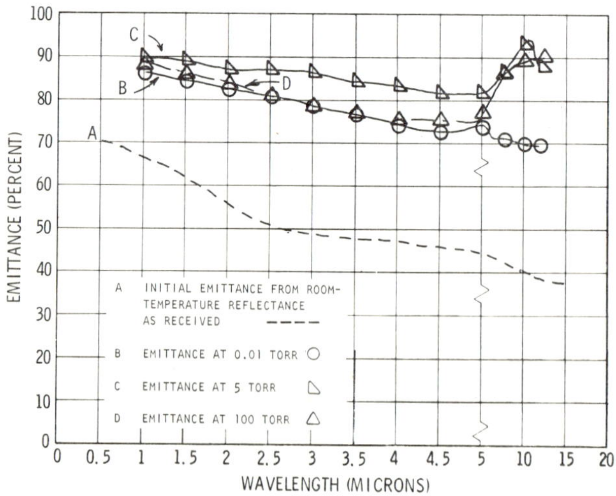

Fig. 9 Spectral emittance of Sylvania Si-20Cr-20Fe coated Cb-752 after 1 hr at 2500° F and various pressures.

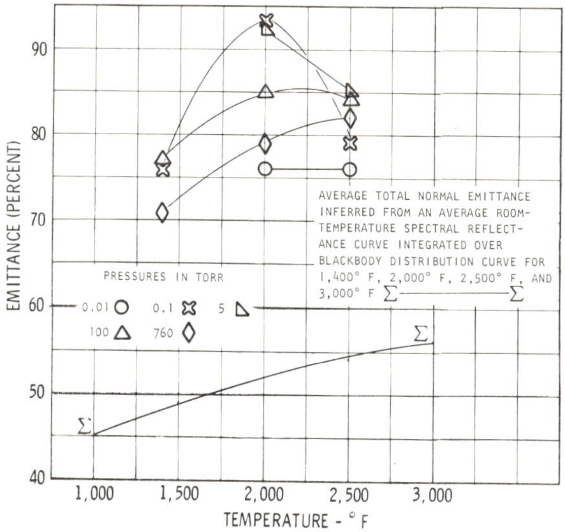

Fig. 10 Effect of temperature and pressure at 1 hr on total normal emittance of Si-20Cr-20Fe coated Cb-752.

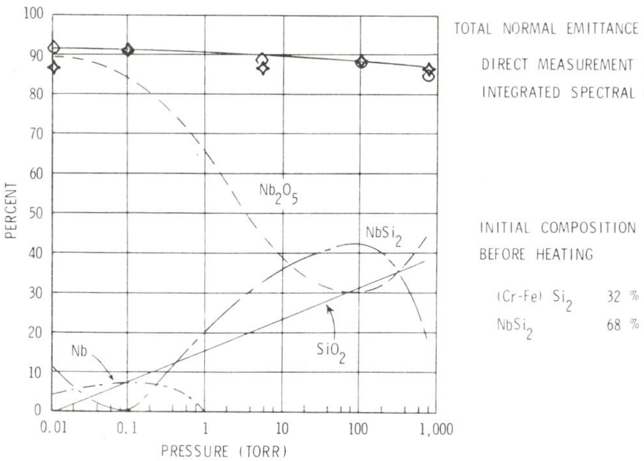

Fig. 11 Percent composition versus pressure for Sylvania Si-20Cr-20Fe coated Cb-752 exposed for 3 hr at 2500° F.

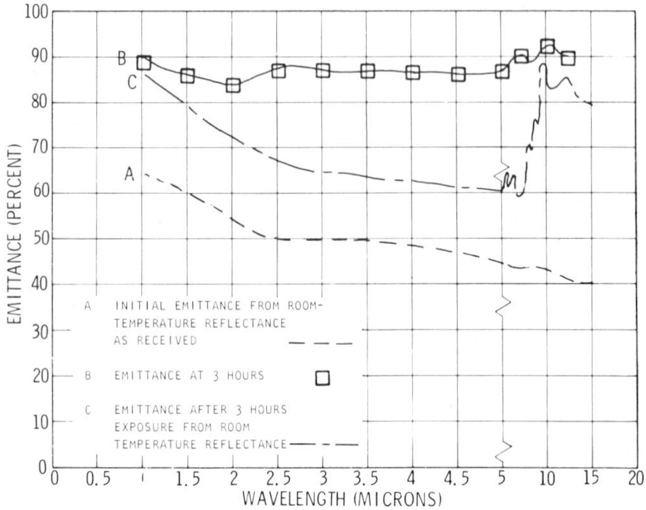

Fig. 12 Spectral emittance of Sylvania Si-20Cr-20Fe coated Cb-752 before exposure - at 3 hr, 2500° F, and 5 torr - and after exposure.

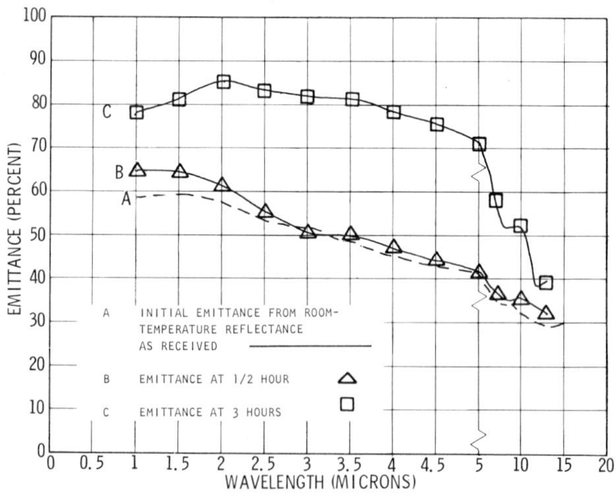

Fig. 13 Spectral emittance of TRW Cr-Ti-Si coated Cb-752 at 2000° F and 5 torr.

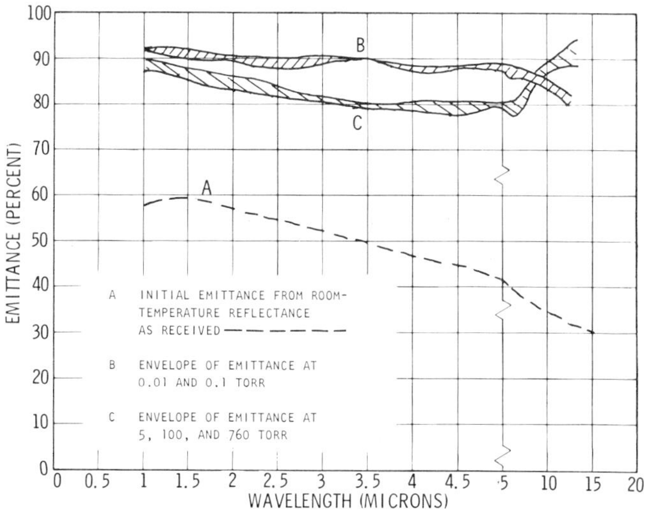

Fig. 14 Envelope of spectral emittance of TRW Cr-Ti-Si coated Cb-752 after 1 hr at 2500° F and various pressures.

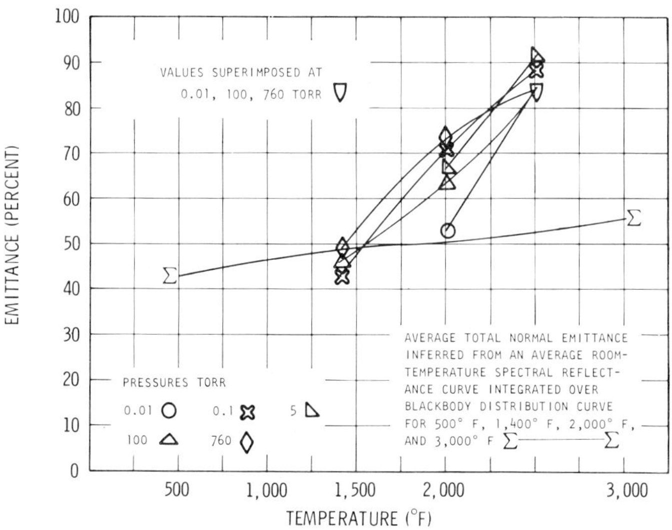

Fig. 15 Effect of temperature and pressure at 1 hr on total normal emittance of TRW Cr-Ti-Si coated Cb-752.

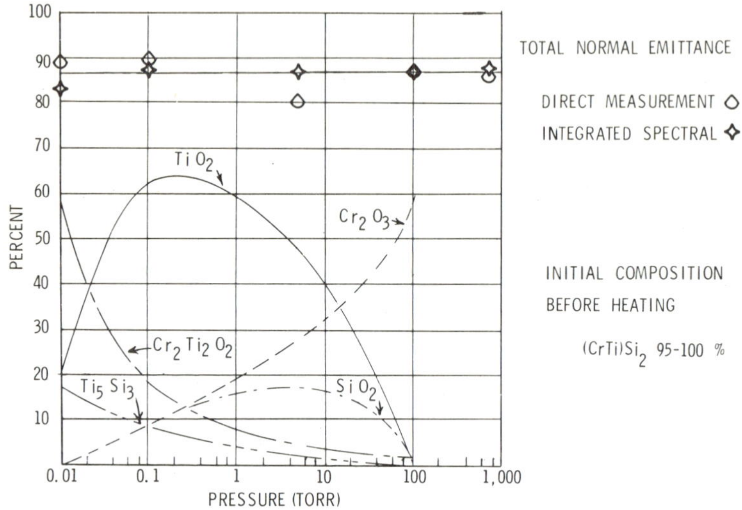

Fig. 16 Percent composition versus pressure for TRW Cr-Ti-Si coated Cb-752 exposed for 3 hr at 2500° F.

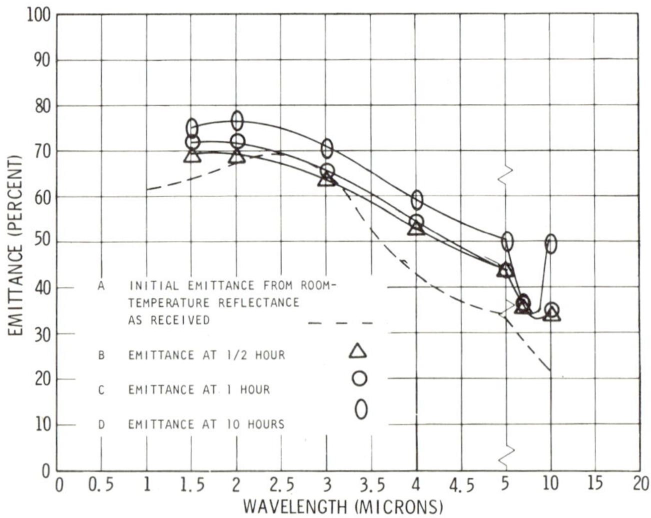

Fig. 17 Spectral emittance of Chromalloy W-3 coated TZM at 2000° F and 5 torr.

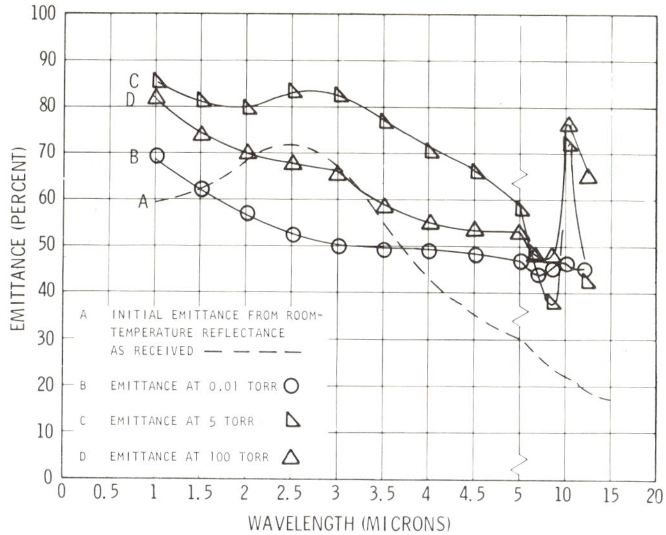

Fig. 18 Spectral emittance of Chromalloy W-3 coated TZM Molybdenum after 1 hr at 2500° F and various pressures.

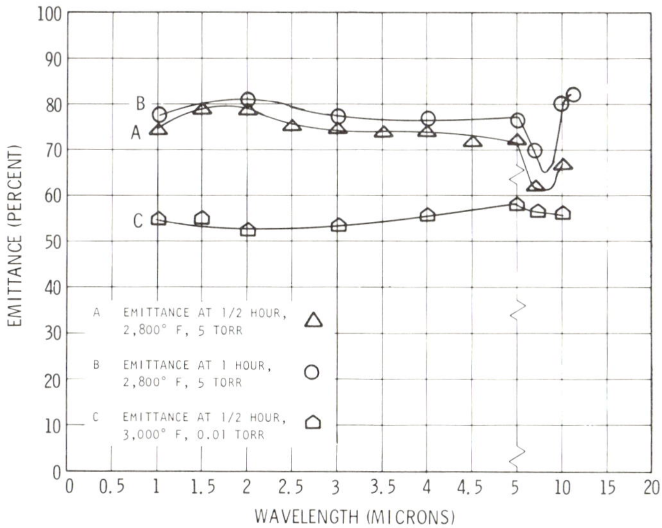

Fig. 19 Spectral emittance of Chromalloy W-3 coated TZM at 2800° F and 3000° F and at 5 and 0.01 torr.

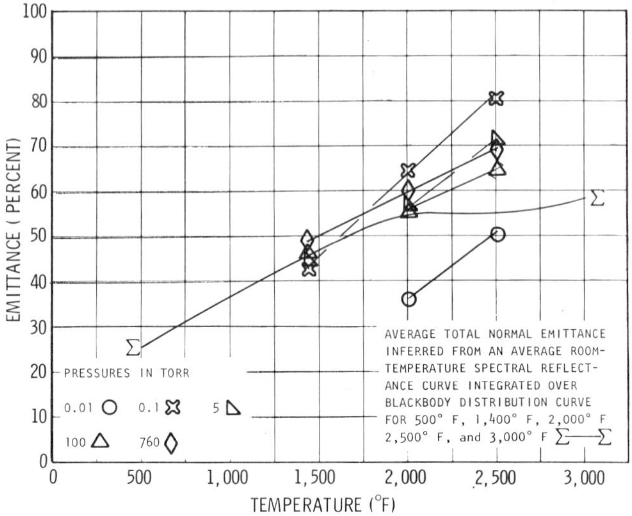

Fig. 20 Effect of temperature and pressure at 1 hr on total normal emittance of Chromalloy W-3 coated TZM.

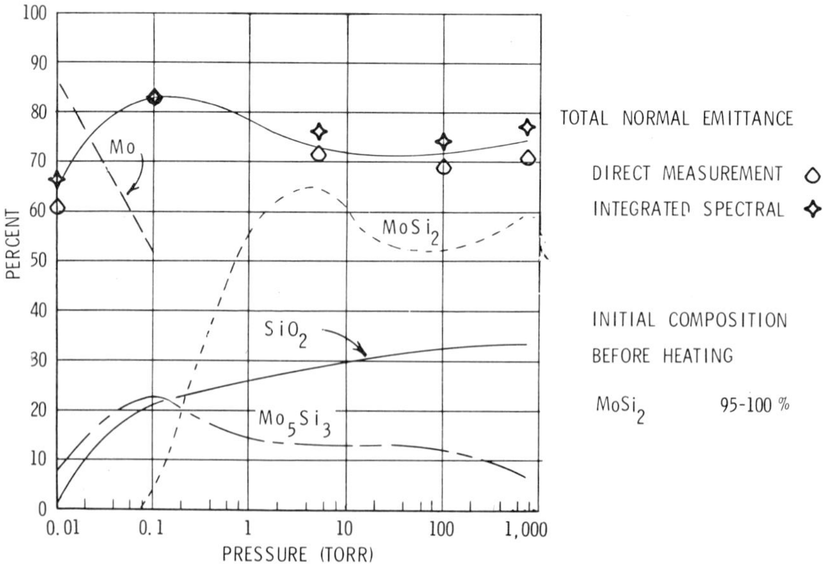

Fig. 21 Percent composition versus pressure Chromalloy W-3 coated TZM exposed for 3 hr at 2500° F.

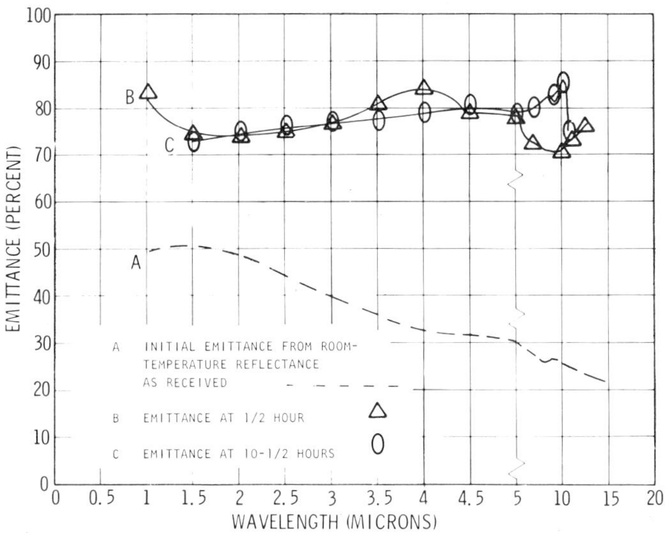

Fig. 22 Spectral emittance of Sylvania Sn-Al coated Ta-10W at 2000° F and 5 torr.

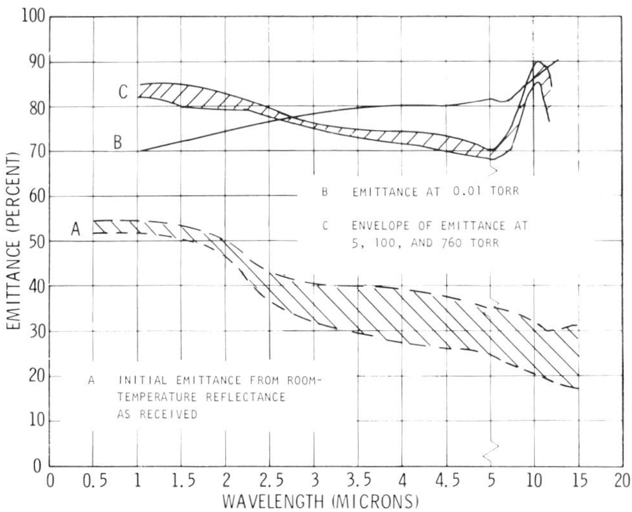

Fig. 23 Envelope of spectral emittance of Sylvania Sn-Al coated Ta-10W tantalum after 1 hr at 2500° F and various pressures.

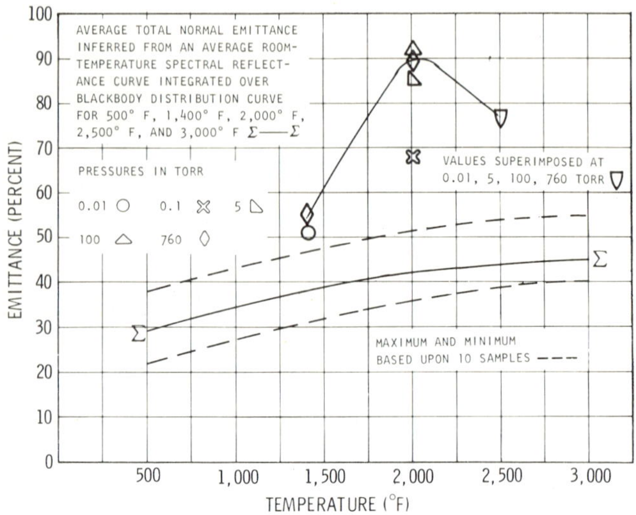

Fig. 24 Effect of temperature and pressure at 1 hr on total normal emittance of Sn-Al coated Ta-10W.

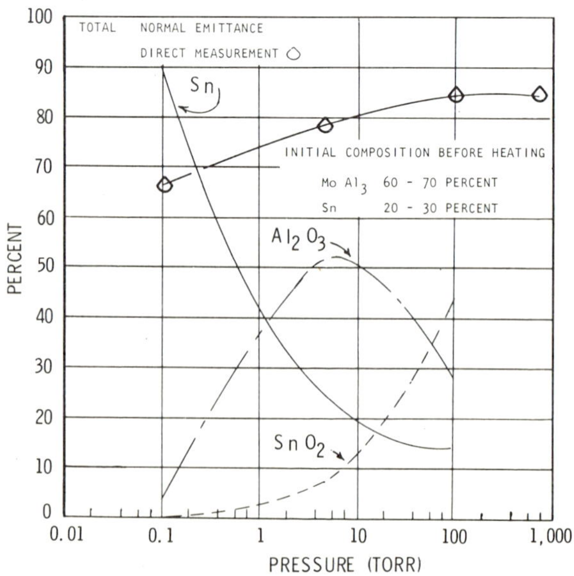

Fig. 25 Percent composition versus pressure for Sylvania Sn-Al coated Ta-10W exposed for 10 hr at 2000° F.

MEASUREMENT OF BIDIRECTIONAL REFLECTANCE USING A PHOTOGRAPHIC TECHNIQUE

J. E. Loehrlein,* E. R. F. Winter,† and R. Viskanta‡

Purdue University, Lafayette, Ind.

Abstract

A photographic technique for measuring the angular distribution of monochromatic radiation reflected from surfaces is described. Relative measurements are reported on both a qualitative and quantitative basis for two specimens of well-characterized V-groove, very smooth aluminum, polycrystalline magnesium oxide and projection screen surfaces in the visible part of the spectrum for a range of polar angles of incidence. Results are compared with the measurements reported in the literature for some of

---

Presented as Paper 70-859 at the AIAA 5th Thermophysics Conference, Los Angeles, Calif., June 29-July 1, 1970. Work supported through NASA Interdisciplinary Institutional Grant NGL 15-005-021. Thanks are due the Thermophysics Properties Research Center of Purdue University and Honeywell, Inc. for supplying some of the surfaces employed in this study; and special thanks are also extended to the Aluminum Company of America in Lafayette, Indiana for the use of their microdensitometer.

*Graduate Research Assistant, Heat Transfer Laboratory, School of Mechanical Engineering; presently Development and Test Engineer, General Electric Company, Schenectady, N. Y.

†Professor of Mechanical Engineering, Heat Transfer Laboratory, School of Mechanical Engineering.

‡Professor of Mechanical Engineering, Heat Transfer Laboratory, School of Mechanical Engineering.

the surfaces using other techniques. They also contribute original information about phenomena observed in the angular distribution of radiation reflected from surfaces. The apparatus measures the angular distribution of reflected radiation very expeditiously and accurately enough for engineering purposes both qualitatively and quantitatively throughout the range of the spectrum for which sensitive films are available.

Nomenclature

- E = radiant energy flux
- e = densitometer reading
- I = intensity of radiation
- R = bidirectional reflectance ratio defined by Equation (3)
- θ = polar angle
- θ_0 = polar angle at which specular peak first appears
- λ = wavelength
- ρ = bidirectional reflectance defined by Equation (2)
- σ = surface roughness
- ϕ = azimuthal angle
- Ω = solid angle

Subscripts

- b = refers to background
- i = refers to incident
- r = refers to reflected
- λ = wavelength

Superscripts

primed = incident direction

I. Introduction

In contemporary engineering applications one usually needs an accurate prediction of radiant heat transfer or equilibrium temperatures. The knowledge of the radiation characteristics of surfaces is essential. The problem of radiant heat exchange between real surfaces can be formulated exactly;[1] however, this type of analysis needs the detailed knowledge of the spectral directional emittance and absorptance as well as the bidirectional reflectance characteristics of surfaces.[§] Of these, the collection of bidirectional reflectance data is rather complicated as well as very time consuming and painstaking because of the large number of variables and parameters involved.

The early measurements of bidirectional reflectance of some metallic and nonmetallic surfaces were made by Eckert.[2] The more recent investigations of bidirectional reflectance reported in the literature are those of Refs. 3-9. All of the results reported were either for only one or at most five azimuthal angles in the reflected direction; therefore, the spectral bidirectional reflectance of a given surface has not yet been completely measured.

A satisfactory theory for predicting the spectral bidirectional reflectance for optically rough surfaces is not available,[10,11] and the needs of present-day technology are only partly met by experimental measurements. Since such an approach to obtain data involves a great deal of resources, a simpler and less time-consuming method than those presently available is needed. In this paper a photographic technique for measuring the angular distribution of reflected monochromatic radiation is described and some results reported.

A review of photographic methods for measuring angle (configuration, view) factors is given by Jakob,[12] and a photographic technique for measuring the mechanical roughness of surfaces has been discussed in some detail by Schmaltz.[13] Recently a photographic method of measuring a bidirectional reflectance of surfaces has been suggested[14,15] and some data reported. Unfortunately, the disadvantage of the method suggested by Maples and Wiebelt[15] is that the radiation reflected from a specimen exposes a film wrapped inside the cylinder. Hence, only the bidirectional reflectance in the plane of incidence can be measured directly instead of in all possible planes.

II. Experimental Apparatus

An apparatus was designed and constructed to measure monochromatically the reflected radiation over

[§]No generally accepted nomenclature exists, and it is also referred to as a biangular, partial and apparent unidirectional reflectance.

the entire hemispherical space above the test specimen in air at room temperature. The basic instrument which facilitated this measurement is a photographic reflectometer. The 10-cm radius hemisphere was constructed from aluminum and then anodized. Seven small circular apertures of 3-mm diam, spaced along a meridian allowed for a change of the polar angle, θ' of the incident beam relative to the test specimen surface normal (see Fig. 1). The surface was located in a lightproof well with its surface situated in the equatorial plane of the hemisphere. The specimen holder was made so that the azimuthal angle ϕ of the test specimen relative to the plane of incidence could be set at convenient angles such as 0°, 45°, and 90°. The interior surface of the hemisphere had a holding mechanism for the attachment of photographic film enveloping the inner contour of the hemisphere.

The hemisphere was mounted on the table of a milling machine indexing head having four degrees of freedom of motion necessary for the optical alignment of the apparatus. These included two horizontal motions, a vertical motion, and one angular motion parallel to the plane of incidence determined by the surface normal and the incident beam. The axis of rotation of the indexing head coincided with the equatorial plane of the hemisphere which in turn coincided with the test specimen surface. Furthermore, the axis of rotation of the indexing head was normal to the optical axis of the monochromator and both intersected at the origin of the hemisphere. By rotating the indexing head, the angle of incidence (polar angle, θ') could be changed from 0° to 85° in intervals of 15° thus exposing at a given time through an opening in the hemisphere the specimen to a monochromatic pencil of light.

The monochromatic light beam was obtained with a Heath Monochromator. A condensor lens and a diaphragm served for the delimitation and the focusing of the incident monochromatic light beam onto the center of the specimen. The aperture size, 3 mm in diameter, was a compromise between the diffraction effects and the energy requirements.

The film was cut by a blade similar to that used in a rolled steel cutter. The cutting die consisted of 3/8-in. wide blades formed to fit the interior contour of the hemisphere and was set into plywood. Resilient material was placed between the blades to keep the film at the edge of the cutting surface. The bearing plate made of aluminum was provided with two one-quarter inch dowel pins to assure positive positioning of the film relative to the cutting surface. The film placed into the die was cut with a hydraulic press, operated manually and capable of developing pressures up to 10,000 psi. (For the films used in this study, a pressure of 4000 psi was usually required to assure a clean cut.) The main advantage of this method was that it produced exact and consistent film profiles throughout the period of experimentation.

The basic components of the apparatus used in the investigations are schematically displayed in Fig. 2a and a photograph is shown in Fig. 2b.

The system was aligned so that the collimating mirror in the monochromator was fully illuminated at maximum light intensity. The monochromator grating G was ruled to 1180 lines per millimeter with a blaze wavelength of 2500 Å. The monochromator was provided with a remote electronic digital control unit to allow for variance of the wavelength from 4000 to 10,000 Å in the zeroth order. The bandwidth of the monochromatic pencil was maintained at $\Delta\lambda$ = 50 Å, while the slit height for both entrance and exit slit was fixed at 5 mm.

III. Test Specimens

The four materials studied in this investigation were purposely selected as being representative of a wide range of engineering surfaces commonly in use. Some of the materials had very well and others poorly defined mechanical surface characteristics. The test specimen included two periodic V-groove aluminum surfaces, a very smooth aluminum surface, a polycrystalline magnesium oxide surface and a woven movie projection screen. A brief description of each of the materials is as follows.

Aluminum V-Groove Surface

The two well-characterized aluminum test specimen consisted of a series of periodic V-shaped grooves.

A summary of the parameters for the samples used is given in Table 1. The specimens were made of blocks of 2024 T-4 aluminum alloy 1 × 1 × 15/16 in. in size with parallel periodic V-grooves cut into one of the 1- × 15/16-in. faces, with the grooves parallel to the 1-in. long side. The samples were ruled on a diffraction grating machine and were first used by Zipin.[16] They were especially well suited for this study because the angular reflectance distributions had previously been reported for these surfaces thus making direct comparison of data possible.

Table 1 Parameters of Samples

Sample	Number of grooves/mm	Included angle	Peak to valley height, μ
1	20	89.9°	25.4
2	100	90°	5.0

Aluminum Foil

The aluminum surface used was of the type produced by Reynolds Aluminum Company in their commercial aluminum foil. While no mechanical surface properties were available, the surface was observed to be very smooth. The root mean square mechanical roughness was estimated to be 0.02μ.

Magnesium Oxide

The fused polycrystalline ceramic was supplied by Honeywell, Inc. and was used "as received." Its chemical composition was 99.9% magnesium oxide. The remaining 0.1% was composed of oxides of various other elements. The root mean square mechanical roughness was estimated by visual inspection with calibrated specimen of identical material to be in the range of 0.5 to 1.5μ.

Woven Movie Projection Screen

The projection screen sample consisted of parallel, almost square, grooves. The peak to peak distance was 0.03 in. and the valley to peak height was 0.02 in.

IV. Measurement Procedure and Calibration

Full illumination of the collimating mirror assured maximum intensity and resolution of the light reflected from the grating. Background light was reduced by delimitation of the light beam entering the monochromator with diaphragms D_1 and D_2. By limiting the exit light beam with diaphragm D_3 and collimating it with lens L_3, the test specimen located at the origin of the hemisphere was illuminated with a uniform pencil of monochromatic light (Figs. 2a and 2b).

Prior to exposure the film was cut and affixed to the inner surface of the hemisphere. Care was taken not to soil or moisten the film during this operation. The hemisphere was then placed on the indexing head, whose center of rotation was so adjusted that it intersected the optical axis of the apparatus. The exposures were made by activating a shutter located on the exit side of the monochromator. This shutter could be operated manually and held open for any specified length of time. Once the film was exposed, it was processed according to manufacturer's specifications.

Because of the wide variance of characteristics of the surfaces studied and, therefore, reflected intensities, two different films were used. For the highly reflective surfaces (aluminum with 20 grooves/mm, aluminum with 100 grooves/mm and aluminum foil) DuPont Cronar Ortho A Litho film was used. This is a high-contrast, low-speed film with very high-dimensional stability. Because of its slow speed, the time of exposure could be determined with good accuracy using a stop watch. Typical exposures were of the order of 10 min. For the other surfaces, such as the magnesium oxide block and the sample of the movie projection screen, a much faster film was required to keep the exposure times reasonably short. For reasons of its speed, dimensional stability, and ease of handling, Kodak Tri-X-Ortho film was selected for the investigation of these surfaces.

In order to interpret the photographic results in terms of reflected intensities, it was necessary to calibrate the film. The characteristic curves were obtained by a substitution technique[17] in which the incident light (energy) was kept constant while the exposure time was varied when exposing different areas of a film sample. The density distribution on the film was measured with a microdensitometer manufactured by the Harry W. Dietert Company. The details of the calibration procedure and its results are described elsewhere.[18]

V. Results

The measurements performed in this investigation lead to two general types of results. Visual examination of the exposed film yields the range of solid angles over which the incident light was reflected, both as a function of incident direction and wavelength, while photodensitometry of the film provides the relative distribution of reflected energy over the hemispherical space.

Qualitative Results

Shown in Figs. 3 and 4 are film exposures typifying the reflectance of monochromatic light (λ = 0.5461μ) incident on the specimen. The photographs of the films have been overexposed for the purpose of illustration.

Figure 3 demonstrates the specular reflection of monochromatic light from some of the surfaces studied. Because of the wavelength at which these exposures were made, and because of the relative characteristic roughness of the surfaces studied, each of these photographs can be analyzed in terms of geometrical optics theory. The reflection from a smooth aluminum surface at an angle of incidence of 30° is shown in Fig. 3a. The reflection is seen to be specular. Reflection from a V-grooved surface with grooves perpendicular to the plane of incidence is illustrated in Fig. 3b. In this case the included angle between the grooves was 90° with 100 grooves/mm. All of the radiation was reflected back within the plane of incidence with peaks at 30° and 54°. This reflectance pattern can be explained with the help of geometrical optics theory. Reflection of radiation incident normally onto a V-grooved surface with the included angle of 89.9°, and 20 grooves/mm is given in Fig. 3c. In agreement with geometrical optics theory, there are several modes of reflection.

The reflection of radiation from a woven movie projection screen for an incident angle of $\theta' = 45°$ is illustrated in Figs. 4a, 4b and 4c with the "grooves" parallel, at 45° and perpendicular to the plane of incidence, respectively. The radiation is reflected fairly uniformly over a large lobe in the forward ($\phi = 0°$) direction with the center of this lobe at $\theta = 30°$ and $\phi = 0°$, Fig. 4a. When the lines of the woven pattern were rotated by 45° relative to the plane of incidence, Fig. 4b, the reflected lobe is unsymmetrical. With the principal threads of the weaving pattern perpendicular to the plane of incidence, the energy is reflected in a well-defined symmetric lobe (Fig. 4c). Superimposed on this relatively diffuse reflection pattern is a specular peak which is reflected back into the direction of the incident beam.

A typical reflectance pattern from a magnesium oxide surface is displayed in Fig. 4d. It is seen from the figure that the energy is reflected quite uniformly over the hemisphere, however, a superimposed specular peak was detected.[18] The existence of a specular peak is in agreement with the work of Toporets[19] for a rough dialectric surface. With increasing polar angle of incidence, θ', the specular peak becomes noticeable when the following condition is satisfied:

$$(\sigma/\lambda)\cos\theta_0' = 1 \qquad (1)$$

Since the intensity of reflected energy in the specular direction is small, compared to that of the diffuse lobe, and since the films used in this study were too fast to record this range of densities linearly, only a qualitative indication can be derived from these results. The result is in general agreement with previous findings by Torrance and Sparrow[6] and Love and Francis.[8] These investigators found that for large angles of incidence the diffuse component of reflection lost prominence and the probability of off-specular peaks appearing in the reflected radiation field increased.

Quantitative Results

For photographic measurements where the relatively fine control of such variables as temperature, age of materials, and processing technique is essential, the comparison of two or more such exposures could lead to gross errors. To eliminate uncertainties related to the photographic processes, it is advantageous to compare exposures relative to some standard which can be determined by the examination of an individual film. Therefore, the quantitative results of this investigation are presented as relative bidirectional reflectance ratio

$$R = \frac{\rho_\lambda(\theta', \phi'; \theta, \phi)}{\rho_\lambda(\theta', \pi; \theta', 0)} \tag{2}$$

where the bidirectional reflectance $\rho_\lambda(\theta', \phi'; \theta, \phi)$ is defined as[5]

$$\rho_\lambda(\theta',\phi';\theta,\phi) = \frac{dI_{r,\lambda}(\theta,\phi)}{dE_{i,\lambda}(\theta',\phi')} = \frac{dI_{r,\lambda}(\theta,\phi)}{I_{i,\lambda}(\theta',\phi')\cos\theta' d\Omega'} \tag{3}$$

and $\rho_\lambda(\theta', \pi; \theta', 0)$ is the reference bidirectional reflectance taken to be in the specular direction $(\theta = \theta', \phi = \phi' - \pi)$. The bidirectional reflectance ratio R can also be interpreted as

$$R = \frac{\Delta E_{r,\lambda}(\theta, \phi)}{\Delta E_{r,\lambda}\ (\theta = \theta', \phi = \phi' - \pi)} \tag{4}$$

where $E_{r,\lambda}(\theta, \phi)$ is the radiant flux reflected into any direction (θ, ϕ) and $E_{r,\lambda}(\theta = \theta', \phi = \phi' - \pi)$ is the radiant flux reflected in the specular direction $(\theta = \theta', \phi = \phi' - \pi)$.

The bidirectional reflectance measurements were obtained with a microdensitometer which provided a direct readout in percent of the amount of light (from an internal source) that passed through the film sample. The transmitted light was detected by a photocell. A suitable background radiation level for each exposure had to be determined. To this end, a sample of film was placed into the equatorial plane of the hemisphere and positioned so that it did not directly intercept light from the monochromatic source. The resulting background level was assumed to be isotropic, and the monochromatic bidirectional reflectance of a specimen relative to the bidirectional reflectance in the specular direction was defined as

$$R = \frac{\rho_\lambda(\theta', \phi'; \theta, \phi)}{\rho_\lambda(\theta', \pi; \theta=\theta', \phi=0)} = \frac{e_{r,\lambda}(\theta, \phi) - e_{b,\lambda}}{e_{r,\lambda}(\theta=\theta', \phi=0) - e_{b,\lambda}} \quad (5)$$

where $e_{r,\lambda}$ and $e_{b,\lambda}$ are the densitometer readings which represent the reflected and background energies, respectively.

The bidirectional reflectance for the aluminum foil showed a mirror-like reflection for angles of incidence of $\theta' = 0$ to $\theta' = 60°$. The V-grooved surface with 20 grooves/mm behaved as an ideal specular reflector when the grooves were oriented parallel to the plane of incidence. For all angles of incidence the angle of reflection was equal to the angle of incidence and all the energy was reflected within the plane of incidence. When the grooves were oriented perpendicular to the plane of incidence the light was also specularly reflected from the walls of the grooves rather than from the surface itself. This behavior is typical for the situation where $(\sigma/\lambda) \ll 1$. Under similar circumstances the reflectance for an aluminum V-groove surface having 100 grooves/mm, with the grooves oriented parallel to the plane of incidence, the reflectance was again specular. On the other hand, with the V-grooves oriented perpendicular to the plane of incidence the specimen reflected radiation within a narrow lobe in a backward direction which depended on the polar angle of the incident beam.[18]

The bidirectional reflectance of a magnesium oxide surface is shown in Fig. 5. The directional distribution of reflected energy is, however, reflected more uniformly with increasing angle of incidence. Identical results were obtained when the surface was rotated by 90°. At $\theta' = 75°$, an off-specular peak appears near $\theta \simeq 85°$ in the plane of incidence.[18] The occurrence of specular and off-specular peaks in the directional distribution of reflected radiation can be attributed to the specular reflection of incident radiation from the microscopic facets with different inclinations, each of which reflects the radiation in accordance to the laws of geometrical optics.[20] The specular peak is considered to be produced by interference of the radiation specularly reflected from the microscopic facets which are parallel to the macroscopic surface while the off-specular peak is found to be caused by Fresnel reflection from microscopic facets that have relatively small inclinations with respect to the surface.

The angular distributions of radiation reflected from a movie projection screen are shown in Fig. 6. For the result of Fig. 6 the grooves were oriented parallel to the plane of incidence and the energy was reflected into several lobes, the most intense of which was centered about the specular direction. Less intense lobes were reflected at $\phi = \pm 50°$ and $\phi = \pm 60°$. Results where the grooves were rotated by an angle of 45° relative to the plane of incidence with nonsymmetrical lobes are shown elsewhere.[21] When the grooves were oriented perpendicularly to the plane of incidence angular distribution of the reflected energy was symmetrical about the plane of incidence for all values of θ' investigated. A marked reduction of the energy reflected out of the plane of incidence is evidenced as the angle of incidence increases from 30° to 60°. These distributions become more uniform with increasing angles of incidence.

VI. Conclusions

Measurements of the monochromatic bidirectional reflectance using a photographic reflectometer were made. A variety of surfaces of practical engineering interest were examined over a range of variables and the feasibility of this technique was demonstrated. For the periodic V-grooved surfaces for which previous data on the bidirectional reflectance were available, the results of the photographic technique were in close agreement. The reflection was predicted well from the geometric optics theory.

With the photographic reflectometer developed the bidirectional reflectance of surfaces was expediently and accurately measured. The utility of the method is that with a single film the complete directional distribution of the reflected radiation for a particular angle of incidence can be obtained. From the visual observation of the exposed film alone one can readily determine the directions and over which range of solid angles the incident energy is reflected. Quantitative distribution of the reflected energy can be obtained from the film with a densitometer. The technique appears particularly well suited in the ultraviolet, visible and near infrared part of the spectrum for use by spacecraft thermal control designers. Practically, the method is presently limited only by the lack of films and emulsions sensitive beyond near infrared radiation.

References

[1]Bevans, J. T. and Edwards, D. K., "Radiation Exchange in an Enclosure with Directional Wall Properties," Transactions of the American Society of Mechanical Engineers, Series C; Journal of Heat Transfer, Vol. 87, No. 3, Aug. 1965, pp. 388-396.

[2]Eckert, E., "Messung der Reflexion von Wärmestrahlen an Technischen Oberflächen," Forschung auf dem Gebiete des Ingenieurwesens, Bd. 7, Heft 6, Nov./Dec. 1936, pp. 265-270.

[3]Münch, B., "Directional Distribution in the Reflection of Heat Radiation and its Effect on Heat Transfer," TT F-497, 1968, NASA.

[4]Birkebak, R. L. and Eckert, E. R. G., "Effects of Roughness of Metal Surfaces on Angular Distribution of Monochromatic Reflected Radiation," Transactions of the Society of Mechanical Engineers, Series C; Journal of Heat Transfer, Vol. 87, No. 1, Feb. 1965, pp. 85-94.

[5]Torrance, K. E. and Sparrow, E. M., "Biangular Reflectance of an Electric Nonconductor as a Function of Wavelength and Surface Roughness," Transactions of the American Society of Mechanical Engineers, Series C; Journal of Heat Transfer, Vol. 87, No. 2, May 1965, pp. 283-292.

[6]Torrance, K. E. and Sparrow, E. M., "Off-Specular Peaks in the Directional Distribution of Reflected Thermal Radiation," Transactions of the American Society of Mechanical Engineers, Series C; Journal of Heat Transfer, Vol. 88, No. 2, May 1966, pp. 223-230.

[7]Herold, L. M. and Edwards, D. K., "Bidirectional Reflectance Characteristics of Rough, Scattered-Metal, and Wire-Screen Surface Systems," AIAA Journal, Vol. 4, No. 10, Oct. 1966, pp. 1802-1810.

[8]Love, T. J. and Francis, R. E., "Experimental Determination of Reflectance Function of Type 302 Stainless Steel," AIAA Progress in Astronautics and Aeronautics: Thermophysics of Spacecraft and Planetary Bodies, Vol. 20, edited by G. B. Heller, Academic Press, New York, 1967, pp. 115-135.

[9]Birkebak, R. C., Dawson, J. P., McCullough, B. A. and Wood, B. E., "Hemispherical Reflectance of Metal Surfaces as a Function of Wavelength and Surface Roughness," International Journal of Heat and Mass Transfer, Vol. 10, No. 9, Sept. 1967, pp. 1225-1232.

[10]Beckmann, P. and Spizzichino, A., The Scattering of Electromagnetic Waves from Rough Surfaces, MacMillan, New York, 1963.

[11]Houchens, A. and Hering, R. G., "Bidirectional Reflectance of Rough Metal Surfaces," AIAA Progress in Astronautics and Aeronautics: Thermophysics of Spacecraft and Planetary Bodies, Vol. 20, edited by G. B. Heller, Academic Press, New York, 1967, pp. 65-89.

[12]Jakob, M., Heat Transfer, Vol. II, Wiley, New York, 1957, pp. 74-84.

[13]Schmaltz, G., Technische Oberflächenkunde, Verlag von Julius Springer, Berlin, 1936, pp. 91-101.

[14]Winter, E. R. F., Viskanta, R. and Wen, A., "A Photographic Bidirectional Reflectometer," Wärme- und Stoffübertragung, Bd. 1, No. 2, 1968, pp. 95-97.

[15]Maples, D. and Wiebelt, J. A., "Experimental and Theoretical Investigation of Electro-Magnetic Scattering from a Known, Rough Surface," AIAA Paper 68-29, New York, 1968.

[16]Zipin, R. B., "The Directional Spectral Reflectance of Well-Characterized Symmetric V-Grooved Surfaces," Ph.D. Thesis, 1965, Purdue University.

[17]James, T. H. and Higgins, G. C., Fundamentals of Photographic Theory, Wiley, New York, 1948.

[18]Loehrlein, J. E., "Measurement of Bidirectional Reflectance Using a Photographic Technique," M.S. Thesis, Jan. 1970, Purdue University.

[19]Toporets, A. S., "Specular Reflection from a Rough Surface," Optics and Spectroscopy, Vol. XVI, No. 1, Jan. 1964, pp. 54-58.

[20]Voishvillo, N. A., "Reflection of Light by a Rough Glass Surface at Large Angles of Incidence of the Illuminating Beam," Optics and Spectroscopy, Vol. XXII, No. 6, June 1967, pp. 417-520.

[21]Loehrlein, J. E., Winter, E. R. F. and Viskanta, R., "Measurement of Bidirectional Reflectance Using a Photographic Technique," AIAA Paper 70-859, New York, 1970.

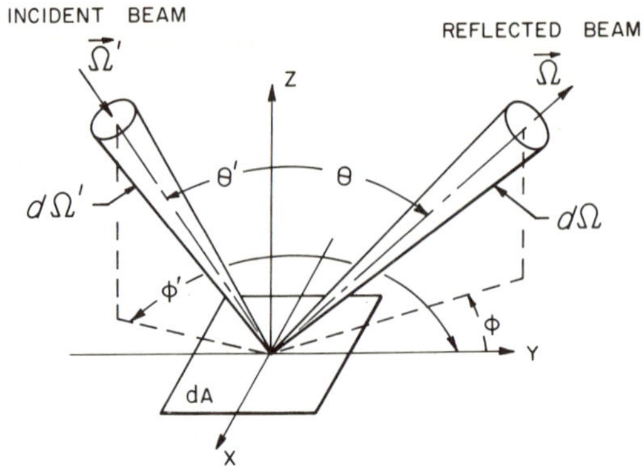

Fig. 1 Geometry of radiation incident and reflected from a differential area.

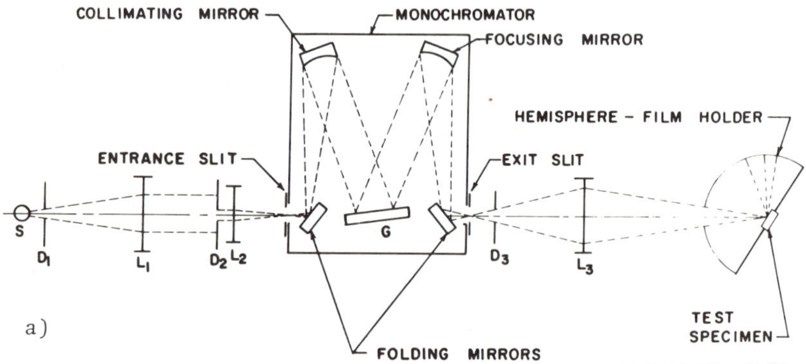

a)

Fig. 2 Schematic diagram of optical apparatus and photograph of facility.

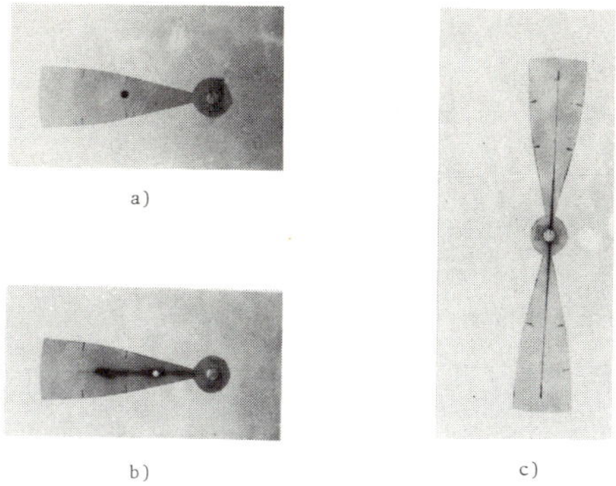

Fig. 3 Reflection from specular surfaces. a) Reflection from an aluminum foil, θ' = 30°. b) Reflection from a V-grooved surface with 100 grooves/mm oriented perpendicularly to the plane of incidence, θ' = 30°. c) Reflection from a V-grooved surface with 20 grooves/mm, θ' = 0°.

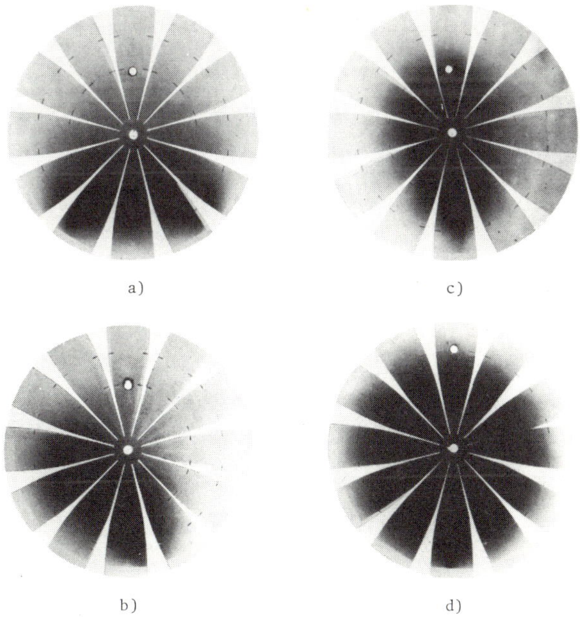

Fig. 4 Reflection from diffuse surfaces. a) Weaving pattern parallel to plane of incidence. b) Weaving pattern at 45° relative to plane of incidence. c) Weaving pattern perpendicular to plane of incidence. d) Reflection from a magnesium oxide surface, θ' = 75°.

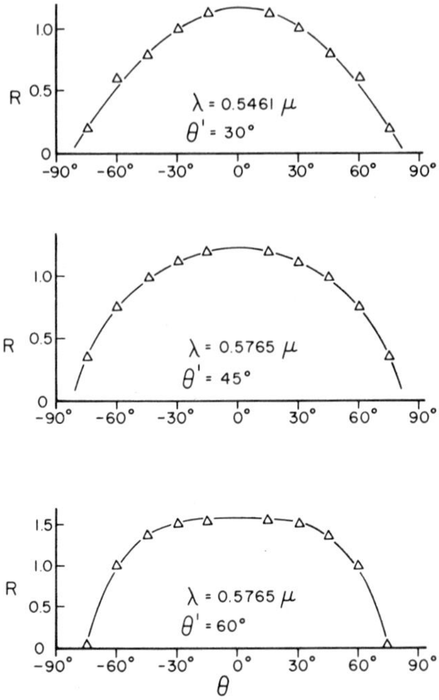

Fig. 5 Bidirectional reflectance from a magnesium oxide surface.

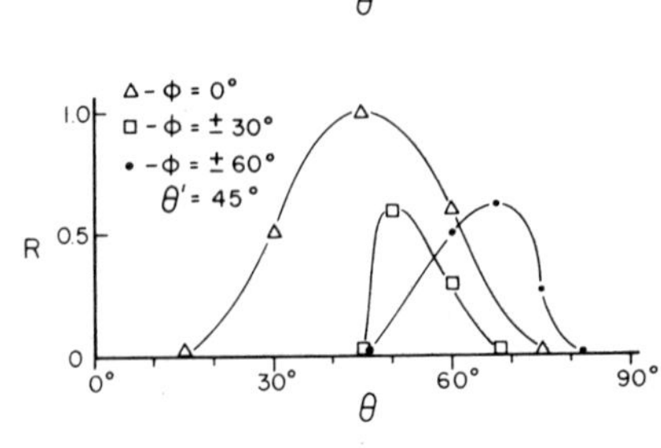

Fig. 6 Bidirectional reflectance from a woven movie projector screen with grooves parallel to the plane of incidence, λ = 0.4355μ.

SUPER- AND SUBSPECULAR MAXIMA IN THE ANGULAR DISTRIBUTION OF POLARIZED RADIATION REFLECTED FROM ROUGHENED DIELECTRIC SURFACES

A. M. Smith,* P. R. Müller,+ and W. Frost‡
ARO, Inc., Arnold Air Force Station, Tennessee, and The University of Tennessee Space Institute, Tullahoma, Tennessee

and

Han M. Hsia§
Cheng Kung University, Tainan, Taiwan, Republic of China

Abstract

Detailed angular distribution measurements were made of the polarized radiant flux reflected from roughened glass samples, and as a result a new type of off-specular peak was discovered. This maximum is termed "subspecular" and occurs for p-polarized radiation provided the wavelength and zenith incidence angle are appreciably less than the dielectric's surface roughness and Brewster angle, respectively. If the incidence angle is greater than the Brewster angle, the maximum in the angular distribution is superspecular and occurs even for a surface roughness smaller than the radiation wavelength. No subspecular maxima are observed for s-polarized radiation, but superspecular maxima occur for all non-normal incidence angles provided the dielectric's surface roughness is significantly greater than the radiation wavelength. For reflected radiant flux containing all components of polarization, only superspecular peaks are observed. These peaks occur if the dielectric surface roughness is larger than the radiation

---

Presented as Paper 70-861 at the AIAA 5th Thermophysics Conference, Los Angeles, Calif., June 29-July 1, 1970. Research sponsored by Arnold Engineering Development Center, Air Force Systems Command, under Contract F40600-69-C-0001 with ARO, Inc.
*Supervisor, Thermal Physics Section and Associate Professor, Aerospace Engineering.
+Research Assistant; presently Assistant Professor, Mechanical Engineering Department, North Dakota State Univ., Fargo, North Dakota.
‡Consultant and Associate Professor of Mechanical Engineering.
§Chairman, Department of Engineering Science.

wavelength and the incidence angle is equal or greater than approximately 30°. A formula is derived for the radiation reflected in the plane of incidence and is used to quantitatively confirm the existence of the super- and subspecular maxima for moderate incidence angles.

Introduction

Off-specular maxima in the directional distributions of radiant flux or intensity reflected from rough dielectric surfaces have been reported by numerous investigators since the early part of this century. Moreover, in recent years, this off-specular peak phenomenon has been experimentally[1,2] and theoretically[3-6] studied in considerable detail for both mixed (containing all polarization components) and plane-polarized reflected radiation. For given zenith incidence angles, the off-specular peaks discussed in these studies occurred in the incidence plane at zenith reflection angles greater than the corresponding specular reflection angles and hence may be referred to as "superspecular" maxima.[#] Similar superspecular maxima have been observed in our laboratory for a dielectric which had a rough surface and significant internal scattering.[7] In order to study this superspecular peak phenomenon for a dielectric with a rough surface but negligible internal scattering, detailed angular distribution measurements were made of the radiant flux reflected from roughened glass samples. These distribution measurements were made in the forward reflection quadrant of the plane of incidence. The objective of this paper is to report on a new type of off-specular maximum which was discovered as a result of the above measurements. This new off-specular maximum is termed "subspecular" since, for a given zenith incidence angle, it occurs in the reflected flux distribution at a zenith reflection angle which is smaller than the specular reflection angle. The effects of polarization, zenith incidence angle, and sample surface properties on the existence and zenith angular location of both the subspecular and superspecular reflection maxima have been investigated and the results are presented. Also, an analytical expression has been formulated for the reflection of radiant flux from a randomly rough dielectric surface having a normal distribution of surface heights. This relation applies to reflection in the plane of incidence and is used to quantitatively predict the existance of the subspecular and superspecular maxima.

#Zenith incidence and reflection angles are measured relative to the normal of the mean surface.

Apparatus and Test Samples

All measurements of the angular distribution of radiation reflected from the roughened glass surfaces were made using the apparatus shown schematically in Fig. 1. With this system, unpolarized light from a tungsten-halogen lamp was collimated by a series of apertures, chopped mechanically at a frequency of 13 Hz, and then reflected from a plane first-surface mirror onto a spherical mirror in the near-normal direction. This mirror focused the radiation on the sample surface at a zenith incidence angle ψ with an incident solid angle Δw_i of 0.017 sterad (see inset, Fig. 1). The sample was vacuum-mounted on a sample holder which was located along with the other aforementioned components on a turntable having an angle indexing device. This sample holder, shown also in Fig. 1, could be adjusted to maintain the roughened surface of the sample on the axis of rotation of the turntable. It also allowed the zenith incidence angle of the irradiance to be set at any desired value within an accuracy of $\pm 0.5^{\circ}$. As noted previously, all zenith angles are measured relative to the outward normal of the mean surface.

The radiant flux reflected from the illuminated area of the sample surface in the direction defined by the zenith reflection angle θ and the plane of incidence was collected by a spherical mirror subtending a solid angle of $\Delta w_r = \Delta w_i =$ 0.017 sterad. This mirror focused the radiation on the entrance slit of a monochromator after reflecting it off a plane first-surface mirror and passing it through a polarizer. Since the over-detection measurement technique was used, the monochromator entrance slit was set to a slightly greater width than the focused image of the illuminated area of the sample. The monochromator was a standard Perkin-Elmer Model 98 equipped with a CaF_2 prism and a 1P28 photomultiplier detector.

All test samples were 6-mm-thick glass disks of 2.5-cm diam. The glass was of optical quality and had a refractive index of $n = 1.51 \pm 0.01$ at the wavelength λ used in this investigation. Initially, both sides of each disk were polished flat by use of a standard optical polishing technique. Then, using a similar technique, one side of the disk was roughened by grinding it with an abrasive. This was done for five samples using different size abrasives for each sample. The resulting rms mechanical surface roughness σ_m of each of the five samples was measured with a profilometer and the following values were recorded: 0.34, 0.63, 1.77, 3.35, and 5.22μ. After these measurements had been performed, the highly polished side of each of the glass disks was coated with a flat black paint of epoxy resin base which had a refractive index of 1.48 ±0.03. Since this refractive index is effectively equal to that of the glass, the amount of radiation internally reflected at the glass-paint interface should be negligible with the radiation transmitted into the paint being absorbed by the pigment particles suspended in the epoxy base. This was confirmed experimentally by irradiating a similarly prepared, but unroughened, glass sample at the Brewster angle ($\psi_{Br} = 56.5^{\circ}$) and observing that no p-polarized reflected radiation was detected at any zenith reflection angle including the specular one. Thus, there was no appreciable back reflection at the glass-paint interface.

Procedure

After alignment and calibration of the irradiation and detection optics, a roughened glass sample was attached to the sample holder and the zenith angle of the incident radiation set to some desired value, ψ. Then, using the polarizer, the radiant flux reflected from the sample surface in a specific θ direction was resolved into components polarized perpendicular and parallel to the plane of incidence. Each of these polarized radiation components was transmitted through the monochromator to the photomultiplier detector. A conventional strip-chart recorder was used to display the detector outputs. Since the monochromator had a 22% greater transmission for parallel-polarized radiation, the detector output recorded for the perpendicular-polarized component was multiplied by 1.22 before comparing it with the detector output obtained for the parallel-polarized component. This unequal transmission of the monochromator for perpendicular- and parallel-polarized radiation is due to differential reflection of these two polarized radiation components at the prism faces.[2]

After the foregoing described measurement was completed, the turntable was rotated and the polarized radiant fluxes reflected from the sample surface in another θ direction were determined. This was subsequently done for values of θ ranging from 0 to 90° in increments of 5°. Also, in the vicinity of all maxima, measurements were made for increments in θ of 0.5 or 1°. Then, the zenith incident angle of the irradiance was set to a new desired value and the procedure repeated. Zenith incidence angles ranging from 10 to 76° were used. All measurements were for a wavelength of 0.5μ.

Experimental Results

Some results of the angular distribution measurements described in the previous section are shown in Figs. 2-4. Additional results are contained in Ref. 8. The data presented here are for glass samples with surface roughnesses σ_m of 0.34, 1.77, and 3.35μ. In each figure, graphs are given for both the parallel (p) and perpendicular (s) polarization components of the reflected radiation. The distribution data shown are presented in the normalized form $\rho_p(\psi,\theta)\cos\theta/\rho_p(\psi,\psi)\cos\psi$ and $\rho_s(\psi,\theta)\cos\theta/\rho_s(\psi,\psi)\cos\psi$ where $\rho_p(\psi,\theta)$ and $\rho_s(\psi,\theta)$ are, respectively, the p- and s-polarized biangular reflectances defined in Ref. 2. This normalization of the reflectance curves by their values at the specular reflection angles $\theta = \psi$ is employed to emphasize the off-specular peaks. The relative magnitude of the off-specular peaks in the p- and s-polarized components of the reflected radiant flux will be shown later.

It is observed in Figs. 2-4 that for some zenith incidence angles ψ, the maxima of the reflected flux distributions occur at zenith reflection angles θ_m greater than the specular reflection angles (superspecular maxima), while for other zenith incidence angles, the maxima of the reflected flux distributions occur at zenith reflection angles θ_m smaller than the specular reflection angles (subspecular maxima). As seen from Figs. 3 and 4, the subspecular maxima occur in the angular distributions of p-polarized reflected flux when the irradiance zenith angles are less than 50° and the sample surface roughness σ_m is significantly larger than the radiation wavelength, $\lambda = 0.5\mu$. Figure 2 shows that no subspecular maxima occur when the sample surface roughness is less than the radiation wavelength. However, for zenith incidence angles greater than about 50° superspecular maxima exist in the distributions of p-polarized reflected flux even though the sample surface roughness is smaller than the radiation wavelength. Similar superspecular maxima are observed in the distributions of p-polarized reflected flux for the rougher sample surfaces as seen in Figs. 3 and 4. Superspecular maxima also occur in the directional distributions of the s-polarized reflected flux when the surface roughness of the sample is significantly larger than the radiation wavelength. These superspecular maxima are observed in the results given in Figs. 3 and 4 and occur for all zenith incidence angles. As can be seen from Fig. 2, superspecular maxima are not observed in the angular distributions of s-polarized reflected flux when the sample roughness is less than the radiation wavelength.

More detailed scrutiny of the p-polarized results shown in Figs. 2, 3, and 4 indicates that the distributions change dramatically as the irradiance incidence angle ψ ranges from below the Brewster angle of the glass to slightly above it. For example, in Fig. 2, the $\psi = 40^\circ$ distribution exhibits only a specular peak but the $\psi = 50^\circ$ distribution contains both a specular peak and an emerging superspecular maximum while the $\psi = 60^\circ$ distribution has only a superspecular maximum. Moreover, in Figs. 3 and 4, the $\psi = 40^\circ$ distributions have a very definite subspecular maximum while the $\psi = 50^\circ$ distributions exhibit an arising superspecular maximum which becomes fully emerged in the $\psi = 60^\circ$ distributions.

Because of the interesting behavior of the p-polarized distributions for irradiance angles near the Brewster angle further directional distribution measurements were made for closely spaced angles of incidence between 40 and 60° The unnormalized results of these measurements for p-polarized reflected flux are shown in Figs. 5 and 6 for surface roughnesses σ_m of 0.34 and 1.77μ, respectively. It is seen in Fig. 5, which is for a σ_m/λ of 0.68, that the specular peak diminishes with increasing incidence angle and vanishes when the incidence angle reaches the Brewster angle. This behavior is consistent with the incidence angle dependence one might expect for the p-polarized specular reflectance of a rough dielectric surface with normally (gaussian) distributed surface heights,[9] $F_p(\psi,n) \exp[-(4\pi\sigma \cos \psi/\lambda)^2]$. Here σ is the rms value of the surface heights and $F_p(\psi,n)$ is the p-polarized component of the Fresnel reflectance for a smooth dielectric surface.[10]

It is further observed in Fig. 5 that for an incidence angle of 46° an emerging superspecular maximum appears in the p-polarized distribution at $\theta_m \simeq 70^\circ$. This superspecular maximum grows in prominence as the incident angle is increased, but remains at essentially the same angular location, $\theta_m=72^\circ$ (see dashed curve C-D). In Fig. 6, which is for a σ_m/λ of 3.54, the subspecular maximum observed in the distribution for $\psi = 40^\circ$ diminishes in the distributions for increasing incidence angles (see dashed curve A-B) while a superspecular maximum emerges and continues to increase in magnitude (see dashed curve C-D). It is seen that for incidence angles below 48°, only subspecular maxima appear in the distributions, while for incidence angles above 54°, solely superspecular maxima are observed. For incidence angles from 48 to 52°, the distributions have both sub- and superspecular maxima.

Figure 7 shows the experimentally determined angular displacements (relative to the specular angle) of the sub- and superspecular maxima in the directional distributions of p- and s-polarized reflected fluxes for roughened glass surfaces. The magnitude of these angular displacements for 0.5μ wavelength radiation is displayed as a function of irradiance zenith angle ψ with sample surface roughness σ_m taken as a parameter. From the results shown, it would appear that when super- and subspecular maxima occur, their angular displacements (relative to the specular direction) will be greater for rougher surfaces and higher incidence angles until the surfaces become rough enough and/or the incidence (and reflection) angles large enough that multiple reflections[4,5,11] and bi-static shadowing[3-6,11-13] begin to modify significantly the distributions and affect the locations of their maxima. Then for larger incidence (and reflection) angles and/or rougher surfaces, the angular displacements of the sub- and superspecular maxima will decrease due to the increased effects of bistatic shadowing[4,13] and multiple reflections[4,11].

Theory and Comparison with Data

In order to confirm theoretically the existence of the subspecular peaks, a simple analytical expression will be formulated for the reflection of radiant flux from a rough dielectric surface into the plane of incidence. To do this, it is first necessary to make several assumptions about the characteristics of the surface. One of these is that the rough surface is isotropic. Another is that the surface is randomly rough. The surface heights $\zeta(x)$ of this surface are shown in Fig. 8 and are considered, after Beckmann,[9] to have a normal distribution

$$w(\zeta) = \frac{1}{\sigma(2\pi)^{1/2}} \exp\left(-\frac{\zeta^2}{2\sigma^2}\right) \tag{1}$$

with standard deviation σ and correlation function $B(\tau)$. Since the mean level of the surface $\langle\zeta\rangle = 0$, the standard deviation σ is also the rms value of the surface heights. Now, as indicated by Beckmann,[9,13] $\zeta'(x)$ also has a normal distribution

$$w(\zeta') = \frac{1}{(2\pi|B''(0)|)^{1/2}} \exp\left(-\frac{(\zeta')^2}{2|B''(0)|}\right) \tag{2}$$

with mean value zero and variance $|B''(0)|$ where $(|B''(0)|)^{1/2}=m$ is the rms slope of the surface. Furthermore, he shows[9] that this distribution function can be transformed to yield the distribution of surface slopes α, defined[9] by $\tan \alpha = \zeta'(x)$. The resulting distribution of the slopes α of a normally distributed surface is given by

$$P(\alpha,m) = \frac{\sec^2\alpha}{m(2\pi)^{1/2}} \exp\left(-\frac{\tan^2\alpha}{2m^2}\right) \qquad (3)$$

Note from Fig. 8 that α is also equal to the angle between the normal to a local surface slope and the normal to the mean plane of the surface.

After determining the distribution of slopes of the normally distributed rough surface, a formula can be derived for the reflection of radiation into the plane of incidence. This has been done using essentially the approach employed in Ref. 3 for the v-grooved facet model of a rough surface. Thus, the wavelength λ of the radiation is considered to be small compared to the rms surface heights, σ, and the derivation is based on geometrical optics.** In addition, the basic reflection model for the random rough surface assumes specular reflection from the local surface slopes, as shown in Fig. 8, plus a contribution due to the multiple reflections[11] which occur when a ray strikes more than one slope before leaving the surface. Also accounted for is "bistatic" shadowing, i.e., the screening of local surface slopes by adjacent surface slopes interrupting the incident and once-reflected radiant flux.[13] There are, however, several differences between the approach used in Ref. 3 and that employed here. One is that the multiple reflections contribution is not considered to always be perfectly diffuse. This is in agreement with previous experimental results for roughened glass.[12] Another difference is that it is not necessary to consider that each slope has the same area as was assumed for the facets in Ref. 3. The relaxation of this assumption is made possible through use of the area relationship given in Refs. 14 and 15. A third difference is that the bistatic shadowing function $S(\psi,\theta,m)$ employed here is for a normally distributed rough surface[13] and depends upon the rms slope of the surface while the $G(\psi,\theta)$ function used in Refs. 3 and 6 for the v-grooved facet surface is independent of the parameter c characterizing the rms slope of the facets.

**The use of geometrical optics to explain the superspecular peak phenomenon for rough glass surfaces was experimentally investigated in Ref. 14 and found to be valid even for rough surfaces not obeying the condition $\sigma/\lambda \gg 1$.

The result of the foregoing described derivation is

$$\rho(\psi,\theta) = \frac{F(\psi+\alpha,n)P(\alpha,m)S(\psi,\theta,m)}{4\cos\theta\cos\psi} + b(\psi,\theta,m) \qquad (4)$$

where $\rho(\psi,\theta)$ is the bidirectional reflectance defined in Refs. 2 and 3 for mixed radiation. Here, $b(\psi,\theta,m)$ is the multiple reflection contribution to the bidirectional reflectance, $P(\alpha,m)$ is the distribution of slopes given in Eq. (3) with $\alpha = (\theta-\psi)/2$ (see Fig. 8), and $F(\psi+\alpha,n)$ is the Fresnel reflectance for local incidence angle $\psi+\alpha$, where $F(\psi+\alpha,n) = [F_s(\psi+\alpha,n)+F_p(\psi+\alpha,n)]/2$ with F_s and F_p being, respectively, the s- and p-polarized components of the Fresnel reflectance.[10] The $S(\psi,\theta,m)$ function in Eq. (4) is Beckmann's bistatic shadowing relation for a normally distributed rough surface[13] and has the form

$$S(\psi,\theta,m) = \begin{cases} S(\psi,m)\ S(|\theta|,m), & -\pi/2<\theta<-\psi \\ S(\psi,m), & -\psi<\theta<0 \\ S(\psi,m)\ S(\theta,m), & 0<\theta<\pi/2 \end{cases} \qquad (5)$$

where

$$S(\psi,m) = \exp\left[-\frac{1}{4}\tan\psi\ \mathrm{erfc}\left(\frac{\cot\psi}{(2)^{1/2}m}\right)\right] \qquad (6)$$

and

$$S(\theta,m) = \exp\left[-\frac{1}{4}\tan\theta\ \mathrm{erfc}\left(\frac{\cot\theta}{(2)^{1/2}m}\right)\right] \qquad (7)$$

It is further noted in Eq. (4) that the multiple reflections contribution $b(\psi,\theta,m)$ has not been assumed perfectly diffuse and is also considered to be a function of the incidence angle ψ and the root-mean-square slope m. This is in agreement with the experimental results of Refs. 11 and 12 and also can be inferred from the theoretical results of Ref. 4. Now, since $\rho(\psi,\theta) = [\rho_s(\psi,\theta) + \rho_p(\psi,\theta)]/2$ for unpolarized incident radiation[2,6] and $F(\psi+\alpha,n) = [F_s(\psi+\alpha,n) + F_p(\psi+\alpha,n)]/2$, Eq. (4) can be resolved into separate expressions for the plane-polarized bidirectional reflectances $\rho_s(\psi,\theta)$ and $\rho_p(\psi,\theta)$. These are

$$\rho_s(\psi,\theta) = \frac{F_s(\psi+\alpha,n)P(\alpha,m)\ S(\psi,\theta,m)}{4\cos\theta\cos\psi} + b(\psi,\theta,m) \qquad (8)$$

and

$$\rho_p(\psi,\theta) = \frac{F_p(\psi+\alpha,n)P(\alpha,m)\ S(\psi,\theta,m)}{4\cos\theta\cos\psi} + b(\psi,\theta,m) \qquad (9)$$

Note that the multiple reflections contribution $b(\psi,\theta,m)$ in Eqs. (8) and (9) is assumed to be unpolarized. This is in agreement with the results of Ref. 12.

Equations (4, 8, and 9) with $b(\psi,\theta,m) = 0$ have been used to predict theoretically the experimental distributions of the mixed and plane-polarized reflected fluxes for the glass sample with roughness $\sigma_m = 1.77\mu$ and refractive index $n = 1.51$. As seen in Fig. 9, this was done for an incidence angle of 30° and an rms slope m of 0.247. The experimental (solid) and theoretical (dashed) curves shown are presented in the normalized forms $\rho_p(\psi,\theta)\cos\theta/\rho(\psi,\psi)\cos\psi$, $\rho_s(\psi,\theta)\cos\theta/\rho(\psi,\psi)\cos\psi$, and $\rho(\psi,\theta)\cos\theta\,/\rho(\psi,\psi)\cos\psi$ for p-polarized, s-polarized, and mixed radiant fluxes, respectively, where $\rho(\psi,\psi)$ is the value of $\rho(\psi,\theta)$ at $\theta = \psi$. This mode of normalization is used to show the relative magnitudes of the subspecular maximum in the p-polarized distribution and the superspecular maximum in the s-polarized distribution. The subspecular maximum seen in the distributions for $\psi = 30^o$ is smaller than the corresponding superspecular maximum but it does have a significant magnitude relative to that of the superspecular maximum. It is further noted in Fig. 9 that there is excellent agreement between the theoretical and experimental distributions for reflection angles in the locality of the peaks and for the smaller reflection angles. The agreement is not as good for large reflection angles and this is attributed to the multiple reflections contribution $b(\psi,\theta,m)$ being neglected. Nevertheless, the excellent agreement between the analytical and experimental results for zenith reflection angles at and around the peaks of the distributions quantitatively confirms the existence of the sub- and superspecular maxima for moderate incidence angles.

Figure 10 presents the theoretical p-polarized flux distributions for zenith incidence angles near the Brewster angle of the glass, 56.5^o. The multiple reflections contribution $b(\psi,\theta,m)$ and the rms slope m of the rough surface are again taken as zero and 0.247, respectively. It is seen in Fig. 10 that as the incidence angle increases the subspecular maximum in the p-polarized distributions diminishes while a superspecular maximum emerges and increases in prominence. For incidence angles below 55°, the subspecular maximum has the greater magnitude but for incidence angles above 55°, the superspecular maximum is larger. The sub- and superspecular maxima are of equal magnitude for $\psi = 55^{o}$. From these results, it can be concluded that the subspecular maximum in the p-polarized distributions does not change continuously to a superspecular maximum as the incidence angle increases but that the superspecular maximum develops separately as the subspecular maximum diminishes. These results corroborate the previous discussion regarding the development and demise of the super- and subspecular maxima in the experimental distributions of Fig. 6.

Figure 11 presents a comparison between the angular locations θ_m of the off-specular maxima of the experimental and theoretical distributions for the glass sample with roughness $\sigma_m = 1.77\mu$. These results were obtained from s- and p-polarized reflected flux distributions for incidence angles of 10 to 70°. As before, the theoretical flux distributions were for an rms slope m of 0.247 and a refractive index n equal to that of the glass, 1.51. Also, the multiple reflections contribution was again neglected and hence $b(\psi,\theta,m)$ in Eqs. (8) and (9) was taken equal to zero. It is observed in Fig. 11 that there is quite good agreement between the theoretical (dashed) and experimental (solid) curves for the angular locations of the superspecular maxima of the s-polarized distributions. This is seen to be true for all incidence angles but especially so for the smaller ones. There also is good agreement between the theoretical and experimental curves for the angular locations of the off-specular maxima of the p-polarized distributions except for the range of zenith incidence and reflection angles where the local incidence angles $\psi + \alpha = (\psi+\theta)/2$ are near the Brewster angle.++ For these local incidence angles, $F_p(\psi+\alpha,n)$ is quite small and the multiple reflections contribution can significantly modify the p-polarized distributions and appreciably affect the location of their maxima. Since the multiple reflections contribution was neglected in obtaining the analytical results presented in Fig. 11, one would not expect good agreement between these results and the p-polarized experimental data for local incidence angles near the Brewster angle.

Conclusions

From the experimental and analytical results presented in the previous sections, it can be concluded that subspecular

---

++The dotted portions of the theoretical curves for p-polarized radiation denote secondary maxima. For examples of such maxima, see Fig. 10.

maxima occur in the angular distributions of parallel-polarized reflected flux for rough dielectric surfaces. These subspecular maxima are observed if the irradiance zenith angle is appreciably less than the dielectric's Brewster angle and the rms mechanical surface roughness of the dielectric is significantly larger than the radiation wavelength. It is also concluded that superspecular maxima occur in the distributions of parallel-polarized reflected flux when the irradiance zenith angle is greater than the Brewster angle of the dielectric. These superspecular maxima in the p-polarized reflected flux distributions are observed even for an rms mechanical surface roughness less than the radiation wavelength. It is further concluded from the aforementioned results that superspecular maxima occur in the angular distributions of perpendicular-polarized reflected flux for roughened dielectric surfaces. These superspecular maxima in the s-polarized reflected flux distributions are observed for irradiance zenith angles ranging from 10° to 76° when the rms mechanical surface roughness of the dielectric is appreciably larger than the radiation wavelength.

References

[1]Torrance, K. E. and Sparrow, E. M., "Off-Specular Peaks in the Directional Distribution of Reflected Thermal Radiation," Journal of Heat Transfer, Vol. 88, Series C, No. 2, May 1966, pp. 223-230.

[2]Torrance, K. E., Sparrow, E. M., and Birkebak, R. C., "Polarization, Directional Distribution, and Off-Specular Peak Phenomena in Light Reflected from Roughened Surfaces," Journal of the Optical Society of America, Vol. 56, No. 7, July 1966, pp. 916-925.

[3]Torrance, K. E. and Sparrow, E. M., "Theory for Off-Specular Reflection from Roughened Surfaces," Journal of the Optical Society of America, Vol. 57, No. 9, Sept. 1967, pp. 1105-1114.

[4]Nelson, H. F., "Ray Reflection from Rough Dielectric Surfaces," AA & ES 66-5, Nov. 1966, Purdue University, Lafayette, Ind.

[5]Nelson, H. F. and Goulard, R., "Reflection from a Periodic Dielectric Surface," Journal of the Optical Society of America, Vol. 57, No. 6, June 1967, pp. 769-771.

[6]Torrance, K. E., "Theoretical Polarization of Off-Specular Reflection Peaks," Journal of Heat Transfer, Vol. 91, Series C, No. 2, May 1969, pp. 287-290.

[7]Smith, A. M., Tempelmeyer, K. E., Müller, P. R., and Wood, B. E., "Angular Distribution of Visible and Near IR Radiation Reflected from CO_2 Cryodeposits," AIAA Journal, Vol. 7, No. 12, Dec. 1969, pp. 2274-2280.

[8]Müller, P. R., "Measurements of Refractive Index, Density, and Reflected Light Distributions for Carbon Dioxide and Water Cryodeposits and also Roughened Glass Surfaces," Ph.D. Dissertation, June 1969, The University of Tennessee, Knoxville, Tennessee.

[9]Beckmann, P. and Spizzichino, A., The Scattering of Electromagnetic Waves from Rough Surfaces, Macmillan, New York, 1963.

[10]Hottel, H. C. and Sarofim, A. F., Radiative Transfer, McGraw-Hill, New York, 1967.

[11]Tanaka, S., "Measurement of Reflection Characteristics of Ground Glass Using Polarized Light," Oyo Butsuri, Vol. 26, No. 3, March 1957, pp. 85-91.

[12]Tanaka, S., "Reflection Characteristics of Nonmetallic Diffusing Surfaces," Oyo Butsuri, Vol. 28, No. 9, Sept. 1959, pp. 508-514.

[13]Beckmann, P., "Shadowing of Random Rough Surfaces," IEEE Transactions on Antennas and Propagation, Vol. AP-13, No. 3, May 1965, pp. 384-388.

[14]Voishvillo, N. A., "Reflection of Light by a Rough Glass Surface at Large Angles of Incidence of the Illuminating Beam," Optics and Spectroscopy, Vol. 22, No. 6, June 1967, pp. 517-520.

[15]Rense, W. A., "Polarization Studies of Light Diffusely Reflected from Ground and Etched Glass Surfaces," Journal of the Optical Society of America, Vol. 40, No. 1, Jan. 1950, pp. 55-59.

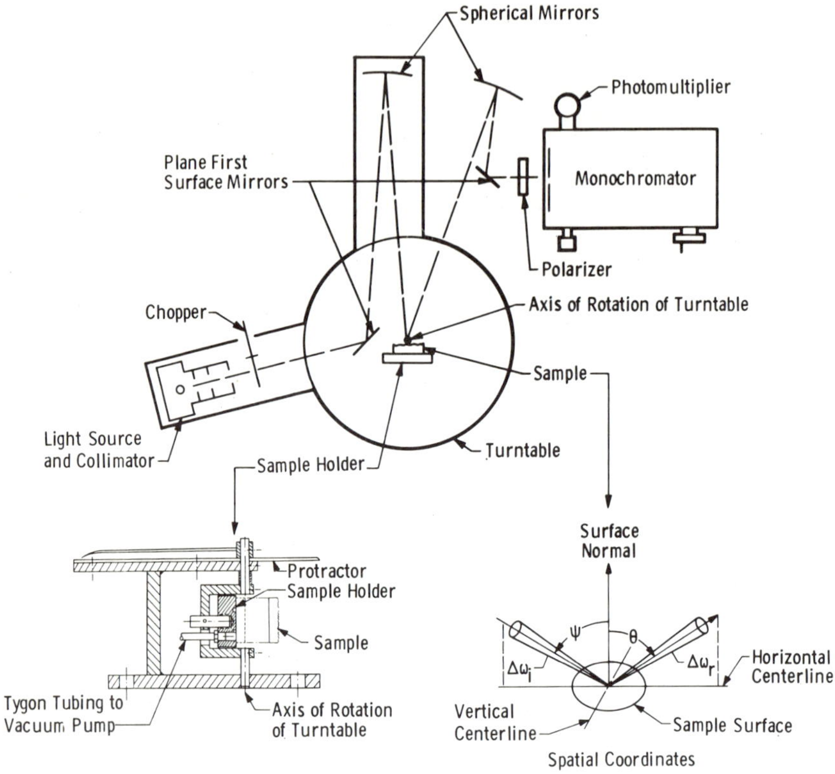

Fig. 1 Schematic of experimental apparatus.

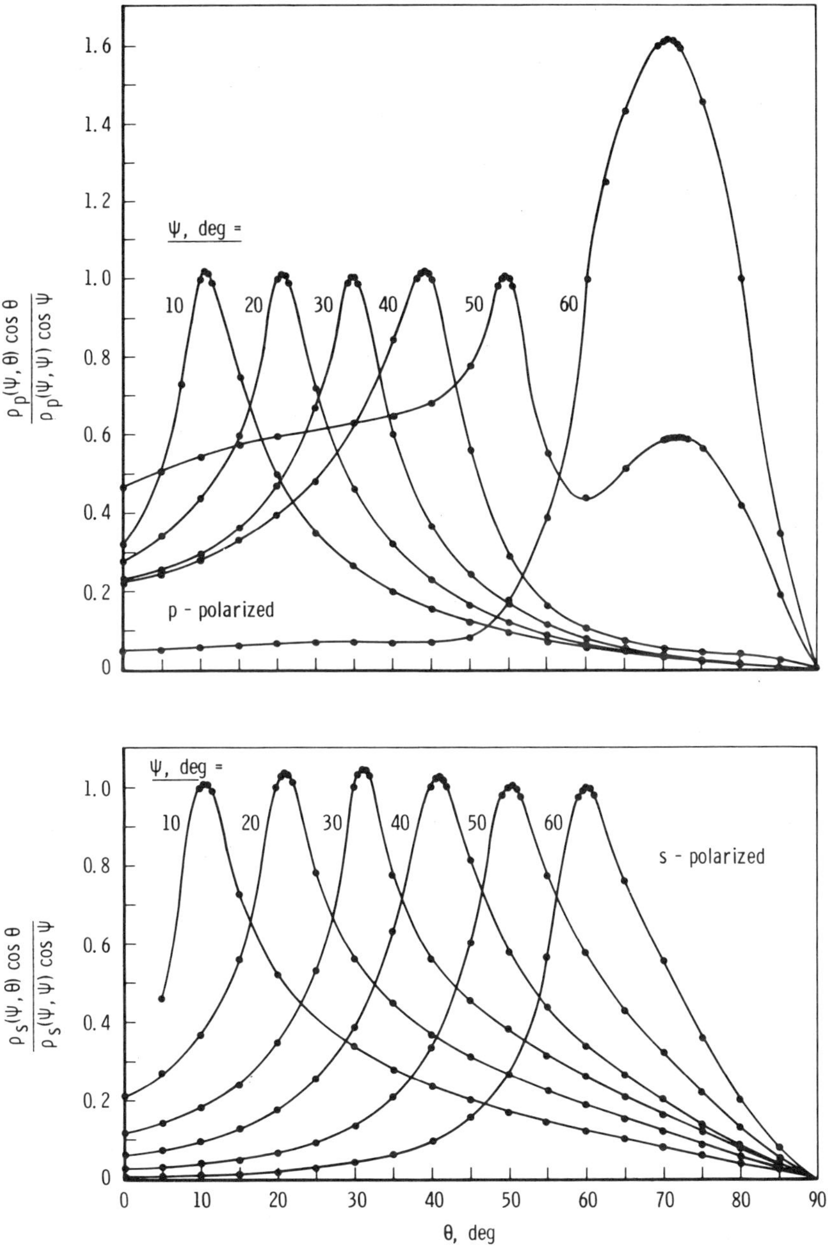

Fig. 2 Directional distributions of plane-polarized radiant flux reflected from roughened glass sample, $\sigma_m = 0.34\mu$, $\lambda = 0.5\mu$, various incidence angles ψ.

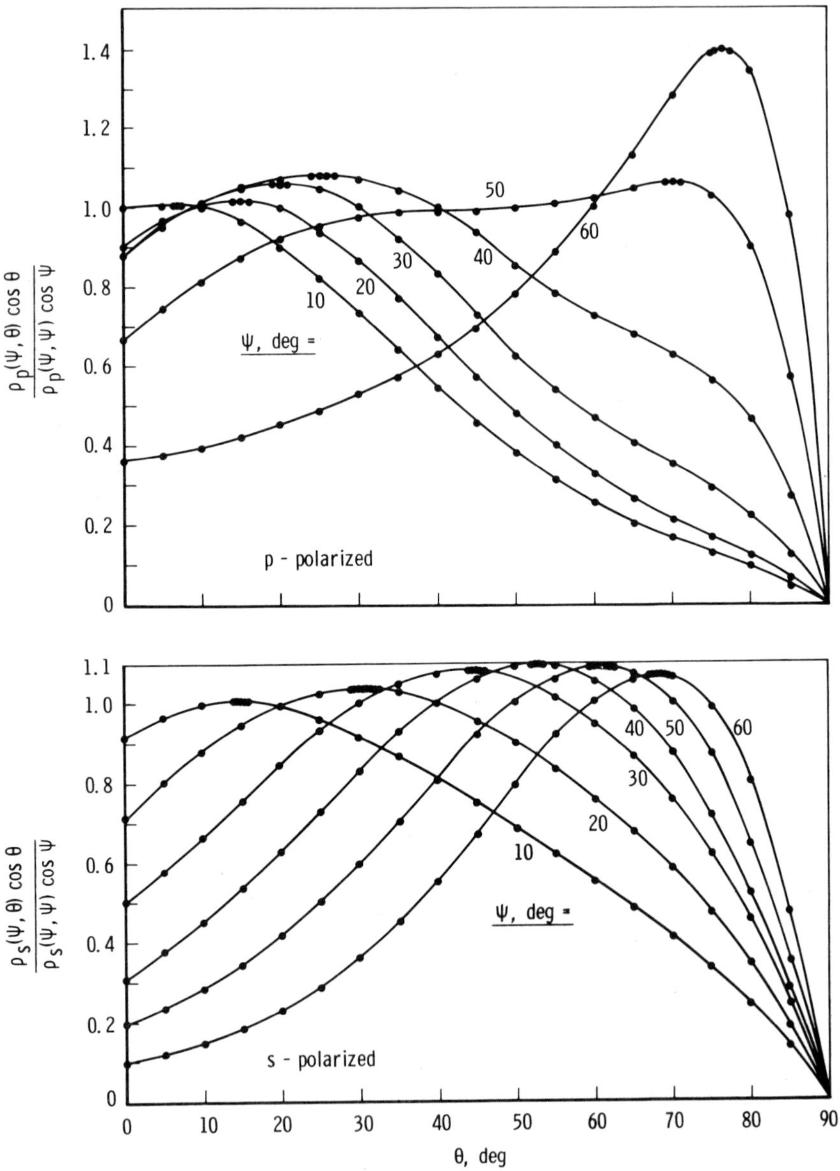

Fig. 3 Directional distributions of plane-polarized radiant flux reflected from roughened glass surface, $\sigma_m = 1.77\mu$, $\lambda = 0.5\mu$, various incidence angles ψ.

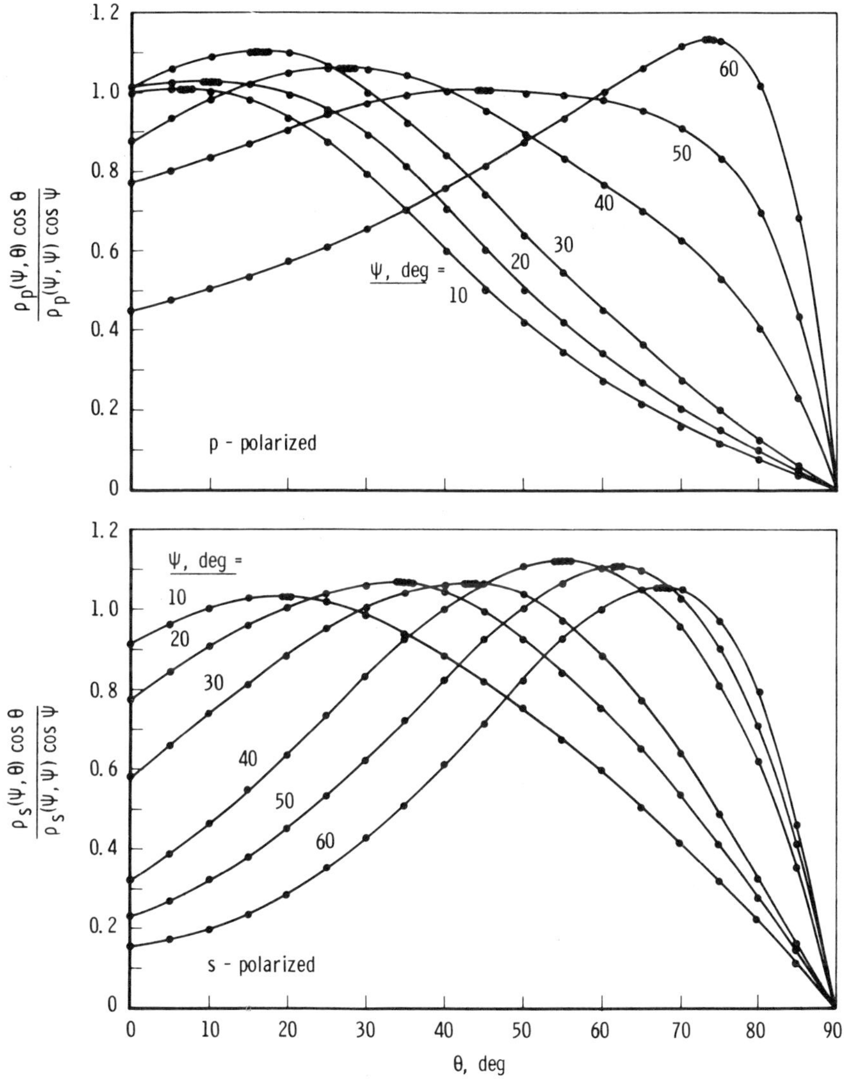

Fig. 4 Directional distributions of plane-polarized radiant flux reflected from roughened glass surface, $\sigma_m = 3.35\mu$, $\lambda = 0.5\mu$, various incidence angles ψ.

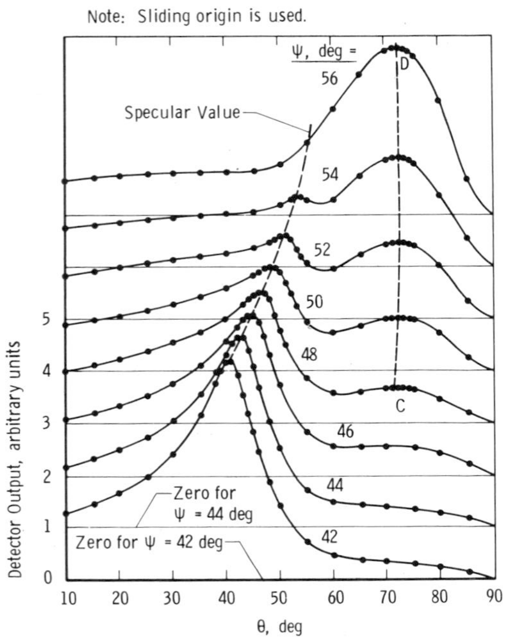

Fig. 5

Directional distributions of p-polarized radiant flux reflected from roughened glass surface, $\sigma_m = 0.34\mu$, $\lambda = 0.5\mu$, incidence angles near the Brewster angle.

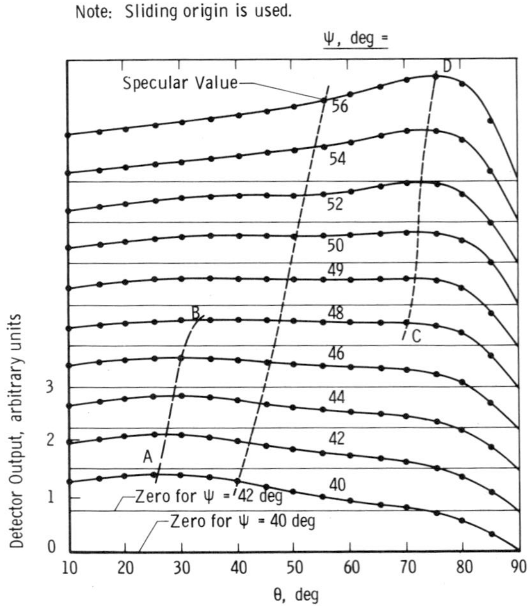

Fig. 6

Directional distributions of p-polarized radiant flux reflected from roughened glass surface, $\sigma_m = 1.77\mu$, $\lambda = 0.5\mu$, incidence angles near the Brewster angle.

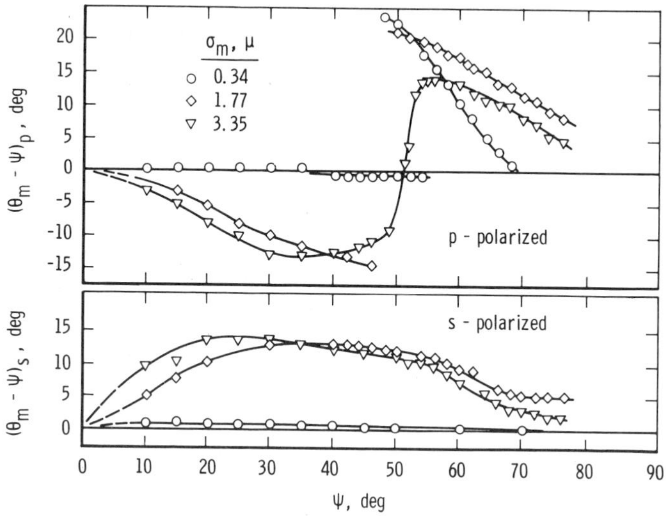

Fig. 7 Measured angular displacements (relative to the specular direction) of off-specular maxima in the distributions of plane-polarized radiant flux reflected from roughened glass surfaces, $\lambda = 0.5\mu$. Various surface roughnesses σ_m and incidence angles ψ.

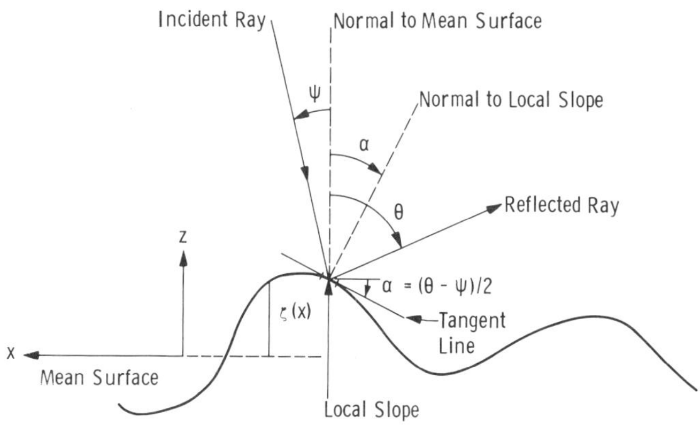

Fig. 8 Sketch of ray reflection from single local slope of rough surface.

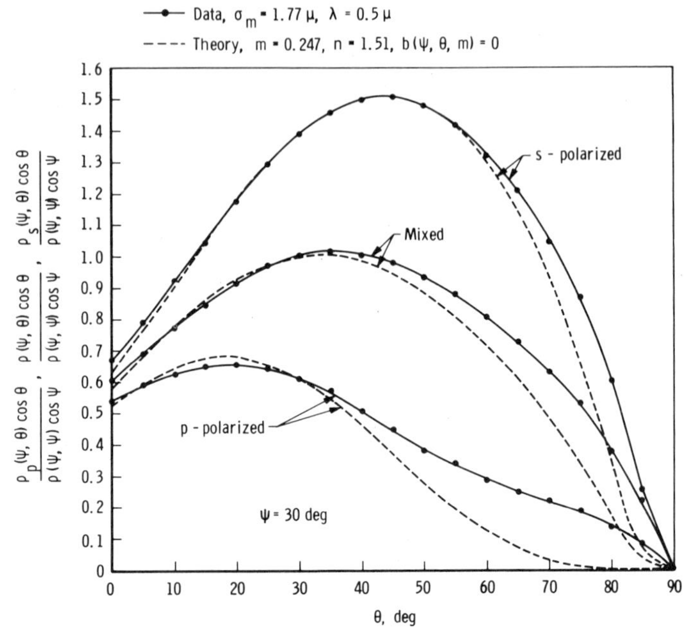

Fig. 9 Comparison between theoretical and experimental distributions of polarized radiant flux reflected from roughened glass surface for incidence angle of 30°.

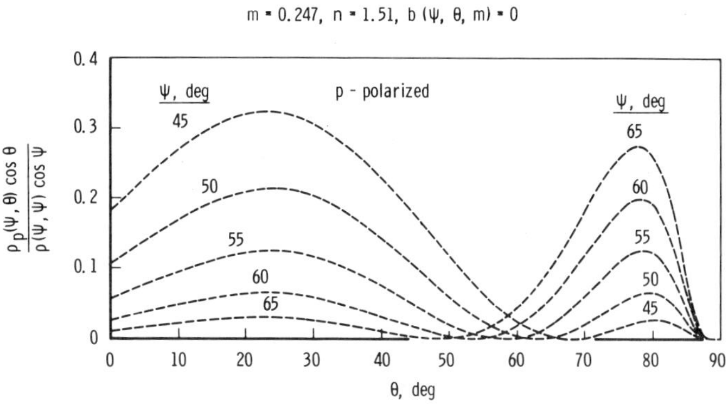

Fig. 10 Theoretical distributions of p-polarized radiant flux reflected from roughened dielectric surface for incidence angles near the Brewster angle.

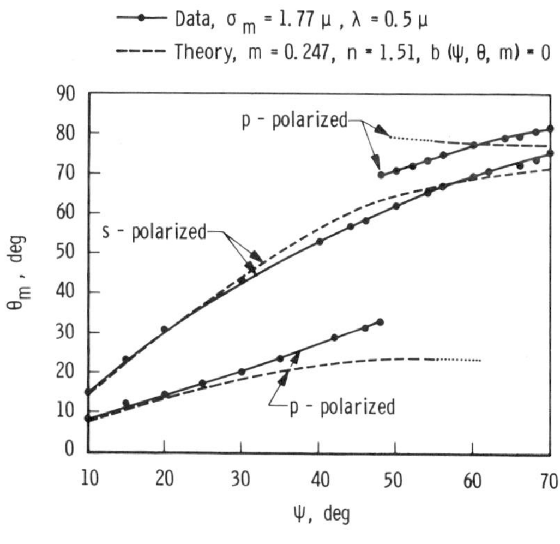

Fig. 11 Comparison between angular locations of the off-specular maxima of theoretical and experimental distributions of plane-polarized radiant flux reflected from roughened glass surface. Various incidence angles ψ.

III. THERMAL JOINT CONDUCTANCE

PREDICTION OF THERMAL CONTACT CONDUCTANCE BETWEEN SIMILAR METAL SURFACES

L.S. Fletcher*
Rutgers University, New Brunswick, N.J.

D.A. Gyorog†
University of Missouri - Rolla, Rolla, Mo.

Abstract

A new expression has been developed for prediction of the thermal conductance of contacting metal surfaces in a vacuum environment. This expression was determined by using the results of an experimental investigation of aluminum, brass, stainless steel, and magnesium over a wide range of test variables. Its dimensionless parameters were formulated in terms of material properties, average surface measurements, and conditions of load and temperature of the metallic contact. These dimensionless parameters correlate the experimental data of this investigation and those of seven other investigators. The expression resulting from this correlation predicted these experimental data within an average over-all error of less than 24%.

Nomenclature

A = area, sq ft
b = specimen radius, ft
C = const
d = surface parameter, μ in.
E = modulus of elasticity, psi
FD = flatness deviation, μ in.
f = function
h_c = contact conductance coefficient, Btu/hr sq ft °F

Presented as Paper 70-852 at the AIAA 5th Thermophysics Conference, Los Angeles, Calif., June 29-July 1, 1970. This research was sponsored in part by NASA Ames Research Center, under Grant NGR 03-001-033, and was conducted in the Mechanical Engineering Laboratories of Arizona State University, Tempe, Ariz.

*Assistant Professor of Aerospace Engineering.
†Associate Professor of Mechanical Engineering.

h_c^* = dimensionless contact conductance coefficient, $h_c \delta / k_m$
k = thermal conductivity, Btu/hr ft °F
m = const
n = const
P = load pressure, psi
P* = dimensionless pressure, P/E
Q = heat flux, Btu/hr
RD = roughness deviation, μ in.
T = temperature, °F
T* = dimensionless temperature, $\beta T_m (T_m$ - °R)
ΔT = temperature difference, °F
α = contact area ratio, A_c/A_a
β = coefficient of thermal expansion, 1/°F
δ = surface parameter; gap thickness between surfaces μ in.
δ* = dimensionless gap thickness, δ/b

Subscripts and Superscripts

a = apparent
c = contact
m = mean
n = exponent
o = initial

Introduction

The development of spacecraft structures to withstand re-entry heating, the thermal isolation of cryogenic storage compartments, the optimum packaging of spacecraft electronic equipment, and the improvement of heat dissipation systems are all problems which are being analyzed continuously by thermal design engineers. Analysis of these problems requires the study of heat exchange between surfaces in contact. In some instances, rather extensive experimental investigations have proved necessary in order to provide the desired design information. Solution of these dilemmas would be far simpler if there existed an improved technique for the thermal analysis of heat exchange between surfaces in contact.

Although theoretical and experimental studies concerning heat-transfer phenomenon in solid materials have resulted in many useful techniques for analysis, studies of energy transfer between contacting solid materials have produced few established procedures for heat transfer prediction. The thermal contact conductance coefficient for a junction between solids has been defined as

$$h_c = \frac{Q/A}{T_1 - T_2} = \frac{Q/A}{\Delta T} \quad (1)$$

where T_1 and T_2 are the temperatures of the bounding surfaces of the contact, and Q/A is the heat flux per unit area. The contact conductance varies considerably, depending upon the mechanical and thermophysical properties of the materials composing the contact, the surface conditions, and the interstitial fluid or filler. Because of the complexity of the phenomenon, a number of analytical and experimental investigations have been conducted, as evidence by the thermal contact conductance bibliographies of Hsieh and Davis[1] and Moore, Atkins, and Blum[2]. In spite of all the published work, however, there have been relatively few attempts directed primarily at the correlation of existing experimental data and the prediction of contact conductance.

This paper presents a new expression for the prediction of thermal contact conductance between similar metal surfaces in a vacuum environment.‡ This expression, based on the single contact model, was developed in terms of dimensionless parameters composed of the material properties, surface measurements, and conditions of load and temperature of the metallic contact. The coefficients were obtained from analysis of contact conductance. The resulting dimensionless parameters were used to correlate the experimental results of this investigation and those of seven other investigators. In order to demonstrate the usefulness of this expression, the thermal contact conductance was predicted for the test conditions of several different investigators, and these predicted values were then compared with the measured results.

Experimental Investigation

The experimental apparatus used in the present investigation was similar to that used by other investigators: a vertical cylindrical column under axial load, with contacting surfaces located at approximately one-third and two-thirds of the column height. The test specimens, which formed the vertical cylindrical column, were one inch in diameter. A band heater with variable transformer control supplied the heat flux, and either water or liquid nitrogen was used as the coolant. The test specimens, band heater, guard heater, radiation shields, and heat sink were all instrumented with thermocouples to provide information on the operation of the system and to measure the specimen axial temperature distribution. A de-

---

‡Since it has been demonstrated that the thermal contact conductance between dissimilar metals evidences a directional effect[3-5], the present study is restricted to similar metallic specimens in contact.

scription of both the apparatus and the experimental techniques employed has been previously reported[6–9].

In order to provide a range of suitable information for the development of a prediction expression, experimental data were obtained for aluminum, brass, stainless steel, and magnesium test specimens. These materials were selected because of their engineering applications and wide range of physical characteristics.

Since the contact surface finish is extremely important in the analysis of thermal contact conductance, care was exercised in the preparation and measurement of the specimen contact surfaces. After all contact surfaces were finished by lapping and polishing, some surfaces were roughened by peening. All surfaces were measured using both a Bendix Micrometrical Proficorder and Profilometer. Since the results given by proficorders and profilometers represent only one trace across the surface, several traces were made on each surface to assure that the measured flatness deviation and roughness would be representative of the surface.[§] The resulting values for the surface finish of the test specimens are presented in Table 1. Tests were made with four contact configurations for each material, designated as smooth (S), medium (M), rough or smooth-wavy (R), and smooth-to-rough (SR).

Table 1 Test specimen surface characteristics

Material	Surface	Flatness (μin.)	Average Roughness (μin.)
Aluminum 2024-T4	S	17	2
	M	465	58
	R	2950	6
Stainless steel 304	S	22	2
	M	165	33
	R	4300	2
Magnesium AZ31B	S	35	4
	M	275	115
	R	4200	6
Brass alloy 271	S	55	2
	M	620	60
	R	4300	6

§The surfaces were measured again after testing to ascertain any deviaitions in the surface finish. These measurements were found to be consistent with the original values.

Test specimen material properties, determinted as a function of temperature, were obtained from engineering handbooks and from representatives of companies providing the material. Thermal conductivity data were verified by laboratory tests of material samples. Armco iron was used as a standard material for ascertaining the reliability of the thermal conductivity measurement facility. The aluminum samples were tested in both the "as received" and "annealed" condition. The resulting thermal conductivity values agreed with the published values within 5%.

The experimental data were obtained in a vacuum environment of 10^{-5} to 10^{-6} torr. The apparent contact pressures, or load pressures, were varied from 20 to 815 psi, and mean junction temperatures ranged from -85°F to 335°F. The length of time required to obtain steady-state conditions for these data ranged from two to twelve hours for each data point. The resulting experimental data, obtained for the elastic contact regime, were found to be repeatable within 5%.

The experimental contact conductance values were determined from Eq. (1). A linear least squares fit of the axial center line temperature measurements was made for each set of data in order to find the interface temperatures. These temperatures were verified by graphical extrapolation of the centerline temperature-axial location curves. The heat flux was calculated from the axial temperature gradient of the centerline thermocouples and the thermal conductivity at the mean temperature of the speciment. This value of heat flux was verified by subtracting the estimated heat losses from the measured heat input.

In order to classify the surface conditions of the experimental data, a surface parameter, d, was defined in terms of the rms roughness and flatness deviation. The total profile height of the surface roughness (peak-to-valley) was specified by Oberg[10] to be four times the rms value. Since the flatness deviation is a measure of the mean surface profile, the total peak-to-valley height could be represented as FD + 2RD. The contact between surfaces was assumed to occur at the mean surface height of the smoother surface such that the contact surface parameter was defined as

$$d = (FD + 2RD)_{\substack{\text{rough}\\\text{surface}}} - \frac{1}{2}(FD + 2RD)_{\substack{\text{smooth}\\\text{surface}}} \qquad (2)$$

This assumption was based on the observation that the thermal contact conductance for rough-to-smooth surface contacts was less than that for rough-to-rough surface contacts.

The experimental conductance data obtained in this investigation were found to compare favorably in both magnitude and trend with those of other investigators. These data, then, were considered suitable for use in the development of a prediction expression for thermal contact conductance.

Analysis

The mathematical analysis of a single thermal contact performed by Fenech and Rohsenow[11] resulted in an expression for the prediction of contact conductance. A model containing macroscopic and microscopic contacts was used by Clausing and Chao[12] as the basis for a conductance prediction expression. Both of these expressions have met with some success in the prediction of contact conductance; however, use of these techniques is limited and impractical for most engineering applications. In order to provide a more useful expression for the prediction of contact conductance, this analysis is directed primarily at the correlation of experimental data, from which a prediction expression is derived.

A theoretical expression for the thermal contact conductance coefficient, developed from a single contact model, may be written in dimensionless form as

$$h_c \delta / k_m = h_c^* = \alpha/(1 - \alpha) \tag{3}$$

where α is the contact area ratio and δ is a surface parameter representing the gap between the contacting surfaces. This expression also evolves from the previously mentioned mathematical analysis[11,13]. As a result of a preliminary analysis of the variables involved in thermal contact conductance, it appeared that a dimensionless correlation of the conductance parameters could be effected.

Earlier attempts at dimensionless correlation and dimensional analysis of the variables effecting contact conductance have yielded very few useful results[14-16]. Hudack[17] and Clausing and Chao[12] obtained some degree of success with the correlating parameters P^*T^* and $h_c b/k_m$, respectively. In his work, Hudack was able to correlate the aluminum conductance data of Bloom[18] for a wide range of temperatures. This correlation indicated that the mean junction temperature exhibits a stronger effect on the contact conductance than can be accounted for by the temperature dependence of the material properties. Clausing and Chao presented a correlation of $h_c b/k_m$ as a function of a dimensionless parameter defined as the elastic conformity modulus. Although this correlation was satisfactory for the experimental data presented, it was not suitable for low mean junction temperatures or for very smooth surfaces. In view of these earlier efforts, the experimental data of the present investigation were graphed as $h_c b/k_m$ as a function of P*T*, as shown in Figs. 1-4. The resulting curves were dimensionless and differed only by the surface conditions.

The plots of $h_c b/k_m$ as a function of P*T* for each specimen material were made coincident at a point corresponding to a load pressure of 20 psi by choosing a value of the no-load surface parameter, δ_o, equal to the value of d for the smooth-to-smooth surfaces and then, by ratio, determining a value of δ_o for each of the other surface configurations. These values of δ_o, however, were found to be dependent upon the magnitude of P*T*. This functional dependence was evaluated by plotting $h_c b/k_m$ as a function of δ_o on semi-log coordinates for specific values of P*T* for each material. Their curves were linear but their slopes varied directly as the magnitude of P*T*, that is

$$\log (h_c b/k_m) = C - f(P^*T^*)$$

After some additional graphical analysis it was found that these curves would coincide on a plot of $h_c b/k_m$ as a function of mP^*T^*/δ_o^* where m, a constant independent of test parameters, was 170. Based on this analysis and the fact that the surface parameter must have the dimensions of length, the effective surface parameter was selected as

$$\delta = \delta_o e^{-170P^*T^*/\delta_o^*} \tag{4}$$

This functional behavior of δ corresponds to the physical observations of a decreasing effective gap thickness with increasing load pressure, increasing junction temperature, smoother surfaces and a decreasing modulus of elasticity.

In order further to establish the validity of Eq. (4) and to determine a reliable value for the initial surface parameter, plots of $h_c b/k_m$ as a function of P*T* for each material were made to coincide at a point corresponding to a load pressure of 800 psi, and values of δ_o were again determined by ratio. These values of δ_o exhibited the same dependence on P*T* as those determined at 20 psi and coincided on a plot of $h_c b/k_m$ as a function of $170P^*T^*/\delta_o^*$. The values of δ_o determined at load pressures of both 20 and 800 psi for all specimen configurations tested were plotted as a function of the contact surface parameter d, as shown in Fig. 5. From this graph it was evident that the surface parameter, δ_o, was related to the contact surface parameter d and, as a result, to the measured quantities of flatness deviation and roughness.

The dimensionless conductance, $h_c\delta/k_m$, was then plotted as a function of P*T* for several values of the surface parameter, δ_o*. The experimental data indicated a correlation between h_c* and P*T* for fixed values of δ_o*. As a result of this correlation, the dimensionless conductance could be written as

$$h_c^* = f(P^*T^*, \delta_o^*) = \alpha/(1 - \alpha)$$

As the load pressure was reduced toward zero, the experimental data indicated that h_c approached a small positive value which appeared to be primarily a function of the surface parameter δ_o*. One expression which satisfies this observed condition is

$$h_c^* = (C_1\delta_o^* + C_2P^*T^*)^n \tag{5}$$

For the finite load pressures at which the present experimental data were obtained, the estimated heat transfer due to radiation at the junction was negligible. If the load pressure were reduced, however, the heat transfer by conduction and by radiation could become comparable. The functional form selected for h_c* in Eq. (5) includes possible radiation effects as the load pressure is reduced to zero.

Since a linear logarithmic relationship of h_c* as a function of P*T* was indicated by the experimental data for large values of P*T*, the C_1 term was neglected and the experimental data were plotted in the form

$$\log h_c^* = \log C_2 + n \log P^*T^*\Big]_{\text{large } P^*T^*}$$

An approximate value for n was selected from the slope and C_2 from the intercept. Then an average value of C_1 was selected from all of the low load pressure experimental data. Because these first values of n, C_1, and C_2 were approximate, the values were refined to predict more closely the experimental data of this investigation by varying these coefficients by small increments and calculating the rms error between the predicted and measured conductance values. The resulting coefficients are given in the following conductance equation:

$$h_c^* = \left[5.22 \times 10^{-6}\ \delta_o^* + 0.036\ P^*T^*\right]^{.56} \qquad (6)$$

or

$$h_c = \frac{k_m}{\delta_o}\ e^{\frac{170\ P^*T^*}{\delta_o^*}} \left[5.22 \times 10^{-6} \delta_o^* + 0.036\ P^*T^*\right]^{.56} \qquad (7)$$

A plot of the present experimental data[8] and published experimental data[6,7,12,18-23] are shown in Fig. 6 to illustrate the close approximation of Eq. (6) with experimental results of widely varying test conditions and contact materials. The curves are calculated for stainless steel test material with δ_o values of 18 (top curve), 200, 1000, and 4000 x 10^{-6} as representative surface finishes. At lower values of P*T*, the curves indicate that the dimensionless conductance is a strong function of surface finish, whereas at higher values of P*T*, the surface finish appears to have little effect. It should be pointed out that all of those data were obtained using one inch diameter test specimens.

The experimental data in Fig. 6 represent mean junction temperatures of -250° to 500°F, apparent contact or load pressures of 10 to 7000 psi, surface flatness deviations of 15 to 4500 μ, and surface roughnesses of 3 to 120 μ in. Data for aluminum, stainless steel, brass, and magnesium are included. It may be seen that for a majority of the data, extremely good correlation exists. Some scatter is to be expected since no two investigators measure or present their data in exactly the same manner. In cases in which all of the experimental characteristics were not listed by the investigator, material properties were selected from the literature in order to complete the data correlation.

Greater deviation occurs below the correlation curve, especially at the higher and lower load pressures. This scatter may be explained in several ways. At higher load pressures, those surfaces with oxide films or other contaminants would yield lower measured conductance values. Also, at high load pressures the interface temperature difference is small and in some instances can be of the same order as the uncertainty, thus permitting deviations of 100% or more. The larger uncertainty in measured conductance values at low load pressures may be due to variations in load application mechanisms used by different investigators. Other explanations may include the uncertainty in measured heat flux, the alignment of the test surfaces, uncertainty in the material properties, and incorrectly reported or assumed surface characteristics.

Results and Discussion

In order to establish the reliability of the expression developed in the previous section for the prediction of conductance values at specific test conditions, comparisons were made with experimental data both for the present investigation and for published data of other investigators.

Equation (7) predicts conductance values of the present investigation within a mean error of 24%. The experimental data for aluminum 2024 are shown in Fig. 1 for comparison with Eq. (7). This comparison indicates that the calculated values are within an average deviation of 20% from the experimental values. The variation of dimensionless conductance for stainless steel 304 is shown as a function of P^*T^* in Fig. 2. The agreement is within 15%, with a maximum deviation of 57% for two experimental test conditions. This good agreement could be a result of the relatively small change in material properties with temperature. Further, stainless steel surfaces are less likely to be oxidized than those of other materials.

Equation (7) is compared with brass conductance data in Fig. 3. These data are predicted within an overall average of 26%. The contact conductance data for magnesium AZ31B are given in Fig. 4 for comparison with calculated values. There appears to be quite poor agreement for the smoother surfaces; however, the agreement for the rougher surfaces is within 25%. The magnesium smooth and medium surfaces were tested even though a visible oxide film remained after thorough cleaning. The deviation in agreement of all the magnesium data with the calculated values averaged about 40%.

The usefulness of Eq. (7) for the prediction of thermal contact conductance at specific test conditions is illustrated in Fig. 7. These published data[6,12,18,22] were selected to represent a wide range of surface finishes and mean junction temperatures, such as might be anticipated in spacecraft structures or electronic equipment. Included are aluminum, magnesium, and stainless steel test materials with mean junction temperatures of −250° to 291°F. In each case the prediction equation approximates the data within a mean deviation of 24% or less. It would appear, then, that Eq. (7) is suitable for the prediction of thermal contact conductance for many materials, surface configurations, and test conditions.

Equation (7) may be used to calculate the thermal conductance of metallic surfaces in contact provided the following characteristics are known: the materials composing the contact, the surface roughness and flatness deviation of each surface composing the junction, the load pressure exerted on the junction, and the mean temperature at the junction. The material properties needed--coefficient of thermal expansion, modulus of elasticity, and thermal conductivity--are temperature dependent and may be obtained from engineering handbooks or material suppliers. These material properties should be selected at a mean temperature appropriate to the system being analyzed.

The first step in calculation of the conductance would be the determination of the surface parameter δ_0 from the surface roughness and flatness deviation. This may be done by use of Eq. (2) and the suggested relationship presented in Fig. 5. Next, the conditions of load pressure, mean junction temperature, and surface parameter should be nondimensionalized with the appropriate material properties and characteristic length. The conductance may then be calculated from Eq. (7).

Summary and Conclusions

Through this analysis, dimensionless parameters were found which would correlate published experimental data as well as data of the present investigation. These data included load pressures to 7000 psi, mean junction temperatures of -250° to 500°F, surface flatness deviations of 15 to 4500 μ in., and surface roughness of 3 to 120 μ in. From this correlation an expression for the prediction of contact conductance was developed in terms of known engineering measurements. This expression, developed by semi-empirical techniques, can be used to predict the contact conductance with an average deviation of 24% or less.

The prediction expression developed in this paper may be used to assist in the solution of thermal design problems involving heat exchange between surfaces in contact. In order to employ this expression to predict a conductance value for a specific application, it is necessary to know the following characteristics of the contact:

1) the metallic materials;
2) the finish of the surfaces;
3) the load pressure;
4) the mean junction temperature.

With knowledge of the material, the appropriate properties may be determined and used to complete calculation of the conductance.

References

[1]Hsieh, C.K. and Davis, F.E., "Bibliography on Thermal Contact Conductance," AFML-TR-69-24, March 1969, Purdue University Thermophysical Properties Research Center.

[2]Moore, C.J., Jr., Atkins, H. and Blum, H.A., "Subject Classification Bibliography for Thermal Contact Resistance Studies," ASME Paper 68-WA/HT-18, Dec. 1968.

[3]Rogers, C.F., "Heat Transfer at the Interface of Dissimilar Metals," International Journal of Heat Mass Transfer, Vol. 2, Nos. 1/2, March 1961, pp. 150-154.

[4]Powell, R.W., Tye, R.P., and Jolliffe, B.W., "Heat Transfer at the Interface of Dissimilar Materials: Evidence of Thermal Comparator Experiments," International Journal of Heat Mass Transfer, Vol. 5, Oct. 1962, pp. 897-902.

[5]Clausing, A.M., "Heat Transfer at the Interface of Dissimilar Metals - the Influence of Thermal Strain," International Journal of Heat Mass Transfer, Vol. 9, No. 8, Aug. 1966, pp. 791-801.

[6]Smuda, P.A., Fletcher, L.S. and Gyorog, D.A., "Heat Transfer Between Surfaces in Contact: The Effect of Low Conductance Interstitial Materials; Part I: Experimental Verification of NASA Test Apparatus," CR 73122, June 1967, NASA.

[7]Fletcher, L.S., Smuda, P.A. and Gyorog, D., "Thermal Contact Resistance of Selected Low Conductance Interstitial Materials," AIAA Journal, Vol. 7, No. 7, July 1969, pp. 1302-1309.

[8]Fletcher, L.S., "Thermal Contact Resistance of Metallic Interfaces: An Analytical and Experimental Study," Ph.D. Dissertation, 1968, Arizona State University, Tempe, Ariz.

[9]Fletcher, L.S., "An Experimental Investigation of Thermal Contact Conductance," N.C. Lind, Ed., Proceedings of the Second Canadian Congress of Applied Mechanics, Waterloo, Canada: University of Waterloo, 1969, pp. 427-428.

[10]Oberg, E. and Jones, F.D., Machinery's Handbook: A Reference Book for the Mechanical Engineer, Draftsman, Toolmaker, and Machinist, 17th Ed., The Industrial Press, New York, 1964.

[11]Fenech, H. and Rohsenow, W.M., "Prediction of Thermal Conductance of Metallic Surfaces in Contact," Journal of Heat Transfer, Vol. 85, No. 1, Feb. 1963, pp. 15-24.

[12]Clausing, A.M. and Chao, B.T., "Thermal Contact Resistance in a Vacuum Environment," Journal of Heat Transfer, Vol. 87, No. 2, May 1965, pp. 243-251.

[13]Henry, J.J. and Fenech, H., "The Use of Analog Computers for Determining Surface Parameters Required for Prediction of Thermal Contact Conductance," Journal of Heat Transfer, Vol. 86, No. 4, Nov. 1964, pp. 543-551.

[14]Cetinkale, T.N. and Fishenden, M., "Thermal Conductance of Metal Surfaces in Contact," General Discussions on Heat Transfer, Conference of the Institution of Mechanical Engineers and ASME, Institution of Mechanical Engineers, London, 1951, pp. 271-275.

[15]Graff, W.J., "Thermal Conductance Across Metal Joints," Machine Design, Vol. 32, No. 19, Sept. 1960, pp. 166-172.

[16]Thomas, T.R. and Probert, S.D., "Thermal Resistances of Some Multilayer Contact under Static Loads," International Journal of Heat Mass Transfer, Vol. 9, No. 8, Aug. 1966, pp. 739-754.

[17]Hudack, L.J., "An Engineering Analysis of Heat Transfer through Metallic Surfaces in Contact," unpublished Master's report, 1965, Arizona State University, Department of Mechanical Engineering, Tempe, Ariz.

[18]Bloom, M.F., "Thermal Contact Conductance in a Vacuum Environment," Rept. SM-47700, Dec. 1964, Douglas Aircraft Company.

[19]Cunnington, G.R., Jr., "Thermal Conductance of Filled Aluminum and Magnesium Joints in a Vacuum Environment," ASME Paper 64-WA/HT-40, Nov. 1964.

[20]Fried, E., "Study of Interface Thermal Contact Conductance, Summary Report," Document 64SD652, May 1964, General Electric Company.

[21]Fried, E., "Study of Interface Thermal Contact Conductance, Summary Report," Document 65SD4395, June 1965, General Electric Company.

[22]Fried, E., "Study of Interface Thermal Contact Conductance, Summary Report," Document 66SD4471, July 1966, General Electric Company.

[23]Hargadon, J.M., Jr., "Thermal Interface Conductance of Thermoelectric Generator Hardware," ASME Paper 66-WA/NE-2, Nov. 1966.

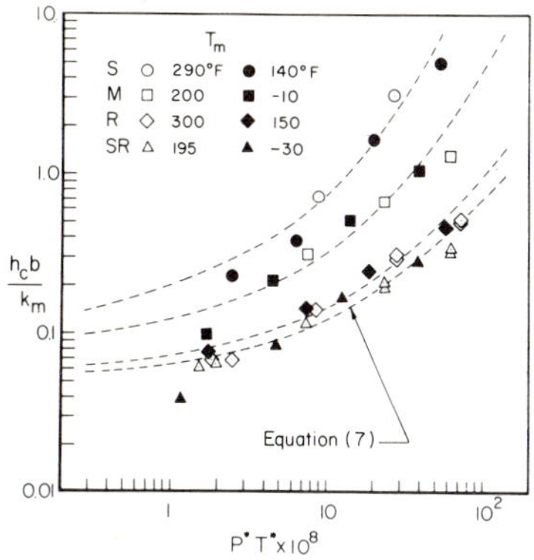

Fig. 1
Variation of conductance with pressure and temperature for aluminum 2024.

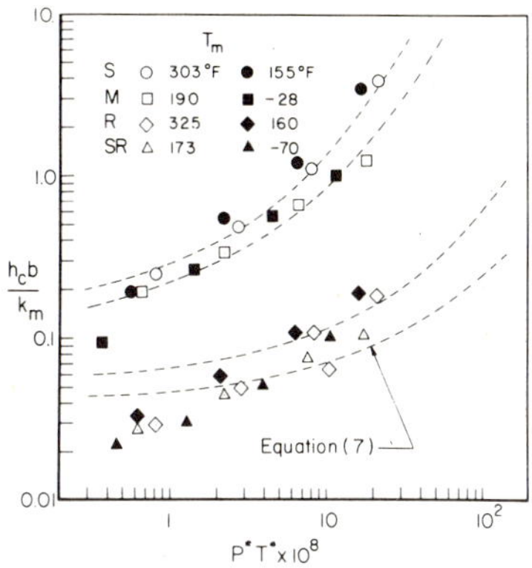

Fig. 2
Variation of conductance with pressure and temperature for stainless steel 304.

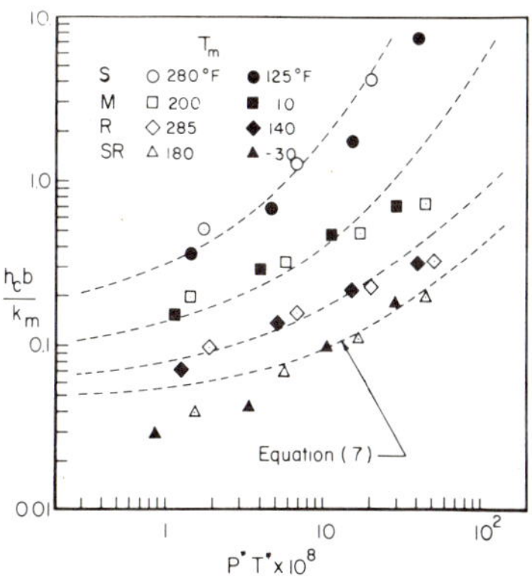

Fig. 3
Variation of conductance with pressure and temperature for brass alloy 271.

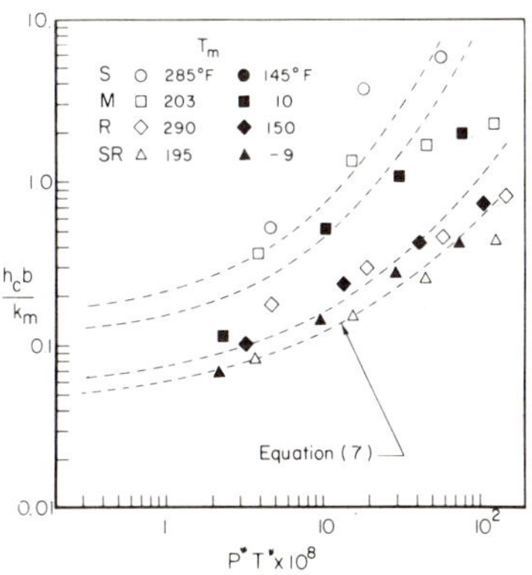

Fig. 4
Variation of conductance with pressure and temperature for magnesium AZ31B.

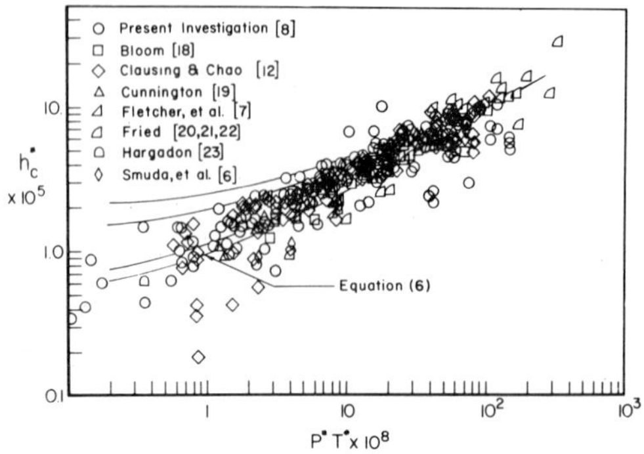

Fig. 5 Variation of surface parameter δ_0 with contact surface parameter d.

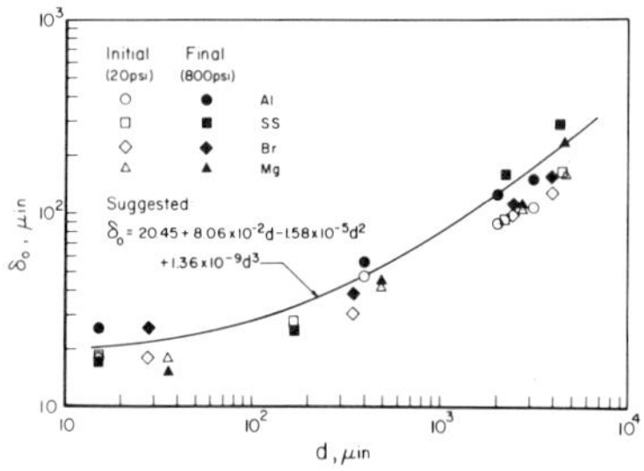

Fig. 6 Variation of thermal conductance number with dimensionless pressure and temperature for all data.

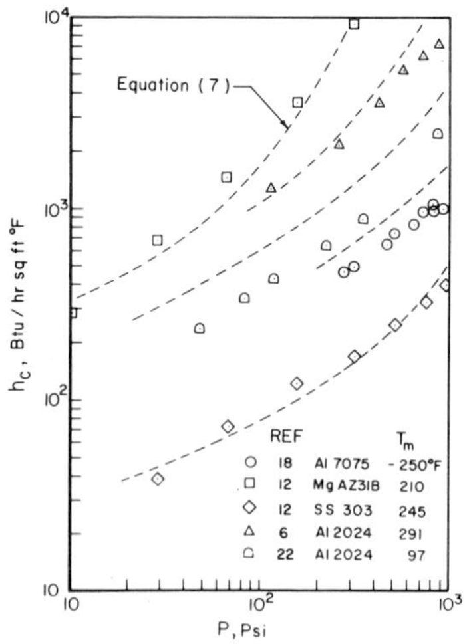

Fig. 7 Variation of contact conductance with apparent interface pressure for selected published data — comparison with the present analysis.

EXPERIMENTAL CONFIRMATION OF CYCLIC THERMAL JOINT CONDUCTANCE

Daniel J. McKinzie Jr. *

NASA Lewis Research Center, Cleveland, Ohio

Abstract

Experimental heat-transfer data obtained from a tungsten-tungsten and an Armco-Armco Iron specimen have confirmed the Bowden and Tabor model of elasto-plastic events which occur during the cyclic engagement of surfaces in the lightly loaded range (4.45×10^5 to 3.45×10^6 N/m^2 or 64.5 to 500 psi). The specimens consisted of cylinders having a diameter of 2.54 cm. Their surfaces of contact were outgassed, and tests were made in a high vacuum environment of approximately 10^{-6} torr. The contact surfaces of the tungsten specimen were nominally flat. Those of the Armco iron specimen were approximated by spherical caps. The surfaces of both specimens were surveyed in roughness profile height, waviness, and total profile before the tests were performed. A recently published simplified approximate theory for calculating the coefficient of thermal contact conductance of plastically deforming nominally flat surfaces was applied to the tungsten specimen and was found to agree well with the data. This theory does not require the use of a computer program or the calculation of any abstract quantities. It does, however, require the use of a surface analyzer and microhardness tester. Clausing and Chao's macroscopic elastic theory of deformation was applied to the Armco Iron specimen, and the agreement was found to be good.

Presented as Paper 70-853 at the AIAA 5th Thermophysics Conference, Los Angeles, Calif., June 29 - July 1, 1970.

*Aerospace Engineer, Electric Propulsion Division.

Nomenclature

A = total cross-sectional area of cylindrical model, or apparent area of contact, m^2
A_c = summation of individual areas of button contacts (results in total area of actual contact), m^2
A_c' = contact area of one asperity, m^2
b = radius of specimen, m
h = coefficient of thermal contact conductance, $kW/(m^2)(K)$
I = microhardness, N/m^2
k = coefficient of thermal conductivity, $W/(m)(K)$
L = length of specimen's upper or lower portion, m
P = total load applied to specimen divided by entire cross-sectional area (apparent contact pressure), N/m^2
$\dot{Q}$ = rate of heat transfer, W
t = temperature, K
Δt = extrapolated temperature difference across the interface of contact, K
W' = load applied to asperity in contact, N
W_0' = upper value of asperity loading in plastic range of softer metal, N
x = length from interface of button contacts along axis of cylinder, m
x_c = constriction ratio or square root of the area ratio of contact, a/b
δ = effective zone of the thermal disturbance or average of the maximum roughness profile heights obtained from the total profile traces of a surface, m

Subscripts

c = plane of button contact
i = interface
0 = condition of maximum loading
1, 2 = stations

Introduction

The thermal control of the structure and instrumentation of a spacecraft can be seriously affected by the use of many lightly loaded fasteners.[1] Thus the prediction of heating rates across the contacting surfaces of such fasteners is important.

In 1950 Bowden and Tabor[2] described the mechanism of metallic contact between surface asperities without heat transfer effects. They considered the loading, work hardening, and unloading of such asperities which deform according to elastic and plastic laws.

In the sixties Fried,[3] Clausing and Chao,[4] and others presented data demonstrating the variation of the coefficient of thermal contact conductance with pressure in a vacuum environment for one or several complete loading cycles, but they did not demonstrate theoretical agreement for a complete cycle. Mikic and Rohsenow,[5] and Mikic, Yovanovich, and Rohsenow[6] have presented approximate solutions to the conductance problem; however, their analyses require the use of computerized programs which some investigators have found to require greater effort and cost than simply obtaining test data directly from sample specimens.

The purposes of the present work are: 1) to describe the Bowden and Tabor model of metallic asperity contact; 2) to present data from a tungsten-tungsten and an Armco iron-Armco iron cylindrical specimen which show the cyclic variation of the coefficient of thermal contact conductance with the apparent joint contact pressure in the lightly loaded range (4.45×10^5 to 3.45×10^6 N/m^2 or 64.5 to 500 psi); 3) to present several surface analyzer traces of the Armco iron specimen's surfaces of contact which were extensively surveyed to determine the variation of their surface roughness profile heights, waviness, total profile, and arithmetical average heights before the tests were conducted; 4) to compare the cyclic test data with the mechanism of the contact phenomena described in the Bowden and Tabor model; 5) to compare the specimens' test data where applicable with the macroscopic elastic deformation theory of Clausing and Chao,[4] and Clausing,[7] and the recently published simplified plastic deformation theory of Ref. 8.

Theoretical Considerations

Bowden and Tabor Description of the Contact Phenomena

Bowden and Tabor in Ref. 2 describe in detail the mechanical aspects of the loading and unloading of a single work-hardened surface asperity through which no heat transfer takes place. The tip of the asperity is assumed to be smooth and spherical in shape. Further, it is assumed to rest on a softer metal, and, in the region of contact, the surface of the softer metal is considered to be a plane.

Figure 1 shows what happens as the load on the asperity is increased. Initially, elastic deformation of the area of contact occurs; this is predicted by Hertz's classical equations. The planar projection of the area of contact A'_c varies with the load raised to the two thirds power, $(W')^{2/3}$.

At some larger load the onset of plastic deformation takes place, followed by a region of full plasticity where the area of contact varies directly with the load. If the asperity is then unloaded from some higher load W'_0 and area of contact $A'_{c,0}$, the variation of contact area with load is indicated by the dashed curve labeled reversible elastic deformation. If the load is removed, the surfaces separate according to the laws of elastic deformation.

Clausing and Chao Macroscopic Theory

In Ref. 4 Clausing and Chao present their macroscopic theory of constrictive resistance in a vacuum environment. The constrictive resistance is defined as the reciprocal of the product of the apparent contact area A and the coefficient of thermal contact conductance h. The apparent contact area is equal to the total cross-sectional area of their cylindrical model. The theory considers the contact region between two cylinders having identical radii placed end to end. A single large, macroscopic, circular contact area is assumed to be concentric with that of the apparent contact area. Thus the noncontact region is annular. Microscopic contact areas are located inside the macroscopic contact; they are assumed circular, of the same radius, and uniformly distributed over the single macroscopic spot. Because of the annular noncontact region, the surfaces cannot be characterized as being nominally flat. Thus, the Clausing and Chao macroscopic theory can be used to predict the coefficient of thermal contact conductance between wavy surfaces with the spherical contour representing one-half wave length. The form of the solution taken from Ref. 7 and used in this paper is

$$hb/k = 1/2\left[10^{f(x_c)}\right] \qquad (1)$$

where the limits of its application are

$$\left\{\begin{matrix} L/b > 0.8 \\ 0.16 < x_c < 0.84 \end{matrix}\right\}$$

L is defined as the length of the specimen's upper or lower piece, b is its radius, and x_c is the constriction ratio or equivalently the square root of the area ratio of contact between two spherical surfaces.

Simplified Method of Reference 8

Reference 8 presents a simplified approximate relation for calculating the coefficient of thermal contact conductance between two rough, nominally flat, outgassed surfaces in contact in a high vacuum. The surface asperities are assumed to be undergoing plastic deformation. The analysis is based on the steady-state Fourier one-dimensional heat conduction law. It was applied to the so-called "button" model of contact (described in Ref. 8).

Figure 2 (modified from Ref. 8) is a view of the assumed contact interface showing a contact region labeled the effective zone of the thermal disturbance. The length of this zone (defined later in this paper) is assumed to be a function of the sum of the largest asperity heights from each of the surfaces. This length is used as the best measurable dimension approximating the physical length of the thermal discontinuity. Because it only approximates the actual length over which the thermal discontinuity at the interface takes place, it is referred to as the effective zone of the thermal disturbance.

The assumptions that apply to the theoretical model and approximate the test specimen are presented in Ref. 8.

When the assumptions are applied to the model, the axial temperature distribution in the vicinity of the interface is as shown in Fig. 3. The basic steady-state one-dimensional Fourier heat conduction law written for the effective zone of the thermal disturbance δ (Fig. 3) is given by

$$\dot{Q}/A_c = -k(dt/dx) \qquad (2)$$

where A_c is the sum of the individual cross-sectional areas of the idealized "button" contacts, k is the coefficient of thermal conductivity, and dt/dx is the axial temperature gradient across the disturbance zone δ which is assumed to be very small.

If Eq. (2) is applied to the effective zone of the thermal disturbance, the heat-transfer rate per unit area at the interface of contact may be approximated by

$$\dot{Q}/A_c = -[k(t_2 - t_1)/\delta] \tag{3}$$

The coefficient of thermal contact conductance h is defined in the equation

$$\dot{Q}/A \equiv -h\,\Delta t \tag{4}$$

where A is the cross-sectional area of the specimen. If the temperature gradients in the specimen outside the effective zone of the thermal disturbance are linearly extrapolated to the interface $x = 0$, as shown in Fig. 3, then Δt at the interface is given by

$$\Delta t = t_{i,2} - t_{i,1} \tag{5}$$

Substitution of Eq. (5) into Eq. (4) gives

$$\dot{Q}/A = -h(t_{i,2} - t_{i,1}) \tag{6}$$

Equate the heat-transfer rate $\dot{Q}$ across the interface, Eqs. (3) and (6), to obtain

$$kA_c[(t_2 - t_1)/\delta] = Ah(t_{i,2} - t_{i,1}) \tag{7}$$

Since δ is assumed to be very small, it is reasonable to assume that

$$t_2 - t_1 \cong t_{i,2} - t_{i,1} \tag{8}$$

Equation (7) thus reduces to

$$h\delta/k = A_c/A \tag{9}$$

From plastic theory for nominally flat surfaces in contact,[9]

$$A_c/A = P/I \tag{10}$$

where P is the pressure acting over the apparent contact area A, and I is the average of the random microhardness values of the softer of the two interface surfaces. Substituting Eq. (10) into Eq. (9) results in the desired equation

$$h(\delta/k) = P/I \tag{11}$$

Unloading the Contacting Surfaces

Reference 8 presents a derivation for the variation of the coefficient of thermal contact conductance during the unloading of the contacting surfaces. The surfaces are assumed to be unloaded from some upper plastically deforming pressure P_0. The following two equations present forms of the results :

$$h_1 = (P_1/P_0)^{2/3} h_0 \tag{12}$$

or substituting Eq. (11) into Eq. (12) yields

$$h_1 = P_1^{2/3} P_0^{1/3} (k/\delta I_0) \tag{13}$$

Thus Eqs. (12) and (13) provide a method for calculating the coefficient of thermal contact conductance for any point below the upper limit of P_0 and h_0. This is true if the mean interface temperature of the contacting surfaces and heights of the plastically deformed asperities remain constant or very nearly constant during the unloading process. It does not apply to flat surfaces or those surfaces which plastically deform under load to flat surfaces.

Experimental Apparatus and Procedure

Tungsten Specimen Test Apparatus

The tungsten specimen was tested in the facility shown in Fig. 4. The test was performed in a bell jar containing an induction heating coil and a column composed of the test specimen, a OFHC copper heat meter, and water-cooled heat sink. A pneumatic loading cylinder was located on top of the bell jar and was used to vary the pressure at the contact interface of the test specimen. Electric currents were induced in the extreme upper portion of the test specimen. The bell jar operates in the 10^{-6} torr pressure range.

Heating coil. Because an induction coil was used to heat the specimen, a discussion of its effect on the specimen's data is needed. Although the magnetic field produced by the coil had a maximum value during the entire test of approximately 3.06×10^{-5} tesla at the center of the coil's axis, its effect at the specimen's interface of contact was negligible since tungsten is nonferrous and has a magnetic permeability of approximately 1.000068 times that of free space. The effect of the magnetic field on the thermocouple data was also negligible because the leads from each of the thermocouples were twisted for some distance from the coil and specimen and had no loops in them.

The use of the magnetic induction coil created the possibility of electric current flowing through a circuit composed of the test column and the bell jar. The formation of this circuit was prevented by isolating the heat sink and its water cooling lines from the base plate and instrument ring of the bell jar.

Specimen. The tungsten test specimen was metallurgically prepared from high-purity pressed and sintered powder. The cylindrical specimen was 2.54 cm in diameter, its upper portion was 25.4 cm long so that it passed completely through the induction coil, and its lower portion was 2.54 cm long.

Surface finish. The specimen's surface lay was produced by grinding. The average value of the maximum roughness profile height obtained from total profile traces across the lay of the upper and lower surfaces was 5.35 and 5.51 μm, respectively. The definition of this length and its measurement is included later in this section. The arithmetical average roughness height of the upper and lower surfaces was 0.538 and 0.943 μm, respectively. This length is defined in Ref. 10. The test was performed with the lay of the top and bottom surfaces approximately parallel. It was found that the maximum waviness height of the surfaces was less than the average value of their maximum roughness profile heights. Thus the tungsten specimen's surface could be considered nominally flat.

Thermometry. Type 30 gage, special error limits, chromel-alumel thermocouple wire twisted at its ends was selected. These thermocouples have an accuracy of ±1.1 K as guaranteed by the manufacturer in the temperature range of their use here. They were peened into holes 0.32 cm deep in the surface of the cylindrical specimen. Six thermocouples were located on the upper and five on the lower portion of the specimen. Also, three sets of two approximately 180° apart were located on the OFHC copper heat meter.

Armco Iron Specimen Test Apparatus

The Armco iron specimen was tested in the same facility as that used in the tungsten test with the following exceptions. The magnetic induction coil was removed and replaced with a tungsten resistance heated cylindrical source. The upper portion of the specimen was appropriately made shorter to allow for the heater in the test column. The OFHC heat meter was removed and replaced with an elongated lower portion of the specimen which was instrumented and calibrated to replace the heat meter. With these revisions and the addition of a supporting structure for the new heat source, the flow of heat remained the same as for the tungsten specimen.

Specimen. The Armco iron cylindrical specimen was 2.54 cm in diameter. Its upper portion was 6.97 cm long and its lower portion was 12.7 cm long.

Surface finish. The specimen's contacting surfaces were produced by lapping and had no lay. Figure 5 shows several typical traces of the Armco iron specimen's top surface of contact. The average values of the maximum roughness profile heights of the upper and lower surfaces of contact were 0.64×10^{-6} and 0.97×10^{-6} m, respectively; their arithmetical average roughness heights were 0.038×10^{-6} m and 0.102×10^{-6} m, respectively. Each surface was found to approximate a section of a spherical cap having a one-half wave length. The average crest to trough wave height was approximately 1.52×10^{-6} m for the upper surface and 1.015×10^{-6} m for the lower surface. Thus the surfaces could not be considered as nominally flat, since their wave height was greater than the average value of their maximum roughness height.

Thermometry. The same type of thermocouple wire was used to instrument this specimen as the tungsten specimen; however, the wire was butt welded at its ends. These thermocouples were calibrated in the specimen after the test was completed and were found to be within the manufacturers' tolerance of ±1.1 K for the range of temperatures encountered. They were peened into slots milled into the specimen's circumference. The slots were approximately 0.076 cm long, 0.033 cm wide, and approximately 0.051 cm deep. The thermocouple's insulated lead wires were wrapped around the circumference of the specimen to minimize conduction errors due to the wire. Four couples were located in the upper portion of the specimen and six were located in the specimen's lower portion.

Experimental Determination of h

Tungsten specimen test procedure. The contact conductance of the tungsten specimen was experimentally determined by using Eq. (6) and the test apparatus described above. The rate of heat transferred was determined with the aid of an OFHC heat meter. The test specimen was designed so that a one-dimensional linear heat-transfer analysis would apply. As pointed out in Ref. 8, several power interruptions occurred during the test. It was found, however, that their effects could be eliminated without introducing an error. Thus the data presented here are correct and usable.

The average time needed to establish steady-state conditions after a change was made in pressure was found to be approximately 2 1/2 hr. After a change in interface temperature was made, steady-state conditions were established in a minimum of 3 hr, with the maximum being 45 hr.

The interface surfaces of contact were cleaned with acetone before installation in the test facility.

Armco iron specimen test procedure. The contact conductance of the Armco iron specimen was experimentally determined by using Eq. (6). The rate of heat transfer across the interface of contact was determined by calibrating the Armco iron test specimen against a 99.993 % pure aluminum (by assay) heat meter. The contact conductance tests were performed by varying the interface contact pressure in a monotonic increasing then decreasing manner through two complete cycles while maintaining the mean interface temperature of the test specimen nearly constant.

To gain an idea of the time required to establish equilibrium between changes in pressure and/or temperature, the test specimen was exposed to steady-state test conditions for periods ranging from 50 to 180 hr between each point. Figure 6 shows a typical variation with time of the conductance after a change was made in the apparent contact pressure. It will be noted that the test specimen required approximately 50 hr to reach a stable condition.

Before placing the test specimen in the test column of the bell jar it was cleaned with acetone and baked in a separate vacuum facility at 10^{-6} torr, through eight 20-min heating and cooling cycles. Each cycle varied from room temperature to approximately 450 K, well below the recrystallization temperature of Armco iron.

Data reduction for both specimens. The experimental interface temperature difference Δt [Eq. (4)] was determined by the linear extrapolation of the thermocouple data to the interface of contact (Fig. 3). A simplification in this procedure was made by determining the thermal gradient in each portion of a specimen from two thermocouple readings in the same manner as that used by Fried.[3] The gradients determined by this technique were continually checked by plots including all of the thermocouple data from each specimen. The error in the temperature difference across the interface introduced by this method was less than 1.1 K from a best line fit (drawn) through all the thermocouple data from each specimen. The coefficient of thermal conductivity, k, was evaluated at a temperature equal to $(t_{i,1} + t_{i,2})/2$.

Measurements of Parameters for Theoretical Determination of h

The theoretical value of h was determined from values of δ, k, and I together with Eq. (11). The quantities δ and I were evaluated from the test specimens. The thermal conductivity, k, of tungsten was estimated from the literature as noted in Ref. 8. The variation of k with temperature for the Armco iron specimen was determined from calibration data obtained in the test facility under vacuum conditions from a 99.993 % (by assay) pure aluminum heat meter. The results of this calibration indicate agreement with data from the literature[11] within 1 1/2%.

Measurement of δ. A Brush surfanalyzer was used to determine the magnitude of δ. It consists of a control unit, a stylus drive unit, a 50-mg stylus probe, and a recorder unit. This apparatus records the surface roughness profile (width and height), total profile, waviness, and arithmetical average or roughness height. All of these quantities are defined in Ref. 10.

During the approach of two rough, nominally flat surfaces the maximum roughness height would make contact first. Therefore, it is appropriate to determine the average value of the maximum roughness profile heights from the total profile traces δ_1 (upper) and δ_2 (lower) of the test specimen before the test is begun. Since it is not possible to obtain surface analyzer traces which include all the points of contact on a surface, a representative number of total profile traces must be taken. From each of these total profile traces, a single value of the maximum roughness profile height is obtained.

These values are then averaged to obtain the value of the maximum roughness profile height used in the calculations. The value of δ, where $\delta = \delta_1 + \delta_2$, is referred to as the effective value of the thermal disturbance. The values of δ_1 and δ_2 determined for the tungsten specimen were obtained by averaging the maximum roughness profile heights obtained from six total profile traces that were made at different positions on each face of the specimen perpendicular to its ground lay. The values determined for the Armco iron specimen were obtained by averaging the maximum roughness profile heights obtained from ten total profile traces, five taken perpendicular to the other five, on each face of the specimen.

Measurement of I. As mentioned in Ref. 8, several authors have found that the microhardness of their specimens typically increase as the indenter load decreases. The tungsten specimen microhardness reported here also indicated an increase in surface microhardness as the indenter load was decreased. The data were obtained with a Knoop indenter at room temperature and are shown in Fig. 7. The increase in the surface microhardness is attributed to surface work hardening caused by the method of preparation. The value of the microhardness I, to be used in Eq. (11) for the tungsten specimen, was based on data from Fig. 7 in the light load range.[8] These data are from the softer of the two contact surfaces.

Discussion of Results

The discussion is separated into first the presentation of the tungsten specimen's results, and then those of the Armco iron specimen.

Tungsten Specimen

Figure 8 presents the Tungsten data from Ref. 8. During the loading portion of the cycle the data were obtained by loading the specimen's interface of contact from zero to 2.65×10^6 N/m^2 so that the surface asperities deformed plastically at a constant interface temperature of 390 K (solid square symbols in Fig. 8). The interface was then loaded to 5.30×10^6 N/m^2 while its temperature was kept constant. At 5.30×10^6 N/m^2 the interface temperature was raised in three increments from 390 K to a high of 666 K (solid triangles). Finally the pressure was lowered in two steps to a low value of 4.82×10^5 N/m^2 (open symbols). At each step three changes in interface temperature were made.

The dashed curve and points labeled plastic theory were calculated from Eq. (11). A second dashed curve labeled elastic theory indicates the theoretical variation of the contact conductance with apparent contact pressure during the disengagement of two nominally flat surfaces. If the interface temperature remains constant, the theoretical prediction of this variation is dependent only on the highest apparent contact pressure that the surfaces experience. Thus agreement is indicated if the data display the same slope as that of the theoretical variation. The symbols containing numbers in Fig. 8 indicate the sequence in which the data were obtained. They also separate the data into distinct groups which, as mentioned in Ref. 8, were bounded by power failures. Because of the likelihood that microscopic lateral asperity misalignment took place, due to large interface differential temperature effects, it is assumed that the data for each of these groups are independent of the others. The data represented by the solid symbols (plastic data) were obtained during the loading portion of the test while the open symbols indicate elastic data obtained during the unloading portion. The curves of Fig. 8 result by joining like temperature data in each of the subgroups. Separating the data in this manner has resulted in curves having slopes that agree well with plastic or elastic theory. The curve and points labeled as plastic theory show the effects of temperature. This effect appears in the theoretical representation of the coefficient of thermal contact conductance, Eq. (11), through the coefficient of thermal conductivity k and the surface microhardness I. The variation with temperature of I was obtained by the new method described in Ref. 8 and was verified by experimental data from the literature. The experimental data (solid symbols) obtained during the first half of the test cycle, shown in Fig. 8 and labeled $t_i = 390$ K, are approximately 20% lower than predicted from plastic theory and calculated from Eq. (11). This agreement is considered good.

Armco Iron Specimen

Figures 9a–9d present data for two loading and unloading cycles obtained from the Armco iron specimen at a constant mean interface temperature. The solid reference curves labeled elastic and plastic variation are shown for comparison with the data. The functional relationships for these curves come from elastic theory Eq. (9) and the Bowden and Tabor model.[2] The data shown in Fig. 9a indicate that the specimen's interface surfaces of contact did not plastically deform during the loading portion of the first cycle. Thus an elastic cyclic variation would be expected to manifest itself during the remainder of the test and, indeed, it has, as shown in Figs. 9b-9d. In Fig. 9a the dashed curve labeled theory of Clausing was determined by application of the method of Ref. 7 and was calculated from Eq. (1). The experimental data are approximately 40% greater than predicted by the theory of Clausing. This agreement is considered good. As noted in comparing Figs. 9b-9d, the data display a distinct reversibility.

Comparison of Experimental Results with the Bowden and Tabor Model

The Bowden and Tabor model of the contact mechanism[2] describes an ideal contact composed of two surface irregularities that are similar to those making contact on real surfaces. It differs from a real surface in the number of contacting asperities. Thus a comparison of the sequence of events that take place during the cyclic contact of two real surfaces and those predicted by the Bowden and Tabor model can be helpful to an understanding of the contact phenomena. Fig. 8 shows that during the initial loading phase the tungsten conductance data vary according to the plastic deformation theory of Ref. 8 as calculated from Eq. (11). During the unloading portion of the cycle the data vary according to the elastic deformation theory as calculated from Eq. (12). Thus the tungsten data indicate good agreement with the Bowden and Tabor model. The Armco iron data of Fig. 9a show that during the initial loading phase the conductance varies according to the elastic deformation theory of Ref. 7 as calculated from Eq. (1). Figs. 9b-9d indicate reversibility which confirms the existence of elastic deformation. Therefore, the Armco iron conductance data also show good agreement with the Bowden and Tabor model.

The data obtained from these two specimens not only confirm that the Bowden and Tabor model describes the contact phenomena, but they point up the necessity of varying test conditions in a specific manner so that the elasto-plastic history of events that occur during the engagement and disengagement of the surfaces may be observed. A lack of consideration of the distinction between elastic and plastic deformation phenomena and the resulting improper scheduling of an experimental test sequence has led to confusion in earlier experimental results.

Hysteresis Effects

Data presented in Ref. 3 were gathered during the initial loading, unloading,and cyclic reloading of contact surfaces. These data show what experimenters frequently refer to as hysteresis effects. The data obtained herein indicate that these effects may be caused by the unintentional use of nonsteady-state data. Specifically, most experimentalists attempt to establish steady-state conditions before they take their data. This is generally done by observing the emf output of the specimen's thermocouples at random points in time. Once these outputs appear to be steady over a prescribed interval of time, the data are taken. The experimental data presented here are the result of recording data at a prescribed test condition for times longer by an order of magnitude than previously done by many investigators. From a plot of these data vs time, an example of which is shown in Fig. 6, a steady-state average value of the coefficient of thermal contact conductance was determined for each prescribed testing condition. The resulting data for the Armco iron specimen (Figs. 9a-9d) indicate that no significant hysteresis effects took place. Rather small shifts in the data occurred which were caused by the inexact matching of test conditions from one point to the next.

Summary of Results

A recently published simplified approximate theoretical method for predicting the coefficient of thermal contact conductance[8] has been applied to a tungsten-tungsten specimen. The specimen's interface of contact consisted of two rough, nominally flat, outgassed surfaces in contact and under load in a high vacuum environment. Good agreement was obtained between the theory and test data. The method of calculation does not require the use of a computer for the determination of any abstract quantities; it does, however, require the use of a surface analyzer and microhardness tester. The surface analyzer provides profile traces that are used only to confirm that the specimen's surfaces of contact are nominally flat and to determine the average value of their maximum roughness profile heights. Thus a major simplification has been made in the computation.

Good agreement was obtained between the macroscopic elastic deformation analysis of Clausing,[7] and the contact conductance data vs apparent contact pressure obtained from two smooth nonflat, outgassed Armco iron surfaces placed in contact and under load in a high vacuum environment. The specimen was run through two complete loading cycles,and the test data showed reversible effects which were anticipated from theoretical considerations.

Cyclic test data presented in this paper show that the coefficient of thermal contact conductance of two distinct and dissimilar specimens, tungsten and Armco iron, vs apparent contact pressure in the light load range (4.5×10^5 to 5.7×10^6 N/m^2) vary according to the sequence of elastic and plastic deformation phenomena described by the Bowden and Tabor model of asperity contact. These results point up the necessity of varying test conditions in a special way so that the elasto-plastic history of events that occur during the engagement and disengagement of the surfaces may be observed.

A steady-state testing procedure used here exposed a specimen to long-time testing conditions which have shown that the hysteresis effects referred to in the literature did not occur.

References

[1]Fletcher, L. S., Smuda, P. A., and Gyorog, D. A., "Thermal Contact Resistance of Selected Low-Conductance Interstitial Materials," AIAA Journal, Vol. 7, No. 7, July 1969, pp. 1302-1309.

[2]Bowden, F. P. and Tabor, D., The Friction and Lubrication of Solids, Clarendon Press, Oxford, 1950, pp. 5-24.

[3]Fried, E., "Study of Interface Thermal Contact Conductance," R66SD4471, NASA CR-82778, July 1966, General Electric Co., Philadelphia, Pa.

[4]Clausing, A. M. and Chao, B. T., "Thermal Contact Resistance in a Vacuum Environment," ME-TN-242-1, 1963, University of Illinois Engineering Experiment Station, Urbana, Ill.

[5]Mikic, M. M. and Rohsenow, W. M., "Thermal Contact Resistance," Tech. Rept. No. 4542-41, Sept. 1966, Dept. of Mech. Engr., Massachusetts Institute of Technology, Cambridge, Mass.

[6]Mikic, M. M., Yovanovich, M. M. and Rohsenow, W. M., "The Effect of Surface Roughness and Waviness upon the Overall Thermal Contact Resistance," Tech. Rept. No. 76361-43, Oct. 1966, Dept. of Mech. Engr., Massachusetts Institute of Technology, Cambridge, Mass.

[7] Clausing, A. M., "Heat Transfer at the Interface of Dissimilar Metals - The Influence of Thermal Strain," International Journal of Heat and Mass Transfer, Vol. 9, No. 8, Aug. 1966, pp. 791-801.

[8] McKinzie, D. J., Jr., "Simplified Method for Calculating Thermal Conductance of Rough, Nominally Flat Surfaces in High Vacuum," TN D-5627, 1970, NASA, Cleveland, Ohio.

[9] Fenech, H. and Rohsenow, W. M., "Prediction of Thermal Conductance of Metallic Surfaces in Contact," Journal of Heat Transfer, Vol. 85, No. 1, Feb. 1963, pp. 15-24.

[10] "Surface Texture, Surface Measurement," American Standard ASA B 46. 1-1962.

[11] Thermophysical Properties Research Center Data Book, Vol. 1 - Metallic Elements and Their Alloys, 1960, Purdue Univ., Lafayette, Ind., Fig. 1011, Curves 2 and 12.

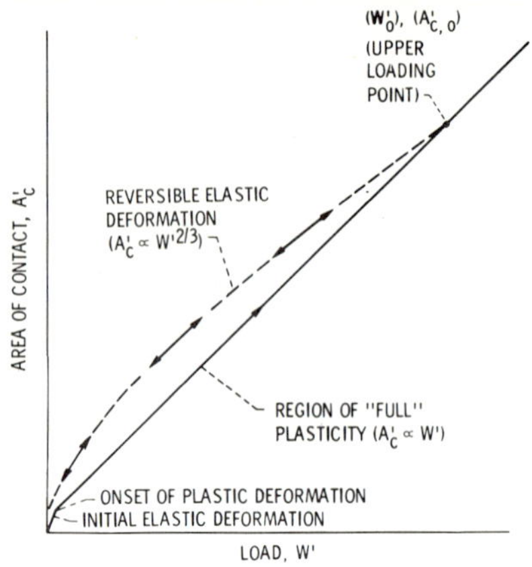

Fig. 1 Area of contact as function of load for increasing and decreasing load application to a single asperity indenting a flat plate. [2]

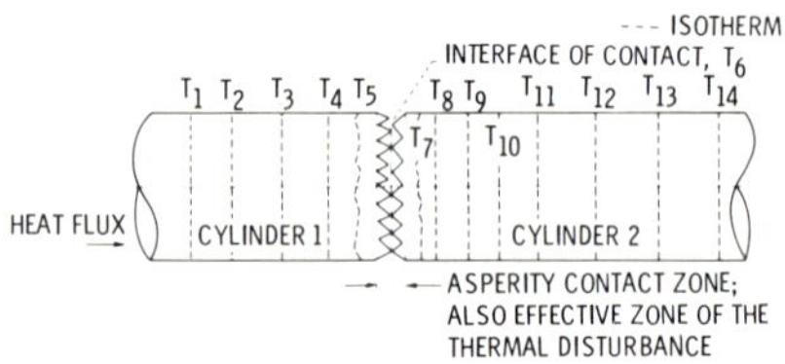

Fig. 2 View of nominally flat contact interface showing the effective zone of the thermal disturbance. $T_1 > T_{14}$.

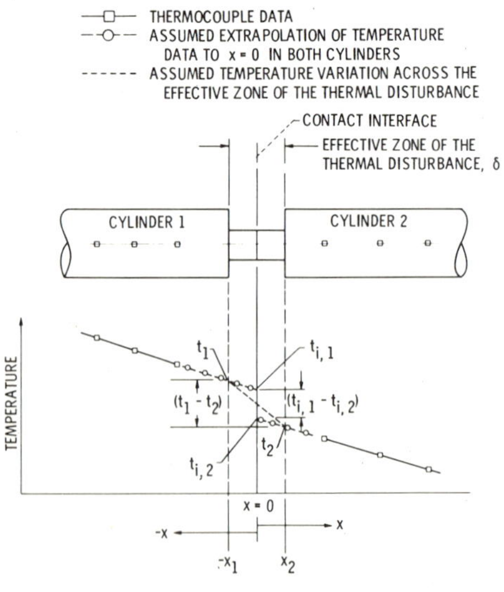

Fig. 3 Temperature profile along specimen.

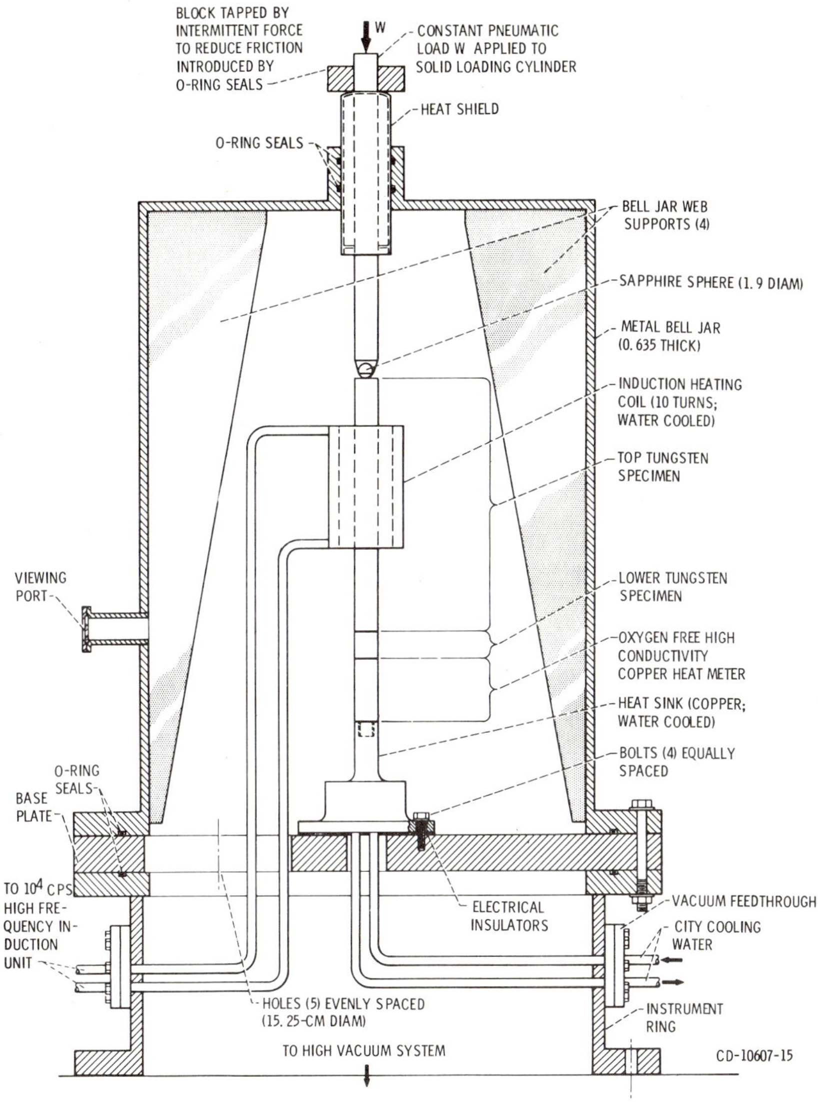

Fig. 4 Internal view of bell jar (all dimensions are in cm).

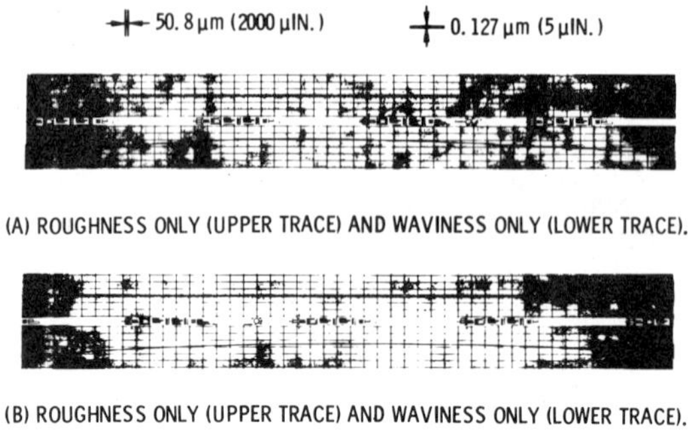

Fig. 5 Brush Surfanalyzer traces of the Armco iron specimen's top surface of contact. The tracks of the traces (a) and (b) are perpendicular to each other along diameters of the specimen's circular area of contact. Shown are typical traces of the roughness only and waviness only.

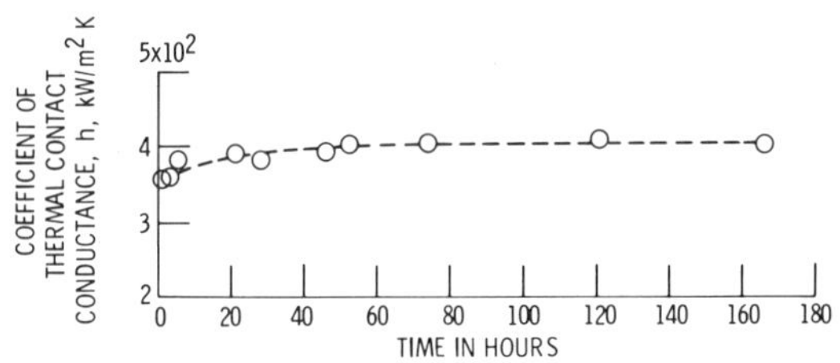

Fig. 6 Variation of thermal conductance with time for the Armco iron specimen after a change was made in pressure from 8.0×10^5 to 1.07×10^6 N/m^2 at an interface temperature of 367 K.

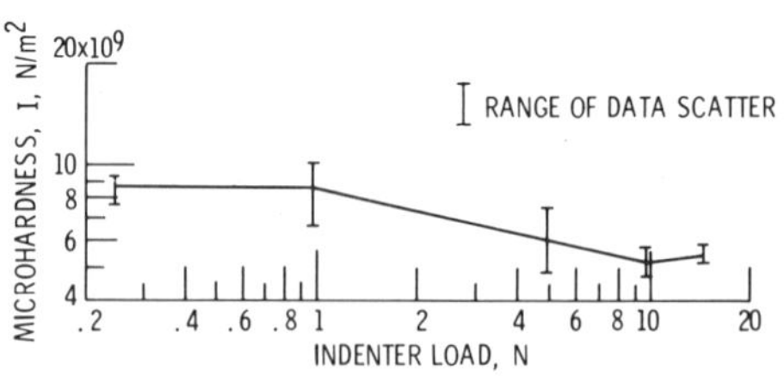

Fig. 7 Microhardness of interface surface of tungsten specimen as function of indentor loadings at approximately 294 K.

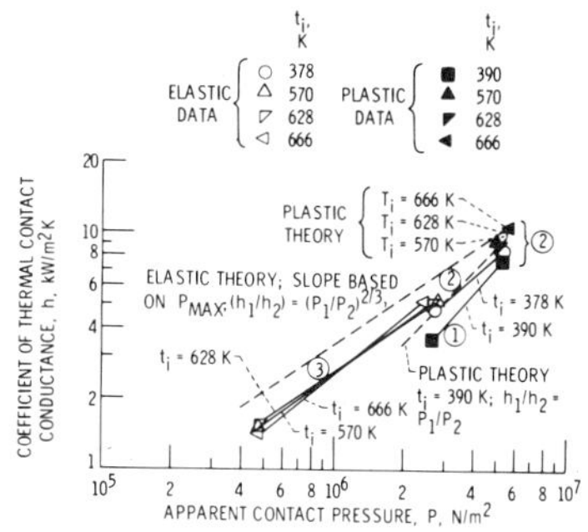

Fig. 8 Selected steady-state tungsten-tungsten data from Ref. 8. Lays of surfaces in contact were approximately parallel.

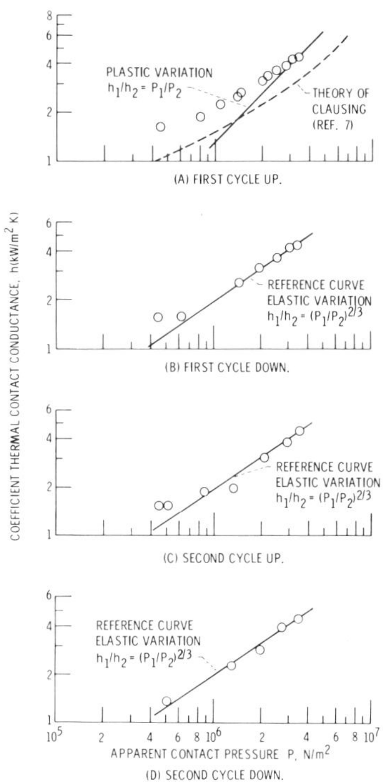

Fig. 9 Coefficient of thermal contact conductance vs pressure for Armco Iron - Armco Iron specimen at a constant interface temperature of 367 K. $\delta_1 = 0.64$ μm; (crest to trough height)$_1$ = 1.52 μm. $\delta_2 = 0.97$ μm; (crest to trough height)$_2$ = 1.02 μm.

INVESTIGATION OF THERMAL ISOLATION MATERIALS FOR CONTACTING SURFACES

Donald A. Gyorog*
University of Missouri - Rolla, Rolla, Mo.

Abstract

Thermal isolation characteristics of various interstitial materials were experimentally evaluated in a vacuum environment. The materials were inserted between contacting metal surfaces and test conditions ranging from 100 to 300 psi load pressure and -60°F to +260°F mean junction temperature were applied. In addition, applications of selected insulating materials were demonstrated by tests with two antenna support flanges. Although thermal resistance is generally a result of the combined effects of surface contact and material conductivity, some materials, notably mica and carbon paper, evidence a primary dependence on only one of these effects. For prediction purposes, pyrotex and teflon offer the advantage of a small variation in thermal resistance with load pressure. Although the analysis of thermal resistance with wire screens is complicated by the wire configuration, a dimensionless correlation was effected based on the functional relationships of load, screen dimensions, and material properties.

Nomenclature

A = apparent area, sq ft
d = diameter, in.
h = junction thermal conductance, Btu/hr sq ft°F
k = thermal conductivity, Btu/hr ft°F

Presented as Paper 70-13 at the AIAA 8th Aerospace Sciences Meeting, New York, January 19-21, 1970. The investigation reported herein was supported by NASA Grant NGR-03-001-033, and the experimental study was conducted at Arizona State University. The cooperation of the University personnel, those organizations who provided the sample materials and especially the laboratory assistance of P. A. Smuda is gratefully acknowledged.

*Associate Professor of Mechanical Engineering.

ℓ = thickness, in.
M = mesh size, per in.
n = number of layers
P = pressure, psi
Q = heat transfer, Btu/hr
R = thermal resistance, hr°F/Btu
r = thermal resistance per contact, hr°F/Btu
T = temperature, °F
ΔT = junction temperature difference, °F
x = length, in.
ρ = density, lb/cu ft

Subscripts

o = initial
c = cooled surface, contact, or conduction
h = heated surface
i = internal conduction
j = junction, or contact
M = metal specimen
m = mean
s = screen

Introduction

The successful operation of a spacecraft is, to a great extent, dependent upon the temperature control of its components and surfaces. As missions become more demanding, the use of protective shields and other insulation techniques take on increased importance. For most structures the thermal analysis requires design information regarding the thermal conductance of joints. A great number of examples from current space craft technology can be listed for which the thermal contact conductance is an important design parameter. Some of these are the thermal isolation of components such as reflective shields, antenna struts, and cryogenic storage compartments. In these particular cases the insulating effect of low conductance interstitial materials would have a direct benefit.

Although the measurement and prediction of thermal contact resistance has received considerable attention, the results have been somewhat disappointing because of the lack of reproducible data and the difficulty of predicting joint conductance under varying conditions of manufacture, load and surface conditions.[1] One proposed solution to enable the prediction of thermal resistance is the use of contact filler materials which are more or less insensitive to loads and surface conditions. The insertion of interstitial materials can serve to increase or decrease the thermal resistance of the junction. However, specific results on the thermal conductance of metallic contacts with interstitial fillers are limited. Further, design information for thermal isolation is even more limited since the majority of the previous investigations on this subject have emphasized the improvement of the contact conductance with such materials as high conductivity metal foils. Thus the primary purpose of this investigation, earlier phases of which were reported in Refs. 2 and 3, was to provide information, based on experiment, for thermal insulation of contacts with low conductance interstitial materials.

In a vacuum environment the heat transfer at the contacting surface is a result of conduction through the interstitial materials and for some materials possible radiation heat transfer. Calculations for the metal test specimen temperatures encountered during the investigation indicate that the contribution due to radiation is less than 0.1%. Hence, the primary mode of heat transfer is conduction. The effective thermal resistance at the junction can therefore be increased by use of a low thermal conductivity material and/or by reduction of the contact area. Junction materials which provide the desired increase in thermal resistance can be classified into the following general categories: Geometric, Powders, and Insulation Sheets. The thermal contact conductance of the junction is defined as

$$h = Q/A(T_1 - T_2)$$

where T_1 and T_2 are the extrapolated temperatures of the bounding surfaces of the contact and Q/A is the apparent heat flux of the total cross-sectional area of the contact. Thus the thermal contact resistance, R, is defined as the reciprocal of the product hA.

Experimental Apparatus and Procedure

Although specific details and design criteria concerning the experimental apparatus were discussed previously[1-3] a schematic diagram is repeated in Fig. 1 for easy reference. The test section was a 1-in. diam column with the contact surface at approximately mid-height. Each metal test specimen was instrumented with five centerline and three surface thermocouples.

The centerline thermocouples were 1/2 in. apart starting 1/4 in. from the junction. These temperatures were used to calculate the temperature gradient and resulting heat flux and, in addition, the temperature at the junction was found by extrapolation. Surface thermocouples were placed in 1/16 in. deep holes diametrically opposite the centerline thermocouple. Thermocouple leads were wrapped around the specimen to reduce the heat transfer from the junction. The thermocouples, lead wires, and thermocouple switches were calibrated with reference to a platinum-platinum 10% rhodium thermocouple and an uncertainty of 0.5% was established.[1]

For load application to the test specimens, high pressure nitrogen gas was admitted to the stainless steel bellows chamber (Fig. 1). The nitrogen bellows was selected since it simplified specimen loading and provided a relatively accurate means of adjusting and maintaining a constant load. The load pressure P is defined as the load force divided by the cross-sectional area of the cylindrical test specimen. The load cell measurements were corrected for the deflection of the stainless steel load pass-through bellows and the atmospheric pressure differential.

A band heater was clamped around the end of the heated test specimen (Fig. 1). This portion of the test specimen was insulated and surrounded by a radiation shield to reduce the heat transferred to the adjacent apparatus. Aluminum foil radiation shields were also installed around the thermocouple instrumented portion of the heated and cooled test specimens. Because a large temperature difference (200°-400°F) existed between the two specimens, with the interstitial materials in place, the radiation shields were separated at the contact plane. Chilled water and liquid nitrogen coolant fluids were supplied to the copper heat sink. All tests were conducted in a vacuum environment of 10^{-5} torr.

To minimize deviations in contact resistance between different sets of test specimens the same surface finishing technique was followed for all specimens. The contacting surfaces were finished on a lathe and then lapped and polished. Measurements with a profilometer and proficorder indicated a 3- to 6-μ-in. roughness and approximately a 20- to 25-μ-in. flatness deviation.

Many bare junction runs were interspersed between the interstitial material runs to test for changes in the surface conditions.† In some cases, large variations in h were encountered, but these were generally found to be the result of some

†Experimental results with the Stainless Steel 304 specimens for this phase of the investigation are tabulated in Table 1. The earlier results for AL2024 and Stainless Steel 304 were reported in Refs. 2 and 3.

Table 1 Tabulated experimental results

Run	P, psi	T_m, °F	ΔT_j, °F	h, $\frac{BTU}{hr\ ft^2\ °F}$
Stainless steel 304 barejunction				
6-BJ-35	100	257	55.6	93.8
6-BJ-38	91	259	73.6	69.3
6-BJ-47	105	264	74.6	64.4
7-BJ-1	105	327	91.8	54.5
7-BJ-2	308	359	17.4	381
7-BJ-3	104	298	82.0	65.6
7-BJ-4	310	310	15.3	457
7-BJ-5	105	296	80.6	68.7
7-BJ-20	104	324	90.5	57.6
7-BJ-21	209	345	29.9	216
7-BJ-22	305	376	21.7	347
7-BJ-24	105	326	79.3	67.4
7-BJ-42	105	277	77.5	64.9
Interstitial materials				
Carbon paper				
6-CA-30	319	211	172.1	9.98
Textolite				
6-TX-31	104	247	165.3	18.8
6-TX-32	314	247	122.1	32.0
7-TX-44	103	244	233.0	8.8
7-TX-45	104	239	131.4	24.0
7-TX-46	306	236	103.8	33.9
Pyrotex 23 RPD				
6-PI-33	94	248	203.1	12.2
6-PI-34	310	249	196.7	13.8
Rutile powder				
6-RP-36	91	237	139.1	22.5
6-RP-37	310	237	96.1	47.8

Table 1 (Cont'd)

Run	P	T_m	ΔT_j	h
Mica				
6-MICP-39	101	240	195.9	12.3
6-MICP-40	307	243	170.2	17.6
7-MIC-6	103	255	119.9	30.1
7-MICP-7	103	236	147.9	17.7
7-MIC2-9	100	253	166.3	17.2
7-MIC3-10	103	241	162.0	15.8
7-MIC2-26	104	256	136.0	23.4
7-MIC3-27	106	254	190.7	14.4
7-MIC2-34	20	239	230.6	7.9
7-MIC2-35	105	254	190.3	15.0
7-MIC2-36	306	262	123.0	34.8
7-MIC2-37	105	252	190.1	15.3
7-MIC3-38	23	233	236.5	6.7
7-MIC3-39	106	244	195.6	13.6
7-MIC3-40	306	258	143.1	27.3
7-MIC3-41	102	244	194.8	13.5
Stainless steel shim -- 0.001-in.				
6-SH-41	103	252	77.5	59.5
6-SHP-42	103	249	76.8	56.9
Stainless steel screen -- 100 mesh				
6-SS100-43	104	249	118.5	32.1
6-SS100L-44	103	251	165.1	18.7
7-SS100-49	100	-35	162.0	21.9
7-SS100-50	304	-17	100.8	46.2
7-SS100-53	105	293	77.6	32.8
7-SS100-54	100	236	112.2	32.8
7-SS100-55	304	254	77.9	64.1
Stainless steel screen -- 10 mesh				
6-SS10-45	103	234	218.2	7.4
6-SS10L-46	104	217	221.8	5.1
7-SS10-51	106	-66	280.8	6.2
7-SS10-52	305	-53	236.5	10.3
7-SS10-56	103	217	215.6	7.0
7-SS10-57	307	219	174.6	12.8

Table 1 (Cont'd)

Run	P	T_m	ΔT_j	h
Laminate T-30LR				
7-LA1/16-11	107	219	215.1	5.04
7-LA2/16-12	108	213	232.3	2.90
7-LA2/16[a]-13	106	213	233.1	2.90
7-LA3/16-14	104	204	227.4	2.25
7-LA3/16[a]-15	105	203	222.8	2.61
7-LA4/16-16	106	197	229.7	1.38
7-LA4/16[a]-17	113	200	236.0	1.35
7-LA3-18	102	206	248.5	1.42
7-LA3[a]-19	104	205	241.5	1.82
Teflon-TFE				
7-TFE-25	106	286	135.7	28.5
7-TFE-47	104	243	121.6	28.8
7-TFE-48	306	246	123.6	29.9
Special multilayer test				
7-$\frac{\text{MIC2}^{b}}{\text{SH}}$-28	102	251	188.5	14.8
7-$\frac{\text{MIC2}^{b}}{\text{SHP}}$-29	104	256	238.9	9.3
7-$\frac{\text{MIC2}}{\text{MICP}}$-30	104	241	193.5	12.6
7-$\frac{\text{MIC2}^{c}}{\text{MICP}}$-31	104	240	193.9	12.8
7-$\frac{\text{MIC2}}{\text{SS100}}$-32	103	241	221.6	9.0
7-$\frac{\text{MIC2}^{c}}{\text{SS100}}$-33	102	234	198.9	11.1

[a]Material initially compressed to 800 psi, initial thickness 2/16, 3/16, and 4/16 and three 1/16 in. layers, respectively.
[b]Two mica layers and intermediate layer of 0.001-in. shim and perforated (P) shim, respectively.
[c]Material initially compressed to 800 psi intermediate layers of perforated mica and 100-mesh screen, respectively.

surface effect such as oxidation, contamination by the previous material, or a scratch or imperfection. The bare junction experimental results for Aluminum 2024 were shown to indicate the same trends and relative magnitudes as those of other investigators.[1,2] Thermal contact conductance values for repeat test runs were shown to agree within 5%. The uncertainties in these values of h for the bare junction tests were estimated to vary from 7.2% at lower values of h to 11.6% at higher values. This variation is due primarily to the reduction in junction temperature difference with increasing h.

The experimental procedure with interstitial materials was essentially the same for all tests. A 1-in. diam disk of the test material was positioned between the metal specimens, and during evacuation of the test chamber the column alignment was maintained by a slight contact pressure. After setting the test conditions (heater power and coolant flow), the guard heater (Fig. 1) was monitored and adjusted to maintain a temperature difference of less than 5°F across the axial insulator. Generally, from 6 to 8 hr were required to obtain a steady-state condition. The results for this phase of the investigation are tabulated in Table 1. Results for earlier tests with other materials were reported in Refs. 2 and 3.

Since the initial thickness of the different interstitial material test samples varied and because the thickness generally varies with load pressure, the comparison of junction thermal conductance should account for the material thickness.[3] Thus, a better comparison of one material to another would be the product $h\ell$ as illustrated in Fig. 2[‡] For this comparison, the value of ℓ was calculated from compression strain $(\ell_0-\ell)/\ell_0$ vs load pressure data obtained from compression tests of similar size material samples. The technique and compression apparatus were described in Ref. 3.

Two antenna assembly bolted flange joint configurations were selected to demonstrate thermal isolation with the interstitial insulating materials (Fig. 3). In a typical application, the flange would be attached to a relatively large surface; therefore, the base flange for the test installation was constructed somewhat larger than the flange dimensions (in a rectangular shape, 0.75-in. thick) to provide a closer simulation of the actual installation. The flanges (and base

---

[‡]Results from Refs. 2 and 3 are also included to provide a summary of various interstitial materials. The material symbols are explained in Table 2.

Table 2 Manufacturer's property data

Material	Run symbol	Supplier	k, Btu/hr ft°F	ρ, lb_m/ft	ℓo, in.
Asbestos board	AB	Johns-Manville	0.04 - 100°F 0.06 - 500°F	136	0.121
Asbestos tape (No. 2074)	AT	Atlas Asbestos Company	0.12 - 300°F	36	0.081
Carbon paper F-907	CA	Fiberite Corp.	0.2 - 300°F	9	0.055
Ceramic paper 970-J	CE	Carborundum Company	0.03 - 400°F 0.10 - 1600°F	12	0.108
WRP-X-AQ felt	FE	Refractory Products Co.	0.04 - 500°F 0.12 - 2000°F	18	0.183
Laminate T-30LR	LA	Carborundum Company	0.37 - 250°F 0.13 - 2000°F	50	0.124
Magnesia 25	MA	Degussa Inc.	15.7 - 112°F	196	0.099
Mica	MIC	Industrial Mica	0.21 - 100°F 1.24 - 1000°F	13	0.003
Pluton B-1 cloth	PL	3M Company	0.02 - 80°F 0.03 - 180°F	87	0.027
Pyroid	PY	Pyrogenics Inc.	1.00 - 70°F 0.30 - 5000°F	162	0.116
Pyrotex 23RPD	PI	Raybestos-Manhattan	1.39 - 70°F	...	0.112
S.S. screen 10/in.	SS10	...	9.30 - 212°F	500	0.050
S.S. screen 100/in.	SS100	...	9.30 - 212°F	500	0.009

Table 2 (Cont'd)

Material	Run symbol	Supplier	k, Btu/hr ft°F	ρ, lb_m/ft	ℓo, in.
S.S. shim	SH	Shop Grade	9.30 - 212°F	500	0.001
Teflon TFE	TFE	Bel-Art Products	... ...	...	0.062
Textolite	TX	General Electric Co.	1.40 - 250°F	...	0.062
Titanium screen, 10/in.	TI10	Newark Wire Cloth Company	11.6 - 200°F	276	0.050
Tungsten screen, 20/in.	W20	Newark Wire Cloth Company	96 - 32°F 70 - 932°F	1210	0.014
Zirconia 23	ZI	Degussa Inc.	1.08 - 212°F	355	0.105

sections) were machined from aluminum 6061 stock. In the vacuum chamber, the flange and base sections were each fastened to 1-in. diam metal support rods. The ends of the 1-in. diam shanks for the flange and base sections were machined to a smaller diameter and inserted into a cylindrical cavity drilled into the end of the respective support rod. They were held in position by set screws. To reduce the thermal resistance between the support rod and the flange (or base), the cavity mating surfaces were coated with a mixture of aluminum powder and vacuum grease. One support rod was threaded into the heat sink, and the other was machined with an enlarged diameter (1.5-in.) at one end to which the band heater was fastened. For the purpose of calculating the heat transfer, three centerline (0.5-in. apart) and two surface thermocouples were installed in each support rod. The thermocouple installation procedure and technique was the same as that used for the interstitial material tests. During the flange tests, the lower section of the test column (load bearing section) was disengaged from the heated section except to provide initial precompression of the joint. The other details of the support rods and radiation shields were the same as for the metal test specimens used in the earlier interstitial material tests.

To measure the heat-transfer rate over a wide range of magnitudes, two sets of instrumented support rods were constructed, one of Armco Iron and the other of Stainless Steel 304. For the uninsulated flange (higher heat flux), the Armco Iron set of heat metering rods was installed, and for low-heat flux runs the stainless steel set was substituted. Two holes, one on either side of the 1-in. diam shank, were drilled (No. 56 drill) from the back face of the flange and the base to within 0.005-in. of the mating surfaces. Copper-constantan thermocouples were inserted to obtain some measure of the temperature difference across the flange junction. The two thermocouples on each flange face agreed within 2°F of one another for all of the test runs. The average of these readings was taken as the flange (or base) temperature, and subsequent values of the temperature difference between flange and base, ΔT, and the mean temperature T_m were calculated (Table 3). The heat transfer rates were calculated by the product of the thermal conductivity and temperature gradients for the heat metering support rods.

In Table 3, the value of $Q/\Delta T$ is an average of the calculated heat transfer for the heated side (flange) and the cooled side (base) of the flange installation. The range of these values is shown for the four-bolt flange in Fig. 4. Run numbers in Table 3 specify the flange test set (two sets were used for the two-bolt flange) and the test run in chronological order for that flange set. Comparison tests with various combinations of bolt and flange face insulations were conducted with the two-bolt flange configuration. Following these runs, carbon paper and 100 mesh stainless steel screen insulations were tested with the four-bolt flange (Fig. 4). The symbols in Table 3, used to describe the type of insulation, are defined in Table 2.

Experimental Results

The test results for the interstitial material and bare junction runs which were conducted during this phase of the investigation are listed in Table 1. The test runs were designated by several symbols to specify the metal test specimen set, the interstitial material, and the run number in chronological order for a particular specimen set. For example, run 7-MICP-7 refers to the seventh run with specimen set 7 and a single layer of perforated clear grade mica, and 6-SS100L-44 is run 44 with set 6 and two layers of 100 mesh stainless steel screen. The screens were oriented at 45° to one another for the two-layer test. Further explanation of the run symbols and conditions are given in Tables 1 and 2.

Interstitial Materials

Comparative tests for textolite, pyrotex, and TFE teflon sheets were run in preparation for their possible use as bolt insulating materials for the flange installation. The supplier and initial thickness of the materials used for the thermal tests are listed in Table 2. The effective thermal conductivities, $h\ell$, are plotted for comparison with the other materials in Fig. 2. Although the magnitudes of $h\ell$ are relatively large for all three materials, the teflon and pyrotex demonstrate the desirable feature of a small dependence of h

Table 3 Two-bolt flange comparison

Run[a]	T_m	$Q/\Delta T$[b]	Insulation	
	°F	Btu/hr°F	Face	Bolt
1-12	104	1.58	none	TX
1-13	151	1.36	none	TX
1-14	218	2.46	none	TX
1-15	106	0.05	CA	TX
1-16	287	0.07	CA	TX
1-17[c]	299	0.14	CA	TX
1-18	286	0.10	CA	PI
1-19[c]	287	0.20	CA	PI
1-20	284	0.14	CA	TFE
1-21	305	0.23	CA	TFE
1-22[c]	292	0.25	CA	TFE
2-5	167	0.42	CA	none
2-6	260	0.45	CA	none
2-7	92	0.41	CA	none
2-8[d]	356	0.29	CA	MIC/SH
2-9	328	0.14	CA	TX
2-10[d]	330	0.29	CA	SH/SS100
2-11	320	0.33	TX	SH/SS100
2-12	154	0.66	SS100	SH/SS100
2-13	244	0.87	SS100	SH/SS100
2-14	335	0.87	SS100	SH/SS100

[a]Run numbers specify flange No. 1 or flange No. 2 and the sequence of test runs with that flange in chronological order.
[b]The average value $Q/\Delta T$ from the values calculated for the heated and cooled sides of the flange installation is tabulated.
[c]Initial bolt torque of 50-in.-lb all other 25-in.-lb.
[d]Axial load of 500 lb applied before setting torque.

on the load pressure for the range 100 to 300 psi. Repeat runs with textolite and teflon agreed very closely except for run 7-TX-44. Test samples for textolite, which is a hard glass laminated sheet, had to be cut on a mill. Without extreme care and a sharp cutting tool, the edges break apart slightly. Some evidence of this existed with the sample for run 7-TX-44. Another sample was carefully machined to avoid this problem, and the results of 7-TX-45 and -46 agreed more closely with the initial runs 6-TX-31 and -32. Disagreement at the lower pressures could be attributed to the surface effect which is reduced with increased load pressure. Thus it can be concluded that the thermal resistance of textolite is a combination of surface contact and bulk material effects.

In an earlier study,[4] it was found that dusting contacting stainless steel surfaces with manganese dioxide power resulted in a large increase in the thermal resistance over that for the bare surfaces. As another comparison, a very thin layer of rutile power was placed on the stainless steel specimen surfaces. To provide a more definitive measure of the powder, it was separated between 0.0024-in. and 0.0017-in. sieves. Approximately a single grain thickness layer was spread on the specimen surface, this powder was then collected and weighed (0.15 g). The same weight of powder was then placed on the specimen surface for testing. The powder is crushed into a layer with increased loads with a resulting large increase in h. Bare junction runs were intermixed with the material runs to test for variations in the metal test specimen surfaces. Run 6-BJ-38 followed the rutile powder tests. The surfaces were wiped clean, but the approximate 30% reduction in h can be directly attributed to the contamination of the surface by the two previous runs with rutile powder, since the same effect is also noted with the later runs with specimen set 6. Thus dusting of the surfaces can easily produce sizeable but not necessarily predictable variations in the junction thermal resistance.

Variable thickness samples of laminate T-30LR were tested to compare the results to a multilayer sample of the same initial thickness and weight. The calculated values of 1/h are plotted vs ℓ in Fig. 5. As a simplified analogy, the junction resistance can be supposed to be a sum of the contact resistance between the metal specimen and the interstitial material and a bulk material resistance of the form ℓ/kA. Thus the linear function shown in the graph, $1/h = 2AR_c+(1/k)\ell$, provides a measure of $1/k$ and R_c at the mean test temperature of 206°F and load pressure of 106 psi. The three-layer sample increased the thermal resistance by 50% over that for a continuous piece of the same thickness as a result of the two intermediate contacts, i.e., an increase of approximately 0.12 hr ft^2 °F/Btu in 1/h for each contact. The value for the nominal 0.25-in. sample is 0.14 hr ft^2 °F/Btu greater than the linear plot would predict. Laminate is a hard laminated board, and the samples were machined from a sheet approximately 0.31-in. thick. The major laminations, which form the sheet, are about 0.06-in. thick. After the 0.25-in. sample test run, one of these laminations was found to have separated. By comparison of the predicted increase of 0.12 with the measured increase of 0.14 in 1/h, the internal contacting surface was apparently present during the test.

Since wire screens are capable of withstanding large compressive loads, tests with stainless steel screens and stainless steel specimens were conducted to compare with the previous results[3] using Aluminum 2024 specimens. The effective values of the junction thermal resistance 1/h are illustrated in Fig. 6. All of the data are for the higher junction termperature runs (chilled water coolant). Results for the aluminum surfaces with stainless steel, titanium, and tungsten screens denoted as -AL in Fig. 6 were reported in Ref 3. The thermal resistance for a single layer (0.055-in.) carbon paper (CA) is also included for comparison from Ref. 3. Slight indentations were noticed in the stainless steel surfaces after the test runs with 10 mesh screens. The same type of indentation was quite noticeable in the aluminum surfaces; therefore, part of the increase in thermal resistance for stainless steel surfaces can be attributed to a reduction in the contact area, but the primary effect seems to be the large decrease in thermal conductivity, a factor of 10 from aluminum to stainless steel. The increased thermal resistance for lower thermal conductivity wire, smaller mesh size (fewer contacting points), smaller wire diameter (diameter decreases with increasing mesh size), lower temperatures, and stainless steel surfaces is apparent.

The prediction of the thermal resistance due to screen wires is difficult because of the complexity of the wire configuration. However, the geometric similarity of wires of different sizes suggests that a graphical correlation of the experimental results could be effected based on the functional relationships of load, screen dimensions (mesh and wire diameter), and wire and contact surface material properties. A simplified model of the screen contact is illustrated in Fig. 7. For the analogy of Fig. 7b, it is assumed that contacts occur at the points of overlap of the screen wire, or approximately M^2 contacts per unit area. Thus, r_{ch} and r_{cc} correspond to the thermal resistance at the points of contact with the flat surfaces, r_{ic} indicates the internal resistance of the wire, and r_{ij} represents a resistance resulting from the internal contacts. In terms of the measured thermal conductance h this assumption gives

$$[Q/AM^2][r_{ch} + r_{cc} + r_i] = \Delta T$$

where r_i represents the equivalent resistance of the possible conduction paths within the screen, and since $Q = hA\Delta T$, the expression reduces to

$$1/h = [r_{ch} + r_{cc} + r_i]/M^2$$

The relation is somewhat complicated by the addition of the term r_i/M^2. However, if $r_{ij} > r_{ic}$, and further if r_{ic} is of the form $\ell/kA \sim 1/Md^2k$, then r_i as an approximation can be assumed in the form

$$r_i/M^2 = [1/k_m M]\Phi_i \text{ (Md, material properties)}$$

where Φ_i is a function of the parameter Md and the screen properties. Since the contact resistances r_{ch} and r_{cc} are macroscopic, the functional relation obtained by Roess[5] and used with success by Clausing and Chao[6] was selected as an approximation for the contact resistances. The constriction function X_L was then defined as proportional to the square root of the ratio of the contacting area per contact A_c and the cross-sectional area of the metal surface per contact $1/M^2$ or in equation form,

$$X_L = M[A_c]^{1/2}$$

With these simplifications for r_c (either r_{ch} or r_{cc}), the expression was reduced to

$$r_c/M^2 = [1/k_m M]\Phi_c(1/X_L)$$

These relations suggest that the parameter $k_m M/h$ should be functionally related to $M^2 A_c$ and Md. With the contact model shown in Fig. 7a and the method outlined in Chapter 11 of Ref. 7, the contact area ratio $M^2 A_c$ for elastic compression is given by

$$M^2 A_c = 2.805\ [1.456(Md)(P/E_m)]^{2/3}$$

where E_m is the harmonic mean of the modulus of elasticity for the screen and the metal surface as defined by

$$1/E_m = [1/E_M + 1/E_s]/2$$

Although the equation for M^2A_c is not exact for the screen contact, it does suggest the functional dependence of k_mM/h on the parameters P/E_m and Md. The value of k_m was determined as the mean of the thermal conductivities for each contact given by

$$1/k_c = [1/k_M + 1/k_s]/2$$

and the thermal conductivity of the wire screen k_{sm} at the temperature T_m. In equation form,

$$1/k_m = [1/k_{ch} + 1/k_{sm} + 1/k_{cc}]/3$$

Since there are two contacts, one at the temperature T_h and the other at T_c, the value of E_m was taken as the average value of E_m for both contacts. Graphs of k_m/h vs P/E_m for the 100 mesh and the 10 mesh stainless steel screens indicated excellent correlation for all of the screen data with both aluminum and stainless steel contacting surfaces and for mean temperatures from 293° to -106°F.§ For the stainless steel screen cases, the dependence of the parameter k_mM/h on $(Md)^3$ was derived to provide the single functional relation shown in Fig. 8. Material property data for the calculations were obtained from Refs. 8-13.

The multilayer test results for mica discussed in Ref. 3 showed that the thermal conductance was greatly dependent upon a surface contact effect in agreement with Fry's[14] earlier observation. To demonstrate this dependence, tests with one, two, and three layers of mica sheet were conducted. The repeated runs generally agreed within the experimental uncertainty except for the two-layer case at 100 psi. At a particular pressure, each test run was conducted with a new test sample. Since the thermal resistance for mica is a strong function of the surface contact, the most probable reason for this disagreement is the possibility of differences in the surface conditions of the mica layers. Two runs, 7-MIC2-37 and 7-MIC3-41, are return pressure test points with the same test samples as runs 7-MIC2-35 and 7-MIC3-39, respectively. Nearly the same values of h were obtained in both cases which indicate that a precompression of mica does not seem to materially affect the thermal conductance.

If the surface contact resistance is predominant with mica, run 7-MIC-6, with a single layer of mica sheet, indicates that

---

§ The test results for these runs are reported in Table 1 and in Ref. 3.

run 7-MIC2/SH-28 (which had an intermediate layer of 0.001-in. stainless steel shim and, therefore, double the number of contact surfaces) should have a thermal conductance of 15 BTU/hr $ft^2\,°F$. This agrees with the measured value of 14.8 BTU/hr $ft^2\,°F$. Also the h values found with three layers of mica (MIC 3) compare closely with run 7-MIC 2/SH-28. Thus it can be concluded that the thermal contact resistance for mica sheets is primarily a surface effect and that the resistance between mica layers is at least as large as that between mica and stainless steel surfaces. The thermal resistance 1/hA for these tests with mica and the resistances reported in Ref. 3 were graphed vs the load pressure. From this graph, a cross plot (Fig. 9) at constant pressure was constructed to illustrate the linear behavior, $1/hA \doteq 2R_c + R_i\,(n-1)$, where R_c is the mica-to-metal surface contact resistance, and R_i represents the mica-to-mica surface resistance.

The effects of reducing the contact area by perforating single layers of mica and 0.001-in. stainless steel shim are shown by runs 6-MICP-39 and -40, 7-MICP-7, and 6-SH-41 and 42. Approximately 25% of the area was removed with a No. 45 drill. The disk was divided into three equal area segments, and the same number of holes were drilled in each region. The lower value of h for run 6-MICP-39 as compared with run 7-MICP-7 may be a result of the same difficulty encountered for the bare junction run 6-BJ-38. (The possible surface contamination with rutile powder.) Another very possible explanation could be the presence of rough edges left by the drilling operation. Nevertheless, a reduction of at least 40% in the magnitude of h for mica at 100 psi can be obtained by the area reduction. On the other hand, the reduction in area for the stainless steel shim seems to have almost a negligible effect.

Insertion of a 100 mesh stainless steel screen between mica sheets (7-MIC2/SS100-32) reduces the thermal conductance to one-half of that obtained with two layers of screen oriented at 45° to one another (6-SS-100L-44). Although the precompression to 800 psi (7-MIC2/SS100-33) has a small effect on mica, the stainless steel screen penetrates the mica layers with a resultant increase in h.

For each run, a maximum and minimum possible slope (dT/dx) was read from the plot of metal specimen temperature vs length. This variation was well within ± 1°F/in. for the runs with interstitial materials, so as an estimate, an uncertainty of 0.5°F/in. was selected for dT/dx. In addition to the slope, the variation in the junction temperature difference ΔT was estimated to be between 2° and 4°F. With the stainless steel screen, the junction temperature difference was as small as

100°F, so that uncertainty in ΔT was on the order of 4%. Combining these estimates, the uncertainty in h was found to be from 4% to 7% for the interstitial material results tabulated in Table 1.

Flange Tests

In order to compare the performance of different insulation combinations, an effective thermal conductance Q/ΔT was defined for the flange tests. The insulating materials were the same thickness as the samples for the earlier interstitial material tests, and the face insulation layer was cut to fit the flange dimensions. Several runs with the two-bolt flange were conducted to compare bolt and flange face insulation to select combinations for the four-bolt flange tests. The results of the comparison tests are listed in Table 3. After preliminary tests with uninsulated flanges (runs 1-1 through 1-11 and 2-1 through 2-4, not tabulated), the bolt attachment holes in flange set 1 were chamferred at 45° and insulating washers of textolite, pyrotex, and teflon were machined to fit the chamfer. Comparison of runs 1-15 through 1-22 indicates that the textolite is the better of the three materials. Values of Q/ΔT for the uninsulated tests ranged from 2Btu/hr°F at a mean temperature of 100°F to a maximum of 6Btu/hr°F at 370°F. Thus the insulating effect is very evident. Although the uninsulated conductance of flange set 2 was considerably smaller, by 30%, than for set 1, the addition of insulation, run 2-9 for example, appears to provide close agreement between flanges of different manufacture. The bolt breaking torque was checked after each run. For runs with the insulating materials, the torque was reduced from 25 in.-lb to 10 to 15 in.-lb as a consequence of the thermal stresses encountered during the test runs. This particular problem was alleviated for the four-bolt flange configuration by precompression of the insulated joint and the addition of compression washers rather than the flat washers used in runs 2-10 through 2-14.

The effective junction thermal conductance for the four-bolt flange is illustrated in Fig. 4. The magnitudes of Q/ΔT for the heated and for the cooled side of the flange joint are plotted in the figure as vertical lines. Face insulations of a single sheet of carbon paper and a single layer of 100 mesh stainless steel screen were selected to demonstrate the two extremes of the insulation materials tested during this investigation. For all runs the bolt insulation was composed of a washer of 100 mesh stainless steel screen and a compression washer. The uninsulated tests (BJ) were conducted with a bolt torque of 25 in.-lb and all others with a torque of 15 in.-lb. After the tests with the stainless steel screen and the carbon paper, without disturbing the torque, the flange installation was removed and attached to a shaker table in a vertical position. The shaker oscillation frequency was varied slowly from zero to 2000 cps over a period of two min. A thermal test was then rerun, and the close agreement with the previous results is shown in Fig. 4. A final check of the bolt torque following this run showed no reduction in the torque. An additional test, with the face insulation screen wire sandwiched between stainless steel shim, demonstrates the benefits of the increased number of contacts and the stainless steel surface. In contrast to the screen the carbon paper provides the desirable trait of a small thermal conductance which is essentially unaffected by the mean temperature.

Summary

The thermal insulation characteristics of the interstitial materials listed in Table 2 were investigated and results reported herein and in previous reports.[2,3] On the bases of an effective conductivity $h\ell$ which also accounts for the material thickness (Fig. 2), and the ability to withstand large cyclic loads,[3] the carbon paper and mica can be selected as the better materials. However, if weight is also a consideration[3] then the carbon paper is substantially the better material.

Some materials, notably mica, evidence a strong dependence on the surface contact effect (Figs. 5 and 9). By contrast, the thermal resistance of carbon paper is solely due to the material resistance.[3] The insulation of aluminum surfaces with wire screen can be measurably improved by separating the screen from the surface with a stainless steel shim (Fig. 4). As illustrated in Fig. 8, the thermal resistance $k_m M/h$ of wire screen for various loads, temperatures, and stainless steel and aluminum surfaces can be determined as a function of $(Md)^3/(P/E_m)$.

Dusting the mating surfaces with a powder, although difficult to control, will also improve the thermal isolation. Some materials offer the advantage, for prediction purposes, of a small dependence of $h\ell$ on load pressure. Several examples from Fig.2 are pyrotex, asbestos board, and TFE teflon. The insulation of bolted flange joints can be effectively accomplished by a single layer of carbon paper inserted between the mating surfaces and textolite bolt insulating washers. The difficulty of reduction in bolt torque, hence the mechanical strength of the joint, as a result of thermal stress or vibration can be avoided by precompression of the insulated joint and the use of tension washers.

References

[1]Smuda, P. A., Fletcher, L. S., and Gyorog, D. A., "Heat Transfer Between Surfaces in Contact: The Effect of Low Conductance Interstitial Materials; Part I: Experimental Verification of NASA Test Apparatus," CR-73122, June 1967, NASA.

[2]Fletcher, L. S., Smuda, P. A., and Gyorog, D. A., "Thermal Contact Resistance of Selected Low Conductance Interstitial Materials," AIAA Journal, Vol. 7, No. 7, July 1969, pp. 1302-1309.

[3]Smuda, P. A. and Gyorog, D. A., "Thermal Isolation with Low Conductance Interstitial Materials Under Compressive Loads," AIAA Paper 69-25, New York, 1969.

[4]Mikesell, R. P. and Scott, R. B., "Heat Conduction Through Insulatory Supports in Very Low Temperature Equipment," Journal of Research of the National Bureau of Standards, Vol. 57, December 1956, pp. 371-379.

[5]Roess, L. C., "Theory of Spreading Conductance," Appendix A, unpublished report, Beacon Labs. of Texas Co., Beacon, N. Y., 1949.

[6]Clausing, A. M. and Chao, B. T., "Thermal Contact Resistance in a Vacuum Environment", Paper 64-HT-16, American Society of Mechanical Engineers, August 1964.

[7]Seely, F. and Smith, J., Advanced Mechanics of Materials, 2nd Ed., Wiley, New York, 1952.

[8]Touloukian, Y. S., ed., Metallic Elements and Their Alloys, Vol. I, Lafayette, Ind., Thermophysical Properties Research Center, Purdue Univ., 1966.

[9]Eldridge, E. A. and Deem, H. W., eds., "Report of Physical Properties of Metals and Alloys from Cryogenic to Elevated Temperatures", ASTM Special Technical Publication No. 296, April 1961.

[10]Aerospace Structural Metals Handbook, Vol. II, Non-Ferrous Alloys, Syracuse University Press, Syracuse, N. Y., revised March 1963.

[11]Goldsmith, A., Waterman, T. E. and Hirschkorn, H. G., Handbook of Thermophysical Properties of Solid Materials, Vols. I and 2, Pergamon, New York, 1961.

[12] Garofalo, F., Malenock, P. R. and Smith, G. V., "Influence of Temperature on the Elastic Constants of Some Commercial Steels", ASTM Special Technical Publication No. 129, 1952.

[13] Birdsall, G. W., ed., The Aluminum Data Book, Reynolds Aluminum Co., Richmond, Virginia, 1965.

[14] Fry, E. M., "Measurements of Contact Coefficients of Thermal Conductance", AIAA Progress in Astronautics and Aeronautics: Thermophysics and Temperature Control of Spacecraft and Entry Vehicles, Vol. 18, edited by G. B. Heller, Academic Press, New York, 1966, pp. 719-734.

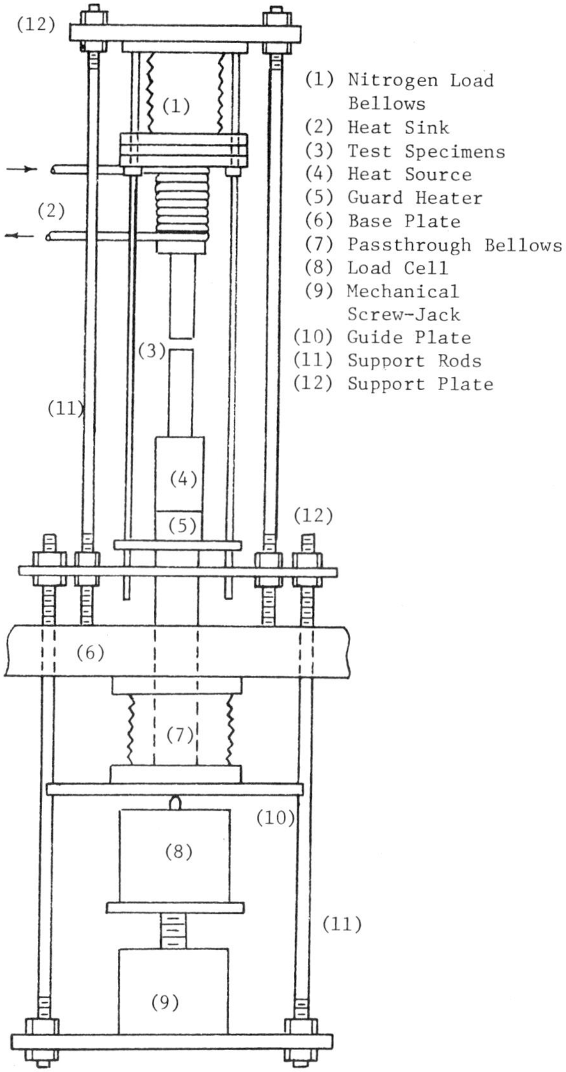

Fig. 1 Schematic of test apparatus.

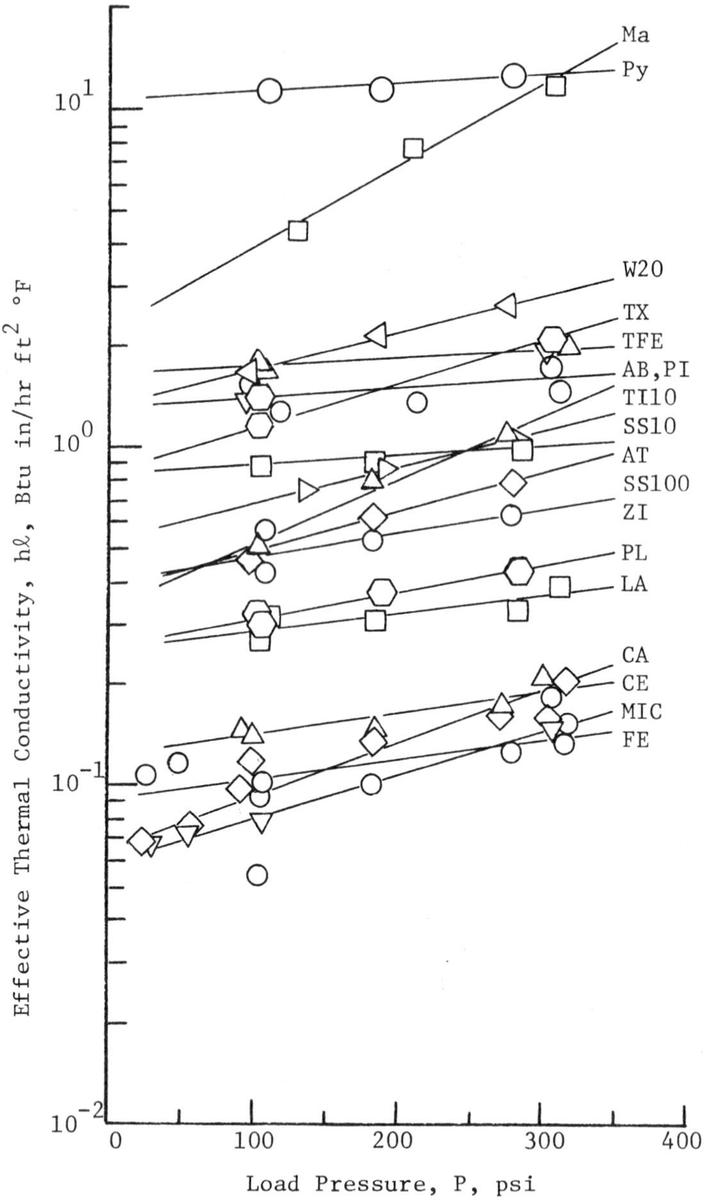

Fig. 2 Effective thermal conductivity comparison (T_m from 180°F to 250°F).

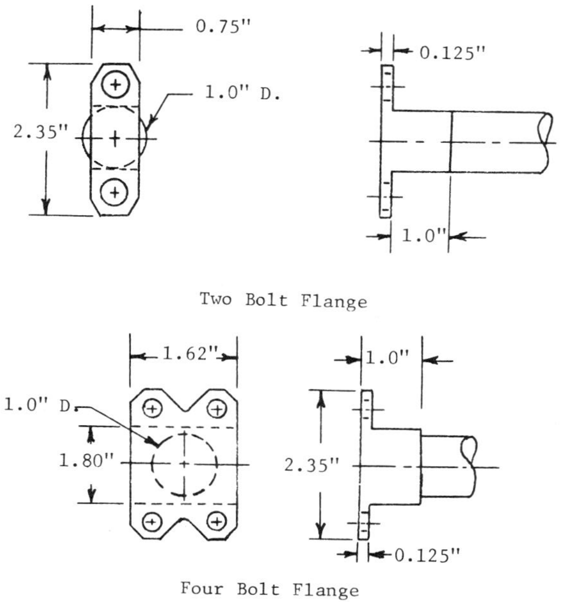

Fig. 3 Flange dimensions.

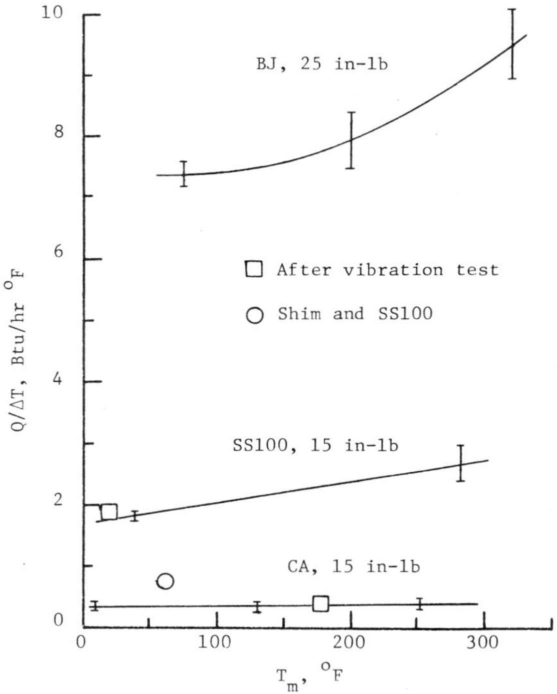

Fig. 4 Thermal conductance for the 4-bolt flange.

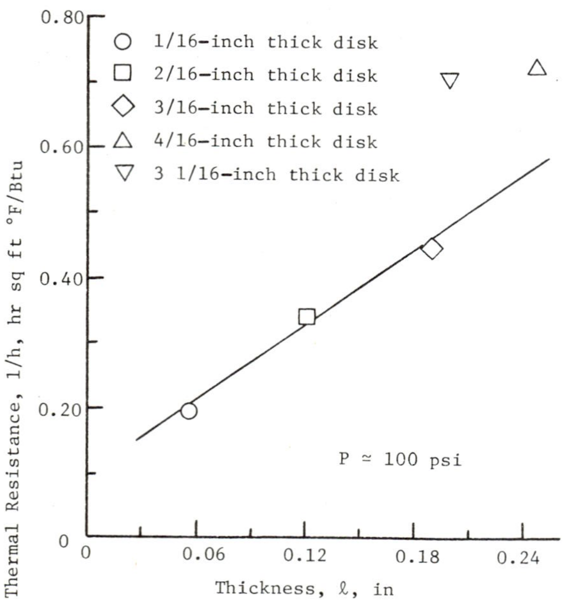

Fig. 5
Laminate T-30LR variable thickness results.

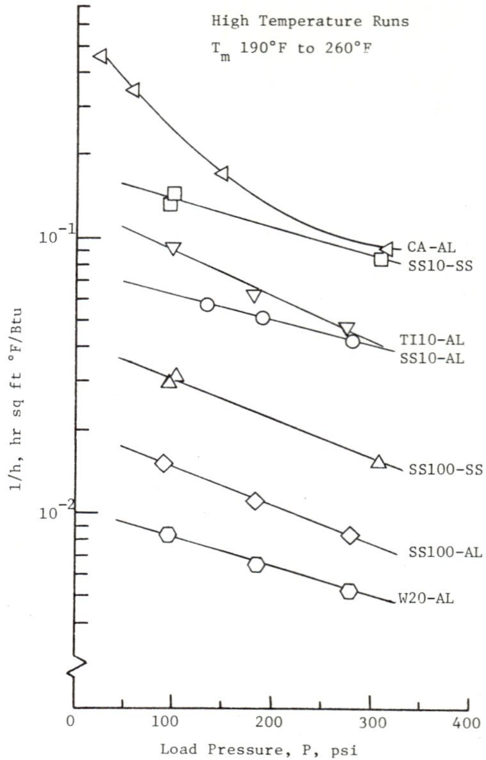

Fig. 6
Thermal resistance of wire screens compared with single layer of carbon paper.

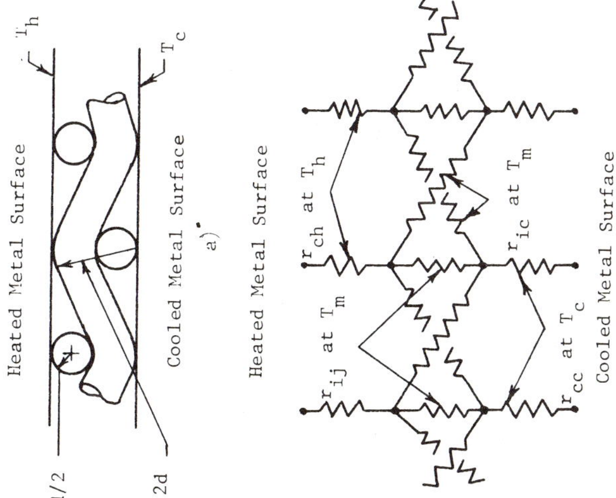

Fig. 7 Wire screen contact.

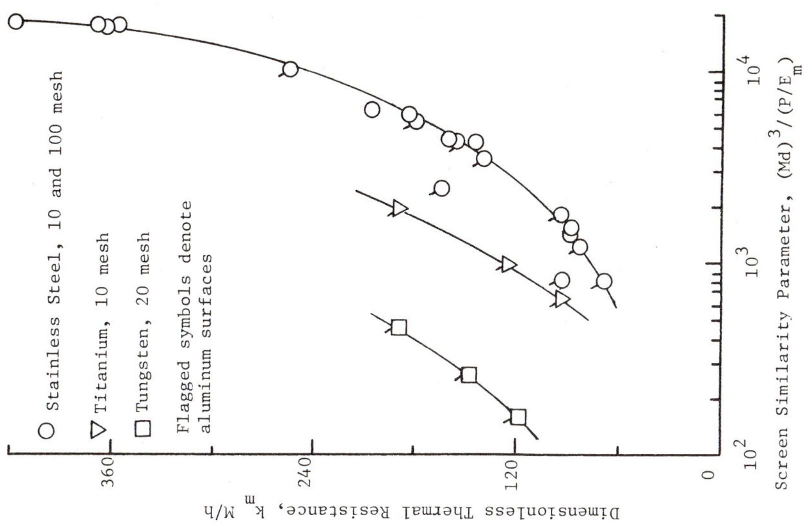

Fig. 8 Wire screen thermal resistance correlation.

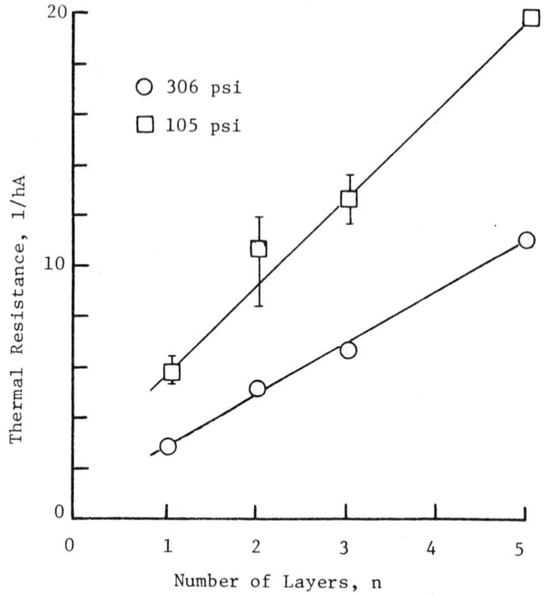

Fig. 9 Multilayer mica test results.

THERMAL CONSTRICTION RESISTANCE BETWEEN CONTACTING METALLIC PARABOLOIDS: APPLICATION TO INSTRUMENT BEARINGS

M. Michael Yovanovich*

University of Waterloo, Waterloo, Ont., Canada

In this paper the thermal constriction resistance between contacting paraboloids is considered, and the results of the theory are applied to the thermal analysis of a typical bearing element. In this analysis the complex thermal problem is resolved by the use of ellipsoidal coordinates which are intrinsic to both the shape of the contact area and the mixed boundary conditions: a) temperature prescribed over the contact which is isothermal and b) temperature gradient prescribed over the remainder of the contact plane which is impervious to heat transfer. An expression is developed for the total constriction resistance of a typical bearing element. The resistance is shown to be directly proportional to a geometric factor, which is a function of the ratio of the semimajor to semiminor axes, and inversely proportional to the product of the thermal conductivity and the semimajor axis. The theory is compared with test results, and the agreement is good.

Nomenclature

a = semimajor axis of elliptic contact area
A = constant
b = semiminor axis of elliptic contact area
B = constant
E = modulus of elasticity

Presented as Paper 70-857 at the AIAA 5th Thermophysics Conference, Los Angeles, Calif., June 29-July 1, 1970. Work performed at the Instrumentation Laboratory of Massachusetts Institute of Technology under NASA contract NAS9-6823. The author wishes to express his thanks to Arthur Grossman, Group Leader, Thermal Laboratory, Bedford Flight Facility; Ronald L. Morey; and Robert L. Henderson for their cooperation during all phases of this work.

*Associate Professor, Department of Mechanical Engineering.

F = force between contact solids
k = thermal conductivity
m = constant
n = constant
N = number of balls in a bearing
q = heat flux vector
Q = total heat flow
r = radial distance
R = thermal constriction resistance defined by Eq. (31)
T = temperature
α = contact angle
β = coefficient of expansion
ν = Poisson's ratio
Γ = contact area
Δ = coefficient defined by Eq. (3)
ρ = radius of curvature of contacting solids
ψ = constriction coefficient defined by Eq. (34)
ϕ = angle between planes containing principal curvatures

Subscripts

1,2 = solids 1 and 2, respectively, in contact
i,o = inner and outer races, respectively

Superscripts

unprimed = minimum radii of curvature
primed = maximum radii of curvature

Introduction

Many modern precision instruments are quite sensitive to temperature levels or temperature changes. When these electromechanical or optical instruments are mounted on gimbaled platforms, the total thermal resistance between the platform and the available thermal sink becomes very important. In a gimbaled system the instrument bearing will dominate the transfer of heat away from the source (the precision instruments). Since many systems operate in a vacuum, there are essentially two modes of heat transfer available: conduction through the solids and the metal-to-metal interfaces, and radiation across the gaps. For small temperature differences and low operating temperature levels the second mechanism will be small relative to the conduction mode.

A survey of the literature shows that only two studies[1,2] have dealt with the problem of heat transfer across a bearing. The first investigation[1] was experimental and showed the magnitude of the thermal resistance of a typical bearing as a function of the bearing load and the influence of air and a lubricant upon the resistance. The investigator considered the case of a dry bearing in a vacuum, a dry bearing in air, as well as a lubricated bearing in a vacuum. He observed that a dry bearing in a vacuum offered from 10 to 5 times more resistance than the same dry bearing placed in air or the same lubricated bearing in a vacuum. The factor of 10 corresponds to a very light bearing load, while the factor of 5 corresponds to a higher bearing load (approximately 10 times the light load). In the first case the thermal resistance was dependent upon the applied load (approximately to the 1/3 power), while for the other two cases, the resistance was fairly independent of the load. The investigator[1] did not attempt an analysis of the thermal resistance, but did conclude that bearings different in size should have similar resistance if the number of balls is the same.

The second investigation[2] was analytical and showed which parameters were important and the influence of these parameters on the total resistance of the bearing. The analysis was based upon a symmetrically load bearing with heat flowing uniformly through each ball. The races were assumed to be flat and the contact areas were determined to be circular and very small relative to the radius of the balls. Assuming a particular thermal model, the investigator[2] showed that the total thermal resistance for a dry bearing in a vacuum is inversely proportional to the product of the number of balls, the radius of the contact area and the thermal conductivity of the balls. Since the contact area radius depends upon the ball radius, the material properties and the applied load, the effect of these parameters could be ascertained. The analysis showed that the resistance should vary as the load to the 1/3 power and is verified by the experimental results of the first investigation[1]. It was concluded that since both the inner and outer race of a bearing are not flat, but possess radii of curvature which can significantly influence the shape of the contact area which determines the magnitude of the thermal resistance, a more exact analysis should be performed to take into consideration the race curvatures.

An interface formed by the elastic contact of smooth solids (angular contact instrument bearing) will consist of a very large density of individual contacts which are said to comprise the contour area. For very smooth surfaces (roughness less than 5 μ in rms value) these small contacts (real contact area) will cover over 90% of the contour area. For such smooth surfaces it has been shown[3-5] that the total resistance can be predicted by considering the resistance due to the contour area only. For this reason, no distinction will be made between the real and contour areas when considering the elastic contact and the thermal resistance.

In this paper it is assumed that heat transfer between contacing smooth paraboloids occurs through the contour area whose shape and size can be predicted by the Hertzian theory. The much more complex problem of heat transfer between contacting rough paraboloids will not be considered.

Elastic Contact Between Smooth Paraboloids

When two elastic solids having well-defined curved surfaces press against each other, an area of contact develops, the shape of which depends upon the principal curvatures of the surfaces near the point of contact, and the extent of which depends upon the force with which the solids press against each other. If the force is zero, the contact is a mathematical point. Whenever one solid exerts a force against another, an area of contact is formed which is generally elliptical. This area increases in size with increases in the force, but the shape remains invariant.

Since the shape and size of the contact area is of primary interest because the thermal constriction is a function of these contact area parameters, a few words will be said about the results of the Hertzian theory which predicts these parameters.

The complete Hertzian solution is available in standard texts on elasticity[6,7] and need not be repeated here. The solution must be evaluated in terms of elliptic integrals of the first and second kind. The semimajor and minor axes of the elliptic contact area are given by

$$a = m\ [(3\pi/4)F\ \Delta]^{1/3} \qquad (1)$$

and

$$b = n\ [(3\pi/4)F\ \Delta]^{1/3} \qquad (2)$$

in which

$$\Delta = \left(\frac{1 - \nu_1^2}{\pi E_1} + \frac{1 - \nu_2^2}{\pi E_2}\right) /(A + B) \qquad (3)$$

The coefficients m and n are numbers depending on the ratio A/B (Fig. 1). A and B are determined from the following equations:

$$A + B = \frac{1}{2}\left(\frac{1}{\rho_1} + \frac{1}{\rho_1'} + \frac{1}{\rho_2} + \frac{1}{\rho_2'}\right) \tag{4}$$

and

$$B\text{-}A = \frac{1}{2}\left[\left(\frac{1}{\rho_1} - \frac{1}{\rho_1'}\right)^2 + \left(\frac{1}{\rho_2} - \frac{1}{\rho_2'}\right)^2 + 2\left(\frac{1}{\rho_1} - \frac{1}{\rho_1'}\right)\left(\frac{1}{\rho_2} - \frac{1}{\rho_2'}\right)\cos 2\phi\right]^{1/2} \tag{5}$$

The constants A and B of Eqs. (4) and (5) have the same sign.

<u>Contact Between Sphere and Paraboloid</u>

The expressions which depend upon the radii of curvature of the contacting solids can be reduced considerably when one of the solids is a sphere. Let us designate the paraboloid as solid 1 and the sphere as solid 2. Since the radii of curvature of the sphere are identical in all planes through its center, the angle ϕ can have any value. But this has no influence upon the results since the term 2ϕ in Eq. (5) is multiplied by $(1/\rho_2 - 1/\rho_2')$ which is zero. Thus Eqs. (4) and (5) simplify, for the paraboloid/sphere contact, to

$$A + B = \frac{1}{2}\left(\frac{1}{\rho_1} + \frac{1}{\rho_1'} + \frac{2}{\rho_2}\right) \tag{6}$$

and

$$B - A = \frac{1}{2}\left(\frac{1}{\rho_1} - \frac{1}{\rho_1'}\right) \tag{7}$$

If we first add Eqs. (6) and (7), we obtain an expression for A, and if we then subtract one from another, we get an expression for B. Taking the ratio of A to B we obtain

$$\frac{A}{B} = \frac{1/\rho_1' + 1/\rho_2}{1/\rho_1 + 1/\rho_2} \tag{8}$$

<u>Contact Area Between Ball and Race</u>

Since the curvatures of the inner and outer race are different (Fig. 2), the expressions for A+B and B/A, will therefore differ for the ball/inner and ball/outer race combinations. For the inner race where ρ_i is negative and ρ_i' is positive, the expressions become

$$A + B = \frac{1}{2}\left(-\frac{1}{\rho_i} + \frac{1}{\rho_i'} + \frac{2}{\rho_2}\right) \tag{9}$$

and

$$\frac{A}{B} = \frac{1/\rho_i' + 1/\rho_2}{-1/\rho_i + 1/\rho_2} \tag{10}$$

At the outer race where both ρ_o and ρ_o' are negative, the expressions become

$$A + B = \frac{1}{2}\left(-\frac{1}{\rho_o} - \frac{1}{\rho_o'} + \frac{2}{\rho_2}\right) \tag{11}$$

and

$$\frac{A}{B} = \frac{-1/\rho_o' + 1/\rho_2}{-1/\rho_o + 1/\rho_2} \tag{12}$$

Illustrative Example

Consider as an example a bearing consisting of an inner and outer race, 42 spherical balls of 3/16 in diameter. All parts are fabricated from 440C stainless steel. The minimum radius of curvature of both the inner and outer race is 52% of the ball diameter. The maximum curvatures of both inner and outer race are 49/16 and 62/16 in, respectively. The physical characteristics of races and balls are $\nu_1 = \nu_2 = 0.25$ and $E_1 = E_2 = 29{,}000{,}000$ psi. It can be readily determined that for these two types of contacts the principal characteristics are as shown in Table 1.

Table 1 Bearing Geometric Characteristics

	Inner race	Outer race
A + B	5.825	5.24
A/B	30.0	27.4
m	3.37	3.25
m/n	8.70	8.20
Δ	3.54×10^{-9}	3.94×10^{-9}
a	$0.68 \times 10^{-2} B^{1/3}$	$0.71 \times 10^{-2} B^{1/3}$

It will be noted that there is only four % difference between the linear dimensions of the elliptic contact area at the inner and outer race. Consider the smaller of the two values as being typical of the contact area. For a bearing load of 17.3 lbs/ball the semimajor axis is $a = 1.84 \times 10^{-2}$ in and the semiminor axis b will be 8.70 times smaller. Thus the contact area is very small relative to the characteristic dimensions of the ball or races.

Thermal Model

As discussed in the previous section, the contact area between two smooth elastic paraboloids is always elliptic and it was shown that the semimajor and minor axes of the elliptic contact are quite small relative to the characteristic dimensions of the contacting solids. In the light of these facts it is assumed that the heat flow pattern in either solid will be fairly symmetrical about the contact plane, and thus it can be argued that the contact area is isothermal. The following discussion will be limited to the determination of the thermal constriction resistance when all the heat flows between the solids via the elliptic contact, i.e., there is no heat flow across the contact plane outside the contact area. This means that the surfaces of the contacting solids are adiabatic outside the contact area.

It is assumed that one half of the thermal constriction resistance problem can be adequately represented by an isolated, isothermal area either supplying or receiving heat from an other-wise insulated conducting half-space of thermal conductivity k_1. It is necessary to obtain a solution for Laplace's equation, $\nabla^2 T = 0$, in the half-space, $z > 0$, subject to the boundary conditions

$$T(x,y) = T_o \quad \text{in } \Gamma \tag{13}$$

and

$$[\partial T(x,y,z)/\partial z]_{z=0} = 0 \quad \text{outside } \Gamma \tag{14}$$

These equations form a so-called "mixed" boundary condition over the surface of the half space, $z > 0$, i.e., a temperature condition over the contact area and a temperature gradient condition over the rest of the surface. In general such boundary conditions are difficult to satisfy by means of simple functions, and this problem is even more difficult because of the shape of the contact area.

The boundary conditions on $z = 0$ are mixed, Eqs. (13) and (14) and not amenable to solution unless ellipsoidal coordinates are used. In ellipsoidal coordinates the Laplace equation becomes[5,8]

$$\nabla^2 T = \partial/\partial\lambda \, [\sqrt{f(\lambda)} \, (\partial T/\partial\lambda)] = 0 \tag{15}$$

where

$$\sqrt{f(\lambda)} = \sqrt{(a^2 + \lambda)(b^2 + \lambda)\lambda} \tag{16}$$

and a, b are the semimajor and semiminor axes, respectively, of the elliptic contact area.

The boundary conditions corresponding to Eqs. (13) and (14) are

$$\lambda = 0, \qquad T = T_o, \text{ constant} \tag{17}$$

$$\lambda \to \infty, \qquad T \to 0 \tag{18}$$

The Dirichlet problem in ellipsoidal coordinates automatically satisfies the mixed boundary value problem in Cartesian coordinates.

Integrating Eq. (15) twice we obtain as solution

$$T = A \int_o^\lambda \frac{d\lambda}{\sqrt{f(\lambda)}} + B^1 \tag{19}$$

where A and B^1 are constants of integration.

Equation (19) can be written as

$$T = A \left\{ \int_o^\infty \frac{d\lambda}{\sqrt{f(\lambda)}} - \int_\lambda^\infty \frac{d\lambda}{\sqrt{f(\lambda)}} \right\} + B^1 \tag{20}$$

where the integral from zero to infinity is constant. Writing

$$B = B^1 + A \int_\lambda^\infty \frac{d\lambda}{\sqrt{f(\lambda)}} \tag{21}$$

the solution becomes

$$T = B - A \int_{\lambda}^{\infty} \frac{d\lambda}{\sqrt{f(\lambda)}} \tag{22}$$

The boundary condition at infinity requires that B = 0 and the Dirichlet boundary condition over the elliptic area requires that we put

$$-A = \frac{T_o}{\int_{o}^{\infty} \frac{d\lambda}{\sqrt{f(\lambda)}}} \tag{23}$$

The temperature distribution in the half-space can now be written as

$$\frac{T}{T_o} = \frac{\int_{\lambda}^{\infty} \frac{d\lambda}{\sqrt{f(\lambda)}}}{\int_{o}^{\infty} \frac{d\lambda}{\sqrt{f(\lambda)}}} \tag{24}$$

The heat flux[5,8] is

$$q_{\lambda} = \frac{-k\ 2\sqrt{f(\lambda)}}{[(\lambda - \mu)(\lambda - \nu)]^{1/2}} \frac{\partial T}{\partial \lambda} \tag{25}$$

and upon taking the derivative of Eq. (24) with respect to λ and substituting into Eq. (25) we get

$$q_{\lambda} = \frac{2\ k\ T_o}{\int_{o}^{\infty} \frac{d\lambda}{\sqrt{f(\lambda)}}} \frac{1}{[(\lambda - \mu)(\lambda - \nu)]^{1/2}} \tag{26}$$

Equation (26) reduces to

$$q_{\lambda} = \frac{2\ k\ T_o}{r^2 \int_{o}^{\infty} \frac{d\lambda}{\sqrt{f(\lambda)}}} \tag{27}$$

for $\lambda \to \infty$ because $\lambda \to r^2$, and $\lambda >> \mu$ and $\lambda >> \nu$. At large distances from the elliptic area, the heat flux is practically radial and therefore one can write

$$q_\lambda = \frac{2\ k\ T_o}{r^2 \int_o^\infty \frac{d\lambda}{\sqrt{f(\lambda)}}} \doteq q_r = \frac{Q}{2\pi r^2} \tag{28}$$

The temperature distribution now becomes

$$T = \frac{Q}{4\pi k} \int_o^\infty \frac{d\lambda}{\sqrt{f(\lambda)}} \tag{29}$$

and by the definition of thermal constriction resistance

$$R = [T_o - T(\lambda \rightarrow \infty)]/Q \tag{30}$$

we have

$$R = \frac{1}{4\pi k} \int_o^\infty \frac{d\lambda}{\sqrt{(a^2 + \lambda)(b^2 + \lambda)\lambda}} \tag{31}$$

Holm[9] has shown that Eq. (31) can be transformed into an equivalent form

$$R = \psi/4ka \tag{32}$$

where

$$\psi = \frac{2}{\pi} \int_o^{\pi/2} \frac{d\theta}{\{1 - \kappa^2 \sin^2 \theta\}^{1/2}} \tag{33}$$

and $\kappa^2 = 1 - b^2/a^2$.

k is the thermal conductivity of the half-space, a is the semi-major axis of the elliptic contact and ψ is a geometric factor depending upon the size of the contact. This geometric factor can be written as

$$\psi = \frac{2}{\pi} K \tag{34}$$

where K is the complete integral of the first kind of modulus $\kappa^2 = 1 - b^2/a^2$.

Equation (33) can be expanded and integrated term by term to yield the following useful expression:

$$\psi = 1 + \left(\frac{1}{2}\right)^2 \kappa^2 + \left(\frac{1.3}{2.4}\right)^2 \kappa^4 + \left(\frac{1.3.5}{2.4.6}\right)^2 \kappa^6 + \ldots + \left[\frac{(2n)!}{2^{2n}(n!)^2}\right]^2 \kappa^{2n} \quad (35)$$

Table 2 Typical Values of Geometric Factor ψ

a/b	ψ	a/b	ψ
1.0	1.000	6.0	2.030
1.5	1.212	6.5	2.082
2.0	1.372	7.0	2.136
2.5	1.402	7.5	2.172
3.0	1.610	8.0	2.214
3.5	1.702	8.5	2.252
4.0	1.784	9.0	2.286
4.5	1.856	9.5	2.322
5.0	1.920	10.0	2.352
5.5	1.980		

Thermal Constriction of a Typical Element

We can now determine the thermal constriction resistance of a typical bearing element (Fig. 2). We will make the analysis general by assuming that the thermal conductivities of the races differ from that of the balls. When all the heat flows from the inner to the outer race via the balls, there will be a constriction resistance at the inner race/ball contact as well as the outer race/ball contact. Thus the total resistance at the inner race/ball contact is

$$R_i = \psi_i/4k_1a_i + \psi_i/4k_2a_i \quad (36)$$

where k_1 and k_2 are the thermal conductivities of the inner race and ball, respectively. The constriction factor ψ_i is taken to be the same because of symmetry. A similar argument for the outer race/ball contact allows us to write the resistance as

$$R_o = \psi_o/4k_3a_o + \psi_o/4k_2a_o \quad (37)$$

The total constriction resistance of a typical bearing element is the sum of Eqs. (36) and (37)

$$R = \frac{\psi_i}{4k_1a_i} + \frac{\psi_o}{4k_3a_o} + \frac{1}{4k_2}\left(\frac{\psi_i}{a_i} + \frac{\psi_o}{a_o}\right) \quad (38)$$

For most bearings the linear dimensions of the elliptic contact areas at the inner and outer races are practically the same, i.e., $a_i = a_o$, and, therefore, $\psi_i = \psi_o$. Also the balls and races are usually fabricated from the same material so that $k_1 = k_2 = k$, and this permits us to write Eq. (20) as

$$R = \psi/ka \tag{39}$$

Thermal Resistance of a Typical Bearing

The results of the elastic deformation and thermal constriction analyses will now be used to obtain an expression for the total constriction resistance of a typical bearing. When a bearing is uniformly loaded, each ball in the assembly supports the same unit load. This means that the contact area will be identical at each ball/race contact. Since there is only 4% difference between the semimajor axes corresponding to the inner and outer races contacts, respectively, it will be assumed that ψ is the same for all contacts. Furthermore we will take the smaller value of ψ corresponding to the inner race contacts.

When the heat flow pattern through the bearing is uniform, we will assume that the constriction resistance is identical for each bearing element. This means that the total constriction of the bearing can be based upon the assumption that all the elements are thermally connected in parallel.

When these conditions of uniform loading and uniform heat flow are satisfied, the total constriction resistance becomes

$$R_b = \psi/Nka \tag{40}$$

where N is the total number of balls in the bearing. Upon substitution of Eq. (1) into the expression for the total constriction resistance we get

$$R_b = \frac{\psi(m/n)}{k\ m\ N^{2/3}\ [(3\pi/4)\Delta]^{1/3}\ [F/\sin\ \alpha]^{1/3}} \tag{41}$$

We note that Eq. (41) depends upon a number of geometric and physical parameters as well as the axial load. For a particular bearing specified by its principal radii of curvature, number of balls, Young's modulus and Poisson's ratio, the parameters m, n, ψ, N and Δ are fixed being independent of the applied load. This means that some of the parameters in Eq. (41) can be lumped together and treated as a bearing constant.

The bearing resistance can be rewritten in the following simple form:

$$R_b = C_1/k\ B^{1/3} \tag{42}$$

where B = F/sin α and C_1 is given by

$$C_1 = \frac{\psi(m/n)}{m\ N^{2/3}\ [(3\pi/4)\ \Delta]^{1/3}} \tag{43}$$

For the two bearings used in the test program the bearing constant C, is as shown in Table 3.

Table 3 Test Bearing Properties

Type	3TAR 33-42U	3TAR 49-62U
Bore, in.	2.0625	3.0625
OD, in.	2.6250	3.8750
Balls	44	42
Ball diameter, in.	1/8	3/16
Clearance, in.	0.0007-0.0012	0.0007-0.0012
Ball/race material, stainless steel	440C	440C
Young's modulus, psi	29×10^6	29×10^6
Coeff. of expansion, $1/^{\circ}F$	5.6×10^{-6}	5.6×10^{-6}
Hardness, Rockwell "C"	60-65	60-65
Thermal conductivity, BTU/HR $FT^{\circ}F$	13.8	13.8
C_1	1255	1130

The effect of applied load, bearing clearance, contact angle and thermal strains will be discussed in the following section utilizing the simple expression given by Eq. (42).

Factors Influencing the Bearing Load

There is a complex relationship between the axial load, the bearing load, the initial bearing clearance and the heat flow rate. The complex relationship arises from the thermal expansion of the balls and races because there is a temperature difference between the races, and the elastic compression of the balls and races. In other words, the clearance under load is equal to the initial clearance less the expansion due to thermal effects plus the displacement due to the elastic compression. With the aid of the nomenclature shown in Fig. 3, the relationship discussed above can be expressed in the following manner:

$$2(\rho_i' - \rho_2)(1 - \cos\alpha) = \delta_o - C_2\Delta T + \frac{C_3 F/\sin\alpha}{C_4 + C_5(F/\sin\alpha)^{2/3}} \tag{44}$$

The constants in Eq. (44) are given by[5,6]

$$C_2 = \beta(\rho_o' - \rho_2) \tag{45}$$

$$C_3 = \rho_i'/S_i + \rho_o'/S_o \tag{46}$$

$$C_4 = 2\pi E \tag{47}$$

$$C_5 = \frac{4}{\pi}\frac{C_\delta}{C_b}\left(\frac{N}{\Delta}\right)^{1/3}\left(\frac{1-\nu^2}{E}\right)\frac{1}{N} \tag{48}$$

The constants C_δ and C_b in Eq. (48) depend upon the principal radii of curvature through the geometric ratio A/B and values are found in Seely[6].

Equation (44) can be cast into a form which clearly shows the relationship between the axial load, the bearing load, heat flow and the initial bearing clearance. For the temperature difference in Eq. (44) write

$$\Delta T = QR_b = QC_1/B^{1/3} \tag{49}$$

Replace the contact angle dependence by writing

$$\cos\alpha = F/\sqrt{B^2 - F^2} \tag{50}$$

We now have

$$2(\rho_i' - \rho_2) = \delta_o - C_1 C_2 Q/B^{1/3} + C_3 B/[C_4 + C_5 B^{2/3}] \tag{51}$$

Knowing the initial clearance, Eq. (51) can be solved for B as a function of the heat flow rate Q after having solved Eqs. (43), (45-48) for the constants appearing in Eq. (51). The initial clearance is not known. The bearing manufacturer specifies a range of values for a typical bearing. Therefore, the best that one can do analytically is to solve Eq. (51) with δ_o as a parameter. Otherwise a series of experiments can be performed to determine the functional relationship between B, F, and Q.

Experimental Verification of the Theory

A systematic test program was conducted to obtain data which was compared with the values predicted by the theory. Two different sizes of angular contact instrument bearings were tested. Dry bearings were used and all tests were conducted in a vacuum of about 5 x 10^{-5} Torr. The test variables were the total axial load and the total heat flow rate. The experimental apparatus consisted of a vacuum chamber and vacuum pumps, a loading device, a heat source and heat sink, radiation shields and an instrument console.

A 15-watt (nominal) button heater was mounted on a copper block called the round centered hot block (Fig. 4). This hot block sat on the copper inner race. The interfaces between the hot block and keeper as well as the keeper and inner race were coated with Dow Corning 340 Silicone Heat Sink Compound. This insured that the heat path from the button heater to the inner race offered negligible thermal resistance. Thermocouples placed in the hot block and the keeper verified this. The outer race sat snuggly in the aluminum bearing support which stood on the base plate which acted as the heat sink. The base plate was cooled by a steady supply of city water. All interfaces were well greased with the heat sink compound. A 15-watt button heater was attached to an aluminum radiation shield which was placed under the inner race keeper (Fig. 4). This insured that all the heat from the source went to the inner race. Thermocouples were welded to the inner and outer race at diametrically opposite locations. They were also epoxied to insure that they would not be detached when handled.

The variable compression load on the bearing was supplied through a calibrated horseshoe-shaped force gauge mounted above the supermica insulating block. The insulating block completely enclosed the heater and rested on the hot block.

The bearings used in the tests were guaranteed by the supplier to be dry. They were kept away from oil and other contaminants when not being tested. All tests were performed with the bearings fixed in place and the total heat supplied by the button heater flowed from the inner to the outer race via the balls.

After mounting the test bearing, the chamber was pumped down to about 5 x 10^{-5} Torr. and these three procedures were followed to obtain the necessary data. In one test program, the load was fixed and the heat to the bearing was increased from

2.5 watts to a maximum of 15 watts in increments of 5 watts. In the second test program, the heat to the bearing was fixed while the total axial load was increased up to a maximum of about 500 lb. The third test program consisted of a mixture of load and heat flow changes from a minimum load and heat flow up to a maximum load and heat flow.

The outer race temperature varied from 70°F up to 81°F while the inner race temperature varied from 91°F up to a maximum of 245°F.

Correlation of Test Data and Discussion

The thermal resistance of a bearing given by Eq. (42) is not amenable to solution because the bearing load depends implicitly upon the temperature difference which is a function of the resistance. This can be seen by examining Eq. (44). If an initial bearing clearance of 0.001 in. is used in Eq. (44) or Eq. (51) one can determine the contact angle directly or the bearing load. Once these are known for a particular heat flow rate, then the thermal resistance can be calculated as a function of the axial load. The theory predicts values of the thermal resistance which show the same trend as the empirical data.

It was decided to use the second alternative described above to determine the relationship between B, F and Q. The results of the preliminary tests are shown in Figs. 5 and 6 where the bearing load is plotted against the axial load with heat flow as a parameter. At low axial loads the thermal expansion effect is important. Here it is seen that for a fixed axial load an increase in the heat flow results in an increase in the bearing load. This effectively reduces the thermal resistance. As the axial load increases, the effect of thermal expansion decreases. For loads above 600 lb the thermal effects are essentially negligible.

The relationship between bearing load and axial load as a function of heat flow rate was subsequently used to correlate all the experimental data obtained following the three procedures outlined in the previous section. There was excellent agreement between the experimental data and the predicted values of the resistance. This is clearly seen in Figs. 7 and 8. At low axial loads the resistance varies from a low value of about 10°F/watt up to a high value of about 22°F/watt for the 2.5-watt and 15-watt heat flow rates, respectively. The difference is due to thermal expansion effects as described above.

The thermal analysis was based upon the assumption that radiation effects are negligible. For the maximum temperatures observed (245-81) it is estimated that the thermal resistance due to radiation heat transfer would be about $300^{o}F$/watt. This is based upon the assumption that the inner race radiates directly to the outer race and both surfaces have emissivities of 0.8. The presence of the balls should increase the radiation resistance considerably. Thus the assumption of neglibible radiation transfer is valid for the low loads with temperature differences of $164^{o}F$ and temperature levels of $245^{o}F$ at the inner race. In the thermal analysis it was implicity assumed that all the resistance of a bearing can be attributed to the constriction of heat flow lines in the neighborhood of the contact areas at the ball/race interfaces. This in effect neglects the resistance to heat flow within the races and across the balls. The assumption used should become less true at high axial loads when the contact areas are large and the constriction resistances are small. In other words the thermal resistance through the races and balls, if important, should influence the total resistance at high axial loads. This was not observed and so it is concluded that this assumption is valid.

Conclusions

The analysis presented suggests that heat transfer through a bearing can be correlated extremely well provided that the surfaces are smooth, the contacts are formed elastically, the bearings are dry and placed in a vacuum, there is no radiation heat transfer across the gaps, and the bearing does not rotate. An expression for the total constriction resistance of a typical angular contact bearing is presented which shows that several physical and geometric parameters play a part in determining the resistance. It is also shown how thermal expansion and elastic compression of the balls and races influence the resistance. For a particular bearing having fixed geometric and physical characteristics, some of the parameters can be lumped together such that the bearing resistance can be correlated with the axial load and heat flow.

It is recommended that further work be done analytically and experimentally to determine the effects of lubricants, gases and rotation of the bearing.

References

[1]Jansson, R.M., "The Heat Transfer Properties of Structural Elements for Space Instruments, Report E-1173, Instrumentation

Laboratory, Massachusetts Institute of Technology, Cambridge, June 1962.

[2]Yovanovich, M.M., "Thermal Contact Resistance Across Elastically Deformed Spheres," Journal of Spacecraft and Rockets, Vol. 4, No. 1, Jan. 1967, pp. 119-122.

[3]Clausing, A.M. and Chao, B.T., "Thermal Contact Resistance in a Vacuum Environment," ASME Journal Heat Transfer, Vol. 87, May 1965, pp. 243-251.

[4]Yovanovich, M.M. and Rohsenow, W.M., "Influence of Surface Roughness and Waviness upon Thermal Contact Resistance," EPL Report 76361-48, June 1967, Mechanical Engineering Department, Mass. Inst. Technology.

[5]Yovanovich, M.M., "Analytical and Experimental Investigation of the Thermal Resistance of Angular Contact Instrument Bearings," Dec. 1967, Instrumentation Laboratory Report E-2215, Mass. Inst. Technology.

[6]Seely, F.B. and Smith, J.O., Advanced Mechanics of Materials, Wiley, New York, 1963, chap. 11.

[7]Love, A.E.H., Mathematical Theory of Elasticity, Dover, New York, 1944.

[8]Jeans, J.H., Mathematical Theory of Electricity and Magnetism. Cambridge University Press, Cambridge, 1963, pp. 244-245.

[9]Holm, R., Electric Contacts, Springer-Verlag, Berlin, 1958.

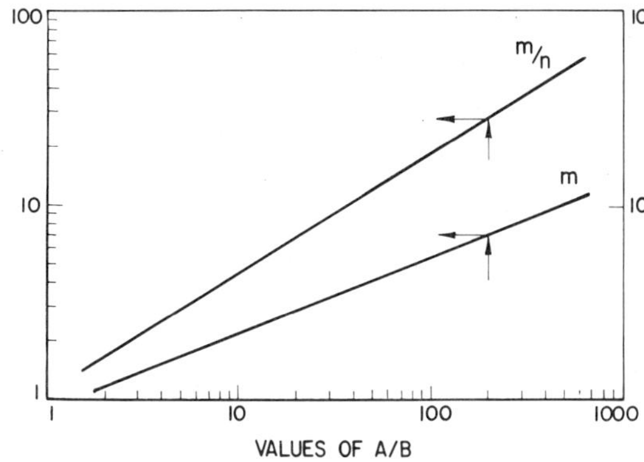

Fig. 1 Geometric parameters.

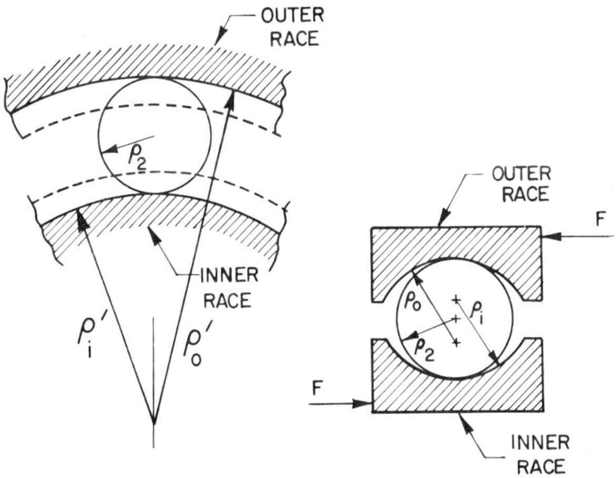

Fig. 2 Typical bearing element.

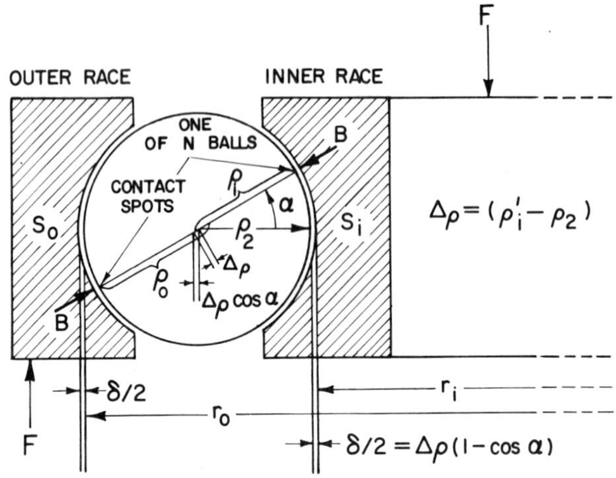

Fig. 3 Schematic of Ball and Races.

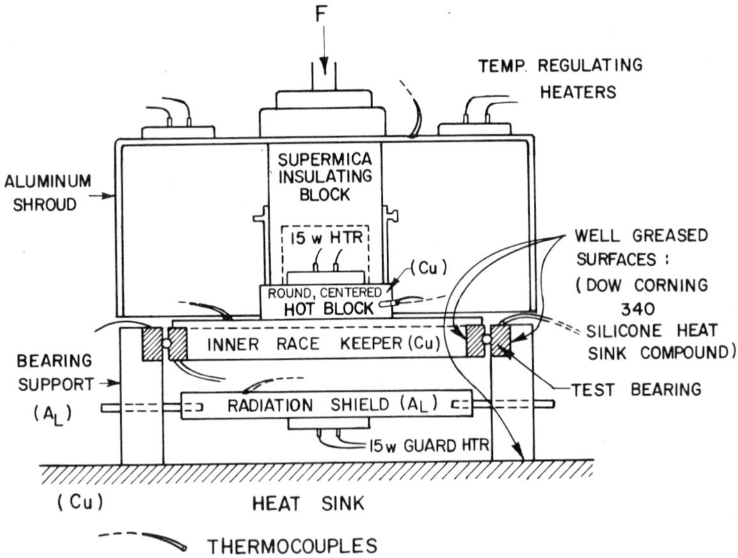

Fig. 4 Test apparatus.

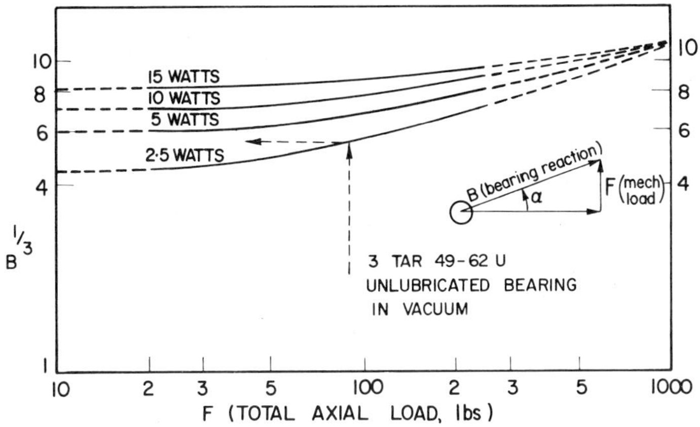

Fig. 5 Bearing load versus axial load.

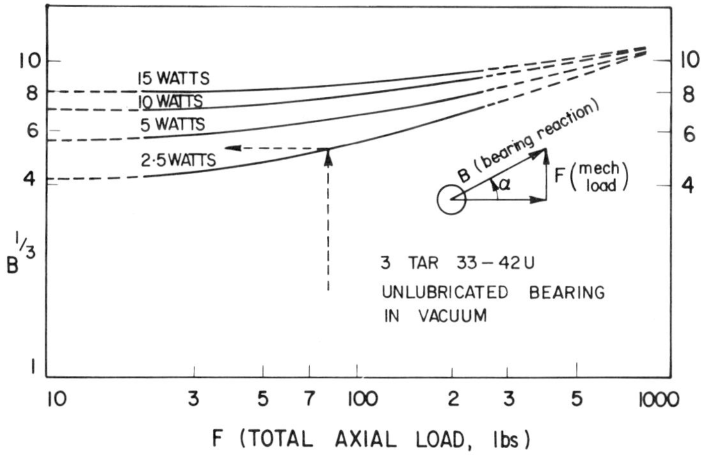

Fig. 6 Bearing load versus axial load.

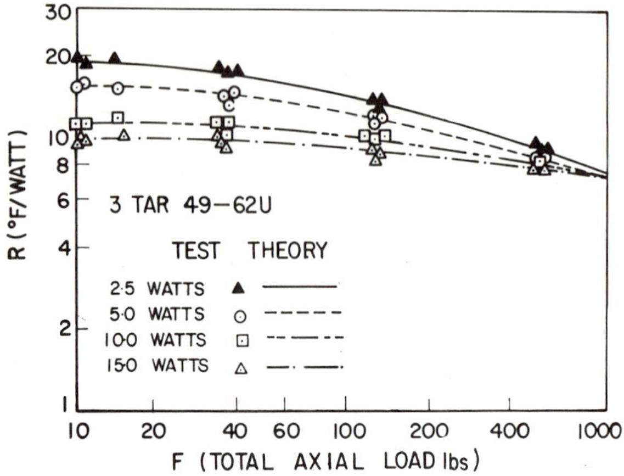

Fig. 7 Correlation of thermal resistance.

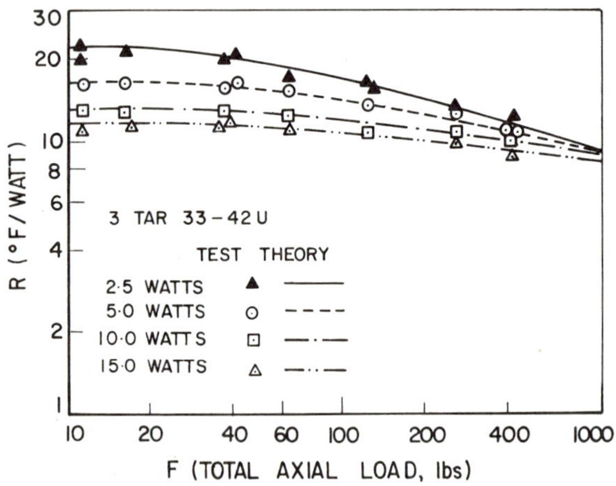

Fig. 8 Correlation of thermal resistance.

IV. HEAT TRANSFER ANALYSIS

RADIANT HEAT TRANSFER IN REFLECTOR SYSTEMS

Robert C. Pfahl Jr.*

Western Electric Engineering Research Center, Princeton, N. J.

Abstract

This paper presents a mathematical model for calculating the distribution of radiant flux absorbed by a heated part in a radiant heating system. The system consists of a point energy source, a reflector, and the heated part. The model considers spectral, directional, and polarizing effects. Reflectors of parabolic, hyperbolic, and elliptic cross section are analyzed with the model. These applications serve to demonstrate the use of the model and to illustrate several features of these common reflector shapes.

Nomenclature

A = part area, m^2
a = distance (Fig. 1), m
b = distance (Fig. 1), m
c = distance (Fig. 1), m
d = distance to source (Fig. 1), m
E = heat flux, W/m^2
I = intensity, W/sr
j = indexing variable for component of polarization
k = extinction coefficient
l = distance to part (Fig. 1), m
m = arc length (Fig. 3), m
n = index of refraction
p = parallel component of polarization
s = perpendicular component of polarization
u = arc length of reflector (Fig. 4), m
w = arc length (Fig. 4), m
x = coordinate (Fig. 1), m
y = radial coordinate of reflector (Fig. 1), m
z = radial coordinate of part (Fig. 1), m

Presented as Paper 70-818 at the AIAA 5th Thermophysics Conference, Los Angeles, Calif., June 29-July 1, 1970.

*Member of Research Staff.

α = absorptance
β = incident angle of direct radiation on part (Fig. 1), rad
γ = incident angle of indirect radiation on part (Fig. 1), rad
ζ = angle (Fig. 4), rad
θ = polar angle (Fig. 1), rad
λ = dummy variable - Eq. (36)
ν = frequency, Hz
π = irrational number 3.14159...
ρ = reflectance
ϕ = power, W
Ψ = incident angle of radiation on reflector (Fig. 1), rad

Subscripts

p = part
r = reflector
s = solid angle
ν = spectral - see footnotes for discussion
o = source
1 = component radiated through central opening
2 = component absorbed by reflector
3 = component reflected by reflector and part
4 = component directly reflected by part
5 = component reflected by reflector and absorbed by part
6 = component directly absorbed by part
7 = component radiated through opening between reflector and part

I. Introduction

Radiant heating systems consisting of a single, compact energy source and a reflector are used for aerospace thermal simulation[1] and for other applications, including: a) manufacturing in the electronics industry,[2] b) photolysis research,[3] c) manufacturing of electrical appliances.[4] Many of the industrial applications have resulted from applying technology developed for thermal simulation.[5]

To design properly a reflector system for heating applications, one must determine the heat flux absorbed by the heated part. The magnitude of the heat flux can vary strongly with position on the part. The purpose of this paper is to present a mathematical model for calculation of the heat flux absorbed by the part. The calculated heat fluxes permit analytical evaluation and optimization of reflector systems without constructing reflectors. Figure 1 illustrates the geometry.

The following sections of this paper discuss the assumptions of the analysis, derive the equations describing the system and discuss their solution, and apply the model to reflectors of parabolic, hyperbolic, and elliptic cross section.

II. Assumptions

This work restricts its attention to a system with rotational symmetry. The heated part is assumed to have a plane surface oriented normal to the axis of symmetry (Fig. 1). The heated "part" may be a physical object that is being heated, or it may be a plane in space where one wishes to calculate the flux distribution exiting from the system. The energy transfer in the system is analyzed by considering only those paths of radiant energy illustrated in Fig. 2. Spectral, directional, and polarizing effects[6] are treated in the model.

Any assumptions required in the analysis restrict the physical situations for which the model is valid. Before the model may be applied to a physical situation the validity of the assumptions must be verified. Following are the assumptions of the analysis followed by brief discussions of the range of validity of the last three: 1) a specular reflector; 2) a nonparticipating gas; 3) a point energy source; 4) negligible emission from the reflector, surroundings, and part; 5) energy transfer only along those paths illustrated in Fig. 2.

No source satisfies assumption three, but experimental measurements verify that many do approximate a point source. The size of the source relative to its distance from the reflector and heated part is important. The effect of source size also depends on the shape of the reflector. An ellipsoidal system with the source at one focal point produces a blurred image of the source at the second focal point; thus, a model which assumes a point source will give poor flux predictions near the second focal point. Other reflector shapes, such as paraboloids, distribute energy rather than imaging the source; for these systems the size of the source is less important.

Assumption four implies that the only important radiator of energy in the system is the source. In verifying the validity of this assumption one must consider the temperature, area, and emittance of each surface. The emission from the reflector can be neglected if the temperature of the reflector is sufficiently low. Reflectors normally are operated at low temperatures (often by water cooling) to maintain surface quality. The emission from room temperature surroundings should also be small; however, the emission from the heated part can be important if the part's temperature and emittance are large.

Assumption five requires that the components of power reflected from the part, ϕ_3 and ϕ_4 (Fig. 2), not strike the reflector. The validity of this assumption is dubious if the part is a good, specular reflector. For parts which are good, diffuse absorbers (qualities which are desirable for effective radiant heating) the assumption should be valid. Assumption five also restricts the energy to traveling either directly or with only one reflection from the source to the part. This restriction limits the generality of the reflector shapes which may be treated; however, most shapes of practical interest meet this restriction. Almost all reflectors of conic cross section (ellipses, parabolas, and hyperbolas) will satisfy the restriction. The reflector may have concave and/or convex cross section.

III. Analysis

Energy Equation

We wish to calculate the radially symmetric heat flux, $E_p(z)$ W/m^2, where the subscript p indicates "part" and the variable z indicates that this flux distribution is a function of the radial coordinate (Fig. 1). $E_p(z)$ is the radiant flux absorbed by the heated part and, thus, depends on the surface characteristics of the part. The total power, ϕ_o, radiated by the source in the reflector system illustrated in Fig. 2 can be divided into seven distinct components: ϕ_1 = total power radiated through the central opening in the back of the reflector,† ϕ_2 = total power absorbed by the reflector, ϕ_3 = total power reflected by both the reflector and the part, ϕ_4 = total power directly reflected by the part, ϕ_5 = total power reflected by the reflector and absorbed by the part, ϕ_6 = total power directly absorbed by the part, ϕ_7 = total power radiated through the opening between the outer edge of the reflector and the part.

At each point z on the part there will be two components to $E_p(z)$: 1) the flux, E_5, which is reflected by the reflector and then absorbed by the part, 2) the flux, E_6, which is directly absorbed by the part. Thus,

†This opening is included to make the analysis more general. The opening may be eliminated as is done for all of the examples presented in a later section.

$$E_p(z) = E_5 + E_6 \quad (W/m^2) \tag{1}$$

In this section we trace polarized spectral power along the two paths from the source to the part using our previous assumptions. These components of power are then integrated over frequency and summed over components of polarization to obtain the total fluxes, E_5 and E_6. Substitution into Eq. (1) will yield the final form of the energy equation.

The polarized spectral radiant intensity of an ideal point source in the direction θ may be denoted as

$$I_\nu(\theta,j) \quad [W/(sr\text{-}Hz)] \tag{2}$$

where θ denotes the polar angle (Fig. 1), ν denotes the frequency,‡ and j denotes the component of polarization.§ Note, that while the intensity is assumed symmetric about the axis of rotation, it may vary with polar angle, θ. The energy distribution from many arc lamps is described by this model since the lamps radiate symmetrically about the axis of the two electrodes.

Consider the differential solid angle, $d\theta_s$, generated by rotating the differential plane angle, $d\theta$, about the x axis. The polarized spectral power, $d\phi_{o\nu}(j)$, radiated in $d\theta_s$ is

$$d\phi_{o\nu}(j) = I_\nu(\theta,j)d\theta_s \quad (W/Hz) \tag{3}$$

‡The subscript ν following a quantity means that it is a spectral quantity; or more precisely that it is the magnitude of the original quantity per unit frequently evaluated at the frequency ν. Thus, the units of I and I_ν are not the same: the first being W/sr and the second being W/(sr-Hz).

§The two components of polarization are the "s" and "p" components. j=s is the intensity component polarized perpendicular to the plane of incidence at the surface of the reflector or part; j=p is the intensity component polarized parallel to the plane of incidence. The plane of incidence is the plane which includes the surface normal and the incident ray. Thus, for the typical cross section of the axially symmetric system illustrated in Fig. 1, the planes of incidence for the reflector and part are identical, namely, the plane of the cross section. Therefore, the s direction is perpendicular to the cross section and the p direction lies in the cross section, perpendicular to the given ray path. The axially symmetric geometry simplifies the handling of polarizing effects.[6]

This power will strike the reflector at the point x,y where a portion will be absorbed and the remainder reflected in the specular direction. The reflected power is

$$d\phi_{r\nu}(j) = \rho_r(\nu,\Psi,j) \quad d\phi_{o\nu}(j) \tag{4}$$

where $\rho_r(\nu,\Psi,j)$¶ is the spectral specular reflectance for the j component of polarized power and Ψ is the angle of incidence of the incoming ray.

The reflected power strikes the part where a portion is absorbed and the remainder is reflected. The portion of power absorbed by the part is

$$d\phi_{5\nu}(j) = \alpha_p(\nu,\gamma,j)d\phi_{r\nu}(j) \tag{5}$$

where α (ν,γ,j) is the spectral directional absorptance of the part and γ is the angle of incidence of the incoming ray. No limitations are placed on the form of the directional absorptance; thus, the part is not restricted to being diffuse or specular.

The polarized spectral radiant flux on the differential annular area

$$dA = 2\pi \ z \ dz \tag{6}$$

due to the energy reflected by the reflector and absorbed by the part is

$$E_{5\nu}(j) = d\phi_{5\nu}(j)/dA \qquad [(W/(m^2\text{-}Hz)] \tag{7}$$

Substitution from Eqs. (3) - (5) yields

$$E_{5\nu}(j) = \rho_r(\nu,\Psi,j) \ \alpha_p(\nu,\gamma,j) \ I_\nu(\theta,j)d\theta s/dA \tag{8}$$

The total flux on the differential annular area due to the energy reflected by the reflector and absorbed by the part is obtained by integrating Eq. (8) over frequency and summing over components of polarization;

¶The frequency, ν, has been shown as a function rather than as a subscript because the units of spectral reflectance and total reflectance are identical (dimensionless). The present notation is used to contrast spectral surface properties from the other spectral quantities discussed in a preceeding footnote.

$$E_5 = \left[\sum_{j=s,p} \int_0^\infty \rho_r(\nu,\Psi,j)\ \alpha_p(\nu,\gamma,j)\ I_\nu(\theta,j)\ d\nu \right] \frac{d\theta s}{dA} \tag{9}$$

We now consider the radiant flux which travels directly from the source to the same differential annular area dA. The polarized spectral power radiated in the differential solid angle $d\beta_s$ is

$$d\phi_{o\nu}(j) = I_\nu(\beta,j)\ d\beta_s \tag{10}$$

The polarized spectral power directly absorbed by the part is

$$d\phi_{6\nu}(j) = \alpha_p(\nu,\beta,j)\ d\phi_{o\nu}(j) \tag{11}$$

The polarized spectral radiant flux on the differential area dA due to the energy directly absorbed by the part is

$$E_{6\nu}(j) = d\phi_{6\nu}(j)/dA \tag{12}$$

Substitution from Eqs. (10) and (11) yields

$$E_{6\nu}(j) = \alpha_p(\nu,\beta,j)\quad I_\nu(\beta,j)\ d\beta_s/dA \tag{13}$$

Using geometrical relationships we now proceed to express $d\beta_s/dA$ in terms of (1-d) and z. From Fig. 3

$$d\beta_s = 2\pi\ z\ dm/c^2 \tag{14}$$

Thus, dividing Eq. (14) by Eq. (6),

$$d\beta_s/dA = C^{-2}\ dm/dz \tag{15}$$

Again, from Fig. 3,

$$dm/dz = \cos\ \beta = (1-d)/c \tag{16}$$

From the Pythagorean Theorem

$$c = \left[(1-d)^2 + z^2\right]^{1/2} \tag{17}$$

Substitution of Eqs. (16) and (17) into Eq. (15) yields

$$d\beta_s/dA = (1-d)/\left[(1-d)^2 + z^2\right]^{3/2} \tag{18}$$

This geometrical relationship is substituted into Eq. (13) yielding

$$E_{6\nu}(j) = \frac{\alpha_p(\nu,\beta,j)\ I_\nu(\beta,j)\ (1-d)}{\left[(1-d)^2 + z^2\right]^{3/2}} \tag{19}$$

The total flux on the differential area, dA, due to the energy directly absorbed by the part is obtained by integrating Eq. (19) over frequency and summing over components of polarization.

$$E_6 = \frac{\left[\sum_{j=s,p}\int_0^\infty \alpha_w(\nu,\beta,j)\ I_\nu(\beta,j)\ d\nu\right](1-d)}{\left[(1-d)^2 + z^2\right]^{3/2}} \tag{20}$$

The total fluxes [Eqs. (9) and (20)] are now substituted into Eq. (1) yielding the energy equation

$$E_p(z) = \left[\sum_{j=s,p}\int_0^\infty \rho_r(\nu,\Psi,j)\ \alpha_p(\nu,\gamma,j)\ I_\nu(\theta,j)d\nu\right] d\theta s/dA$$

$$+ \frac{\left[\sum_{j=s,p}\int_0^\infty \alpha_p\ (\nu,\beta,j)\ I_\nu(\beta,j)\ d\nu\right](1-d)}{\left[(1-d)^2 + z^2\right]^{3/2}} \tag{H 21}$$ **

Geometrical Equations

The remaining six equations required to complete the mathematical model are geometrical and are developed in this subsection. From Fig. 1

$$\theta = \arctan\ \left[(y/d - x)\right];\ 0 \le \theta \le \pi \tag{A 22}$$

From Fig. 4

$$d\theta_s = dArea/b^2 = 2\pi\ y\ dw/b^2 \tag{23}$$

$$d\theta = dw/b \tag{24}$$

$$y/b = \sin\ \theta \tag{25}$$

Substitution of Eqs. (24) and (25) in Eq. (23) yields

** Lettered equations are used in the final model.

$$d\theta_s = 2\,\pi\,\sin\theta\,d\theta \tag{26}$$

Integration of Eq. (26) results in a relationship between θ_s and θ,

$$\theta_s = 2\,\pi\,(1 - \cos\theta) \qquad \text{(B 27)}$$

From Fig. 4

$$\Psi = \theta - \zeta \tag{28}$$

but

$$\tan\zeta = dx/dy \tag{29}$$

Thus,

$$\Psi = \theta - \arctan\left[1/(dy/dx)\right];\;\; 0 \leq \zeta \leq \pi \qquad \text{(C 30)}$$

Applying the law of cosines to the triangle with sides a, b, c yields

$$c^2 = a^2 + b^2 - 2\,a\,b\,\cos 2\Psi \tag{31}$$

where from Fig. 1

$$a^2 = (1-x)^2 + (z-y)^2$$

$$b^2 = (d-x)^2 + y^2$$

$$c^2 = (1-d)^2 + z^2$$

Rearranging Eq. 31 yields

$$z = f(z) = \qquad \text{(D 32)}$$

$$\frac{x^2+y^2+1(d-x)-xd-\sqrt{\left[(1-x)^2+(z-y)^2\right]\left[(d-x)^2+y^2\right]}\,\cos 2\Psi}{y}$$

Note that Eq. 32 is an implicit equation for z. The area of a part of radius z is

$$A = \pi\,z^2 \qquad \text{(E 33)}$$

From Fig. 1

$$\beta = \arctan\left[z/(1-d)\right]; \quad 0 \le \beta \le \pi/2 \qquad \text{(F 34)}$$

Also from Fig. 1

$$\gamma = \left| \theta - 2\Psi \right| \qquad \text{(G 35)}$$

Solution of Equations

The preceding two subsections have developed the seven equations, (A) - (H), which form the model. This subsection outlines the numerical solution of the equations. The model has been programed for numerical solution in Fortran IV.

The problem may be stated in functional form as

Given: x, y(x), dy/dx: The Reflector Shape

Find: $E_p(z)$: The Flux Distribution by Solving:

$$\theta = \theta(x, y) \qquad \text{(A)}$$

$$\theta_s = \theta_s(\theta) \qquad \text{(B)}$$

$$\Psi = \Psi(\theta, dy/dx) \qquad \text{(C)}$$

$$z = z(x, y, \Psi, z) \qquad \text{(D)}$$

$$A = A(z) \qquad \text{(E)}$$

$$\beta = \beta(z) \qquad \text{(F)}$$

$$\gamma = \gamma(\theta, \Psi) \qquad \text{(G)}$$

$$E_p = E_p(\Psi, \gamma, \theta, \beta, z, d\theta_s/dA) \qquad \text{(H)}$$

Inspection of the equations shows they have been formulated so that, knowing the values of x, y, and dy/dx at a point on the reflector, it is possible to evaluate numerically Eqs. (A)-(G) in sequence solving for: θ, θ_s, Ψ, z, A, β, and γ. All of the equations are explicit formulas except Eq. (D). Equation (D) must solved by an iterative technique; the technique used is Wegstein's method.[7] $d\theta_s/dA$ must be calculated before Eq. (H) may be evaluated to determine the value of $E_p(z)$. For most reflectors this derivative is a slowly varying function and may be calculated numerically with reasonable accuracy. $d\theta_s/dA$ is computed from a table of θ_s vs A by calculating the Lagrangian interpolation polynomial of degree two relevant to three successive tabulated points.[8††] The integrals in Eq. (H) are evaluated numerically when spectral effects are considered.

During the numerical calculations several quantities, in addition to $E_p(z)$, are computed. The calculations determine the seven components of the total power and further subdivide each component into its two polarized components. One other quantity, the power distribution,

$$\phi_p(z) = \int_0^z 2\pi \lambda E_p(\lambda)\, d\lambda, \tag{36}$$

is also calculated.

IV. Applications

This section applies the preceding model to reflectors of parabolic, hyperbolic, and elliptic cross section. These examples have been chosen both to demonstrate the use of the model and to illustrate several features of these common reflector shapes. In all of the examples spectral effects have been ignored to illustrate other important features of the system. The data for the nine examples are summarized in Table 1 and discussed in the following paragraphs.

Point Source

The energy emitted by the source is assumed to be unpolarized and independent of frequency and polar angle. Thus, the intensity of the source is

$$I(\theta,s) = I(\theta,p) = \phi_o/(8\pi) \ (W/sr) \tag{37}$$

The total radiated power, ϕ_o, is 16.5 W for all examples.

†† The derivative, $d\theta_s/dA$, may be eliminated by an alternative formulation of the problem; however, this formulation requires x, y, dy/dx, and d^2y/dx^2 to be given. In many problems only x and y are known, and dy/dx and d^2y/dx^2 must be calculated numerically. While the numerical value of dy/dx will be subject to limited numerical error, the calculated value of d^2y/dx^2 may be quite inaccurate; thus, the present approach appears preferable.

Table 1 Summary of data for examples

	Reflector			Heated Part	Geometry	
Ex.	Shape	n	k	Surface Characteristics	d, in.	l, in.
1	Parabolic	0.47[a]	4.84	Blackbody	0.5	5.0
2	Parabolic	0.47	4.84	Blackbody	0.5	20.0
3	Hyperbolic	0.47	4.84	Blackbody	0.5	5.0
4	Elliptic	0.47	4.84	Blackbody	0.902	4.439
5	Elliptic	0.47	4.84	Blackbody	0.902	4.389
6	Elliptic	0.47	4.84	Blackbody	0.902	4.339
7	Elliptic	0.47	4.84	Blackbody	0.902	4.289
8	Elliptic	1.75[b]	8.50	$n = 1.75$ $k = 8.5$[b]	0.902	4.289
9	Elliptic	$\rho = 0.912$		$\rho = 1 - \alpha = 0.912$	0.902	4.289

[a] Experimental measurements on aluminum at 0.436μ.[10]
[b] Experimental measurements on aluminum at 0.95μ.[10]

Reflector Shapes

The cross sections of the three reflectors studied are shown to scale in Fig. 5. The equations which describe the conic sections are

$$\begin{aligned} &\text{Parabola:} && y = \pm(2x)^{1/2} \text{ in.} \\ &\text{Hyperbola:} && y = \pm 0.6009(6x+x^2)^{1/2} \text{ in.} \\ &\text{Ellipse:} && y = \pm 0.75(5.34x-x^2)^{1/2} \text{ in.} \end{aligned} \tag{38}$$

The required values of y and dy/dx are computed analytically from Eq. (38). In every example the source is located at a focus of the conic cross section. The distance to the source from the rear of the reflector, d, is tabulated in Table 1. The second focus of the ellipse is located at $x = 4.439$ in.

Surface Characteristics of the Reflector

The reflectance, $\rho_r(\Psi,j)$ must be specified in Eq. (H). It is assumed that the dependency on direction and component of polarization is described by Fresnel's equations,[9]

$$\rho(\Psi,s) = \frac{a^2 + b^2 - 2a\cos\Psi + \cos^2\Psi}{a^2 + b^2 + 2a\cos\Psi + \cos^2\Psi}$$

$$\rho(\Psi,p) = \rho(\Psi,s)\left(\frac{a^2 + b^2 - 2a\sin\Psi\tan\Psi + \sin^2\Psi\tan^2\Psi}{a^2 + b^2 + 2a\sin\Psi\tan\Psi + \sin^2\Psi\tan^2\Psi}\right) \qquad (39)$$

where

$$2a^2 = \sqrt{(n^2 - k^2 - \sin^2\Psi)^2 + 4n^2k^2} + (n^2 - k^2 - \sin^2\Psi)$$

$$2b^2 = \sqrt{(n^2 - k^2 - \sin^2\Psi)^2 + 4n^2k^2} - (n^2 - k^2 - \sin^2\Psi)$$

and where Ψ is the angle of incidence, n is the refractive index, and k is the extinction coefficient. Experimental data are available for n and k as functions of frequency for many reflector materials. The values of n and k used for each example are tabulated in Table 1.

Figure 6 shows $\rho(\Psi,s)$ and $\rho(\Psi,p)$ for aluminum at 0.95μ[10] as calculated from Eq. (39). The figure indicates that the reflectance is constant only if Ψ is less than 45°; thus, if many of the rays are reflected at greater angles, the variation of reflectance with both Ψ and j may become significant in evaluating Eq. (H). Examples 8 and 9 show the importance of these directional and polarizing effects. Example 8 is based on the aluminum data presented in Fig. 6. Example 9 assumes the reflectance is independent of direction and component of polarization and uses the value of reflectance from Example 8 for $\Psi = 0$.

<u>Surface Characteristics and Location of the Heated Part</u>

Examples 1 - 7 assume the heated part to be a blackbody; thus,

$$\alpha_p(\beta,j) = 1 \qquad (40)$$

This value was selected to approximate the black, diffuse surface of a calorimeter. Examples 8 and 9 assume the heated part to have the same surface characteristics as the reflector; the absorptance is expressed in terms of the spectral reflectance as

$$\alpha_p(\beta,j) = 1 - \rho_p(\beta,j) \tag{41}$$

The distance to the heated part from the rear of the reflector, l, is given in Table 1.

Results

Parabolic Reflector. Figure 7 presents the flux distributions for Examples 1 and 2 illustrating the effect of part location, l, and of radial position, z, on the flux. The numerical calculations, in agreement with geometrical optics, predict that the reflected flux, E_5, is collimated; thus, the distribution of the reflected flux is independent of l. The direct flux, E_6, is not collimated, and its axial value is proportional to $(l-d)^{-2}$; thus, its magnitude decreases rapidly with increasing l. In Example 2 (l = 20 in.) the contribution of the direct flux is negligible (0.72% of the total power incident on a 3.16-in. radius circle). In a parabolic reflector system the energy is not completely collimated; the flux distribution is insensitive to l; and the flux is a strong function of z.

Hyperbolic Reflector. Figure 8 presents the flux distribution for Examples 3 and 1, illustrating the difference in distribution between hyperbolic and parabolic reflectors of similar geometry. The numerical calculations show that the reflected rays are divergent for the hyperbolic reflector. The diverging rays cause the flux distribution from the hyperbolic reflector to be lower and more dispersed than that from the parabolic reflector. This effect is more pronounced at large values of l.

Elliptic Reflector. Figure 9 presents the flux distributions for Examples 4-7, illustrating the effect of part location, l. In Example 4 the part is located at the second focus; the flux distribution at this plane is the two-dimensional Dirac delta-function[11] if the direct radiation is neglected. This unrealistic prediction results from the idealized assumption of a point source. In each succeeding example the part is moved an increment of 0.050 in. toward the source. Succeeding distributions are less peaked and more dispersed than preceding ones. Figure 10 presents the power distributions calculated from Eq. (36). In Example 4 all of the power is incident at z = 0 if the direct radiation is neglected. As the part is moved toward the source an increasing area is required to absorb a prescribed amount of power.

Directional and Polarizing Effects. In the preceding examples directional and polarizing effects were now shown.‡‡ Figure 6 has illustrated that the reflectance of aluminum depends on Ψ, and j only when Ψ is greater than 45°. Unpolarized energy must strike two such surfaces at angles greater than 45° before polarizing effects must be treated; the first surface polarizes the reflected energy, the second surface then selectively absorbs or reflects the two different components of polarized energy. Directional and polarizing effects are minimal in parabolic and hyperbolic reflectors (when the source is at a focus) since most of the energy strikes both the reflector and part at angles less than 45°. For a parabolic system, exactly one half of the energy radiated by the source strikes the reflector at an angle less than 45°, and all of the reflected energy strikes the part at normal incidence.

In an elliptical system with the part near the second focus many rays will strike the reflector and then the part at angles greater than 45° causing directional and polarizing effects to become significant. These effects are important for the geometry of Examples 8 and 9; these effects are considered in Example 8; they are neglected in Example 9. (See Table 1). Neglecting directional and polarizing effects in Example 9 causes an error in the calculated absorbed flux which ranges from +0.3 percent at z = 0 in. to -27% at z = 0.464 in. It should be noted, however, that the magnitude of the absorbed flux at z = 0.464 in. is only 0.06 percent of the magnitude at z = 0 in. The integrated effect is illustrated by comparing the total power [Eq. (36)] absorbed by a 1.600 in. radius part; neglecting directional and polarizing effects results in an error in the power of -3.5%. Thus, while directional and polarizing effects frequently can be ignored, their effect should be investigated when the energy strikes both the reflector and part at large angles of incidence.

V. Conclusions

The following conclusions have been reached from the cases analyzed:

---

‡‡Since the absorptance of the part was independent of j, the summations in Eq. (H) could have been eliminated by using a directional spectral reflectance for the reflector which was averaged with respect to j.

1) The model, in agreement with geometrical optics, predicts that a parabolic reflector collimates the reflected energy. However, the direct energy (ϕ_6 of Fig. 2) is not collimated. The heat flux absorbed by the part is a strong function of radial position, but insensitive to axial position.

2) An hyperbolic reflector results in an uncollimated beam of more uniform distribution than the parabolic reflector.

3) The model, in agreement with geometrical optics, predicts that an elliptic reflector will refocus energy from a point source located at one focus to form a point image at the other focus. Thus, the predicted heat flux distribution on a plane passing through the second focal point is the two-dimensional Dirac delta-function.

4) Directional and polarizing effects should be considered when the energy strikes both the reflector and part at large angles of incidence (on the average).

References

[1]Anderson, J. W., LaBlanc, E. A., and Cohan, H., "Experimental and Analytical Assessment of Space Thermal and Vacuum Environment Simulation Requirements," Journal of Spacecraft and Rockets, Vol. 3, No. 7, July 1966, pp. 976-982.

[2]Schoenthaler, D., "Reflow Soldering with Radiant Heating," Symposium Record, International Electronic Circuit Packaging Symposium, Calif., Aug. 20-21, 1969.

[3]Jacob, S. M., Dranoff, J. S., "Light Intensity Profiles in an Elliptical Photoreactor," AIChE Journal, Vol. 15, No. 1, Jan. 1969, pp. 141-144.

[4]Jeffery, R. W., "Infrared Heating in Appliance Manufacturing," IEEE paper #34CP 67-301, presented at the IEEE 1967 18th Annual Appliance Technical Conference, 1967.

[5]Schalk, E. N., "Radiant Heating - High Power Density," IEEE Special Publication T-148, April 1963, pp. 102-116.

[6]Edwards, D. K. and Bevans, J. T., "Effect of Polarization on Spacecraft Radiation Heat Transfer," AIAA Journal, Vol. 3 No. 7, July 1965, pp. 1323-1329.

[7]Wegstein, J. H., "Accelerating Convergence of Iterative Processes," Communications of the Association for Computing Machinery, Vol. 1, No. 6, June 1958, pp. 9-13.

[8]Hildebrand, F. B., Introduction to Numerical Analysis, McGraw-Hill, New York, 1956, pp. 64-68.

[9]Sparrow, E. M. and Cess, R. D., Radiation Heat Transfer, Brooks/Cole, Belmont, Calif., 1966, pp. 63-64.

[10]Hass, G., "Mirror Coatings," Chap. 8, Applied Optics and Optical Engineering, Vol. III, edited by R. Kingslake, Academic Press, New York, 1965, pp. 309-330.

[11]Courant, R. and Hilbert, D., Methods of Mathematical Physics, Vol. II, Interscience, New York, 1962, p. 791.

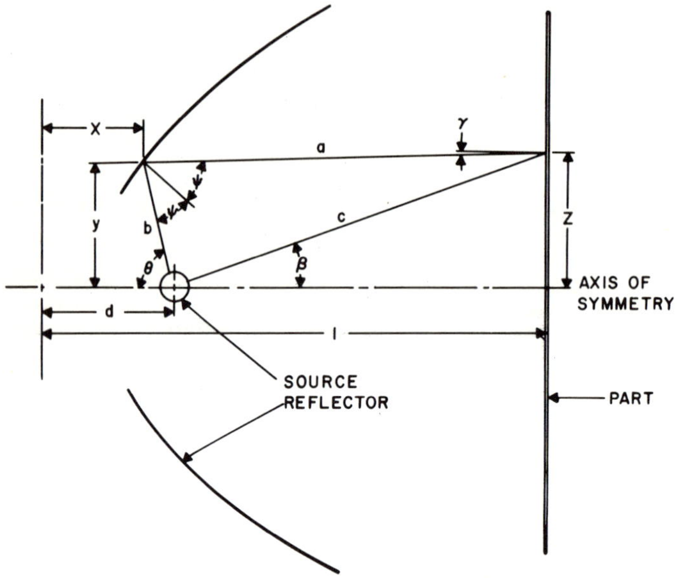

Fig. 1 System geometry.

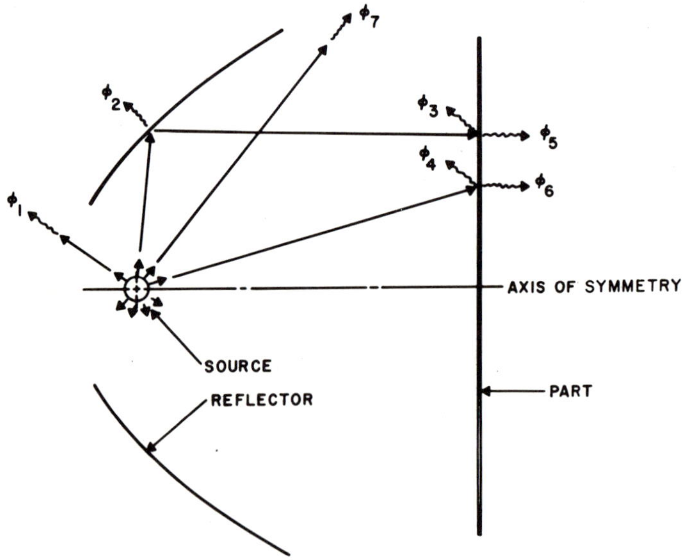

Fig. 2 Components of power.

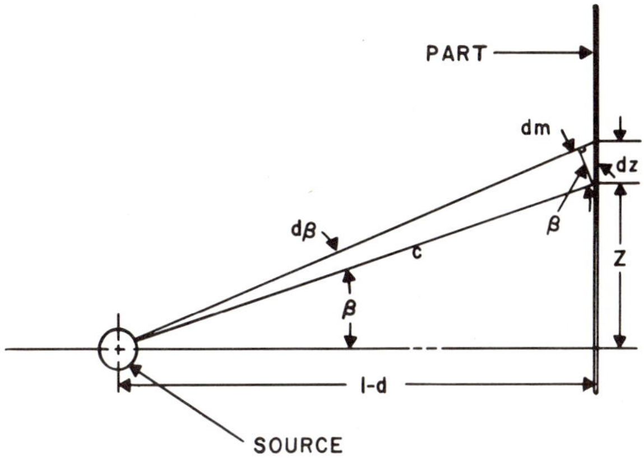

Fig. 3 Differential geometry for part.

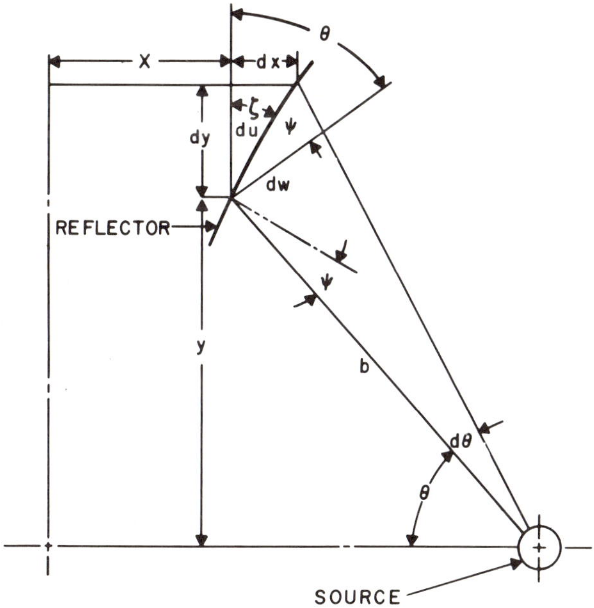

Fig. 4 Differential geometry for reflector.

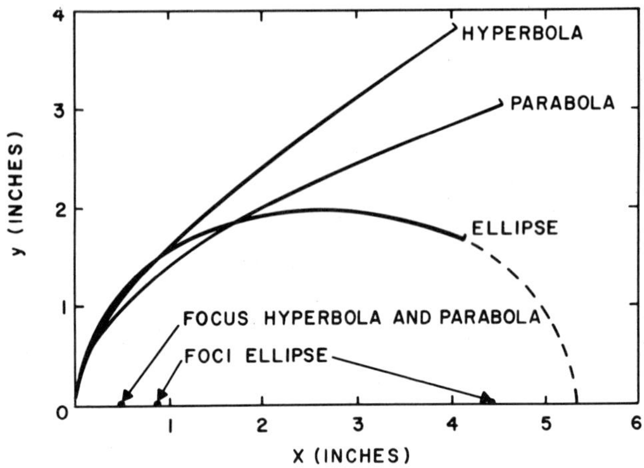

Fig. 5 Reflector shapes.

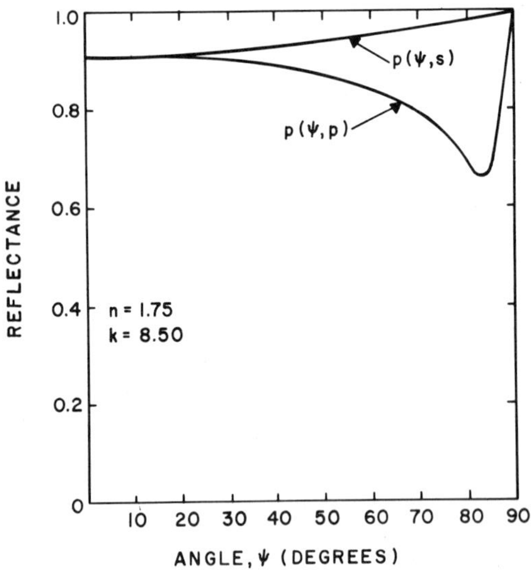

Fig. 6 Reflectance of aluminum at 0.95μ.

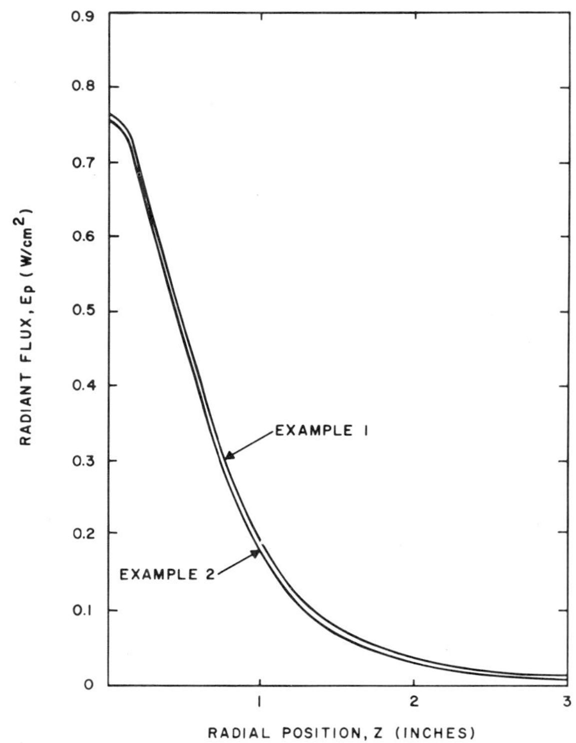

Fig. 7
Flux distributions for parabolic reflectors.

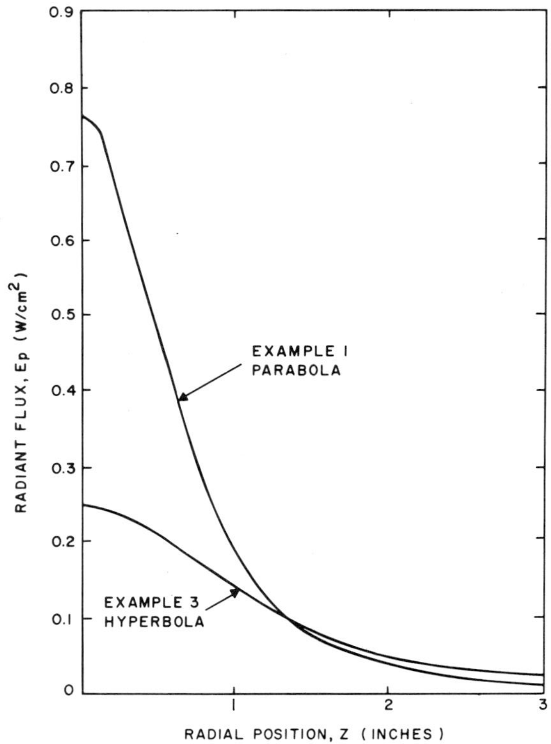

Fig. 8
Comparison of flux distributions for hyperbolic and parabolic reflectors.

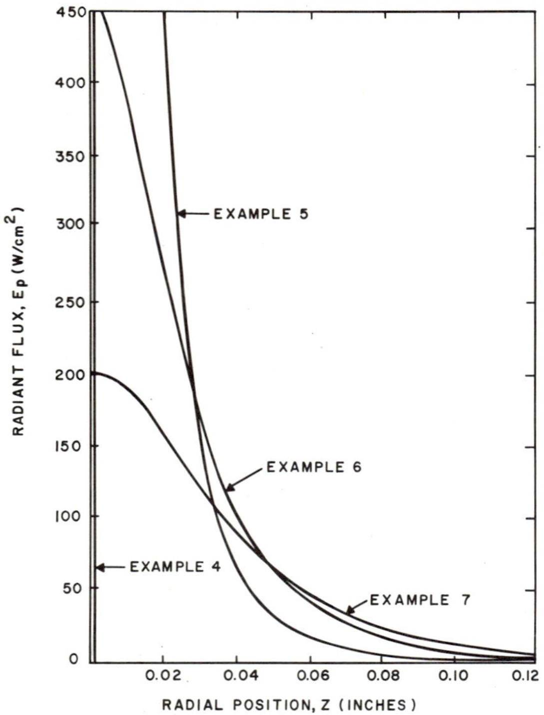

Fig. 9
Flux distributions for elliptic reflectors.

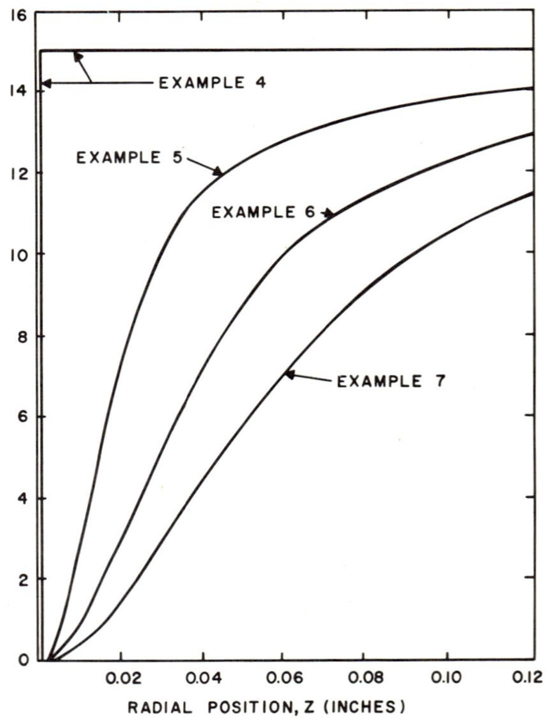

Fig. 10
Power distributions for elliptic reflectors.

THERMAL NETWORK CORRECTION TECHNIQUES

Takao Ishimoto* and Henry M. Pan†

TRW Systems Group, Redondo Beach, Calif.

Abstract

Predictions from the use of a lumped parameter system (thermal network) to describe a complex physical system accurately are subject to the inaccuracies of the parameters, which include conduction and radiation coupling considerations. Quite often the parameter inaccuracies are large, and the true parameter values are difficult to determine. As a result, thermal tests are conducted in ground-based facilities to verify the thermal math-models. Thermal math-model verification represents a very difficult undertaking because of the interaction of the parameter values on the temperatures. The present paper reviews a number of techniques that may evaluate the parameters of the network directly, given experimental temperature measurements. A detailed development of the Kalman filtering method and application to a thermal model also are presented.

1. Introduction

Design and projected operational characteristics of a spacecraft thermal control system in a heat facility and in space require an accurate mathematical model of the physical system. The more sophisticated the physical system, the more complex the mathematical model; as a result, the more difficult it is to generate an accurate representation of the physical system.

Presented as Paper 70-821 at the AIAA 5th Thermophysics Conference, Los Angeles, California, June 29-July 1, 1970. This work was sponsored by NASA/MSC under Contract NAS 9-8289. The authors are grateful to J.T. Bevans for his guiding encouragement, to R. Dotts for his sympathetic understanding of the problems as the NASA/MSC technical monitor, to J.D. Gaski for his generation of the Kalman filtering program, to E.B. Stear for helpful suggestions, and to S. Franklin for her effort in the typing of this paper.

*Staff Engineer, Heat Transfer and Thermodynamics Department.

†Member of the Technical Staff, Heat Transfer and Thermodynamics Department.

Confidence in a thermal design can be achieved either by a direct verification of the thermal control system in an accurate space simulation (near duplication) test or by the use of mathematical model verification. Since the former is extremely difficult to achieve, the latter has been receiving more and more attention.

Model verification requires a comparison between analytical predictions and test data with correction of the mathematical model when analytical and test results are not in agreement. When the thermal network is large, the usual trial and error corrective procedure which relies exclusively on engineering judgment is totally inadequate; a direct technique would represent a significant analytical tool that would have considerable utility in the area of thermal control design.

The types of errors that make up the difference between measured and predicted temperature are mainly three in nature: 1) functional form of the model; 2) measurement errors; and 3) incorrect parameter values. If the functional form of the model is not sufficiently accurate, a new model must be constructed. Thus, parameter estimation methods will be applicable only if errors are confined to measurement errors and parameter inaccuracies.

The physical system of concern is one that can be represented by a thermal network of resistances and capacitances which can be expressed mathematically as follows:

$$C_i \frac{dT_i}{dt} = Q_i + \sum_{j=1}^{r} a_{ij}(T_j - T_i) + \sum_{j=1}^{r} \sigma b_{ij}(T_j^4 - T_i^4) \qquad (1)$$

$$i = 1,2,\ldots,n, \quad T_j = \text{const for } n<j\leq r$$

where C_i is the capacitance of the ith node, T_i is the temperature of the ith node, t is the time, Q_i is the net heat input to the ith node, a_{ij} is the conductance, b_{ij} is the radiation coefficient, and σ is the Stefan-Boltzmann constant.

2. Methods of Parameter Estimation

Correction of thermal model parameters from noisy temperature measurements is a class of problem designated estimation problems; the term noisy is defined here to mean random errors. In general, an estimation problem involves the identification and estimation of the parameters and states of a physical system from noisy measurements of some observable quantities of the system.

2.1 Literature Search

A literature search revealed papers on regression analysis,[1-6] on least squares,[7-19] on Bayesian estimation,[6,17,20] on maximum likelihood,[1,6,17] on invariant imbedding,[21-23] on quasilinearization,[24,25] on Kalman filtering,[16,17,26-29] and on sensitivity analysis which may be an integral part of any parameter correction scheme.[30,31] These methods have been categorized here into two different classes. One class, denoted as one-pass methods, involves the estimation of parameters and states over a period of time, whereby all the measurements are processed in one pass; the methods in this class include regression analysis, least squares, maximum likelihood, and quasilinearization. The other, denoted as sequential methods, involves the estimation of parameters and states sequentially, whereby the measurement data are processed sequentially and new estimates of the parameters are obtained after each set of measurements is processed. Methods of sequential estimation include Kalman filtering, invariant imbedding, and Bayesian estimation. These methods are discussed below, but selected ones are presented in more detail.

2.2 One-Pass Methods

The method of least squares is a classical method that in its simplest form involves the formation of a criterion function ϵ which is the sum of the squares of the difference between the computed model response and a set y of m experimental observations of a scaler output. These observations can be weighed if desired. Least squares estimate of the parameters p and states x are defined as the values of p and x that minimize ϵ.

Estimation of the temperature requires the use of the heat balance equations with the measured temperatures as boundary conditions; this represents a multi-point boundary value problem that can be solved by the method of quasilinearization.[24,25] Quasilinearization in essence is a method whereby the governing nonlinear heat balance differential equations are linearized by a 1st-order Taylor series expansion about nominal temperature profiles and nominal parameter values. Nominal values are defined by an a priori estimate of the initial conditions. The linearized heat balance equations are solved assuming unknown parameter values and initial conditions. Improved temperature profiles are obtained as a function of knowns, and the unknowns are then determined in the least squares sense by minimizing the square of the difference between the improved temperature profiles and the measured temperatures. After all the data have been processed, a new estimate of the unknowns is obtained. The results are then used to obtain new temperature profiles, and the cycle is repeated. This process continues (generally several times) until the change in these unknowns is small. This method can also be applied when the temperature is not measured at every nodal location.

Toussaint[12] applied the method of least squares for the correction of thermal modal parameters. However, in lieu of using temperature as the measurement vector, Toussaint minimized the net heat flux residual at each nodal location over some specified time span.

2.2.1 Deterministic. The deterministic method requires accurate (if not perfect) temperature measurements at each and every nodal location; if transient temperatures are used, the temperature derivative must also be known accurately. If the number of unknown parameters (excluding capacitances) are less than or equal to the number of nodal temperatures, steady-state heat balance equations may be used; otherwise transient data must be utilized.

Consider the model heat balance equations (1); if the measured temperatures were used in lieu of model temperatures, the heat flow at the ith node would be zero only if the measured temperatures and the coefficients were perfect. If all of the coefficients were not exact, the net heat flux at the ith node would be a finite value; $\tau_i(t)$, Eq. (1) would be expressed as

$$\tau_i(t) = C_i \frac{dT_i}{dt} - Q_i + \sum_{j=1}^{r} a_{ij}(T_i - T_j) + \sum_{j=1}^{r} \sigma b_{ij}(T_i^4 - T_j^4) \qquad (2)$$

$$i = 1, 2, \ldots, n$$
$$T_j = \text{const for } n < j \leq r$$
$$a_{ij} = a_{ji},\; b_{ij} = b_{ji}$$

where all the coefficients are not exact.

Setting $\tau_i(t) = 0$ and using the temperature data at a sufficient number of time instants, a set of linear equations can be generated to obtain the parameter values; this method assumes, of course, that the temperature derivatives can be determined with arbitrary accuracy and hence can be assumed to be perfect. Note that since multiplication of each parameter by a constant for fixed i gives the same equation, only ratios of the parameters to one parameter can be determined for each node. It should also be noted that, if the time slices are too closely spaced, the parameter coefficient matrix will be almost singular - that is, the set of equations are ill-conditioned. This leads to numerical solution difficulties. If the temperature data are not accurate, the parameter values would necessarily be forced to compensate for the inaccuracies.

2.2.2 Method of Least Squares (Net Heat Flux Residual). One formulation of least squares is the use of the net heat flux residual at every node i as indicated in Eq. (2). This method also requires that the temperatures be measured at every nodal location. If the temperature at every node were perfect, then the residual at a given node is the result of inaccuracies of the parameter values. If the temperature measurements were not accurate, then the heat flux residual represents combined effects of temperature and parameter inaccuracies.

Using Eq. (2), the mean squared residue R may be expressed over a time period $t_f - t_o$ as follows:

$$R = \frac{1}{(t_f - t_o)} \int_{t_o}^{t_f} \sum_{i=1}^{n} [\tau_i(t)]^2 dt \qquad (3)$$

The set of normal equations is obtained by taking the derivative of Eq. (2) with respect to parameters Q_k, $a_{k\ell}$, $b_{k\ell}$, C_k, and T_k. Solution of resultant equations yields the unknown parameter values if the temperatures are known accurately at every node. Since this least squares method represents, in essence, smoothing of the information from time t_o to t_f, the need for accurate temperature derivatives and the problem of ill-conditioned equations are not considered to be as severe as for the deterministic method discussed previously.

2.2.3 Method of Least Squares (Measured and Model Temperature Difference). Another formulation of the least squares method is the use of the temperature difference between measurement and model. With this formulation, random noise can be superimposed on the temperature measurement and temperature sparsity can be accommodated provided of course that physical considerations are satisfied.

Consider a rearranged form of the heat balance Eq. (1).

$$\frac{dT_i}{dt} = \frac{Q_i(t)}{C_i} + \sum_{j=1}^{r} \frac{a_{ij}}{C_i}(T_j - T_i) + \sigma \sum_{j=1}^{r} \frac{b_{ij}}{C_i}(T_j^4 - T_i^4) \tag{4}$$

$$i = 1,2,\ldots,n$$

If the heat input to each node is a function of time and temperature, an additional set of n heat input equations of the following form must be considered:

$$Q_i(t) = f_i(T_i, T_j, t, \beta) \tag{5}$$

where β represents a heating constant such as the solar constant.

For a typical thermal model with n nodes, with (p-n) model parameters and m measured nodal temperature, where $m \leq n$ and p are the sum of the number of nodes and the number of model parameters, it is desired to obtain the least squares estimate of the n nodal temperatures and the (p-n) model parameters over some period, using the m noise corrupted nodal temperature measurements.

Let the measurement vector $\underline{y}_t^*$ at time t be an m by 1 vector, whose elements are the m noise corrupted measured temperatures at time t. Then

$$\underline{y}_t^{*T} = (T_{1,t}^*, T_{2,t}^*, \ldots, T_{m,t}^*) \tag{6}$$

(Note that the superscript T indicates transpose.)

Let the state vector $\underline{x}_i$ at time t be a p by 1 vector, whose elements are the n nodal temperatures and the (p-n) model parameters at time t. Then

$$\underline{x}_t^T = [T_{1,t}, T_{2,t}, \ldots, T_{n,t}, (Q_i/C_i)_t, \ldots, (b_{ij}/C_i)_t] \tag{7}$$

The relation between the measurement vector and the state vector at time t is given by the following observation equation:

$$\underline{y}_t^* = M_t \underline{x}_t + \underline{w}_t \tag{8}$$

where M_t is the m by p measurement matrix at time t, and $\underline{w}_t$ is the m by 1 noise vector whose elements are the noise associated with the measured temperatures at time t.

$$M_t = \left[\begin{array}{c|c} I & \\ (m \times m) & 0 \end{array} \right] \tag{9}$$

Assuming that linear relations‡ can be used to express the updating of the n nodal temperatures and the n nodal heat input from time t to a later time $t+\Delta t$, and assuming the remaining (p-2n) model parameters are constant, the following updating equation for the state vector can be generated:

$$\underline{x}_{t+\Delta t} = U_t \underline{x}_t \tag{10}$$

where U_t is the updating matrix evaluated at time t.

Assuming that k sets of noisy temperature measurements were obtained for the m nodes from time t=0 to time t=k, and these k sets are $\underline{y}_o^*, \underline{y}_1^*, \ldots, \underline{y}_k^*$, we wish to obtain the least squares estimate of $\underline{x}_k$, whose elements are the n nodal temperatures and the (p-n) model parameters at time t=k. In other words, we want to obtain $x_{i,k}$, the elements of $\underline{x}_k$ as a function of the k sets of temperature measurements such that they minimize the following criterion function:

$$\epsilon_k = \sum_{t=o}^{k} \left| \left| M_t \underline{x}_t - \underline{y}_t^* \right| \right|_{W_t} \tag{11}$$

where W_t, an m by m diagonal matrix, is an arbitrary positive weighting matrix for the measurement vector $\underline{y}_t^*$. This implies that we have some idea of the relative magnitudes of the random measurement noises involved, otherwise W_t can be taken to be an identity matrix.

Upon expressing $\underline{x}_t$ as a function of $\underline{x}_k$, $t < k$, Eq. (11) can be expressed as

$$\epsilon_k = \sum_{t=0}^{k} \left| \left| K_{k,t} \underline{x}_k - \underline{y}_t^* \right| \right|_{W_t}^2 \tag{12}$$

$$\text{where} \quad K_{k,t} \triangleq M_t \, \Phi_{k,t}^{-1} \tag{13}$$

‡The linear relations can usually be obtained by proper linearization of the heat balance equations and the equations governing the heat input; e.g., truncated Taylor series expansion.

$$\Phi_{k,t} \triangleq \prod_{i=t}^{k-1} U_i \tag{14}$$

The estimation problem is reduced to the minimization of ϵ_k with respect to $\underline{x}_k$. The set of optimal $\underline{x}_k$ that minimizes the criterion function ϵ_k is called the least squares estimate of $\underline{x}_k$ and is denoted $\hat{\underline{x}}_k$. $\hat{\underline{x}}_k$ can be obtained by solving the set of normal equations obtained by differentiating Eq. (12) with respect to $\underline{x}_k$. The set of normal equations is given by

$$\sum_{t=0}^{k} [K_{k,t}\hat{\underline{x}}_k - \underline{y}_t^*]^T W_t K_{k,t} = 0 \tag{15}$$

Solving for $\hat{\underline{x}}_k$ gives

$$\hat{\underline{x}}_k = [\sum_{t=o}^{k} K_{k,t}^T W_t K_{k,t}]^{-1} [\sum_{t=o}^{k} K_{k,t}^T W_t \underline{y}_t^*] \tag{16}$$

Equation (16) gives the least squares estimate of the n nodal temperatures and the (p-n) thermal model parameters at time t=k using the noisy temperature measurements for time t=o to time t=k.

Available a priori estimates can be incorporated into the criterion function as additional measurements. Let the a priori estimate to the state vector at time t=k be $\underline{x}_{a,k}$. Incorporating this a priori estimate into the criterion function as defined in Eq. (12), we get a new criterion function given by

$$\epsilon_{k,a} = \left|\left| \underline{x}_k - \underline{x}_{a,k} \right|\right|_P^2 + \sum_{t=o}^{k} \left|\left| K_{k,t}\underline{x}_k - \underline{y}_t^* \right|\right|_{W_t}^2 \tag{17}$$

where the p by p matrix [P] , is a weighting matrix for the a priori estimates.

Upon differentiating $\epsilon_{k,a}$ with respect to the state vector $\underline{x}_k$, and solving for the least squares estimate $\hat{\underline{x}}_k$ from the resultant normal equations, we get

$$\hat{\underline{x}}_k = [P + \sum_{t=o}^{k} K_{k,t}^T W_t K_{k,t}]^{-1} [P\underline{x}_{a,k} + \sum_{t=o}^{k} K_{k,t}^T W_t \underline{y}_t^*] \tag{18}$$

2.2.4 Method of Quasilinearization. The method of quasilinearization as applied to thermal model parameter correction may be developed as follows:

Consider the heat balance equation (4). This set of n equations can be represented by the following vector differential equation:

$$\dot{\underline{T}} = \underline{f}(\underline{T},\underline{p},t) \tag{19}$$

where $\underline{T}$ is a vector that consists of the n nodal temperatures and $\underline{p}$ is a vector that consists of the k thermal model parameters. By making transient temperature measurement of all or part of the n isothermal nodes, it is desired to determine the vector $\underline{p}$ and a complete set of initial conditions $\underline{T}(t_o)$ for the n nodes, such that the solution to Eq. (19) is in closest agreement with the measured data. The intent is to determine the thermal model parameter vector $\underline{p}$ and the initial temperature vector $\underline{T}(t_o)$ such that the sum of the squares of the deviations between the computed temperatures and the measured temperatures for the measured nodes is minimized.

Let the mth approximated solution to Eq. (19) be given by

$$\underline{T}^{(m)} = \underline{g}\ [\underline{T}^{(m)}(t_o),\ \underline{p}^{(m)},\ t\ ] \tag{20}$$

By the method of quasilinearization, the m+1 approximated solution can be obtained by solving the following n linearized differential equations:

$$\dot{T}_i^{(m+1)} = f_i\ [\underline{T}^{(m)},\underline{p}^{(m)},t] + \sum_{j=1}^{n} \frac{\partial f_i[\underline{T}^{(m)},\underline{p}^{(m)},t]}{\partial T_j}\ x$$

$$[T_j^{(m+1)}-T_j^{(m)}] + \sum_{j=1}^{k} \frac{\partial f_i[\underline{T}^{(m)},\underline{p}^{(m)},t]}{\partial p_j}\ x$$

$$[p_j^{(m+1)}-p_j^{(m)}]\ ,\quad i = 1,2,\ldots n \tag{21}$$

The improved thermal model parameter vector $\underline{p}^{(m+1)}$ and the improved initial temperature vector $\underline{T}(t_o)^{(m+1)}$ are determined in the least squares sense by minimizing the following criterion function:

$$\epsilon = \sum_{t=o}^{t} \left|\left|\ \underline{T}^{*}(t) - \underline{T}(t)^{(m+1)}\right|\right|_{W_t} \tag{22}$$

where $\underline{T}^*(t)$ = measured temperature for the measured nodes, $\underline{T}(t)^{(m+1)}$ = computed temperature for the measured nodes obtained by solving the system of differential equations as given by Eq. (21), and w_t = some weighting function for the measured temperature if the relative magnitude concerning the measurement noise is known.

Since $\underline{T}(t)^{(m+1)}$ is the solution of a system of linear differential equations, it may be represented in the form

$$\underline{T}(t)^{(m+1)} = \underline{c}\ [t,\underline{p}^{(m+1)}] + \sum_{j=1}^{n} \underline{h}_j(t)T_j(t_o)^{(m+1)} \tag{23}$$

where the vector $\underline{c}\ [t,\underline{p}^{(m+1)}]$ is the particular solution of Eq. (21) that involves terms consisting of the improved parameter vector $\underline{p}^{(m+1)}$. The vector $\underline{h}_j(t)$ is the solution of the homogeneous form of Eq. (21). These vectors, $\underline{c}\ [t,\underline{p}^{(m+1)}]$ and $\underline{h}_j(t)$, are determined computationally on the measurement interval, time t_o to time t.

The improved initial nodal temperatures, $\underline{T}(t_o)^{(m+1)}$, and the improved thermal parameters, $\underline{p}^{(m+1)}$, are determined by minimizing the following:

$$\epsilon = \sum_{t=t_o}^{t} \left|\left| \underline{T}^*(t)-\underline{c}\ [t,\underline{p}^{(m+1)}] - \sum_{j=1}^{h} \underline{h}_j(t)T_j(t_o)^{(m+1)} \right|\right|_{W_t} \tag{24}$$

In this way, the problem of thermal model parameters correction is reduced to the solution of systems of linear initial value problems and of linear algebraic equations.

2.3 Sequential Methods

Kalman,[26,28] Kalman and Bucy[27] were the first to propose a scheme whereby the states and parameters of a linear dynamic system can be estimated sequentially. This is the method of Kalman filtering which is a minimum variance estimate whereby the variances of the states and parameters to be estimated are minimized. Kalman filtering can also handle nonlinear systems by linearizing the nonlinear equations about some nominal trajectory (temperature history for the present study).[16] Browne[29] has applied the Kalman filter to correct the parameters of a thermal model. In his study, he neglected the radiation coupling terms and thus eliminated the nonlinearity in the governing heat balance differential equations. This approximation allowed the direct application of the Kalman filter.

Other sequential schemes have been developed for the estimation of states and parameters. These are the recursive least squares and invariant imbedding. The recursive least squares is a scheme whereby the states and parameters of a system can be estimated in a least squares sense without recalculating the entire least squares problem when additional data are available. Invariant imbedding is essentially an approximation method whereby the states and parameters of nonlinear dynamic systems can be estimated using the method of least squares.

2.3.1 Kalman Filtering Method. The functional model, as assumed for the one-pass method, is again considered to represent the physical system accurately and it is the intent to correct the model parameters using the measured temperatures. Since the measured temperatures generally include random observation noises, it is desirable to formulate a technique whereby the random noises are filtered. Consider the heat balance equation (4) with the heat input function of each node expressed as in Eq. (5). An applicable method currently being used in the field of optimal control is the method of Kalman filtering. In this study, the method of Kalman filtering is applied to correct the various thermal model parameters using noisy temperature measurements.

For a thermal model that contains n nodes with m nodal temperatures measured, where $m \leq n$, the random noise corrupted measurement vector $\underline{y}^*$ is an m by 1 vector whose elements are given by the m noise corrupted measured temperatures. This is given by Eq. (6).

In the present problem, the sum of the number of model parameters and the number of isothermal nodes is p. The state vector is a p by 1 vector whose elements are the n nodal temperatures and the (p-n) model parameters. The (p-n) model parameters are the (Q_i/C_i)'s, (a_{ij}/C_i)'s, and the (b_{ij}/C_i)'s. The state vector is given in Eq. (7). The relation between the measurement vector and the state vector is given by the observation Eqs. (8) and (9). The random measurement noise vector $\underline{w}$ is composed of elements which are associated with the m measured temperatures. This is given by

$$\underline{w}^T = (w_1\ w_2\ \dots\ w_m) \tag{25}$$

It is assumed that $\underline{w}$ has a zero mean, that is $E\{w\} = 0$, and the covariance matrix of $\underline{w}$ is given by

$$W = E\ [\underline{w}\ \underline{w}^T] \tag{26}$$

Let $\underline{x}_a$ be an unbiased a priori estimate of $\underline{x}$, where unbiased means $E[\underline{x} - \underline{x}_a] = 0$, and let the p by p error covariance matrix of $\underline{x}_a$ be given as

$$A = E\,[(\underline{x} - \underline{x}_a)\,(\underline{x} - \underline{x}_a)^T] \tag{27}$$

Furthermore, it is assumed that the random measurement error and the a priori estimation error are uncorrelated; that is,

$$E\,[\underline{w}(\underline{x}-\underline{x}_a)^T] = 0 \tag{28}$$

It is now desired to obtain a new and "better" estimate of $\underline{x}$ using the measurement $\underline{y}^*$ and the a priori estimate $\underline{x}_a$. Let the new estimate of $\underline{x}$ be $\hat{\underline{x}}$, and let $\hat{\underline{x}}$ be a linear combination of $\underline{y}^*$ and $\underline{x}_a$ such that the following relation is satisfied:

$$\hat{\underline{x}} = B\underline{y}^* + C\underline{x}_a \tag{29}$$

The matrix B and C are called the measurement weighting matrix and the a priori weighting matrix, respectively.

For $\hat{\underline{x}}$ to be an unbiased estimate, $E\{\hat{x}\}$ must be equal to $E\{x\}$. Thus from Eq. (29), we get

$$E\{\hat{\underline{x}}\} = E(B\underline{y}^* + C\underline{x}_a) \tag{30}$$

Upon substituting Eq. (8) into Eq. (30), we get

$$E\{\hat{x}\} = BME\{x\} + BE\{w\} + CE\{x_a\} \tag{31}$$

since $\quad E\{x\} = E\{x_a\}$ and $E\{w\} = 0$

Equation (31) reduces to

$$E\{x\} = (BM + C)E\{x\}$$

Thus, in order for $\hat{\underline{x}}$ to be an unbiased estimate, we must have

$$BM + C = I \tag{32}$$

where I is a p by p identity matrix.

Similar to the error covariance matrix for $\underline{x}_a$, we define the error covariance matrix of the estimate $\hat{\underline{x}}$

$$J = E\,[(\underline{x} - \hat{\underline{x}})\,(\underline{x} - \hat{\underline{x}})^T] \tag{33}$$

By using Eqs. (20), (8), and (32), we obtain

$$J = E\ [\{(I-BM)(x-\underline{x}_a) - B\underline{w}\}\ \{(I-BM)(\underline{x}\ \underline{x}_a) - B\underline{w}\}^T]$$

With Eqs. (26) and (27), Eq. (33) reduces to

$$J = (I-BM)A(I-BM)^T + (BWB^T) \tag{34}$$

In order to obtain a minimum variance estimate of $\underline{x}$, the matrix B must be chosen such that the quadratic form associated with J is minimized. The quadratic form of J is given by

$$S = \underline{z}^T J \underline{z} \tag{35}$$

where $\underline{z}$ is any arbitrary p by 1 vector.

Taking the variation of S gives

$$\delta S = \underline{z}^T\ [-(I-BM)AM^T + BW]\ \delta B^T \underline{z} + [\underline{z}^T\ [-(I-BM)AM^T + BW]\ \delta B^T \underline{z}]^T \tag{36}$$

Since both terms in the right-hand side are scalers, each of which is the transpose of the other, and because the transpose of a scaler is that scalar itself, we obtain

$$\delta S = 2\underline{z}^T\ [-(I-BM)AM^T + BW]\ \delta B^T \underline{z}$$

S will be minimized for all $\underline{z}$ if

$$(I-BM)AM^T = BW \tag{37}$$

Solving for B gives

$$B = AM^T(MAM^T + W)^{-1} \tag{38}$$

Combining Eqs. (29) and (32), we get

$$\hat{\underline{x}} = \underline{x}_a + B(\underline{y}^* - \underline{y}_a) \tag{39}$$

Equation (39), together with Eq. (38), gives the minimum variance estimate $\hat{\underline{x}}$ of $\underline{x}$.

Upon substitution of Eq. (38) into Eq. (34), the expression for the optimal error covariance matrix is obtained to give

$$J = (I-BM)A \tag{40}$$

The equations developed form the basis for the Kalman filter equations. The remaining task is to show how $\underline{x}_a$ and A for a new time t+Δt can be obtained from $\hat{\underline{x}}$ and J for the time t. In other words, one wishes to use estimates at time t to form the a priori estimate for time $t+\Delta t$. The time t is taken to mean the time at which a measurement or measurements are made, and time $t+\Delta t$ is taken to mean the time at which the next measurement or measurements are made.

Let the state vector at time $t+\Delta t$ be expressed by some linear function involving‡the value of the state vector at time t as given by Eq. (10). Using Eq. (10), the new a priori estimate for the state vector is obtained to give

$$\underline{x}_{a\ t+\Delta t} = U_t \hat{\underline{x}}_t \tag{41}$$

and the associated a priori error covariance matrix is given by

$$A_{t+\Delta t} = E\ [U_t(\underline{x}_t - \hat{\underline{x}}_t)\ (\underline{x}_t - \hat{\underline{x}}_t)^T U_t^T]$$

Using Eq. (33), we get

$$A_{t+\Delta t} = U_t J_t U_t^T \tag{42}$$

Equations (41) and (42) give the relationship whereby the new a priori estimate of the state vector and its associated error covariance matrix can be obtained.

2.4 Comments on the One-Pass and Sequential Methods

The essential difference between the one-pass estimate and sequential estimate is that the former estimates the unknown over a time period by expressing the unknowns at any time point in terms of the unknowns for the particular chosen period. The latter estimates the unknowns at some time t and uses the new estimate at time t to project it forward to some new t+1; this projected estimate is used as the a priori estimate for the processing of data at time t+1. This process continues until all data are processed.

The nature of the estimation problem governs the choice between the one-pass estimate and the sequential estimate. In general, the sequential estimation scheme is primarily used for on line estimation where the current estimate of the states and parameters are desired rather than the states and parameters at some epoch (time) is desired. In problems where large numbers of parameters are involved (for example, thermal model parameter estimation), it is preferable to use the sequential estimate since it only requires the inversion of an m x m matrix where m is the number of nodes in the model. The one-pass estimate necessitates the inversion of p x p matrix where p is the sum of the number of parameters and the number of nodes in the model (p is often time very much greater than m).

Additional differences between the various methods studied involved the use of a priori knowledge and the use of statistical assumptions concerning the measurement noise and the status of the system. The various methods that require the use of some form of least squares formulation require no a priori knowledge of the states and parameters. However, if a priori knowledge of the unknown states and parameters and measurement noise statistics are available, they can be incorporated in the calculation, respectively as additional measurements and measurement weighting functions.

The method of Kalman filtering requires the assumption of a priori knowledge of the unknown states and parameters and their corresponding covariances; the measurement noise statistics are also required.

2.5 Application of the Kalman Filtering Method to Parameter Correction of Thermal Models

The similarity of the methods discussed in the previous section required the need to study only a single method to evaluate the extent to which a thermal network can be corrected. As a result, the Kalman filtering method was selected because of the inherent advantage of a convergence trace and the need for the inversion of a smaller matrix. Application of this technique to the correction of thermal model parameters requires that the heat balance equations be arranged in particular Kalman filtering format.

If spacecraft can be divided into n isothermal nodes, and if transient temperature histories for all the isothermal nodes can be obtained from a thermal test, an artificial filtering problem can be constructed. The governing heat balance equations are arranged in the following form:

$$\underline{y}^{*} = M\underline{x} + \underline{w}$$

This is done by placing those terms that involved the known (hard) model parameters on the left-hand side; these terms are combined to form an artificial measurement vector $\underline{y}^*$. The remaining terms that involve the unknown (soft) model parameters are placed on the right-hand side to form $M\underline{x}$, where $\underline{x}$ is the state vector formed by the soft model parameters (parameters that needed to be corrected) and M is the artificial measurement matrix that involves the coefficients of the soft parameters. The vector $\underline{w}$ is the noise associated with measurement vector $\underline{y}^*$. If all of the unknown parameters are constant, the updating matrix U is given by an identity matrix. With the vectors $\underline{x}$, y^*, and the matrices M and U formulated, the Kalman filter equations can be used to correct the thermal model parameters sequentially as soon as the problem is initialized by assuming a priori knowledge for the soft parameters. Figure 1 gives a schematic diagram for the sequential correction scheme.

The artificial filtering problem was applied to two mathematical models of a physical system identified as a hollow cube, one consisting of five isothermal nodes and the other consisting of twenty isothermal nodes as shown in Figs. 2 and 3. The nominal parameter values employed in this study are tabulated in Tables 1 and 2.

Several of the known model parameters were perturbed, and the exact transient temperature histories were used as test temperatures to correct these perturbed (soft) parameters.

Table 1 Five-node model nominal parameter values

Parameters	Value
a_{12}, a_{14}, a_{15}, a_{23} a_{25}, a_{34}, a_{35}, a_{45}	0.5 Btu/hr^{o}F
b_{12}, b_{13}, b_{14}, b_{15} b_{23}, b_{24}, b_{25}, b_{26} b_{34}, b_{35}, b_{45}, b_{46}	0.2 (dimensionless)
b_{16}, b_{36}, b_{56}	1.2 (dimensionless)
C_1, C_2, C_3, C_4, C_5	0.2 Btu/^{o}F
Q_1	100 Btu/hr
Q_2	150 Btu/hr
Q_3, Q_4	0 Btu/hr
Q_5	300 Btu/hr

Table 2 Twenty-node model nominal parameter values

$a_{12}, a_{14}, a_{16}, a_{17}, a_{23}, a_{2,15}, a_{34}, a_{36}, a_{45}, a_{4,20}, a_{56}, a_{58}, a_{5,20} a_{67}, a_{78}, a_{7,10}, a_{89}, a_{8,19}, a_{9,10}, a_{9,12}, a_{9,19}, a_{10,11}, a_{11,12}, a_{11,14}, a_{12,13}, a_{12,18}, a_{13,14}, a_{13,16}, a_{13,18}, a_{14,15}, a_{15,16}, a_{16,17}, a_{17,18}, a_{17,20}, a_{18,19}, a_{19,20}$	= 0.5 Btu/hr^{o}F
$b_{1,21}, b_{2,21}, b_{3,21}, b_{4,21}, b_{9,21}, b_{10,21}, b_{11,21}, b_{12,21}, b_{17,21}, b_{20,21}$	= 1 (dimensionless)
$b_{7,21}, b_{15,21}$	= 0.295 (dimensionless)
$b_{18,21}, b_{19,21}$	= 1.2 (dimensionless)
$b_{8,21}, b_{16,21}$	= 0.105 (dimensionless)
$b_{7,15}, b_{8,16}$	= 0.048 (dimensionless)
$b_{7,16}, b_{8,15}$	= 0.036 (dimensionless)
$b_{7,18}$	= 0.03 (dimensionless)
$b_{7,19}, b_{8,18}$	= 0.035 (dimensionless)
$b_{8,19}, b_{15,18}, b_{16,19}, b_{16,19}$	= 0.2 (dimensionless)
$b_{15,18}$	= 0.04 (dimensionless)
$C_1, C_2, \ldots C_{20}$	= 0.05 Btu/^{o}F
Q_1, Q_2, Q_3, Q_4	= 100 Btu/hr
Q_5, Q_6, Q_7, Q_8	= 150 Btu/hr
$Q_{17}, Q_{18}, Q_{19}, Q_{20}$	= 300 Btu/hr

Initially, the five-node model was examined with the measurement noise vector $\underline{w}$ set to zero; this meant that the Kalman filtering scheme was used to solve a deterministic problem. The correction of five soft parameters, two linear and three radiation, was attempted. The results were quite unsatisfactory; a convergence trend appeared after two sets of data were processed, but divergence from the true values occurred when additional data were processed.

A detailed examination of the computer printouts revealed that the state error covariance matrix, which should be a symmetric and positive semi-definite matrix, became asymmetric after having processed the first set of data; the asymmetry became more and more pronounced when additional data were processed. The asymmetry of the state error covariance matrix caused the parameters to diverge from the true values. The cause of this asymmetry appeared to be numerical round-off errors even though specific sources could not be isolated. Symmetry of the off-diagonal elements was forced by averaging the values of the elements. With this forced symmetrical matrix, the case with the two linear and three radiation parameters was re-examined. The results (not shown here) were excellent.

The Kalman filter was examined further with a number of other cases: 1) four linear and four radiation conductors with no parallel sets; 2) one capacitor, three linear, and four radiation conductors with no parallel set; 3) four linear and four radiation conductors with one parallel set; 4) four linear and four radiation conductors with all sets parallel; and 5) three linear and five radiation conductors with three parallel sets. Excellent correlation between the calculated parameter values and their true value was obtained for cases 1-3 typified by Table 3. Case 4 (Table 4) shows divergence parameters. Case 5 (Table 5) shows that all the perturbed parameters converged to values within fractions of 1% of their true values after having processed three sets of temperature data, and then diverged to undesirable results.

Detailed examination of the Kalman filter equations indicated that numerical errors probably caused the divergence and oscillations. The nature of the Kalman filter equations is such that by setting the measurement noise vector to zero, it is necessary to invert a matrix given by MAM^T [refer to Eq. (38)]. When the unknown parameter values approach their true values, the elements in the state error covariance matrix become very small. The inversion of matrices whose elements are very small poses a numerical problem. This problem was eliminated by employing an artificial measurement noise vector whose elements are very small, stationary and uncorrelated;

case 5 was then rerun. The results are in Table 6. It can be observed that excellent correlations between the calculated values and their true values were obtained; convergence to the true parameter values were obtained as more and more temperature data were processed.

For the twenty-node model, twenty linear conductors were perturbed. Excellent results were obtained; the results are not shown here. A further study was made by perturbing all the parameters with the exception of the heat input Q's. The number of soft parameters totaled eighty-four (84) as shown in Table 2. The results were excellent as shown in Table 7. As a note of interest, this case required approximately 20,000 core locations and a run time of approximately 4 min (nine passes).

Table 3 Five-node model, Kalman filter, 4 linear and 4 radiation conductors (one parallel set, no noise, forced symmetry of state error covariance matrix)

Unknown parameters	True values	A priori estimates	Estimates after 1st set of data	Estimates after 4th set of data
a_{12}	0.5	0.9	0.418620	0.500014
a_{23}	0.5	0.1	0.446879	0.500001
a_{34}	0.5	0.7	0.529639	0.500000
a_{45}	0.5	0.3	0.354025	0.500004
σb_{12}	$4.0x10^{-10}$	$7.0x10^{-10}$	$5.28459x10^{-10}$	$3.99978x10^{-10}$
σb_{24}	$4.0x10^{-10}$	$1.0x10^{-10}$	$5.40279x10^{-10}$	$3.99988x10^{-10}$
σb_{35}	$4.0x10^{-10}$	$6.0x10^{-10}$	$5.09572x10^{-10}$	$4.00001x10^{-10}$
σb_{46}	$4.0x10^{-10}$	$2.0x10^{-10}$	$3.99998x10^{-10}$	$3.99998x10^{-10}$

Table 4 Five-node model, Kalman filter,
4 linear and 4 radiation conductors
(four parallel sets, no noise,
forced symmetry of state error covariance matrix)

Unknown parameters	True values	A priori estimates	Estimates after 1st set of data	Estimates after 6th set of data
a_{12}	0.5	0.9	0.423799	0.500680
a_{23}	0.5	0.1	0.443865	0.788397
a_{34}	0.5	0.7	0.456180	0.501029
a_{45}	0.5	0.3	0.535847	0.499602
σb_{12}	$4.0\text{x}10^{-10}$	$7.0\text{x}10^{-10}$	$5.27431\text{x}10^{-10}$	$3.98752\text{x}10^{-10}$
σb_{23}	$4.0\text{x}10^{-10}$	$1.0\text{x}10^{-10}$	$2.03844\text{x}10^{-10}$	$-1.66470\text{x}10^{-10}$
σb_{34}	$4.0\text{x}10^{-10}$	$6.0\text{x}10^{-10}$	$4.73948\text{x}10^{-10}$	$3.97814\text{x}10^{-10}$
σb_{45}	$4.0\text{x}10^{-10}$	$2.0\text{x}10^{-10}$	$3.39281\text{x}10^{-10}$	$4.00786\text{x}10^{-10}$

Table 5 Five-node model, Kalman filter,
3 linear and 5 radiation conductors
(three parallel sets, no noise,
forced symmetry of state error covariance matrix)

Unknown parameters	True values	A priori estimates	Estimates after 3rd set of data	Estimates after 4th set of data
a_{12}	0.5	0.9	0.496027	-37.3930
a_{23}	0.5	0.1	0.501868	-12.6194
a_{25}	0.5	0.7	0.499040	9.44752
σb_{12}	$4.0\text{x}10^{-10}$	$7.0\text{x}10^{-10}$	$4.01928\text{x}10^{-10}$	$193.759\text{x}10^{-10}$
σb_{23}	$4.0\text{x}10^{-10}$	$1.0\text{x}10^{-10}$	$3.99058\text{x}10^{-10}$	$72.2106\text{x}10^{-10}$
σb_{24}	$4.0\text{x}10^{-10}$	$1.0\text{x}10^{-10}$	$4.00005\text{x}10^{-10}$	$4.0008\text{x}10^{-10}$
σb_{25}	$4.0\text{x}10^{-10}$	$6.0\text{x}10^{-10}$	$4.00478\text{x}10^{-10}$	$-42.2503\text{x}10^{-10}$
σb_{26}	$4.0\text{x}10^{-10}$	$2.0\text{x}10^{-10}$	$4.00000\text{x}10^{-10}$	$4.00071\text{x}10^{-1}$[illegible]

Table 6 Five-node model, Kalman filter
3 linear and 5 radiation conductors with measurement noise
(three parallel sets, artificial noise,
forced symmetry of state error covariance matrix)

Unknown parameters	True values	A priori estimates	Estimates after 3rd set of data	Estimates after 8th set of data
a_{12}	0.5	0.9	0.499031	0.499782
a_{23}	0.5	0.1	0.500374	0.500031
a_{25}	0.5	0.7	0.499763	0.499950
σb_{12}	$4.0x10^{-10}$	$7.0x10^{-10}$	$4.01568x10^{-10}$	$4.00360x10^{-10}$
σb_{23}	$4.0x10^{-10}$	$1.0x10^{-10}$	$3.99339x10^{-10}$	$3.99941x10^{-10}$
σb_{24}	$4.0x10^{-10}$	$1.0x10^{-10}$	$3.99994x10^{-10}$	$3.99995x10^{-10}$
σb_{25}	$4.0x10^{-10}$	$6.0x10^{-10}$	$4.00416x10^{-10}$	$4.00106x10^{-10}$
σb_{26}	$4.0x10^{-10}$	$2.0x10^{-10}$	$3.99999x10^{-10}$	$3.99998x10^{-10}$

3. Method of Using Experimental Data with Noise

The numerical results presented in the computer simulation of the Kalman filtering scheme were all based upon the use of computer-generated temperatures in lieu of experimental data, which are subject to inaccuracies. The Kalman filtering method can accommodate noisy data, but the parameter correction scheme must be treated as a regular filtering problem; this means that the statistical nature of the noise must be known. Often the statistical characteristics of the noise are not known, but even if known a major consideration is the program complexities that would arise from the need to generate and use the statistical noise in a proper sequence.

For the purpose of correcting the parameters of a thermal model, perhaps a better alternative is to employ experimental data that has been "smoothed" by the use of a method such as least squares. Naturally, the corrected parameters will reflect the "smoothed" temperature data.

4. Summary

A number of correlation methods have been reviewed and details of several selected techniques presented. The Kalman filter was chosen for further examination and the results indicate that if the temperature measurements are near-ideal,

Table 7 Twenty-node model, Kalman filter, 36 linear and 28 radiative conductors, 20 capacitances (artificial noise, forced symmetry of state error covariance matrix)

Unknown param-eters	True value	Estimate after 9th set of data	Unknown param-eters	True value	Estimate after 9th set of data	Unknown param-eters	True value	Estimate after 9th set of data
C_1	0.05	0.0498935	a_{45}	0.5	0.502054	$\sigma b_{1,21}$	a	5.04613×10^{-10}
C_2	0.05	0.0499921	$a_{4,20}$	0.5	0.489397	$\sigma b_{2,21}$	a	4.99538×10^{-10}
C_3	0.05	0.0500326	a_{56}	0.5	0.500873	$\sigma b_{3,21}$	a	5.01477×10^{-10}
C_4	0.05	0.0496892	a_{58}	0.5	0.501775	$\sigma b_{4,21}$	a	5.00575×10^{-10}
C_5	0.05	0.0501638	$a_{4,20}$	0.5	0.500681	$\sigma b_{9,21}$	a	4.97538×10^{-10}
C_6	0.05	0.0500891	a_{67}	0.5	0.500894	$\sigma b_{10,21}$	a	5.00803×10^{-10}
C_7	0.05	0.0477995	a_{78}	0.5	0.456812	$\sigma b_{11,21}$	a	4.98179×10^{-10}
C_8	0.05	0.0507621	$a_{7,10}$	0.5	0.501038	$\sigma b_{12,21}$	a	4.83844×10^{-10}
C_9	0.05	0.0497347	a_{89}	0.5	0.496495	$\sigma b_{17,21}$	a	4.99464×10^{-10}
C_{10}	0.05	0.0500998	$a_{8,19}$	0.5	0.511881	$\sigma b_{20,21}$	a	5.04621×10^{-10}
C_{11}	0.05	0.0498834	$a_{9,10}$	0.5	0.499504	$\sigma b_{7,21}$	a	4.87098×10^{-10}
C_{12}	0.05	0.0489263	$a_{9,12}$	0.5	0.496274	$\sigma b_{15,21}$	a	4.89399×10^{-10}
C_{13}	0.05	0.0493641	$a_{9,19}$	0.5	0.499056	$\sigma b_{18,21}$	a	5.24572×10^{-10}
C_{14}	0.05	0.0498457	$a_{10,11}$	0.5	0.50379	$\sigma b_{19,21}$	a	4.77062×10^{-10}
C_{15}	0.05	0.0491943	$a_{11,12}$	0.5	0.496140	$\sigma b_{8,21}$	a	5.04009×10^{-10}

Table 7 (Continued)

Unknown param-eters	True value	Estimate after 9th set of data	Unknown param-eters	True value	Estimate after 9th set of data	Unknown param-eters	True value	Estimate after 9th set of data
C_{16}	0.05	0.0504122	$a_{11,14}$	0.5	0.498470	$\sigma b_{16,21}$	a	$5.15093x10^{-10}$
C_{17}	0.05	0.0499071	$a_{12,13}$	0.5	0.496418	$\sigma b_{7,15}$	a	$3.90605x10^{-10}$
C_{18}	0.05	0.0506099	$a_{12,18}$	0.5	0.470691	$\sigma b_{8,16}$	a	$5.53624x10^{-10}$
C_{19}	0.05	0.0497443	$a_{13,14}$	0.5	0.498318	$\sigma b_{7,16}$	a	$4.17923x10^{-10}$
C_{20}	0.05	0.0493960	$a_{13,16}$	0.5	0.495731	$\sigma b_{8,15}$	a	$4.06697x10^{-10}$
a_{12}	0.5	0.500243	$a_{13,18}$	0.5	0.496732	$\sigma b_{7,18}$	a	$5.02361x10^{-10}$
a_{14}	0.5	0.497493	$a_{14,15}$	0.5	0.498458	$\sigma b_{7,19}$	a	$5.34679x10^{-10}$
a_{16}	0.5	0.497666	$a_{15,16}$	0.5	0.523403	$\sigma b_{8,18}$	a	$5.54961x10^{-10}$
a_{17}	0.5	0.498447	$a_{16,17}$	0.5	0.500182	$\sigma b_{8,19}$	a	$5.20843x10^{-10}$
a_{23}	0.5	0.500335	$a_{17,18}$	0.5	0.496115	$\sigma b_{15,18}$	a	$5.95466x10^{-10}$
$a_{2,15}$	0.5	0.500291	$a_{17,20}$	0.5	0.499229	$\sigma b_{15,19}$	a	$5.10081x10^{-10}$
a_{34}	0.5	0.497182	$a_{18,19}$	0.5	0.515076	$\sigma b_{16,19}$	a	$4.03536x10^{-10}$
a_{36}	0.5	0.500888	$a_{19,20}$	0.5	0.465891	$\sigma b_{16,8}$	a	$4.83587x10^{-10}$

a For convenience, σb_{ij} were all set equal to $5.0x10^{-10}$.

tne parameters of a thermal network are indeed correctable. No attempt was made here to include statistical noise. Problems generated by temperature sparsity (less than total measurements) were considered to be beyond the intent of the present paper; however, this problem is being studied in some depth at the present time. Some preliminary results are contained in Refs. 32 and 33.

5. References

[1] Graybill, F.A., An Introduction to Linear Statistical Models, McGraw-Hill, New York, 1961, Vol. 1, Chap. 6.

[2] William, E.J., Regression Analysis, Wiley, 1959.

[3] Efroymson, M.A., "Multiple Regression Analysis," Mathematical Methods for Digital Computers, edited by A. Ralston and H.S. Wilf, Wiley, 1960.

[4] Hartley, H.O., "The Modified Gauss-Newton Method for the Fitting of Nonlinear Functions by Least Squares," Technometrices, Vol. 3, No. 2, May 1961, pp. 269-280.

[5] Breaux, H.J., "On Stepwise Multiple Linear Regression," Rept. 1369, Aug. 1967, Ballistic Research Labs.

[6] Giese, C. and McGhee, R.B., "Estimation of Nonlinear System States and Parameters by Regression Methods," 1965 Joint Automatic Control Conference, Rensselaer Polytechnic Institute, Troy, New York.

[7] Rauch, H.E., "Least Squares Estimation with a Large Number of Parameters," 1967 Joint Automatic Control Conference, University of Pennsylvania, Philadelphia, Pennsylvania.

[8] Golub, G., "Numerical Methods for Solving Linear Least Squares Problems," Numerische Mathematik, Vol. 7, 1965, pp. 206-216.

[9] Marquardt, D.W., "An algorithm for Least Square Estimation of Nonlinear Parameters," Journal of the Society for Industrial and Applied Mathematics, Vol. 11, No. 2, June 1963, pp. 431441.

[10] Smith, G.A., "The Theory and Application of Least Squares," TM X-63127, Dec. 1967, NASA.

[11] Levenberg, K., "A Method for the Solution of Certain Non-Linear Problems in Least Squares," Quarterly Applied Mathematics, Vol. 2, 1964, pp. 164-168.

[12] Toussaint, M., "Verification of the Thermal Mathematical Model for Artificial Satellites: A New Test Philosophy," AIAA Progress in Astronautics and Aeronautics: Thermophysics of Spacecraft and Planetary Bodies, Vol. 20, edited by G.B. Heller, Academic Press, New York, 1967, pp. 611-629.

[13] Allison, J.S., "The Parameter Estimation Problem in Mathematical Modeling," Rept. SAD 41, Nov. 1966, Department of Supply, Australian Defense Scientific Service, Weapons Research Establishment.

[14] Allison, J.S., "The Parameter Estimation Problem in Model Checking," Proceedings of the Third Australian Computer Conference, Canberra, May 16-20, 1966.

[15] Ho, Y.C., "The Method of Least Squares and Optimal Filtering Theory," RM-3329-PR, Oct. 1962, Rand Corp.

[16] Alford, M.W., Jr., "A Computationally Stable Noise-in-the-State Filtering Algorithm," AIAA Paper 68-887, Pasadena, Calif., 1968.

[17] Aoki, M., Optimization of Stochastic Systems, Topics in Discrete-Time Domain, Academic Press, New York, 1967, Chap. 5.

[18] Nickle, C.E., "Linear Error Analysis and Filtering by Recursive Least Squares," TM HSM-M58-67, June 1967, Space Division, Chrysler Corp., Huntsville Operation.

[19] Mowery, V.O., "Least Squares Recursive Differential-Correction Estimation in Non-Linear Problems," IEEE Transactions on Automatic Control, Oct. 1965, Vol. AC-10, No. 4, pp. 399-407.

[20] Ho, H.C., "A Bayesian Approach to Problems in Stochastic Estimation and Control," IEEE Transactions on Automatic Control, Oct. 1964, pp. 333-339.

[21] Bellman, R.E. et al., "Invariant Imbedding and Nonlinear Filtering Theory," The Journal of Astronautical Science, Vol. 13, No. 3, June 1966, pp. 110-115.

[22] Detchmendy, D.M. and Sridhar, R., "Sequential Estimation of States and Parameters in Noisy Nonlinear Dynamic Systems,"

Transactions of the ASME, Ser. D: Journal of Basic Engineering, June 1966.

[23] Sridhar, R., Bellman, R.E. and Kalaba, R.E., "Sensitivity Analysis and Invariant Imbedding," RM-4039-PR, March 1964, Rand Corp.

[24] Bellman, R., Kagiwada, H.H. and Kalaba, R.E., "Quasilinearization, Systems Identification, and Prediction," RM-3812-PR, Aug. 1963, Rand Corp.

[25] Prasanna-Kumar, K.S. and Sridhar, R., "On the Identification of Control Systems by the Quasi-Linearization Method," IEEE Transactions on Automatic Control, Vol. AC-9, April 1964, pp. 151-154.

[26] Kalman, R.E., "A New Approach to Linear Filtering and Prediction Problems," Transactions of ASME, Ser. D: Journal of Basic Engineering, March 1960, pp. 35-45.

[27] Kalman, R.E. and Bucy, R.S., "New Results in Linear Filtering and Prediction Theory," Transactions of ASME, Ser. D: Journal of Basic Engineering, March 1961, pp. 95-108.

[28] Kalman, R.E., "New Methods and Results in Linear Prediction and Filtering Theory," RIAS TR 61-1, Nov. 1960, Purdue Univ.

[29] Browne, B.H., Jr., "Statistical Estimation and Elimination of Thermal Network Model Errors," Paper 68-HT-48, 1968, American Society of Mechanical Engineers.

[30] Tomovic, R., "Sensitivity Analysis of Dynamic Systems," McGraw-Hill, New York, 1963.

[31] Meissinger, H.F., "The Use of Parameter Influence Coefficients in Computer Analysis of Dynamic Systems," Simulation, Aug. 1964, Vol. 4, No. 10, pp. 53-63.

[32] Ishimoto, T. et al., "Development of Digital Computer Program for Thermal Network Correction, Final Report, Phase I - Investigation/Feasibility Study," Rep. 11027-6001-RP-00, Jan. 1969, TRW Systems.

[33] Ishimoto, T. et al., "Addendum to the Final Report Development of Digital Computer Program for Thermal Network Correction, Phase I - Investigation/Feasibility Study," Rep. 11027-6001-RP, Addendum, June 1969, TRW Systems.

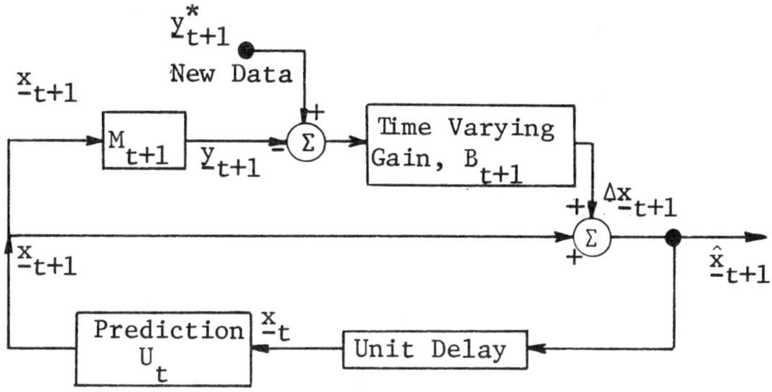

Fig. 1 Kalman filter block diagram.

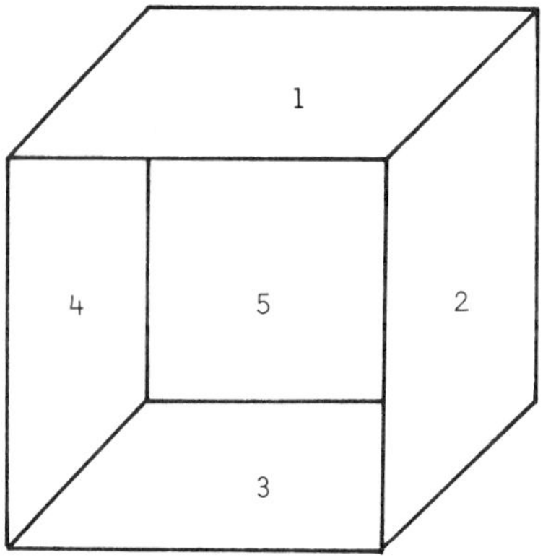

Fig. 2 Five-node hollow cube model.

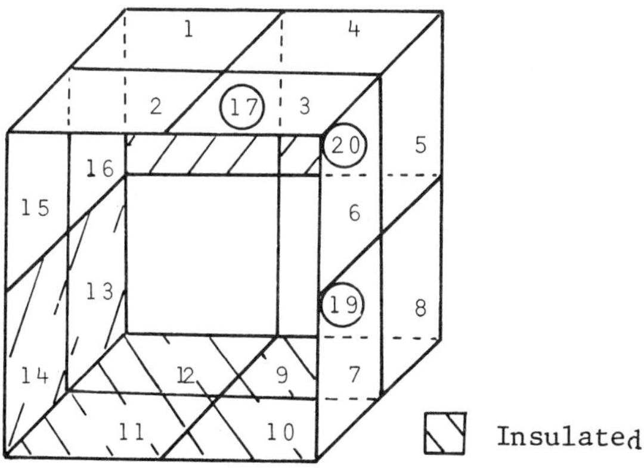

Fig. 3 Twenty-node hollow cube model.

INFLUENCE OF A NONABSORBING GAS IN RADIATION-CONDUCTION INTERACTION

J. L. Novotny* and F. A. Olsofka†

University of Notre Dame, Notre Dame, Ind.

Abstract

The effect of a nonabsorbing gas in radiation-conduction interaction is examined experimentally. Temperature profiles in a 2.55-cm heated-from-above gas layer are obtained with a Mach-Zehnder interferometer for mixtures of CO_2-N_2 and N_2O-N_2 at a total pressure of 200 kN/m^2. The experimental temperature gradients at the bounding surfaces are compared to a nongray analysis based on the exponential wide-band gas model. The agreement between experiment and theory indicates that the wide-band gas model correctly predicts the magnitude and pressure trends of radiation-conduction interaction.

Nomenclature

A_i = total band absorptance of the ith band

A_{oi} = correlation parameter for the ith band

b_i = self-broadening coefficient for the ith band

B_i^2 = correlation parameter for the ith band

C_{oi}^2 = correlation parameter for the ith band

e_ω = Planck function

$\tilde{e}_i$ = $\left(de_{\omega i}/dT\right) / \left(de_{\omega i}/dT\right)_{T_2}$

Presented as Paper 70-836 at the AIAA 5th Thermophysics Conference, Los Angeles, Calif., June 29-July 1, 1970. This work was sponsored by the National Science Foundation under Grant GK-2093.

*Professor, Aerospace and Mechanical Engineering Department.

†Research Assistant; presently Engineer, Clark Brothers Company, Olean, New York.

$f(\beta_i)$ = $2.94\ [1-\exp(-2.6\beta_i)]$

λ = thermal conductivity

L = distance between plates

m = summation index

N_i = $(P_a L^2/\lambda)\ A_{oi} \left(de_{\omega i}/dT\right) T_2 \left(C_{oi}^2\right)$

n_i = pressure broadening exponent

P = pressure

P_a = absorbing gas partial pressure

P_{ei} = effective broadening pressure for the ith band, $[(b_i P_a + P_{N_2})/P_o]^{n_i}$

P_{N_2} = nitrogen partial pressure

P_o = atmospheric pressure

P_t = total pressure, $P_a + P_{N_2}$

q_c = conductive heat-transfer rate

q_{co} = conductive heat-transfer rate for pure conduction

q_r = radiative heat-transfer rate

T = temperature

u_i = $C_{oi}^2\ P_a y$

u_{oi} = $C_{oi}^2\ P_a L$

X = mole fraction

y = coordinate normal to plate surface

β_i = line structure parameter, $B_i^2\ P_e$

ε = fringe shift

η = dummy variable of integration for ξ

θ = T/T_2

κ_ω = absorption coefficient

ξ = y/L

ρ = reflectance

ω = wave number

Subscripts

1 = wall at y = 0

2 = wall at y = L

i = ith band

Introduction

In the last decade, there have been numerous investigations of the influence of gaseous thermal radiation at moderate temperatures in heat-transfer processes. The majority of the publications in this area have been analytical investigations using gas models such as the gray gas, the box or the exponential wide-band model. Examples of analyses using the gray gas approach are given in the text by Sparrow and Cess[1]; good examples of the use of the Edwards and Menard exponential wide-band model[2] and the box model are given in the work of Cess, Mighdoll, and Tiwari.[3]

At the present time, there are very few existing experimental investigations concerned with the ability of the various gas models to predict the interaction of gaseous radiation with the other modes of heat-transfer at moderate temperatures. Gille and Goody[4] examined the stability of a horizontal radiating layer heated from below. In the course of this study, temperature profiles and the total heat flux in a heated-from-above gas layer were also measured. Though the temperature and total heat flux data exhibit quite a bit of scatter, a general agreement was found with a nongray analysis based on the first derivative of the total band emittance. England and Emery[5] measured temperature profiles in a free convection boundary layer on a vertical flat plate; the interaction of gaseous radiation was found to be virtually negligible in this case. Schimmel, Novotny, and Olsofka[6] measured the temperature distribution in a horizontal gas layer heated from above where the gas layer was bounded by two polished aluminum plates. The temperature gradients at each surface were found to agree very well with an analysis based on the exponential wide-band model.

Tests were run for pure CO_2 and N_2O, and mixtures of CO_2-N_2O and CO_2-CH_4. The gray gas and box models were found to be inadequate since they did not predict the proper pressure trends of the data.

Although the ability of the exponential wide-band model to predict local interaction in a pure gas layer has been demonstrated,[6] it was of interest to further test this model in its ability to predict the details of the absorption-emission process in the presence of a nonabsorbing foreign gas. This was accomplished by an experimental investigation of the influence of pressure broadening by a nonabsorbing gas. Temperature profiles in a 2.55 cm heated from above horizontal gas layer were obtained with a Mach-Zehnder interferometer for mixtures of CO_2-N_2 and N_2O-N_2 at a total pressure of 200 kN/m^2. The temperature gradients at the bounding surfaces were compared to a nongray analysis based on the exponential wide-band model.

Experimental Investigation

Since the experimental apparatus and procedures used to gather the information reported here is described by Schimmel, Novotny, and Olsofka,[6] only a brief description of the experiment will be given in this report. The experimental apparatus attempted to reproduce in the laboratory a horizontal gas layer bounded by two infinite parallel plates. Although an approximation to the infinite geometry will introduce edge effects and natural convection into the gas layer, the apparatus should closely approximate a one-dimensional radiating-conducting gas layer if the ratio of plate spacing to plate length is kept small.

The experimental gas layer was bounded by two polished aluminum plates, 50.8 cm square, ground to better than 25-μm flatness. The top plate was heated by strip heaters and the bottom plate was cooled by water from a constant temperature bath. Both plates were thoroughly instrumented with thermocouples. The plates were kept at a plate spacing of 2.55 cm[6] by four Micarta spacers placed in the corners of the plates. The plates were housed in a specially constructed pressure chamber during the experiments. Strips of Styrofoam isolated the edges of the gas layer from the pressure chamber. Two optical flats mounted in the walls of the chamber allowed the passage of the interferometer light beam. The details of the Mach-Zehnder interferometer and the other instrumentation are given in previous investigations.[6,7] All gases used in the experiments were obtained from the Matheson Company. The bone dry grade carbon dioxide and the prepurified grade of nitrogen were used in the experiments. The nitrous oxide had a minimum purity of 98% by mole with the major contaminent being air. The reflectance of the aluminum surfaces, which is a necessary input to the analytical calculations, was obtained from monochromatic measurements with a sample of aluminum prepared in the same manner as the test plates.[7] The reflectance was measured as a function of direction and wavelength from 1μm to 18μm.

The procedures for reducing the fringe patterns into temperature data is described in detail in the investigations of Schimmel, Novotny, and Olsofka,[6] and Olsofka[7]; a brief description will be given here. In this type of apparatus, a reference temperature is not available in the final interferogram. There was also a possibility that the interferometer would shift during the heating period. To solve these difficulties, the temperature difference between the plates, as obtained from thermocouples, was used to calculate the total fringe shift between the plates. The total fringe shift value was then used to correct the data from the interferogram. Additionally, the data from each final fringe pattern were also corrected for optical imperfections in the system and pressure effects on the optical flats by basing the fringe shift on an initial unheated fringe pattern. To accentuate the curvature in the fringe profile, each final fringe pattern was adjusted such that the center portion of the fringe was vertical. An example of the interferograms taken for this investigation is given in Fig. 1; the field of view of the fringe pattern is restricted on the vertical sides by a viewing slot cut into the Styrofoam strips around the edge of the gas layer. The warm wall is at the top of the fringe pattern.

All interferograms (negatives) were analyzed using a Gaertner M2001 toolmakers microscopic. An example of the fringe data is given in Fig. 2 with a sketch of the geometry. By repeated analysis of a negative, it was determined that the data could be reproduced to better than ±1/30 of a fringe. The Gladstone-Dale approximation to the Lorenz-Lorentz equation along with the principle of additivity for the molar refractivity was employed in the data reduction. The properties necessary for evaluation of the Gladstone-Dale constant were taken from the Critical Tables,[8] and the density was evaluated from the perfect gas law. The errors encountered in the experiments are discussed in detail by Olsofka[7]; these will also be mentioned in the comparison of the data to the analysis.

Analysis

The governing energy equation for radiation-conduction interaction in a one-dimensional gas layer is given by

$$d^2\theta/d\xi^2 = \left(L^2/\lambda T_2\right)\left(dq_r/dy\right) \quad (1)$$

where q_r is the spectral radiant flux integrated over all wave numbers. Equation (1) is given on p. 250 of Sparrow and Cess.[1] A portion of the nomenclature associated with Eq. (1) is given by the sketch contained in Fig. 2. The boundary conditions are $\theta = T_1/T_2$ at $\xi = y/L = 0$ and $\theta = 1$ at $\xi = 1$.

To formulate the radiation term on the right-hand side of Eq. (1), it is necessary to designate a gas model for the absorption-emission process. Here, the wide-band model of Edwards and Menard[2] coupled with correlation of Tien and Lowder[9] for the total band absorptance

$$A_i = \int_{\Delta\omega_i} [1-\exp(-\kappa_\omega y)]\, d\omega \quad (2)$$

will be employed. The correlation for the total band absorptance is

$$A_i = A_{oi} \,\ln \left\{u_i f_i [(u_i + 2)/(u_i + 2f_i)] + 1\right\} \quad (3)$$

where $u_i = C_{oi}^2 P_a y$ and $f_i = 2.94\,[1-\exp(-2.6\, B_i^2 P_{ei})]$. The quantities A_{oi}, C_{oi}^2, and B_i^2 are spectroscopic correlation parameters for the ith band and P_{ei} is the effective broadening pressure. The function f_i represents the pressure-broadening effect. Following Edwards, Glassen, Hauser and Tuchscher,[10] the effective broadening pressure is given by

$$P_{ei} = [(b_i P_a + P_{N_2})/P_o]^{n_i} \quad (4)$$

where b_i is the self-broadening coefficient, P_a is the partial pressure of the absorbing gas, P_{N_2} is the partial pressure of the nitrogen and P_o is the standard atmospheric pressure.

The Edwards and Menard wide-band model[2] is an arbitrary model in the sense that the distribution of the mean line intensity to spacing ratio in a vibrational-rotational band is assumed to be given by the exponential function

$$C_{oi}^2 \exp[-(\omega-\omega_{oi})/A_{oi}]$$

where ω_{oi} is the center of the ith band. This distribution is then used in the Goody statistical model for the spectral absorptance of the band.[11]

To incorporate this gas model into the radiative transfer equations, a few assumptions have to be made. First, to simplify the governing radiation equation, it is assumed that the bounding surfaces are diffuse. Although the surfaces in the experiment are specular,[7] it is felt, based on previous investigations,[1,12] that the discrepancy introduced by a comparison of a diffuse surface analysis to the experiments should be small. The previous study dealing with radiation-conduction interaction in a parallel-plate system with one perfectly reflecting specular wall and one black wall[12] shows that the difference between an exact kernel formulation and the diffuse wall formulation used here is negligible. In fact, the two formulations are identical in the logarithmic limit. The surfaces are taken to be gray in the region of each band. Additionally, it is necessary to replace the exponential integral function with the exponential kernel approximation

$$E_3(y) \cong (1/2)\exp(-3y/2).$$

It has been shown that this assumption introduces a very small error in radiation-conduction interaction situations.[12] In the derivation, the surface radiosities contained terms like

$$[1-\rho^2\exp\{-(3/2)\kappa_\omega L\}]^{-1}$$

which were expanded into the form

$$\sum_{m=0}^{\infty} (\rho)^{2m}\exp[-(3/2)\kappa_\omega mL] .$$

To integrate the spectral radiant flux over all wave numbers, the gradient of the Planck function $de_{\omega i}/dT$ was taken as a constant across the spectral width of a band.

By using the basic equations given in Sparrow and Cess,[1] the previous assumptions and the total band absorptance correlation, the energy equation is given by [12]

$$\frac{d^2\theta}{d\xi^2} = -\frac{3}{2}\sum_i N_i \left\{ \int_0^1 \tilde{e}_i \frac{d\theta}{d\eta} \sum_{m=0}^{\infty} (\rho_i)^{2m} \right.$$

$$\Big[\rho_i \bar{A}_i' \left\{ \tfrac{3}{2} u_{oi} [2m+\xi+\eta] \right\} - \rho_i \bar{A}_i' \left\{ \tfrac{3}{2} u_{oi} [2(m+1)-\xi-\eta] \right\}$$

$$+ \rho_i^2 \bar{A}_i' \left\{ \tfrac{3}{2} u_{oi} [2(m+1)-\xi+\eta] \right\} - \rho_i^2 \bar{A}_i' \left\{ \tfrac{3}{2} u_{oi} [2(m+1)+\xi-\eta] \right\} \Big] d\eta$$

$$\left. - \int_0^{\xi} \tilde{e}_i \frac{d\theta}{d\eta} A_i' \left\{ \tfrac{3}{2} u_{oi} (\xi-\eta) \right\} d\eta + \int_{\xi}^{1} \tilde{e}_i \frac{d\theta}{d\eta} \bar{A}_i' \left\{ \tfrac{3}{2} u_{oi} (\eta-\xi) \right\} d\eta \right\} \tag{5}$$

where $\bar{A}'_i(u_i)$ denotes the derivative of A/A_{oi} with respect to u_i and the summation index i refers to the ith band. This formulation assumes that the wall reflectances of the bounding surfaces are equal.

For mixtures of carbon dioxide and nitrogen, the 15-and 4.3-μm bands were included in the calculations; at the temperature level of the experiments, the contribution of the 4.3-μm band is small. The 17-, 7.8-, and 4.5-μm bands were included in the nitrous oxide-nitrogen mixture calculations; the contribution of the 4.5-μm band is small. The wall reflectances were taken as 0.95 for the 17-μm band, 0.94 for the 15-μm band, 0.86 for the 7.8-μm band, and 0.79 for the 4.5-and 4.3-μm bands. As stated before, these values were obtained from experimental monochromatic measurements.[7]

The spectroscopic correlation parameters for carbon dioxide and the integrated band intensities $A_{oi}C_{oi}^2$ for nitrous oxide were taken from Tien.[13] The remaining correlation parameters for nitrous oxide are not available in the literature. The parameters B_i^2 and A_{oi} for nitrous oxide were estimated using the method of Edwards[14] with the line half width and line spacing given in Burch and Williams[15] and the rotational constant of Pliva[16]; the same values of B_i^2 and A_{oi} were taken for each band. The self-broadening coefficient b_i for nitrous oxide was obtained from Burch, Singleton, and Williams,[17] and the pressure-broadening exponent for all nitrous oxide bands was taken as $n_i = 1.0$. The mixture thermal conductivity in the energy equation was calculated as outlined in Novotny and Irvine[18]; the force constant parameters necessary for this calculation were taken from the literature.[18-20] Except for B_i^2 in the N_2O-N_2 case, all parameters and properties were introduced at the mean temperature of the gas layer. Since the temperature dependency of B_i^2 for nitrous oxide was not derived,

B_i^2 was introduced at 300°K. The band absorptance for the nitrous oxide bands was calculated using the previously mentioned parameters and compared to existing literature data; the average deviation for the bands used here was about 14% with the largest deviations appearing for the extremely short and long path lengths.[7]

Numerical solutions to Eq. (5) for mixtures of CO_2-N_2 and N_2O-N_2 were obtained on a UNIVAC 1107 computer for total pressures of 200 and 400 kN/m^2.

Results and Discussion

Before presenting the comparison between experiment and theory, it is of interest to examine the influence of the pressure in Eq. (5). The parameters

$$N_i = (P_a L^2 A_{oi} C_{oi}^2/\lambda)\,(de_{\omega i}/dT)_{T_2}\ ;\ u_{oi} = P_a C_{oi}^2 L$$

contain the partial pressure of the absorbing gas. The equivalent broadening pressure, as given by Eq. (4), contains the partial pressure of the absorbing gas as well as the partial pressure of nonabsorbing broadening gas; this equivalent pressure appears in f_i as the product $\beta_i = B_i^2 P_e$.

For purposes of discussion, the Edwards and Menard relationship[2,10] will be used here rather than the Tien and Lowder continuous representation[9] of the Edwards and Menard model. According to Edwards and Menard, there are two regions of absorption corresponding to $\beta_i \leq 1.0$ or $\beta_i > 1.0$. For $\beta_i \leq 1.0$, the total band absorptance A_i/A_{oi} has three distinct regions in terms of its variation with respect to u_i. These are linear, the square root, and the logarithmic regions. The linear region is valid up to $A_i/A_{oi} = \beta_i$ and the square root variation holds for $\beta_i \leq A_i/A_{oi} \leq (2-\beta_i)$. Above $A_i/A_{oi} = (2-\beta_i)$, the logarithmic relationship is valid. For $\beta_i > 1.0$, the total band absorptance is given by two functions, the linear and logarithmic; the linear relationship holds up to $A_i/A_{oi} = 1.0$.

The left-hand side of Eq. (5) is proportional to P_a, $P_a^{1/2}$ or independent of P_a depending on whether the band absorptance is totally in the linear, square root or logarithmic region, respectively. Therefore, for high absorbing gas pressures, the radiation interaction is virtually independent of the absorbing gas pressure.[6] Also, it is noticed that the square-root region for A_i/A_{oi} decreases with increasing β_i or P_e; in fact, for $\beta_i > 1.0$, the square root region no longer exists. Thus, at a fixed P_a as the partial pressure of the nitrogen increases, the total band absorptance becomes closer to the situation of a direct transition from the linear to the logarithmic relationship. Although the derivative of the band absorptance in Eq. (5) will generally encompass all three band absorptance regions, the radiation interaction for significant values of the nonabsorbing gas partial pressure should increase at a more rapid rate with respect to the situation without a nonabsorbing broadener. From a consideration of the behavior of the band absorptance alone, the radiation interaction in the pressure broadened case should always be larger than for the pure case at equal values of the absorbing gas pressure; the interaction curves would join at $P_t = P_a$. This behavior, with respect to the absorbing gas pressure, is due to the total band absorptance becoming closer to the case of a direct transition from the linear region to the logarithmic region as the effective broadening pressure increases.

However, the CO_2-N_2 and N_2O-N_2 mixtures examined here indicate that although the broadened gas interaction is larger at low pressures ($P < 50$ kN/m^2) the interaction curves cross. The interaction curves intersect again at $P_t = P_a$. This behavior of the radiation interaction is caused by the mixture thermal conductivity in the denominator of the parameter N_i. The pure nitrogen thermal conductivity is quite a bit larger than the thermal conductivity of pure carbon dioxide or nitrous oxide. Thus, the interaction parameter N_i, which characterizes the relative importance of gaseous radiation, decreases with increasing pressure of the nonabsorbing gas causing the crossover of the curves. Although this discussion is based on the Edwards and Menard model rather than the continuous correlation, the conclusions should be the same in both cases.

To estimate the amount of free-convection introduced into the pressure cell by edge effects, nitrogen was used as a test gas.[6] Also, for a basis of comparison, the data given in Schimmel, Novotny, and Olsofka[6] for pure carbon dioxide and nitrous oxide will be reproduced here along with the analytical wide-band results for the pure gases. Figure 3 represents this information for pure carbon dioxide and nitrogen along with the analysis and data for mixtures of carbon dioxide and nitrogen. This figure contains a plot of ratio of the conductive heat transfer at each wall with gaseous radiation to the conductive heat transfer for pure conduction; the ratio q_c/q_{co} is just the ratio of the temperature gradients in the two situations. In the absence of free convection, the nitrogen data for constant properties should be $q_c/q_{co} = 1.0$. The error exhibited by the nitrogen data is due primarily to free convection[6]; this error is larger for the hot wall since there is a heat loss at the edge of the gas layer. Although this error will be slightly different for carbon dioxide due to property changes and radiative transfer errors, the nitrogen data indicate the approximate systematic error[6,7] (the estimated maximum random error is ±1.5%[7]). It should be pointed out that for the temperature level and temperature ratio of the walls in these experiments, the conductive heat transfer at the hot and cold walls should be indistinguishable in this figure; thus, a single analytical curve represents both walls.

The pure carbon dioxide data and the wide-band analytical results taken from the previous work[6] are in good agreement. The pressure trends of the data are correct; the lack of dependency on absorbing gas partial pressure for the larger values of the carbon dioxide partial pressure is clearly shown. The data for the carbon dioxide and nitrogen mixtures at P_t = 200 kN/m^2 and the wide-band analytical results for P_t = 200 and 400 kN/m^2 are also given in Fig. 3. For small values of the carbon dioxide partial pressure, the interaction for the pressure broadened case increases at a faster rate with respect to the absorbing gas pressure than in the pure gas case. At carbon dioxide partial pressures above 50 kN/m^2, the pure gas interaction is larger than that for the carbon dioxide and nitrogen mixtures. This is in agreement with the previous discussion concerning the effect of the nonabsorbing gas. The experimental data for mixtures of carbon dioxide and nitrogen exhibit a fair agreement with the wide-band results. The analytical dependence of the interaction with respect to the partial pressure of carbon dioxide is definitely supported by the data. The rather sharp intersection of the pure and broadened interaction curves is due to the behavior of the mixture thermal conductivity.

Figure 4 presents the results for the case where nitrous oxide is the absorbing gas. The interaction behavior of the mixtures of nitrous oxide and nitrogen as a function of the absorbing-gas partial pressure is slightly different than the carbon dioxide-nitrogen case. At 300^{0}K, B_i^2 for the 15-μm carbon dioxide band is 0.084 whereas it is 0.75 for the nitrous oxide bands. Thus, for equal values of the effective broadening pressure P_e, the β_i for nitrous oxide is almost an order of magnitude greater than the β_i for the important carbon dioxide band. Although β_i for the 15-μm band of carbon dioxide is always less than one for the effective pressures examined here, the β_i for the nitrous oxide bands is greater than one for the nitrogen broadened cases examined here and is greater than one for pressures greater than about 100 kN/m^2 in the pure nitrous oxide case. Therefore, the pure nitrous oxide interaction behavior with respect to the absorbing gas pressure should be closer to the nitrogen broadened nitrous oxide cases. Figure 4 shows this tendency at low pressures. As in the case of the carbon dioxide-nitrogen data, the agreement between the nitrous oxide-nitrogen data and the wide-band analysis is good. However, if the data were corrected for the free convection error, the analysis would tend to overpredict the interaction. Considering the manner in which the majority of the spectroscopic parameters for nitrous oxide were determined, the agreement between the data and the analysis is excellent.

The information presented further confirms the correctness of the exponential wide-band gas model for predicting local interaction effects in a radiating-conducting gas layer. As previously stated in this paper, the requirement of being able to predict the local interaction in a gas layer is a severe test of a gas model. The results presented here, indicate that this model could be extended to other interaction situations with a fair degree of confidence.

References

[1]Sparrow, E. M. and Cess, R. D., Radiation Heat Transfer, Wadsworth, Belmont, Calif., 1966.

[2]Edwards, D. K. and Menard, W. A., "Comparison of Models for Correlation of Total Band Absorption," Applied Optics, Vol. 3, 1964, pp. 621-625.

[3]Cess, R. D., Mighdoll, P., and Tiwari, S. N., "Infrared Radiative Heat Transfer in Nongray Gases," International Journal of Heat Mass Transfer, Vol. 10, 1967, pp. 1521-1532.

[4]Gille, J. and Goody, R., "Convection in a Radiating Gas," Journal of Fluid Mechanics, Vol. 20, 1964, pp. 47-79.

[5]England, W. G. and Emery, A. F., "Thermal Radiation Effects on the Laminar Free Convection Boundary Layer of an Absorbing Gas," Journal of Heat Transfer, Vol. 91, 1969, pp. 37-44.

[6]Schimmel, W. P., Novotny, J. L., and Olsofka, F. A., "Interferometric Study of Radiation-Conduction Interaction," Proceedings of the Fourth International Heat Transfer Conference, 1970.

[7]Olsofka, F. A., "An Interferometric Study of Gaseous Radiation with Conduction in Plane Layers of N_2O and CO_2-N_2O Mixtures," M. S. thesis, 1969, Univ. of Notre Dame.

[8]International Critical Tables, Vols. 2 and 7, McGraw-Hill, New York, 1930.

[9]Tien, C. L. and Lowder, J. E., "A Correlation for Total Band Absorption of Radiating Gases," International Journal of Heat Mass Transfer, Vol. 9, 1966, pp. 698-701.

[10]Edwards, D. K. et al.; "Radiation Heat Transfer in Nonisothermal Nongray Gases," Journal of Heat Transfer, Vol. 89, 1967, pp. 219-229.

[11]Plass, G. N., "Models for Spectral Band Absorption," Journal of the Optical Society of America, Vol. 48, 1958, pp. 690-703.

[12]Schimmel, W. P., Jr., Novotny, J. L., and Kast, S., "Effect of Surface Emittance and Approximate Kernels in Radiation-Conduction Interaction," Wärme-und Stoffübertragung, Vol. 3, 1970, pp. 1-6.

[13]Tien, C. L., "Thermal Radiation Properties of Gases," Advances in Heat Transfer, Vol. 5, 1968, pp. 253-324.

[14]Edwards, D. K., "Absorption of Radiation by Carbon Monoxide Gas According to the Exponential Wide-Band Model," Applied Optics, Vol. 4, 1965, pp. 1352-1353.

[15]Burch, D. E. and Williams, D., "Infrared Absorption by Minor Atmospheric Constituents," AFCRL-TN-60-674, Contract AF 19(604)-2633, 1960, Air Force Cambridge Research Laboratory.

[16]Pliva, J., "Infrared Spectra and Molecular Constants of Isotropic Nitrous Oxides," Proceedings of the International Symposium on Molecular Structure and Spectroscopy, C-222-1, Science Council of Japan, 1962.

[17]Burch, D. E., Singleton, E. B. and Williams, D., "Absorption Line Broadening in the Infrared," Applied Optics, Vol. 1, 1962, pp. 359-363.

[18]Novotny, J. L. and Irvine, T. F., Jr., "Thermal Conductivity and Prandtl Number of CO_2 and CO_2-Air Mixtures at One Atmosphere," Journal of Heat Transfer, Vol. 83, 1961, pp. 125-132.

[19]Westenberg, A. A., "A Critical Survey of the Major Methods for Measuring and Calculating Dilute Gas Properties," Advances in Heat Transfer, Vol. 5, Academic Press, New York, 1968, pp. 253-302.

[20]Irvine, T. F., Jr. and Ibele, W. E., "Certain Thermodynamic and Transport Properties of the Atmospheres of Venus, Mars, and Jupiter," Journal of Heat Transfer, Vol. 82, 1960, pp. 381-386.

Fig. 1 N_2O-N_2 interferogram. P_t=200 kN/m^2, X(N_2O)=0.75, T_1/T_2=1.018, T_2=296°K.

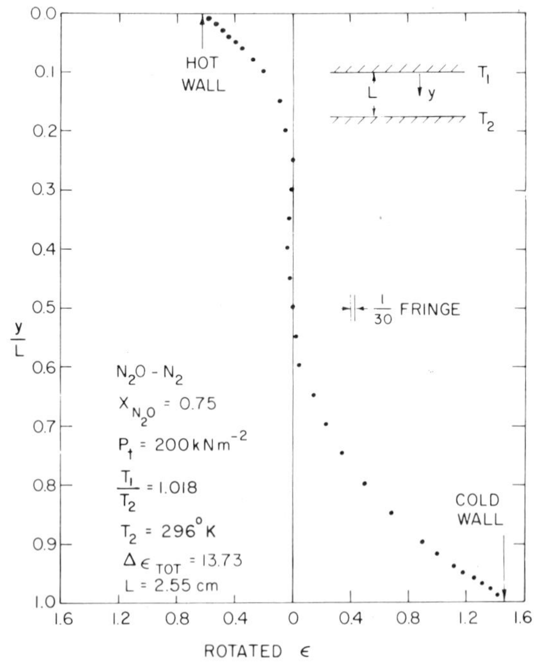

Fig. 2 N_2O-N_2 fringe profile.

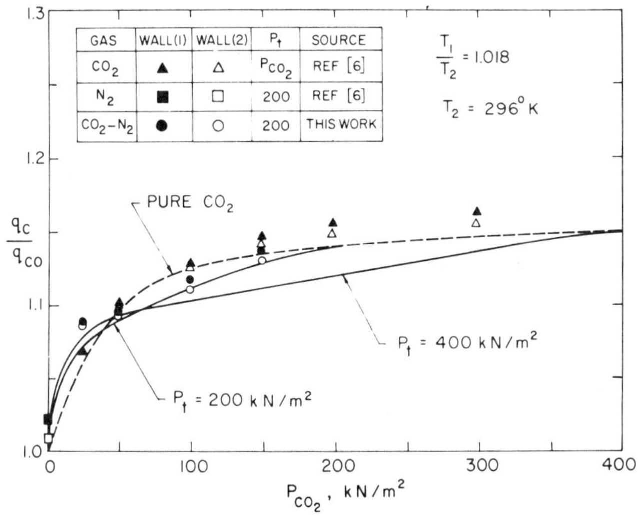

Fig. 3 Conductive heat transfer, CO_2-N_2 mixtures.

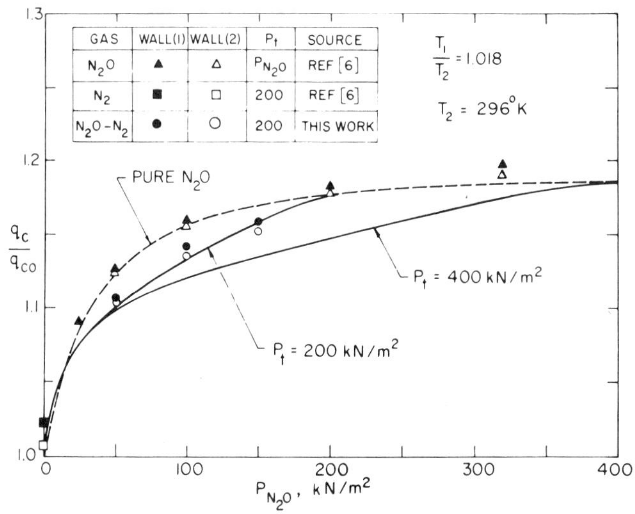

Fig. 4 Conductive heat transfer, N_2O-N_2 mixtures.

TRANSIENT HEAT FLOW IN THREE-DIMENSIONAL SANDWICH PANELS

Bert K. Larkin*

The Aerospace Corporation, El Segundo, Calif.

Abstract

An analytical solution has been developed for transient heat flow in a sandwich panel. Unlike most previous studies which treat only a one-dimensional panel, this study treats both two- and three-dimensional panels. The panel is rectangular in shape, with its core material thick in comparison with the top and bottom laminations. Because of the rather general assumptions under which the solution is developed, this solution may be applied to re-entry vehicles, spacecraft, cryogenic storage vessels, or ocean tankers. Out of the infinite series form of the solution, a criterion is developed. This criterion indicates the dimensionality of the problem; it shows when one-, two-, or three-dimensional aspects of the problem need to be considered. Thus, without actually working the problem, the engineer might use this criterion to judge the complexity of a problem.

Nomenclature

b = $\delta_2 \pi / p\ell_x$

c = specific heat

$\bar{f}(i,j,s) = \int_0^\infty \int_0^\pi \int_0^\pi F(\eta,\zeta,\theta) \cos i\eta \cos j\zeta e^{-s\theta} d\eta d\zeta d\theta$

F = transformed heat flux

h = convection coefficient

Presented as Paper 70-16 at the AIAA 8th Aerospace Sciences Meeting, New York, N.Y., January 19-21, 1970.

*Staff Engineer.

k = thermal conductivity

ℓ = length

m = $(s + i^2 + w^2 j^2)^{1/2}$

M = see Eq. (29)

p = $(k_{2z}/k_{2x})^{1/2}$

q = heat flux to material 1

r = dimensionless distance

s = Laplace transform parameter

t = time

T = temperature

U = transform of temperature

V_n = $(\epsilon_n - 1)(i^2 + w^2 j^2) + \gamma_n$, $n = 1,3$

w = ℓ_x/ℓ_y

x = distance variable

y = distance variable

z = distance variable

α = thermal diffusivity (note that α_2 is based upon k_{2x})

β = dimensionless conductivity

γ = dimensionless convection coefficient

δ = thickness

δ_{ij} = Kronecker delta

ϵ = dimensionless diffusivity

= dimensionless distance

η = dimensionless distance

θ = dimensionless time

λ = eigenvalue

ρ = density

Subscripts

x = x direction

y = y direction

z = z direction

1 = material 1

2 = material 2

3 = material 3

I. Introduction

Because of their unique combination of desirable properties, sandwich panels enjoy wide usage in the aerospace industry. To illustrate some areas of panel application, we cite: insulation for cryogenic tanks, radiatively cooled walls of re-entry vehicles, and solar panels in a spacecraft. Typically, these panels are formed such that the top and bottom layers are solid material and the center or core layer is porous. Materials used in panel construction may be wood, glass, metal, or plastic, depending upon the specific application. It is the intent of this study to consider a very general panel rather than one designed in a specific manner or one intended for a specific area of application.

Because the panel thickness is usually small when compared with its lateral dimensions, most studies in panel heat flow need only account for one-dimensional aspects of the heat flow problem. For this reason virtually all treatments of heat flow in sandwich panels that the author has seen consider only the one-dimensional form of the problem. There are, however, conditions under which two- or even three-dimensional aspects of the problem can be important. These conditions could arise whenever the heat flux to the panel surface varies greatly with lateral distance and when conductivity in the lateral direction is significant. It is to these latter problems that the material presented here is focused.

The problem to be treated is so very complex that a number of simplifying assumptions must be made in order to reduce it to a solvable form. For example, the panel is assumed to be made of three materials (each of possibly different thermo-physical properties) of finite extent in all three directions. The top and bottom layers of the panel are assumed to be very thin and possess no significant temperature gradient in depth. Transient temperatures can be obtained by exact analytical methods since the governing equations are assumed to be linear. The form of the solution is that of an infinite series. In view of the complexity of the problem and the amount of work necessary in solving the problem by alternate means, the solution contained here is fairly easy to use. Since allowance is made for 12 constants to describe the panel, the solution can be made to accommodate a variety of applications.

As an alternate, and possibly more rigorous treatment of the problem, it would be natural to attempt to amplify one of the many excellent one-dimensional solutions of the problem. The technique of Kendall and Gonzalez[1] represents one possible choice. These authors (see their Fig. 1) subdivided a single square cell into five elements or nodes and employed digital computer solution. However, the unit cell considered by Kendall and Gonzalez was only 3/8-in. on a side and the 8-in.^{2} test panel would contain over 450 unit cells. Thus, a three-dimensional treatment of the 8-in.^{2} test panel might result in a model of 2250 nodes. A problem of this magnitude would heavily load even the largest computer presently available. Since an 8-in.^{2} test panel is small when compared to panels actually used in aerospace applications, it is obvious that certain simplifications must be made in the more rigorous cell-by-cell analysis in order to allow even a hint of a solution. Thus, it seems mandatory to lump a number of cells together by employing an "effective" or estimated set of thermal properties in order to obtain a solution with a reasonable amount of effort. Since some lumping of cells is required, it seems natural to lump the properties such that only three layers comprise the panel. By assuming constant thermal properties, it is possible to develop an analytical solution in a comparatively simple form.

A rather useful criterion may be formulated from the infinite series form of the solution. This criterion establishes the dimensionality of the problem. The first term in the series expansion is equivalent to the one-dimensional solution to the heat flow problem. The higher order terms, when compared to the first term, show the contribution to the solution by considering two- and three-dimensional aspects of the problem. Even in cases where the assumptions of constant thermal properties may not justify use of an analytical solution, the analytical solution could be used to define the dimensionality. In these cases the analytical solution could fill the vital role of guiding the engineer in the proper choice of the detail necessary to solve the problem by more exact, but more costly procedures.

II. Formulation of the Problem

The heat balance equations† for the three materials comprising the panel, shown schematically in Fig. 1, are

$$\frac{\partial T_1}{\partial t} = \frac{q(x,y,t)}{\delta_1 \rho_1 c_1} + \alpha_1 \left(\frac{\partial^2 T_1}{\partial x^2} + \frac{\partial^2 T_1}{\partial y^2} \right) - \frac{h_1 T_1}{\delta_1 \rho_1 c_1} + \frac{k_{2z}}{\delta_1 \rho_1 c_1} \frac{\partial T_2}{\partial z} (x,y,0,t) \tag{1}$$

$$\frac{\partial T_2}{\partial t} = \alpha_2 \left(\frac{\partial^2 T_2}{\partial x^2} + p^2 \frac{\partial^2 T_2}{\partial z^2} + \frac{\partial^2 T_2}{\partial y^2} \right) \tag{2}$$

$$\frac{\partial T_3}{\partial t} = \alpha_3 \left(\frac{\partial^2 T_3}{\partial x^2} + \frac{\partial^2 T_3}{\partial y^2} \right) - \frac{h_3 T_3}{\delta_3 \rho_3 c_3} - \frac{k_{2z}}{\delta_3 \rho_3 c_3} \frac{\partial T_2}{\partial z} (x,y,\delta_2,t) \tag{3}$$

Equations (1) and (3) contain the assumption that the thickness of the media which they represent is so small that temperature differences in the z direction are negligible. Carslaw and Jaeger[2] (pp. 133 and 169) show the development of such equations. Each of these equations assumes convective heat exchange with a medium external to the panel and these external media are held at a fixed temperature of 0°. Conductive heat flow to material 2, of finite thickness, is

†The linearization, mentioned above, consists of considering such physical properties as α, h, k and c as constants. Actually, these properties may be functions of temperature as pointed out in Chapter I of Carslaw and Jaeger.[2]

described by the last term on the right-hand sides of Eqs. (1) and (3). The heat flux term q in Eq. (1) is any transformable function of x, y, t. Equation (2) assumes purely conductive heat flow in material 2 and the only comment necessary in describing this equation is to point out that the factor p^2 allows a different conductivity in the z direction from that of the x and y directions.

The assumed initial condition of the panel is that of zero temperature in each of the three materials.

Boundary conditions at the lateral edges, $x = 0$, $x = \ell_x$, $y = 0$ and $y = \ell_y$ assume zero temperature gradient in these lateral directions. Certain boundary conditions in materials 1 and 3 are contained in Eqs. (1) and (3). Consider, for example, the plane $z = 0$ where materials 1 and 2 meet. As the last term on the right-hand side of Eq. (1) shows, it is assumed that heat flux is continuous at the boundary $z = 0$. It is also assumed that temperature is continuous at the boundary, or

$$T_2(x,y,0,t) = T_1(x,y,t) \tag{4}$$

By a similar assumption at $z = \delta_2$, it follows that

$$T_2(x,y,\delta_2,t) = T_3(x,y,t) \tag{5}$$

The final form of the solution has a much neater appearance if the independent variables are changed to the dimensionless forms

$$\eta = x\pi/\ell_x, \quad \zeta = y\pi/\ell_y \tag{6a}$$

$$r = z\pi/p\ell_x \tag{6b}$$

$$\theta = \pi^2\alpha_2 t/\ell_x^2 \tag{6c}$$

$$b = \pi\delta_2/p\ell_x \tag{6d}$$

Dimensionless parameters which define the panel thermal properties are defined by

$$\gamma_n = h_n\ell_x^2/(\pi^2\alpha_2\delta_n\rho_n c_n) \qquad n = 1,3 \tag{7}$$

$$\beta_n = \ell_x k_{2z}/(\pi \alpha_2 p \delta_n \rho_n c_n) \qquad n = 1,3 \tag{8}$$

$$\epsilon_n = \alpha_n/\alpha_2 \qquad n = 1,3 \tag{9}$$

The input heat flux to material 1 is expressed, in units of temperature, by

$$F(\eta,\zeta,\theta) = \frac{q(\eta \ell_x/\pi,\ \zeta \ell_y/\pi,\ \ell_x^2 \theta/\pi^2 \alpha_2)\ell_x^2}{\pi^2 \alpha_2 \rho_1 c_1 \delta_1} \tag{10}$$

If one uses the variables defined by Eqs. (6) and (10), it is possible to write the governing equations as

$$\frac{\partial T_1}{\partial \theta} = F(\eta,\zeta,\theta) + \epsilon_1 \left(\frac{\partial^2 T_1}{\partial \eta^2} + w^2 \frac{\partial^2 T_1}{\partial \zeta^2} \right) - \gamma_1 T_1$$

$$+ \beta_1 \frac{\partial T_2}{\partial r} (\eta,\zeta,0,\theta) \tag{11}$$

$$\frac{\partial T_2}{\partial \theta} = \frac{\partial^2 T_2}{\partial \eta^2} + \frac{w^2 \partial^2 T_2}{\partial \zeta^2} + \frac{\partial^2 T_2}{\partial r^2} \tag{12}$$

$$\frac{\partial T_3}{\partial \theta} = \epsilon_3 \left(\frac{\partial^2 T_3}{\partial \eta^2} + w^2 \frac{\partial^2 T_3}{\partial \zeta^2} \right) - \gamma_3 T_3 - \beta_3 \frac{\partial T_2}{\partial r} (\eta,\zeta,b,\theta) \tag{13}$$

III. Solution of the Problem by Transform Techniques

The solution of the previously formulated equations is possible by transform methods. Let U_n be the transform of T_n such that

$$U_n(i,j,r,s) = \int_0^{-\infty} \int_0^{\pi} \int_0^{\pi} T_n(\eta,\zeta,r,\theta) \cos i\eta \cos j\zeta e^{-s\theta} d\eta d\zeta d\theta \tag{14}$$

for n = 1,2,3. It follows from Churchill's discussion[3] of transform properties (pp. 6 and 273) that the above transform, when applied to Eqs. (11-13), reduces these equations to a rather simple form. The only derivatives in transformed equations are those of r, and it is possible to show that the U's may be written as

$$U_1 = A \tag{15}$$

$$U_2 = B \sinh m\,(b-r) + C \cosh m\,(b-r) \tag{16}$$

$$U_3 = C \tag{17}$$

where $m = (s + i^2 + w^2j^2)^{1/2}$ and coefficients A, B, and C will be determined.

From Eqs. (4,11, and 13) it follows that

$$A - B \sinh mb - C \cosh mb = 0 \tag{18}$$

$$A[s + \epsilon_1(i^2 + w^2j^2) + \gamma_1] + Bm\beta_1 \cosh mb$$
$$+ Cm\beta_1 \sinh mb = \overline{f}(i,j,s) \tag{19}$$

$$[s + \epsilon_3(i^2 + w^2j^2) + \gamma_3]C - m\beta_3 B = 0 \tag{20}$$

The solutions for Eqs. (18,19, and 20) are

$$A = \frac{\overline{f}(i,j,s)\{m\beta_3 \cosh mb + [s + \epsilon_3(i^2 + w^2j^2) + \gamma_3]\sinh mb\}}{\Delta(s)} \tag{21}$$

$$B = \frac{\overline{f}(i,j,s)\ \{s + \epsilon_3(i^2 + w^2j^2) + \gamma_3\}}{\Delta(s)} \tag{22}$$

$$C = \overline{f}(i,j,s)m\beta_3/\Delta(s) \tag{23}$$

where
$$\Delta^{\dagger}(s,i,j) = \Big\{m\beta_3\left[s + \epsilon_1(i^2 + w^2j^2) + \gamma_1\right] + m\beta_1\left[s + \epsilon_3(i^2 + w^2j^2) + \gamma_3\right]\Big\}\cosh mb + \Big\{\left[s + \epsilon_1(i^2 + w^2j^2) + \gamma_1\right] \times \left[s + \epsilon_3(i^2 + w^2j^2) + \gamma_3\right] + m^2\beta_1\beta_3\Big\}\sinh mb \tag{24}$$

Since values for U_1 and U_3 are given by U_2 at $r = 0$, b (respectively), it is only necessary to treat U_2. Since the inverse cosine transforms are written as series expansions, the only remaining work is in finding an inverse Laplace transform. The expression for U_2 [Eq. (16)] can be considered as the product of two Laplace transforms, one of which is $\bar{f}(i,j,s)$. Thus, it is natural to express the inverse Laplace transform of U_2 as a convolution integral. From this it follows that an inverse transform needs to be found for H(s) where

$$H(s) = \frac{B(s) \sinh m\,(b-r)}{\bar{f}(i,j,s)} + \frac{m\beta_3 \cosh m\,(b-r)}{\Delta(s)} \tag{25}$$

From a substitution theorem (see Churchill,[3] p. 12), it follows that the inverse transform for H(s) may be expressed as

$$L^{-1}[H(s)] = e^{-(i^2 + w^2j^2)\theta} L^{-1}\left[H(s-i^2 - w^2j^2)\right] \tag{26}$$

The inverse transform of $H(s-i^2 - w^2j^2)$ may be expressed as an infinite series. Each root of the equation $\Delta(s-i^2 - w^2j^2) = 0$ generates one term in the series expansion and these roots lie along the negative real axis in the s plane.[§] This result allows the solution to be written as

---

[†]For simplicity, $\Delta(s,i,j)$ will be referred to merely as $\Delta(s)$.

[§]The use of residue theory to compute the inverse transform is discussed by Churchill,[3] p. 167, and the fact that the poles of H(s) lie along the negative real axis may be shown by the method of Carslaw and Jaeger,[2] p. 114-119.

$$T_2(\eta,\zeta,r,\theta) = \frac{1}{\pi^2} \sum_{j=0}^{\infty} (2-\delta_{j0})\cos j\zeta \sum_{i=0}^{\infty} (2-\delta_{i0})\cos i\eta \sum_{k=1}^{\infty}$$

$$\times \frac{[(V_3-\lambda_k^2)\sin \lambda_k(b-r) + \lambda_k\beta_3\cos \lambda_k(b-r)]}{M(\lambda_k)}$$

$$\times \int_0^{\theta} \overline{F}(i,j,\tau)e^{-(i^2 + w^2j^2 + \lambda_k^2)\ (\theta-\tau)}d\tau \qquad (27)$$

where λ_k is the kth positive root¶ of

$$\tan(b\lambda_k) = \frac{\lambda_k[\beta_3(\lambda_k^2 - V_1) + \beta_1(\lambda_k^2 - V_3)]}{(\lambda_k^2 - V_1)\ (\lambda_k^2 - V_3) - \beta_1\beta_3\lambda_k^2} \qquad (28)$$

and

$$M(\lambda_k) = \frac{1}{2\lambda_k}\Big\{[3\beta_3\lambda_k^2 - \beta_3V_1 - \beta_1V_3 + 3\beta_1\lambda_k^2$$

$$+ b(V_1 - \lambda_k^2)\ (\lambda_k^2 - V_3) + b\beta_1\beta_3\lambda_k^2]\cos\lambda_k b$$

$$+ [2\lambda_k(V_1 - \lambda_k^2) + 2\beta_1\beta_3\lambda_k + 2\lambda_k(V_3 - \lambda_k^2)$$

$$- b\beta_3\lambda_k(\lambda_k^2 - V_1) - b\beta_1\lambda_k(\lambda_k^2 - V_3)]\sin\lambda_k b\Big\}$$

$$(29)$$

and

$$\overline{F}(i,j,\tau) = \int_0^{\pi}\int_0^{\pi} F(\eta,\zeta,\tau)\ \cos i\eta\ \cos j\zeta\ d\eta\ d\zeta \qquad (30)$$

where V_n is $(\epsilon_n-1)(i^2 + w^2j^2) + \gamma_n$, $n = 1,3$.

¶$\lambda_k = 0$ is not to be counted unless both V_1 and V_3 are zero. In such a case then $\lambda_k = 0$ is a root and one must take the limit of Eq. (27) as $\lambda_k \rightarrow 0$.

IV. Estimate of Lumped or Effective Thermal Properties

A vital part of the process of predicting heat flow in panels is that of establishing the thermal properties. Unfortunately, this is generally an area of considerable uncertainty. Many times conduction, convection, and radiation occur simultaneously and the relative importance of each mechanism may change with both positions in the panel and time. In every case, the best source of properties data would be experimental test data in the particular panel of interest. Sometimes such data are not available and one must resort to the literature for the necessary values. As an aid to the reader, some of the papers providing guidance in this area will be listed.

Although the top and bottom layers are usually a single material, the bonding agent which holds these layers to the core may present a significant thermal resistance. Sauer and Nevins[4] show how the thermal resistance of the bonding agent may be employed, along with the heat capacity of the outer layer, to characterize the outer layers as a single material.

If the core material is a powdered insulation, then the review paper of Glaser[5] would be of interest. The papers of Harding[6] and Hammond[7] should be helpful in those applications in which the core material is cellular foam. For information on transfer through fibrous core materials, the works of Rolinski and Purcell[8] and Christiansen[9] et al., may be helpful. Even multilayer insulations might be contained in the panel core, and the Cryogenic Engineering Conference is one of the more outstanding sources of information on such insulations. For example, see the work of Ruccia[10] et al. Many of the panel cores are lattices of rectangular or hexagonal cells. Heat transfer in such materials has been studied by Swann,[11,12] Sauer and Nevins[4] and Minges[13].

Criterion for Determining the Dimensionality of Panel Heat Flow

Because the study of panel heat flow in two- and three-dimensions can represent a massive technical effort, it seems desirable to possess a criterion showing when such effort is really necessary. Were one to formulate a one-dimensional solution to the governing equations, the resulting solution would depend only on the thickness variable r and the time variable θ. This special case is readily obtained from Eq. (27) by considering only the $i = 0$ and $j = 0$ terms in the summation, and the result appears as

$$T_2(r,\theta) = \frac{1}{\pi^2} \sum_{k=1}^{\infty} \frac{[(V_3-\lambda_k^2)\ \sin\lambda_k\ (b-r) + \lambda_k\beta_3\ \cos\lambda_k\ (b-r)]}{M(\lambda_k)}$$

$$\times\int_0^\theta \overline{f}(0,0,\tau)e^{-\lambda_k^2(\theta-\tau)}d\tau \tag{31}$$

Note that $\overline{f}$ gives the coefficients for the cosine expansion of the input heat flux $F(\eta,\zeta,\theta)$. The natural choice for the one-dimensional approximation to the input flux would be the spatially averaged flux and that is exactly what the factor $\overline{f}(0,0,\tau)$ represents in Eq. (31).

Suppose that a two-dimensional problem is to be considered. Then the appropriate form of Eq. (27) is obtained by only taking the $j = 0$ term in the j summation, and the dependent variable could be considered to be $T_2(\eta,r,\theta)$. In the process of solving this problem by use of Eq. (27), it is necessary to expand the input flux function $F(\eta,\theta)$ into the Fourier coefficients $\overline{f}(i,\theta)$. The first term of the series expansion of Eq. (27) is given by Eq. (31). Let n represent the first non-zero value of index i which produces a non-zero $\overline{f}(i,\theta)$. As an indication of the contribution of the higher terms in the series, one may employ the ratio of the nth term to the first term in the series, which would be

$$P = \frac{2\cos n\eta}{\pi^2 T_2(r,\theta)} \sum_{k=1}^{\infty} \frac{[(V_3-\lambda_k^2)\ \sin\lambda_k\ (b-r) + \lambda_k\beta_3\ \cos\lambda_k\ (b-r)]}{M(\lambda_k)}$$

$$\times\int_0^\theta \overline{f}(n,\tau)e^{-(n^2+\lambda_k^2)(\theta-\tau)}d\tau \tag{32}$$

If P is nearly zero in magnitude, then the heat flow problem is essentially one-dimensional in character. On the other hand, if the magnitude of P is nearly unity, then the problem should receive a two-dimensional treatment. Thus P can be used as a dimensionality criterion.

The criterion for choice between two- and three-dimensions follows in just the same manner. The two-dimensional solution comes from Eq. (27) by choosing $j = 0$. Then the effect of the third-dimension is estimated from the next higher value of j which produces non-zero $\overline{f}(i,j,\theta)$. Again, the first and n^{th}

terms of the series may be used to form the criterion ratio as in Eq. (32).

VI. An Illustrative Problem

To illustrate the application of the above equations, a simple two-dimensional example will be worked. The sandwich panel is made of two aluminum plates separated by a layer of polyurethane foam -- see Table 1 for the detailed properties. The dimensionless values corresponding to these assumed conditions are shown in Table 2. The heat flux input is assumed to be

$$F(\eta,\theta) = 10^4 \eta^{-1/2} (1 - S_{\theta_o}),\ ^{\circ}\mathrm{F} \tag{33}$$

where S_{θ_o} is the unit step function[3] (p. 4). We set θ_o to be 0.017 which corresponds to 0.5 hr. Because of the nature of the step function, Eq. (33) provides an input heat flux for $0 < \theta < \theta_o$, but no heat input for $\theta > \theta_o$. Thus Eq. (33) provides a pulse input of flux. This pulse is of such size that the spatially averaged value would be 115 Btu/hr-ft^2. Because the $\eta^{-1/2}$ spatial flux variation results from a fully developed laminar boundary layer flowing over a flat plate (Ref. 14, pp. 517-526), it arises frequently in aerospace applications.

Table 1 Example problem specifications

δ_1, δ_3	0.1 in.
δ_2	1.0 in.
α_1, α_3	3.36 ft^2/hr
α_2	0.0139 ft^2/hr
ℓ_x	2 ft
p	1
k_1, k_3	120 Btu/hr-ft-°F
k_2	0.0167 Btu/hr-ft-°F
h_1	2 Btu/hr-ft^2-°F
h_3	0 Btu/hr-ft^2-°F

Table 2 Dimensionless parameter list

b	0.131
θ	0.034 t, t in hr
β_1, β_3	2.57
γ_1	196
γ_3	0
ϵ_1	242
ϵ_3	242

As Eq. (27) shows, the solution requires a cosine expansion, which for this problem would be

$$\bar{f}(i,\theta) = 10^4(1-S_{\theta_o}) \int_0^{\pi} \eta^{-1/2} \cos i\eta \, d\eta \qquad (34)$$

The integral in Eq. (34) is a Fresnel integral and is identified by Abramowitz and Stegun[15] as $C_2(x)$. These authors give, on pp. 301-302, convenient formulas for the computation of Fresnel integrals.

Figure 2 shows the spatial heat flux variation of Eq. (33) and (the circled points) the corresponding cosine expansion, truncated to 40 terms. Over the interval shown, the expansion agrees with the $\eta^{-1/2}$ curve well enough for the purposes of illustration. However, for very small values of η, there is considerable deviation since $\eta^{-1/2}$ tends to become infinite while the truncated expansion remains finite. This deviation is really a trivial one since the assumption of fully-developed flow is not valid for small values of η. Thus, the truncated expansion is physically realistic while the $\eta^{-1/2}$ curve is not.

The solution of Eq. (28) for the eigenvalues λ_k is illustrated in Fig. 3. The functions R(x) and L(x) shown in Fig. 3 are the right- and left-hand sides of Eq. (28). Each intersection of L and R gives an eigenvalue (the abscissa value of the intersection point), and Fig. 3 shows only the first three for this sample problem. (These eigenvalues are: 3.99, 13.93, and 26.01.) As λ_k becomes large, the right-hand side of Eq. (28) approaches zero, and the eigenvalues are obtained from the zeros of the tangent function. Digital computation of these eigenvalues is very easy if one takes care to avoid the vertical asymptotes of L and R.

Figure 4 illustrates typical temperature profiles for that side of the panel which receives the heat flux. Note that the temperature variable has units of degrees F, and that the temperature represents the increase with respect to the initial temperature. The spatial distribution for the two higher curves (those at $\theta = 0.0034$ and 0.017) reflect the $\eta^{-1/2}$ heat flux variation shown on Fig. 2. At time θ greater than 0.017 the heat flux is zero and this results in decreasing surface temperatures. The lowest curve, $\theta = 0.0238$, illustrates this point. Besides showing a general decrease in temperature level, the $\theta = 0.0238$ curve shows a decrease in the magnitude of the temperature gradient in the η direction.

Shown in Fig. 5 are temperatures for that panel side opposite the side that receives the flux. In general, these temperatures are lower than those shown on Fig. 4 and the η gradients are also somewhat smaller. The maximum temperature level is attained on this interior side at some time later than that corresponding to pulse removal.

Figure 6 shows the dimensionality criterion P as computed from Eq. (32). Because P is the ratio of the second term to the first term in the expansion [see Eq. (27)], it follows that P will vary with r, η, and θ. Those values shown in Fig. 6 were computed at r = 0, the side receiving the flux. Since only the magnitude of P is of importance and since the η variation is that of the cosine, only the interval from 0 to $\pi/2$ needs to be shown. For this problem to be essentially one-dimensional, it is required that P be small everywhere in the panel. This is clearly not the case as P attains values on the order of 0.5. It was observed, however, that the magnitude of P has an important transient variation. For values of θ prior to heat pulse termination, the problem is two-dimensional. This point is illustrated by the top two curves on Fig. 6. However, as the bottom curve shows, P decreases substantially after the heat pulse is removed. Thus, this problem tends to assume a one-dimensional character as θ increases. Because one-dimensional problems are much easier and more economical to solve than two- or three-dimensional ones, there may be an advantage in switching to a one-dimensional solution form for larger values of θ.

Although the above problem is only two-dimensional, it can be used as a basis for predicting the difficulty in solving a three-dimensional problem. The computer program which produced the results shown on Figs. 2-6 required 52 sec of time on a CDC 6400 system. As Eq. (27) shows, the two-dimensional solution is merely the first term of a three-dimensional solution. To develop a three-dimensional solution to a ten-term approximation would require about 10 times more computation, and this would require about 10 min. To study the dimensionality properties of a three-dimensional solution requires but two terms in the series expansion and this should require only 2 min of computer time. Thus, the equations outlined above should be compatible with the capabilities of modern computing systems even for three-dimensional problems.

VII. Conclusions

An analytical solution can be developed which describes transient three-dimensional heat flow in a sandwich panel comprised of three materials. From the series form of the solution, it is possible to develop a criterion which indicates the need for solution in one, two or three dimensions. A worked example problem suggests that even three-dimensional solutions are within the capability of modern computing machines.

References

[1]Kendall, P.J. and Gonzalez, J.I., "An Analytical Model for Predicting Thermal Response in Honeycomb Sandwich Panels," AIAA Paper 64-258, Washington, D.C., 1964.

[2]Carslaw, H.S. and Jaeger, J.C., Conduction of Heat in Solids, 2nd ed., Oxford University Press, Oxford, 1959.

[3]Churchill, R.V., Modern Operational Mathematics in Engineering, McGraw-Hill, New York, 1944.

[4]Sauer, H.J., Jr., and Nevins, R.G., "Thermal Characteristics of Honeycomb-Core Panels," AIAA Paper 64-257, Washington, D.C., 1964.

[5]Glaser, P.E., "Heat Transfer Mechanisms in Evacuated Power Insulation," International Developments in Heat Transfer, ASME, Vol. IV, 1961, pp. 829-837.

[6]Harding, R.H., "Heat Transfer Through Low-Density Cellular Materials," Industrial and Engineering Chemistry Process Design and Development, Vol. 3, 1964, pp. 117-125.

[7]Hammond, M.B., Jr., "An Analytical Model for Determining the Thermal Conductivity of Closed-Cell Foam Insulation," AIAA Paper 68-766, Los Angeles, 1968.

[8]Rolinski, E.J. and Purcell, G.V., "Thermal Conductivity Measurements of Fibrous Insulations Up to 2500°F," Proceedings of Third International Heat Transfer Conference, American Institute of Chemical Engineers, Vol. IV, 1966, pp. 133-140.

[9]Christiansen, R.M., Hollingsworth, M., Jr., and Marsh, H.N., Jr., "Low-Temperature Insulating Systems," Advances in Cryogenic Engineering, Vol. 5, 1960, pp. 171-178.

[10]Ruccia, F.E., Hinckley, R.B., and Reid, R.C., "Thermal Performance of Tank Applied Multilayer Insulations," Advances in Cryogenic Engineering, Vol. 12, 1966, pp. 218-228.

[11]Swann, R.T., "Heat Transfer and Thermal Stresses in Sandwich Panels," TN 4349, 1958, NASA.

[12]Swann, R.T., "Calculated Effective Thermal Conductivities of Honeycomb Sandwich Panels," TN D-171, 1959, NASA.

[13]Minges, M.L., "Heat Transfer in Structural Honeycomb Composites at High Temperatures," Proceedings of the Third International Heat Transfer Conference, American Institute of Chemical Engineers, Vol. IV, 1966, pp. 89-99.

[14]Kreith, F., Principles of Heat Transfer, 2nd ed., International Textbook Co., Scranton, 1965.

[15]Abramowitz, M. and Stegun, I.A., National Bureau of Standards Applied Math Series, No. 55, U.S. Government Printing Office, Washington, D.C., 1964.

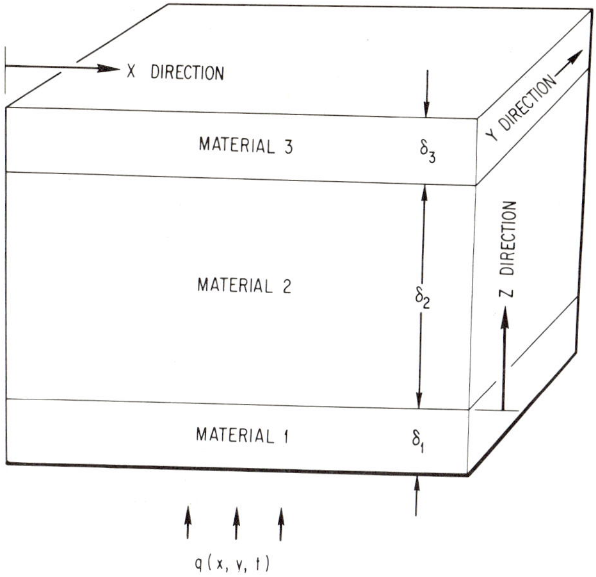

Fig. 1 Sandwich panel schematic.

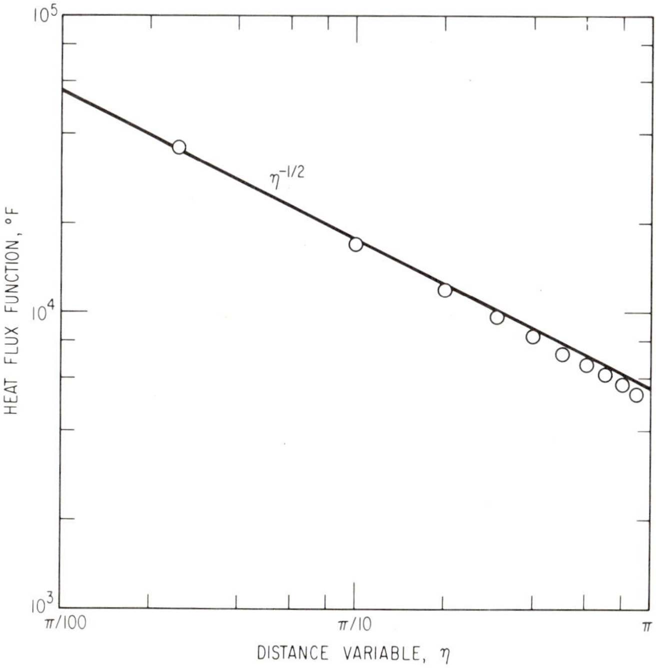

Fig. 2 Heat flux distribution.

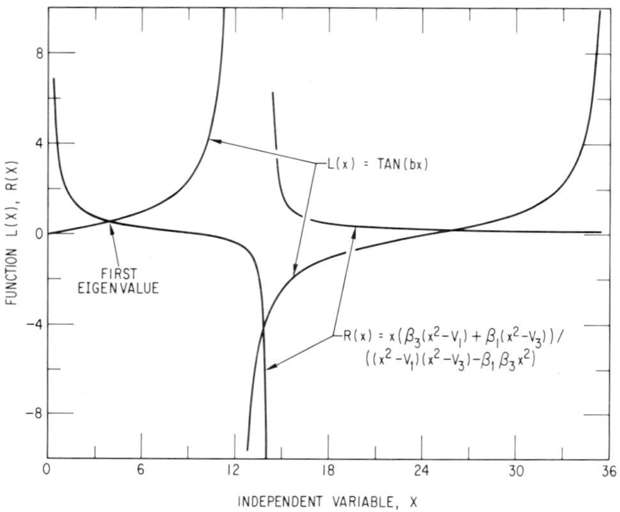

Fig. 3 An eigenvalue calculation.

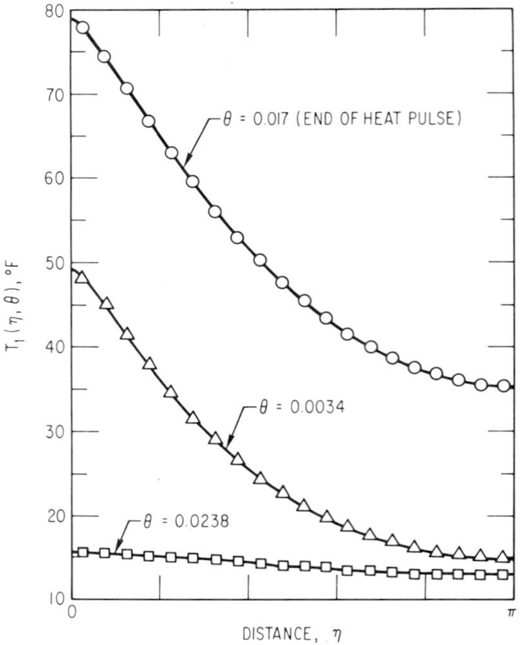

Fig. 4 Temperature of material 1.

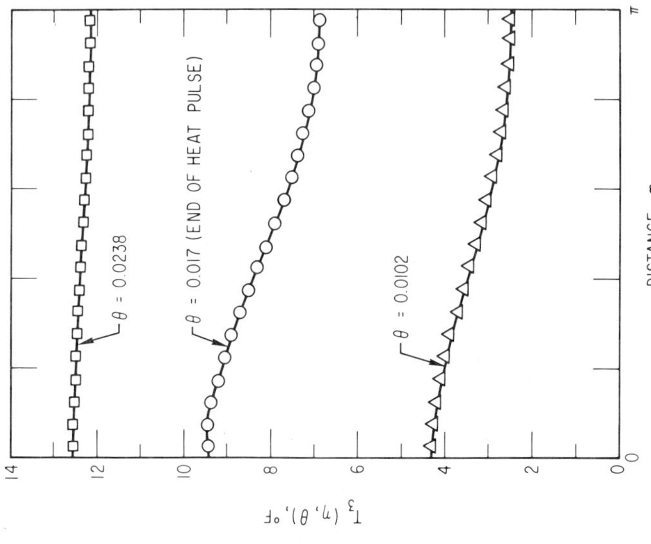

Fig. 5 Temperature of material 3.

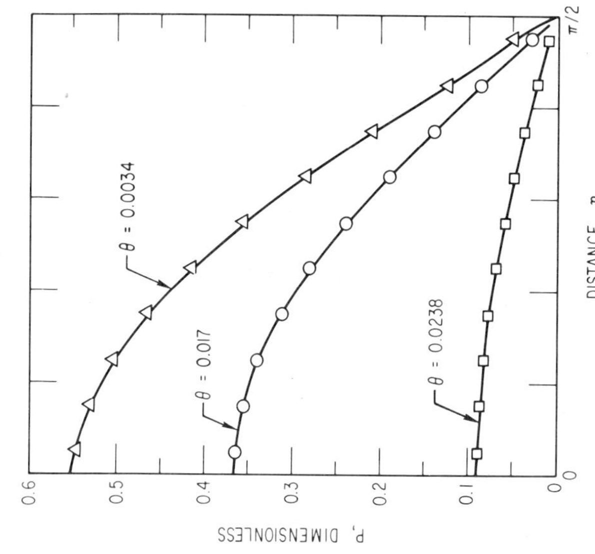

Fig. 6 Dimensionality criterion.

V. MULTILAYER INSULATION

THERMAL CHARACTERISTICS OF MULTILAYER INSULATION

H. CHAU* and H. C. MOY†
Grumman Aerospace Corporation, Bethpage, N.Y.

Abstract

Analytical and experimental techniques are used to determine the magnitudes of radiation, gas conduction, and contact solid conduction of cryogenic multilayer insulation (MLI). The analytical investigation evaluates the broadside venting of entrained gas to the space environment. This analysis, together with a heat-transfer mathematical model, is then used to optimize the performance of the MLI. A 40.6-cm -diam spherical calorimeter and a 40.6-cm -diam spherical tank supported by six tension struts are used to evaluate experimentally the interleaving layer-to-layer MLI's thermal characteristics. Heat leakage comparison between the calorimeter and the supported tank indicated negligible heat loss associated with the penetrations. This was achieved by integrating the wrapping of the blanket and penetrations to minimize parallel conduction. Gross heat flux to the supported tank with 50 layers of perforated MLI was 6.4×10^{-5} watt/cm^2, based on the tank surface area of 0.518 m^2. The contact solid conduction thermal conductivity was within the range of $2.6 - 6.2 \times 10^{-7}$ watt/cm$^\circ$C.

Nomenclature

A = cryogenic tank surface area
A_s = perforated plus unperforated area of a shield
A_h = perforated area of a shield
A_{ij} = mean surface area of layers i and j
A_o = constant related to the available adsorption sites of the outgassing material
H = activation energy of desorption of Van der Waals molecules

---

Presented as Paper 70-850 at the AIAA 5th Thermophysics Conference, Los Angeles, Calif., June 29-July 1, 1970.

This study was supported by the Grumman Advanced Development Program, Project AD 04-02.

*Advanced Development Project Engineer.

†Senior Thermodynamics Engineer.

K = Boltzmann's constant
K_m = free outgassing rate as defined by Eq. (1)
K_o = outgassing parameter as defined by Eq. (3)
M = molecular weight
n = independent variable designating layer number
R = universal gas constant
T = temperature; when subscripted, T refers to the temperature of a specific layer
t = time
$\bar{v}$ = average molecular speed
α = fractional area of the perforations
β = molecular density
ϵ_{ij} = average surface emissivity, $\frac{1}{2}(\epsilon_i + \epsilon_j)$
Γ_n, Γ_{n-1} = net free molecular flux across the two adjacent radiating shields
λ = heat of vaporization of the cryogen, evaluated at the saturation condition of the cryogenic tank pressure
σ = Stefan-Boltzmann's constant
τ = half-life of the outgassing rate

Introduction

High-performance insulation has been extensively investigated over the past decade.[1-7] In general, calorimetric tests are conducted, and the data on thermal performance is interpreted in terms of a radiation contribution of the shields and a contact solid conduction contribution of the spacer between the shields.[1,5,7] The gas conduction contribution, which is significant on thick multilayer blankets (>30 layers), is normally ignored, because a thin blanket geometry which assures negligible gas conduction within the sample is chosen for these calorimetric tests. However, when the theoretical thermal prediction based on this calorimetric data is used for the design of a multilayer insulation blanket, the resulting heat leak attributable to the composite (shield and spacer) is generally higher than predicted.[7] This degradation in thermal performance rests heavily on the failure in the design of thick multilayer insulation blankets to provide adequate molecular venting such that the interstitial spaces between blanket layers can reach a pressure below 10^{-4} torr in a reasonable length of time. In order to minimize gas conduction, the interstitial pressure must be below 10^{-4} torr.[3]

The difficulties of assessing the gas conduction contribution are due to experimental difficulties in measuring gas pressures within the blanket as well as inadequate analytical modeling and the complex nature of the adsorption-desorption phenomena of molecular species from the layers of a MLI blanket. From a design point of view, there is an optimum degree of molecular venting, because as the venting capability is increased (with perforated radiating shields, for example) to lower gas conduction, the radiation heat-transfer contribution through the blanket increases. Therefore, an overall thermal performance optimization method which includes radiation, gas conduction and contact solid conduction must be developed.

This paper presents a general method for incorporating gas conduction into just such an optimization scheme. The results of the analysis are also combined with the experimental data from two MLI-insulated 40.6-cm -diam spherical tanks in order to relate the three heat transfer modes within an MLI blanket.

Analysis

Outgassing Characteristics

In order to describe the interstitial pressure distribution across an MLI blanket, the amount of the entrained gases inside the blanket must be known. Typically, these gases come from the vaporization of water vapor and other species sorbed on the individual layers, outgassing by other blanket components, and residual purge gas. Since for the case to be considered here, the interest is in the noncontinuum flow range of the pressure-time curve, the total gas load on the blanket will be restricted to the outgassing load from the composite (shield and spacer).

The free outgassing rate or desorption rate of molecules from a surface is both time and temperature dependent. In general, the outgassing rate of a material at a given temperature is determined by a vacuum weight-loss measurement technique. However, one of the difficulties in using this raw data for predicting the outgassing load of a composite applied to a cryogenic surface is that the weight-loss equilibrium changes as the temperature and partial pressure of the species above the material being outgassed changes. This situation is prevalent in an MLI blanket system, and the problem of finding an appropriate empirical relationship that takes into account the pressure and temperature effects on the outgassing characteristics of the material is extremely difficult.

For this study, the approximate theoretical outgassing formula[8] is based on a free outgassing rate K_m for the time period when the ambient pressure change is negligible. The equation is

$$K_m = A_o T \exp(-0.69 t/\tau) \cdot \exp(-H/RT) \quad (1)$$

In evaluating the outgassing function for a given period of time, the above equation can be rewritten as

$$K_m = K_o T \exp(-H/RT) \quad (2)$$

where

$$K_o = A_o \exp(-0.69 t/\tau) \quad (3)$$

The outgassing parameter K_o can be approximated from the weight-loss data for a particular material, assuming an average molecular weight for the departed gases. Thus, for a different time period there will be a corresponding value of K_o. As an example, Table 1 presents some of these values for several insulation materials, using the molecular weight of water as the average molecular weight. The outgassing parameter K_o for the composite presented in Table 1, is assumed to be the sum of the individual outgassing parameters.

Table 1 Outgassing parameters

Specimen	K_o, molecules/cm^2-sec-^{o}K		
	2000 sec	7000 sec	12,000 sec
1) 1/8 mil double aluminized mylar	1.75×10^{12}	1.10×10^{12}	6.95×10^{11}
2) 1/4 mil double gold coated mylar	8.32×10^{13}	4.51×10^{12}	. . .
3) Nylon netting (4.9×10^{-4} gm/cm^2)	3.74×10^{13}	1.11×10^{13}	3.29×10^{12}
4) 1/2 mil aluminized H film	1.22×10^{14}	9.05×10^{12}	6.69×10^{11}
5) 1 layer of 1) and 2 layers of 3)	7.66×10^{13}	2.33×10^{13}	7.28×10^{12}

Venting Analysis

An insight into the molecular density distribution which affects gas conduction can be obtained from a steady-state, noncontinuum, molecular venting analysis, using a temperature dependent source term in the governing equation. The source term is the temperature and time dependent outgassing function developed in the previous paragraph, with the time dependence accounted for in the outgassing parameter K_o. In effect, the steady-state analysis represents a frozen-time zone study. It should be noted here that the approximate outgassing function, as well as its corresponding outgassing parameter K_o quantify the interstitial molecular density inside the blanket for a particular outgassing load, but give only an order of magnitude approximation to the outgassing load that exists on a particular blanket.

The analytical model consists of two adjacent infinite, parallel, uniformly perforated shields separated by spacer material. (The model assumed that for any given layer the hole fraction is uniformly distributed so that in effect no molecular density gradient exists throughout a given layer.) The spacer is a netting material and the shields are metallized on both sides. The apparent thickness of the net spacer is such that it occupies the space h between the two shields. The standard formulation for equilibrium venting through an orifice is assumed to be valid here, even though the parallel surfaces are separated by a netting material. The outgassing load of the netting material is added to the surface outgassing load of the shield, and the subsequent theoretical development considers only infinite parallel surfaces.

The approximate molecular balance for the above modified venting model is

$$A_h \Gamma_n = A_h \Gamma_{n-1} + 2(A_S - A_h)\cdot K_m(T)$$

Therefore

$$\Gamma_n - \Gamma_{n-1} = \left[2(1-\alpha)/\alpha\right]\cdot K_m(T) \tag{4}$$

where

$$\alpha = A_h/A_S$$

Equation (4) is an expression for the net molecular transport across the interstitial space. When a system is composed of many shields and the distance between shields is small compared to the total multilayer thickness, the net one-dimensional free molecular flux Γ can be expressed as

$$\Gamma = -\frac{1}{4}\cdot\bar{v}\cdot d\beta/dx$$

Therefore

$$\Gamma_n - \Gamma_{n-1} = \frac{d}{dn}\left[-\frac{1}{4}\,\bar{v}\,\frac{d\beta}{dn}\right]$$

or, neglecting the variation in $\bar{v}$ between adjacent shields,

$$d^2\beta/dn^2 = -\left[8(1-\alpha)/\bar{v}\alpha\right]\cdot K_m(T) \qquad (5)$$

Equation (5) approximates a steady-state free molecular venting process through the perforations of a multilayer insulation blanket. In order to solve Eq. (5) for β as a function of n, the temperature in the outgassing function and the average molecular speed must be expressed as function of n. The temperature of the nth layer (counting from the tank wall) in a N-layer MLI system can be empirically approximated by

$$T(n) \simeq T_N\cdot(n/N)^{\frac{\ln T_N - \ln T_1}{\ln N}} \qquad (6)$$

If we let

$$B = (\ln T_N - \ln T_1)/\ln N$$

$$C = 2\left(\frac{2\pi M}{R}\right)^{\frac{1}{2}}\left(\frac{1-\alpha}{\alpha}\right)K_o\,T_N^{\frac{1}{2}}\,N^{-\frac{B}{2}}$$

$$D = HN^B/RT_N$$

Then Eq. (5) can be rewritten as

$$d^2\beta/dn^2 = -Cn^{\frac{B}{2}}\cdot\exp(-Dn^{-B}) \qquad (7)$$

with boundary conditions

$$d\beta/dn = 0 \qquad \text{at } n = 1$$

$$\beta = \beta_o \qquad \text{at } n = N$$

This boundary value problem can be readily solved by two successive integrations.

Figure 1 represents the results for two different outgassing parameters and a perforation fraction of 0.5%. The calculations were based on taking $T_N = 300^oK$ and $T_1 = 78^oK$ for $N > 45$ and

$$T_1 = 290\left(\frac{1}{N}\right)^{0.3319} \;{}^oK, \text{ for } N \leqslant 45 \qquad (8)$$

Equation (8) is an empirical expression for the first layer temperature, obtained by extrapolating the experimental data. It should be emphasized that since the present venting study aimed at approximating the order of magnitude of the gas density distribution within the MLI, the MLI temperature profile approximated by Eq. (6) is considered sufficient. Knowing the molecular density distribution, the gas conduction contribution for an MLI system can be calculated.

Thermal Analysis

Heat is transferred through an MLI system, consisting of radiating shields and low-conductive spacers, by three additive mechanisms: radiation, gas conduction, and contact solid conduction.

The steady-state thermal radiation transmitted between adjacent perforated radiating shields, i and j, can be estimated from the one-dimensional, infinite parallel plates model[9]

$$Q_{R_{ij}} = \frac{\epsilon_{ij} + \alpha(2-\epsilon_{ij})}{(2-\epsilon_{ij})(1-\alpha)} \cdot \sigma A_{ij}\,(T_j^4 - T_i^4) \qquad (9)$$

The radiation can be separated into two terms:

$$Q_{R_{ij}} = \left(\frac{2}{\epsilon_{ij}} - 1\right)^{-1} \sigma A_{ij}\,(T_j^4 - T_i^4) + \left(\frac{2}{2-\epsilon_{ij}}\right)\left(\frac{\alpha}{1-\alpha}\right) \sigma A_{ij}\,(T_j^4 - T_i^4) \qquad (10)$$

or

$$Q_{R_{ij}} = Q_{RS_{ij}} + Q_{RH_{ij}} \qquad (11)$$

where the first righthand term, $Q_{RS_{ij}}$, represents the radiation transferred through the unperforated portion of the radiating shield, and the second righthand term, QRH_{ij}, is the radiation transmitted through the perforations.

From kinetic theory, the heat-transfer between adjacent radiating shields, i and j, due to the effect of residual interstitial gas can be estimated from

$$Q_{GC_{ij}} = (3K/4)\mathcal{A}\beta_{ij}\bar{v}_{ij}A_{ij}(T_j - T_i) \qquad (12)$$

In the derivation of the previous equation, the specific heat at constant volume is taken to be 3 K, and includes both the translational and rotational internal energies. The accommodation coefficient $\mathcal{A}$ is assumed to be one. β_{ij}, the residual gas concentration within the MLI system, has been discussed in the venting analysis section.

The heat transmitted at the points of physical contact between insulation layers can be defined as:

$$Q_{KC_{ij}} = k_{KC_{ij}} A_{ij}(T_j - T_i)/h \qquad (13)$$

where h is the distance between adjacent layers, i and j. The contact solid conduction thermal conductivity k_{KC} is a function of the thermal conductivities of the insulating materials and the contact resistance between layers, which depends upon the static load on the specimen. Since the contact resistance is difficult to analyze theoretically, k_{KC} may be estimated by utilizing extensive test data. In summary, the total heat-transfer Q_T is the summation of the radiation, gas conduction, and contact solid conduction modes, that is:

$$Q_T = Q_{RS} + Q_{RH} + Q_{GC} + Q_{KC} \qquad (14)$$

Optimization of the Percent of Shield Perforation

The interstitial gases resulting from the outgassing of MLI components can significantly reduce the thermal performance of the MLI if not allowed to vent to the ambient. It has been found that perforated radiating shields provide a controllable means for alleviating this problem in addition to satisfying the boost venting requirement. However, with the introduction of perforations, the radiating shields experience an increase in emittance. Therefore, an optimum perforation percentage should be employed, i.e., a perforation percentage which will allow sufficient venting yet not seriously degrade the MLI emittance. Yet, the optimum hole perforation fraction alone is not sufficient to guarantee lower equilibrium interstitial pressure. In free molecular flow the passage length from a hole in one layer to a hole in the next layer should be as short as possible. Thus, the number of holes per unit area is an important parameter. To have a greater number of holes for the same perforation fraction, the diameter of the holes must be smaller. The smallest hole diameter commercially available in $\frac{1}{4}$ mil aluminized mylar is 0.1168 cm.

For the blanket optimization study, Eq. (14) can be rewritten as

$$Q_T = Q_R(\alpha, \epsilon, T) + Q_{GC}(\alpha, \beta, \bar{v}, T) + Q_{KC}(T)$$

On the assumptions that Q_{KC} is not significantly influenced by perforations, and ϵ, β and $\bar{v}$ are constant, the optimum perforation fraction can then be obtained by setting the first derivative of Q_T with respect to α equal to zero and solving for α. In evaluating the optimization equation, the maximum molecular density, β, was evaluated at n=1, $\bar{v}$ was evaluated at the average temperature of the MLI, and ϵ was evaluated at 300°K [Eq. (17)].

Some typical results for optimum perforation are presented in Fig. 2 for various K_o and total number of radiating layers in an MLI system. The influence of the optimum perforation on heat flux, without the contact solid conduction contribution, is presented in Fig. 3. The heat flux through an MLI with a specified number of layers can be optimized as a function of hole perforation fraction. Figure 4 shows that a 30-layer blanket should have an optimum hole perforation fraction of 0.4% for K_o = 1.47 x 10^{12} molecules/cm^2°K. However, in the 50-layer blanket system, as illustrated in Fig. 5, the optimum hole perforation should be 0.65% for the same outgassing parameter.

Experiment

The experimental program consisted of two phases. The first phase was conducted primarily to determine the thermal performance of the MLI blanket while the second phase emphasized the thermal integration of the MLI with a cryostat supported by six fiberglass tension struts, simulating an actual cryogenic system.

Apparatus

Calorimeter. A thermally guarded calorimeter was designed, for the first phase study, to evaluate the thermal characteristics of the MLI system. It essentially consisted of two components: a cryogenic test tank(Fig. 6) and a boil-off measurement system. The thermal performance of the specimen was evaluated from the boil-off rate of cryogen.

The 40.6-cm -diam test tank was made by welding together two 0.159-cm -thick stainless steel hemispheres. A double walled stainless steel tube served as fill and vent lines. The guard collar, forming a cryogenic envelope surrounding the upper portion of the fill and vent, was designed to minimize any heat conducted through the fill and vent line to the test tank. The two highly polished copper plates extending from the bottom of the guard collar served as radiating shields, further thermally isolating the cryogenic tank. The filling of the guard collar with cryogen through its own fill and vent lines was controlled automatically.

The tank and guard collar assembly was positioned in a brazing rod frame, resembling a bird cage, and to this frame was attached the MLI system.

Preliminary analysis indicated that the cryogen boil-off could be as low as 1 cc/sec. The measurement of such an extremely low boil-off rate was accomplished by measuring the pressure differential across a 122 cm long, 0.14-cm -i.d. capillary tube. A Brooks 70 cc volume calibrator was incorporated into the flow measurement section to provide constant in-line calibration of the differential pressure transducer.

Pressure transducers were used to monitor the cryogenic tank pressure (the saturated cryogen vapor pressure inside the calorimeter) and the pressure drop across the capillary tube.

A manostat was installed in-line to restrict the variation of vapor pressure inside the calorimeter caused by the fluctuating barometric pressure. At the small flow rate encountered in the present study, even small fluctuations in vapor pressure would result in significant changes in the cryogenic internal energy which in turn precipitate an unstable evaporization process.

Strut-Supported Cryostat. In order to evaluate the thermal efficiency of a more realistic cryogenic tank system, a supported cryostat was designed.

This system consisted of a 40.6-cm -o.d. spherical stainless steel tank with penetrations of three pairs of tension support struts, and one fill and one vent line. The support struts were made of 35.6-cm long, 0.318-cm square fiberglass rods. They were fastened along the equatorial perimeter of the cryogenic tank at 120° intervals, as illustrated in Fig. 7. The 0.635-cm -o.d. stainless steel fill and vent lines were located at the top of the tank. Their 274.3-cm extension was formed into a helix before penetrating the vacuum chamber cover.

Similar to the calorimeter design, a "bird cage" frame was provided for supporting the multilayer insulation. The cryogen vapor measurement system, also, was identical to that for the calorimeter.

MLI Composition

One-quarter mil perforated mylar film aluminized on both sides was selected as the radiating shield. The perforation diameter was 46 mils and the percentage of open area was 0.5%. The perforation fraction of 0.5% was selected based on analysis (Fig. 2) of a 30-layer blanket with an outgassing parameter of 1.47×10^{12} molecules/cm^2-sec-$^{\circ}$K. The spacer material was 5.5 mil thick nylon netting of 15 denier, weighing 4.9×10^{-4} gm/cm^2. Two layers of the spacer were used to separate the adjacent radiating shields in order to minimize the direct physical contact of the shields and to reduce the solid conduction heat transfer through the spacer material itself. A single "layer" in the MLI system studied here is thus defined as the composition of one shield and a double layer of spacer.

To simulate an ideal one-dimensional heat-transfer MLI system, as defined by the mathematical model, the insulation layers were installed individually. First, the two nylon spacers were attached to the aluminized mylar shield with the aid of mylar tape, and then different patterns were cut to match the contour of the supporting cage. Finally, the blanket was wrapped layer-by-layer with approximately 1.5 cm overlapping between sections. The circumference of the wrapping was measured and adjusted every 5 layers to assure uniform MLI density (a density of 30 layers per 2.54 cm was maintained). The wrapping scheme of the penetrations was integrated with that of the cryostat to minimize parallel conduction. This included not only matching of the two MLI systems layer-to-layer, but also mating of the local penetration temperature with that of the MLI.

Test Procedure

The calorimeter and supported cryostat tanks were initially wrapped with MLI blankets of 30 and 50 layers, respectively. At the completion of a test series, a specified number of layers were stripped from this initial blanket, and the remaining blanket tested in the next series. These initial blankets were baked out in the evacuated test chamber (10^{-6} torr) for three days at a tank temperature of approximately 82°C.

Before the start of each series, the MLI was baked out under the aforementioned conditions for one day. After this, the tank in conjunction with the rest of the test system was halogen leak tested. The tank was then filled with liquid nitrogen and allowed approximately 24 hr to stabilize before actual data logging commenced.

Daily readings, taken at half-hour intervals, included tank pressure, boil-off (pressure drop across the capillary), and MLI temperatures as well as other temperature records necessary in the MLI performance analysis. A calibration point, using the Brooks Volumeter, for the capillary tube was also obtained at the end of each data logging interval. An interesting point to note here is that the cryogenic system had to have sufficient time to restabilize before another calibration point could be taken. A visual observation of the recovery time was obtained by monitoring the millivolt output of the capillary pressure transducer with a pen recorder.

Since the mass flow rate changes when layers are removed for each successive series, calibration points for the capillary tube vary significantly. Thus, at the end of the total test series all the calibration points were fitted by a least square method to provide an equation for mass flow rate versus ΔP across the capillary tube.

Results and Discussion

From the experimental cryogen boil-off rate ($\dot{m}$, gm/hr) and tank pressure the thermal performance of the testing system is

$$q_T = \dot{m}\lambda/A \quad \text{watt/cm}^2 \tag{15}$$

The values of q_T for the calorimeter and the strut-supported cryostat test series, for various MLI thicknesses, are summarized in Table 2. Initially, for the 30-and 25-layer calorimeter tests, only 10 to 14 data sets (one test

Table 2 Summary of experimental heat leak

Test sequence	Total MLI layer - 30 Layer per 2.54 cm	$Q_T \times 10^4$, watt/cm^2 — Test series no. 1	2	3	4	5	6	7	8	9	Grand average watt/cm^2
Calorimeter test (40.64-cm-diam spherical tank)											
1	30	0.751	...	...	...	...	...	...	...	...	0.751×10^{-4}
2	25	0.814	...	...	...	...	...	...	...	...	0.814×10^{-4}
3	20	1.031	1.012	1.015	1.025	...	...	...	...	...	1.022×10^{-4}
4	15	1.025	0.981	0.889	0.933	0.990	...	...	...	...	0.962×10^{-4}
5	10	1.309	1.362	1.523	1.425	1.315	1.343	1.555	...	...	1.397×10^{-4}
6	5	2.283	2.299	2.160	2.283	2.192	2.267	...	...	...	2.245×10^{-4}
Supported tank test (40.64-cm-diam spherical tank)											
7	50	0.539	0.634	0.694	0.599	0.643	0.650	0.741	0.681	0.624	0.640×10^{-4}
8	40	0.656	0.681	0.710	0.744	0.691	0.697	0.618	0.691	...	0.687×10^{-4}
9	30	0.694	0.738	0.694	0.725	0.725	0.681	...	...	...	0.710×10^{-4}
10	10	1.236	1.353	1.309	1.274	1.302	...	...	...	...	1.296×10^{-4}

series) were taken, at thermal equilibrium, at half-hour intervals. Since additional data yields a better statistical sample, it was decided to increase the number of test series for the subsequent blanket sizes. Figure 8 is a graphical representation of Table 2. The results indicated negligible heat leak associated with the penetrations, and this is believed to be due to the wrapping scheme discussed previously. Gross heat flux to the supported tank with 50 layers of MLI was 6.4 x 10^{-5} watt/cm^2, based on the tank surface area of 0.518 m^2, and would have been about 20% less if the surface area of the innermost MLI layer had been used.

In a layer-to-layer evaluation of the MLI heat transfer modes and the contact solid conduction thermal conductivity, the accuracy of a continuous MLI temperature profile is essential. For this reason, the approximation given by Eq. 6 would not be suitable here. The discrete experimental temperature distribution can, however, be represented by a second-order least-square polynominal fit on log-log scale

$$\ln T_n = \sum_{i=0}^{2} C_i (\ln n)^i, \quad n \geq 5 \tag{16}$$

where T_n is the nth layer temperature, in degrees Kelvin, in an N-layer MLI, and C_i are the empirical constants. A comparison between the experimental data and the temperature calculated from the empirical equation are shown in Figs. 9 and 10. The results of the 5-, 10-and 15-layer MLI calorimeter tests and the 10-layer MLI supported cryostat test are excluded from the presentation because of insufficient or erratic temperature readings.

Once the MLI temperature distribution is accurately described, the layer-to-layer thermal analysis can then be performed using Eqs. 9 - 14. In the detailed computation, it is assumed that the surface emissivity of aluminized mylar is proportional to the 2/3 power of the absolute temperature.[6]

$$\epsilon = 6.337 \times 10^{-4} T^{\frac{2}{3}}, \quad T \text{ in } {}^{\circ}K \tag{17}$$

To evaluate the magnitude of the gas conduction term, the value of K_o was estimated to be 1.47 x 10^{12} molecules/cm^2-sec-°K for the 20-, 25- and 30-layer MLI, and 2.94 x 10^{12} molecules/cm^2-sec-°K for the 40-and 50-layer MLI.

These estimates were arrived at after an examination of the nonradiation heat transfer through the MLI. Figure 11 presents the apparent thermal conductivity k_{NR} of the MLI for the nonradiation contribution Q_{NR} and is defined as

$$k_{NR_{ij}} = Q_{NR_{ij}} \cdot \frac{h}{A_{ij}(T_j - T_i)}$$

$$= Q_{GC_{ij}} \cdot \frac{h}{A_{ij}(T_j - T_i)} + k_{KC_{ij}} \qquad (18)$$

Those portions of the curves bounded by the isotherms of 224°K and 300°K appear to be temperature independent. It is reasonable to assume, therefore, that within this range, the contact solid conduction thermal conductivity k_{KC} of the MLI is also constant. Based on this assumption, Eq. (18) can be combined with the outgassing and venting analyses in order to determine values for K_O, which can then be used in the evaluation of the gas conduction terms for different blanket sizes.

The values of K_O evaluated from this analysis were substituted back into Eq. (18) in order to evaluate k_{KC}. The results presented in Fig. 12 show that k_{KC} seems constant over the range considered above. However, at the innermost layers, k_{KC} appears to be temperature dependent. The calculated values vary between 2.64-6.17 x 10^{-7} watt/cm-°K. This can be attributed to the fact that at this inner section of the blanket there is an extremely large temperature gradient. Unfortunately, the effect of temperature on either k_{NR} or k_{KC} at the inner layers could not be properly evaluated because of insufficient temperature measurements.

Figure 13 presents the heat-transfer characteristics of several total MLI blankets for the calorimeter tests (i.e., taking i = 1 and j = N in the analysis presented previously and dividing Eqs. (9, 12, 13 and 14) by N - 1, the total number of layer gaps in an N-layer MLI). It should be noted that in the overall analysis, the detailed MLI temperature distribution is not required. Only the first layer and the outermost layer temperatures are needed. Since no experimental temperature measurements were made on the first layer, this temperature is approximated by Eq. (8). The calculations were based on K_O = 1.47 x 10^{12} molecules/cm^2-sec-°K and a constant emissivity of 0.028. The nonradiation heat transfer is significant and becomes more pronounced as the number of MLI blanket layers increases. As illustrated in the figure, the nonradiation is only 19% of the total heat transmitted through the 10-layer MLI, however, for the 30-layer MLI the contribution is 55%.

The calculated layer-to-layer thermal characteristics for the different blanket sizes (total layers) is summarized in Table 3 for both the calorimeter and supported cryostat tests. The relative importance of the different heat-transfer modes changes in passing through the MLI system. Radiation dominates at the hot boundary, and conduction dominates at the cold boundary. Almost constant gas conduction prevails throughout the insulation, but then decreases rapidly near the outermost layers. This is a consequence of the calculated residual gas density distribution, illustrated in Fig. 1.

Concluding Remarks

The analytical and experimental program presented here serves as a basis for a more accurate evaluation of MLI system thermal performance. Specifically, the gas conduction contribution, significant in the performance of thick MLI blankets, has been shown to be amenable to analysis. Current efforts in this area, which include provisions for closer monitoring of the temperatures of the MLI layers and measurement of the interstitial pressure distribution in the 10^{-5} torr range, should lead to further refinement of the outgassing parameter in the mathematical model. This should result in a consistent and reliable thermal performance evaluation method for blankets with a high number of layers.

References

[1]Black, I. A., Fowle, A. A. and Glaser, P. E., "Development of High-Efficiency Insulation," Advances in Cryogenic Engineering, Vol. 5, 1960, pp. 181-188.

[2]Kropschot, R. H., Schrodt, J. E., Fulk, M. M. and Hunter, B. J., "Multiple-Layer Insulation," Advances in Cryogenic Engineering, Vol. 5, 1960, pp. 189-198.

[3]Hnilicka, M. P., "Engineering Aspects of Heat-Transfer in Multilayer Reflective Insulation and Performance of NRC Insulation," Advances in Cryogenic Engineering, Vol. 5, 1960, pp. 199-208.

[4]Riede, P. M. and Wang, D. I-J., "Characteristics and Applications of Some Superinsulations," Advances in Cryogenic Engineering, Vol. 5, 1960, pp. 209-215.

Table 3 Summary of layer-to-layer thermal characteristics of MLI

Test sequence	Total MLI layer	Layer no. in MLI		Grand average heat leak, watt				
		i	j	Q_T	$Q_{RS_{ij}}$	$Q_{RH_{ij}}$	$Q_{GC_{ij}}$	$Q_{KC_{ij}}$
1 (Calorimeter test)	30	5	6	0.389	0.090	0.046	0.048	0.205
		10	11	0.389	0.133	0.058	0.032	0.166
		20	21	0.389	0.168	0.063	0.019	0.139
		29	30	0.389	0.177	0.062	0.002	0.148
2 (Calorimeter test)	25	5	6	0.422	0.099	0.050	0.033	0.240
		10	11	0.422	0.155	0.065	0.024	0.178
		20	21	0.422	0.226	0.082	0.014	0.100
		24	25	0.422	0.247	0.087	0.002	0.086
3 (Calorimeter test)	20	5	6	0.530	0.141	0.067	0.023	0.299
		10	11	0.530	0.216	0.087	0.017	0.210
		19	20	0.530	0.305	0.107	0.002	0.116
7 (Supported tank test)	50	5	6	0.332	0.052	0.026	0.155	0.099
		10	11	0.332	0.062	0.029	0.114	0.127
		20	21	0.332	0.082	0.033	0.086	0.131
		30	31	0.332	0.101	0.038	0.075	0.118
		40	41	0.332	0.121	0.043	0.059	0.109
		49	50	0.332	0.139	0.048	0.005	0.140
8 (Supported tank test)	40	5	6	0.357	0.064	0.031	0.108	0.154
		10	11	0.357	0.078	0.034	0.079	0.166
		20	21	0.357	0.103	0.041	0.060	0.153
		30	31	0.357	0.128	0.047	0.048	0.134
		39	40	0.357	0.152	0.053	0.005	0.147
9 (Supported tank test)	30	5	6	0.368	0.091	0.043	0.034	0.200
		10	11	0.368	0.112	0.048	0.025	0.183
		20	21	0.368	0.151	0.057	0.018	0.142
		29	30	0.368	0.184	0.065	0.002	0.117

[5]Leonhard, K. E., Collom, J. E., Kneisel, K. M. and Tatro, R. E., "Superinsulation Research and Development Program," Convair Division of General Dynamics Report GDC-ERR-AN-1201, December, 1967.

[6]Cunnington, G. R. and Tien, C. L., "A Study of Heat-Transfer Processes in Multilayer Insulation, "AIAA Paper 69-607, San Francisco, Calif., 1969.

[7]Caren, R. P. and Cunnington, G. R., "Heat-Transfer in Multilayer Insulation System," Advances in Cryogenic Heat Transfer, Vol. 64, No. 87, 1968, pp. 67-81.

[8]Dayton, B. B., "Relations Between Size of Vacuum Chamber, Outgassing Rate, and Required Pumping Speed," Vacuum Symposium Transactions, American Vacuum Society, Inc., 1959, pp. 101-119.

[9]Emslie, A. G., "Gas Conduction Problems with Multilayer Radiation Shields," Rept. 63270-04-01 (AD-256894), April 1961, Arthur D. Little, Inc.

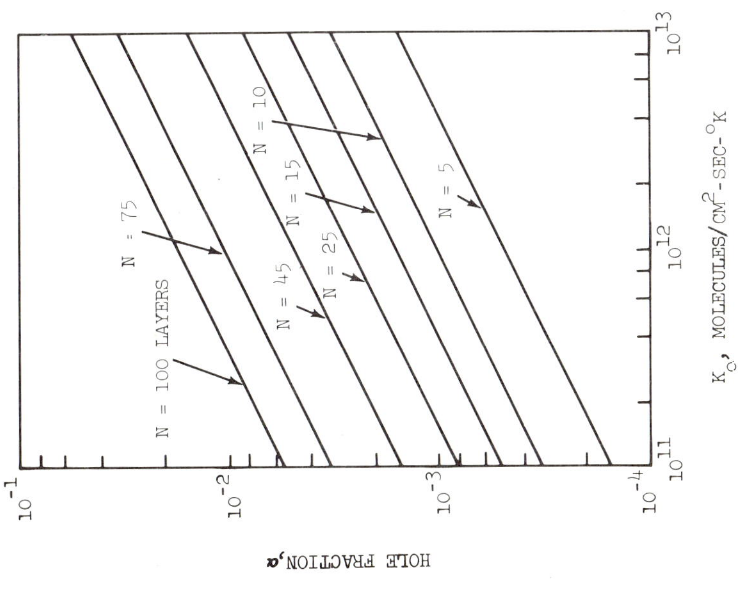

Fig. 1 Gas density distribution in MLI.

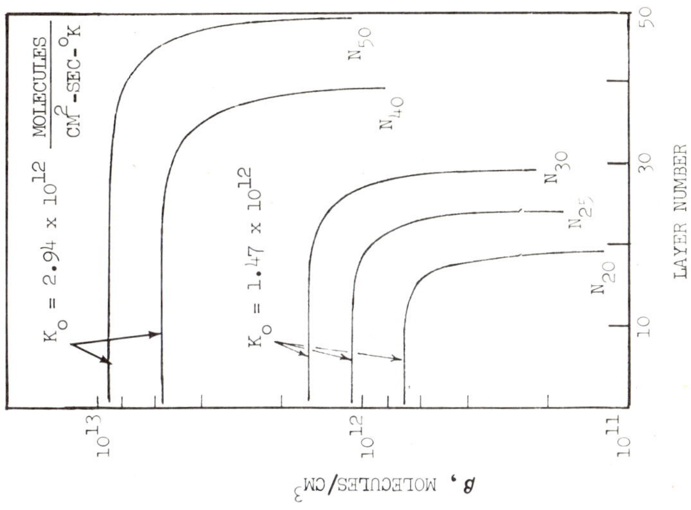

Fig. 2 Optimum radiating shield perforation vs K_o.

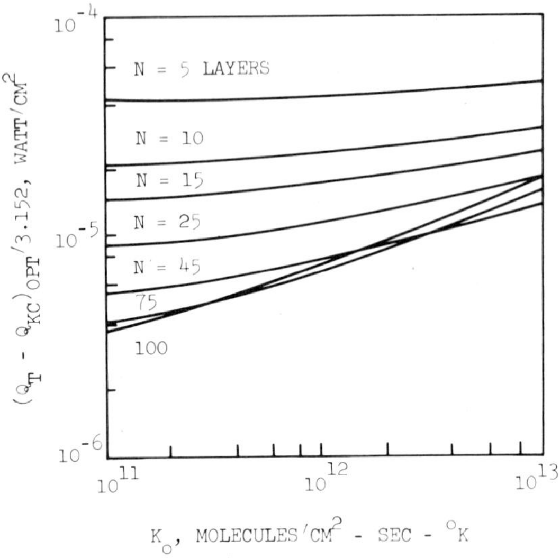

Fig. 3 Optimum performance of MLI vs K_o.

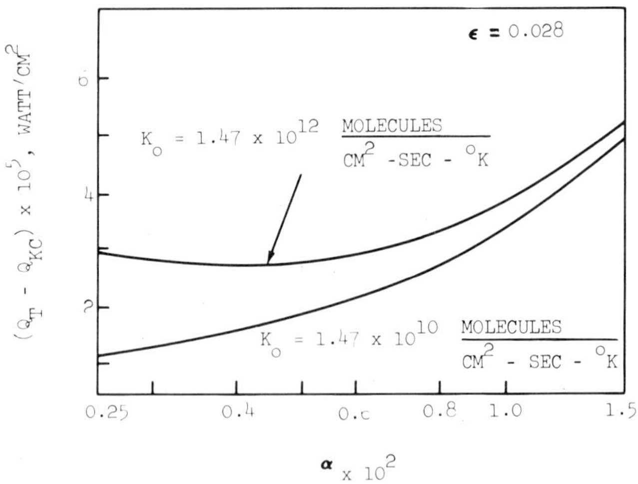

Fig. 4 Thermal performance vs hole fraction of a 30-layer MLI.

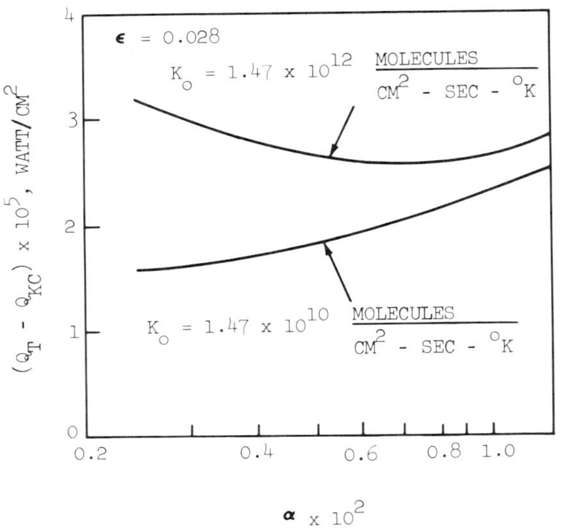

Fig. 5
Thermal performance vs hole fraction of a 50-layer MLI.

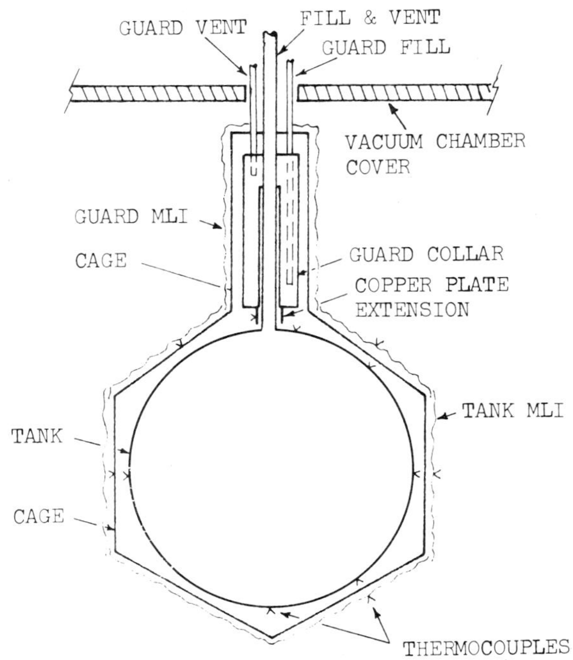

Fig. 6 Cryogenic calorimeter.

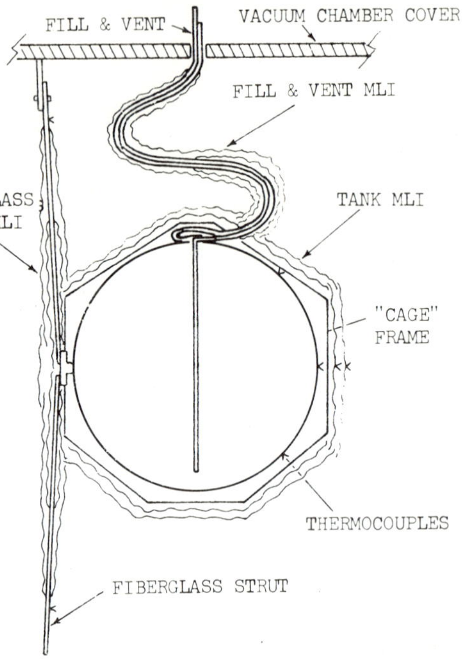

Fig. 7 Supported cryostat.

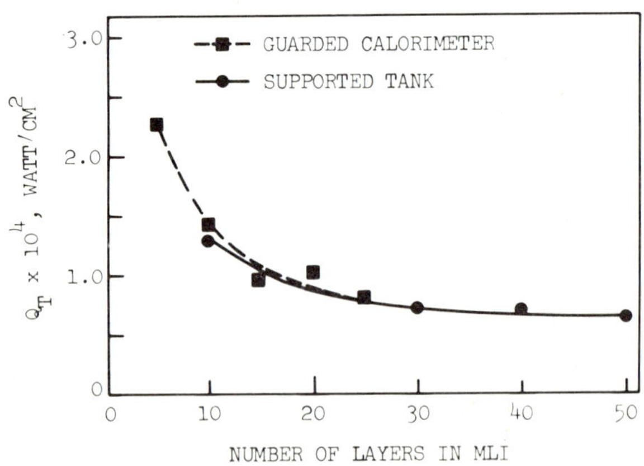

Fig. 8 Thermal performance of MLI.

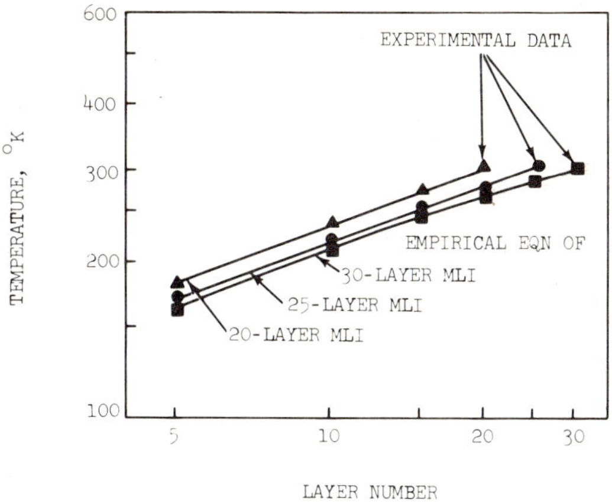

Fig. 9 MLI temperature profiles (calorimeter test).

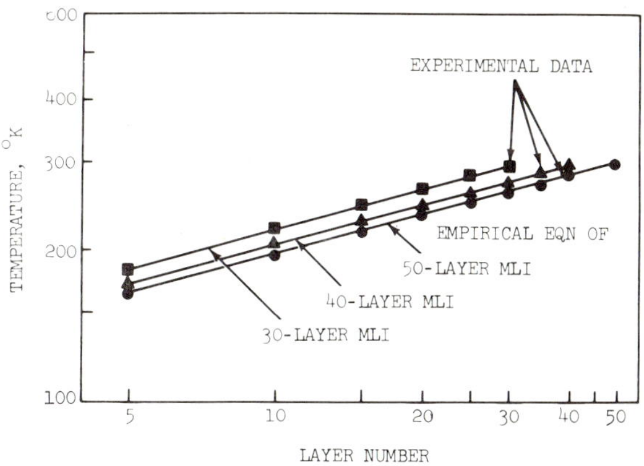

Fig. 10 MLI temperature profiles (supported cryostat test).

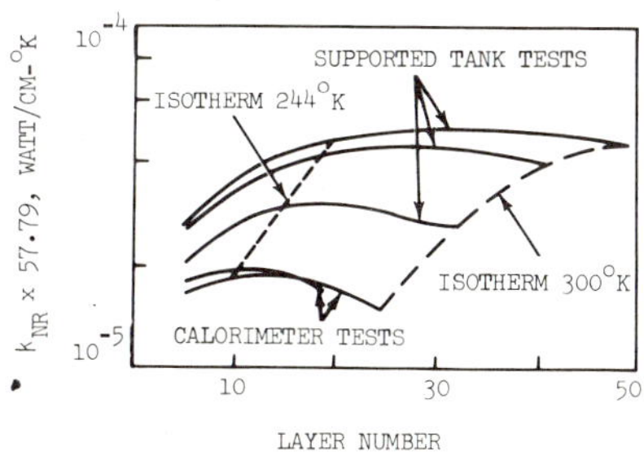

Fig. 11 MLI nonradiation thermal cnductivity.

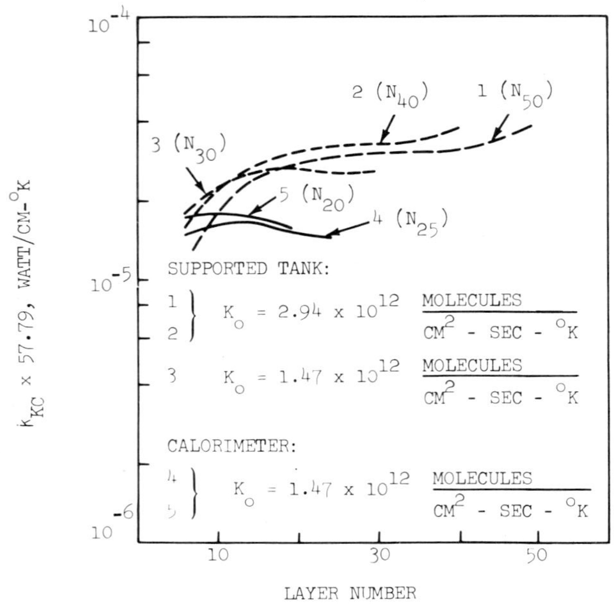

Fig. 12 MLI solid conduction thermal conductivity.

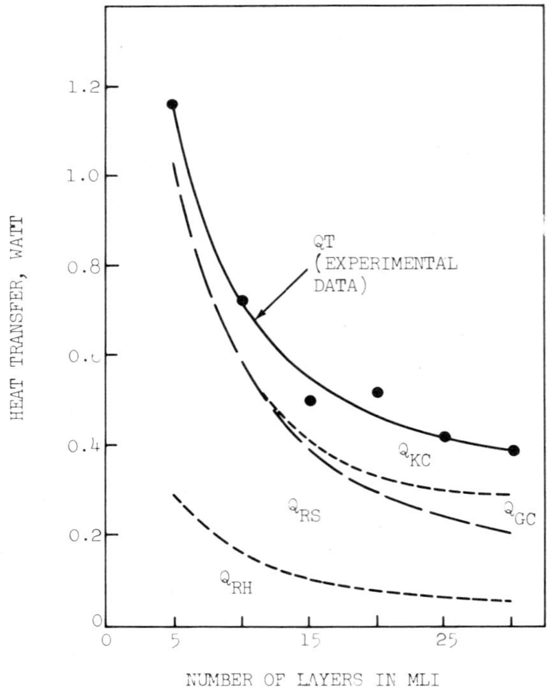

Fig. 13 Thermal characteristics of MLI (calorimeter test).

EFFECTIVE CONDUCTANCE ALONG PARALLEL RADIATION SHIELDS

John T. Pogson* and Robert K. MacGregor*

The Boeing Company, Seattle, Wash.

Abstract

A numerical method is proposed for the calculation of the energy transport parallel to metalized radiation shields. The network model used considers simultaneously both the radiation and conduction between nodes. Results from the computations are presented as an effective thermal conductivity varying with temperature. The effects of emittance, specular reflectance, packing density and spacer material were studied and are illustrated as parametric curves. A comparison was made with experimental data taken from the literature, and good agreement was obtained. The technique is general and can be applied to a wide range of radiation shield and spacer material combinations. The results indicate radiative transport can be the dominate mode of energy transport and can be reduced by diffusing spacers.

Nomenclature

A_c	= node cross-sectional area
A_s	= node surface area
C_{ij}	= conductive thermal conductance between nodes i and j
F_{ij}	= radiation interchange factor
K_f	= shield thermal conductivity
K_{eff}	= lateral effective thermal conductivity
$O(\delta T)$	= temperature difference is of order δT
Q	= rate of heat transfer
R_{ij}	= radiative thermal conductance between nodes i and j
T	= temperature
T_c	= characteristic temperature
ε	= emittance
δ	= gap spacing
$\delta \ell$	= incremental length of radiation shield

Presented as paper 70-847 at the AIAA 5th Thermophysics Conference, Los Angeles, Calif., June 29-July 1, 1970.

*Specialist Engineer, Aerospace Systems Division.

δT = temperature difference
σ = Stefan-Boltzmann constant
ρ_d = diffuse reflectance
ρ_s = specular reflectance
τ_d = diffused transmittance

Introduction

Insulation systems fabricated from highly reflecting radiation shields are light-weight with a relatively large thermal resistance for space conditions. The radiation shields are typically made of thin mylar or kapton films with a vapor deposited metallic coating on one or both sides (i.e., single aluminized mylar or double aluminized mylar). Techniques for spacing the radiation shields range from utilization of a coarse silk net to the use of continuous sheet materials such as foams or borosilicate fiberglass paper (tissuglas) or by "crinkling" of the radiation shields.

Heat transfer normal to the insulation layers is small because conduction between the layers is minimized by the fabrication techniques and radiation occurs between low emissivity metalized films (ε = 0.02 to 0.05). There are, however, relatively larger heat fluxes parallel to the layers in regions of discontinuities in the radiation shields (i.e. seams, joints, and penetrations). This is because the metalized films have a thermal conductivity which is two to three orders of magnitude larger than the effective thermal conductivity normal to the layers. Conduction along the metalized radiation shields is further enhanced by parallel radiant energy transport resulting in an increase in effective lateral thermal conductivity.

The heat transfer to seams, joints, and penetrations affects the temperature distribution over substantial regions of multilayer blankets. This results in an over-all degradation of the insulation system thermal performance. So that methods may be developed to reduce the effects of blanket discontinuities, a detailed understanding of energy transport parallel to radiation shields is needed.

Several investigations,[1,2] have suggested nodal network thermal models to determine the local temperature distributions and heat fluxes within the insulation layers. These nodal models combined the conduction and radiation transport into single nodal connectors, which have experimentally determined conductivity values. In some cases the effect of temperature on the parallel conductivity is not correctly threated by this combined method.

The purpose of this paper, further documented in Ref. 3, is to present the results of an analysis which permits the parallel conductivity to be determined directly as a function of temperature and the lateral radiation conductivity to be treated separately from the thermal conductance of the metalized film.

Analysis

In a recent study a nodal model was selected (Fig. 1) which permits the calculation of radiation heat transfer parallel to two opposing radiation shields.[4] The selection of this nodal model was based on a study of radiation interchange factors from a single nodal area to all other nodal areas representing the radiation shields. It was found that radiation heat transfer to only the nearest nodes need be considered since the interchange factors to the remaining nodes rapidly became negligible as the distance to the node increased. In addition it was determined that the specularity of the metalized coatings was important and must be considered. That is, the use of a diffuse approximation to the reflectance of a metalized coating is not always valid, and can result in large errors.

The same nodal model was used in this study to determine the combined radiative and conductive heat flux parallel to the shields. The value assigned to each radiation connector was calculated from the following expression:

$$R_{ij} = \sigma A_s F_{ij}$$

The interchange factors, F_{ij}, were determined by a Monte-Carlo, ray tracing computer program[5] in which a specular-diffuse reflection model was used.[6] This computer program uses a diffuse emission model for all surfaces. Transmitting surfaces with absorptance and reflectance were also treated by the aforementioned radiation interchange factor computer program allowing the effects of spacer materials to be considered.

The array of radiation interchange factors from a single node to all other nodes required for this study was determined for each packing density, radiation shield emissivity, and spacer material considered. In an attempt to include the effects of the radiation exchange with nodes other than the nearest diagonal and adjacent nodes, the sums of the view factors to the remaining nodal areas on the opposing layer and to the same shield were calculated and lumped with the respective values computed for the diagonal and adjacent nodes. In all cases this represented only a small increase in these interchange factors.

The conduction connectors shown in Fig. 1 were based on an effective thermal conductivity which was weighted according to the respective metal and substrate cross-sectional areas. The variation of the metalized thin film thermal conductivity with temperature was not included in the computation of connector values except in the case of data presented in Table 3. Conduction connector values were calculated from the following equation:

$$C_{ij} = K_f A_c / \delta \ell$$

The network of conduction and radiation connectors were input into a finite difference thermal analyzer program.[7] The thermal network represented two concentric cylindrical radiation shields of prescribed length linked to two constant temperature boundaries, T_H and T_C, Fig. 2. The finite difference solution to this thermal model was used to compute local temperatures and lateral heat fluxes. Effective thermal conductivity values between nodes could then be computed as follows:

$$K_{eff} = \frac{Q}{(T_i - T_j)} \frac{\delta \ell}{A_c}$$

In this equation A_c represents the cross-sectional area of the radiation shield.

The lateral effective thermal conductivity calculated from the above equation is a linear function of the heat flux and inversely proportional to the temperature difference between nodes i and j. The temperature distribution between nodes is a nonlinear function, therefore a characteristic temperature was used to represent the temperature corresponding to each calculated conductivity value. That is, the radiation conductance may be written as

$$\text{Radiation conductance} = \sigma A_s F_{ij} (T_i + T_j)(T_i^2 + T_j^2)$$

Expanding the temperature expression for small differences,

$$(T_i + T_j)(T_i^2 + T_j^2) = 4T^3 + 0(\delta T) + 0(\delta T^2) + \ldots$$

Hence, for small temperature differences, the following definition of characteristic temperature results:

$$T_c = [1/4 \, (T_i + T_j)(T_i^2 + T_j^2)]^{1/3}$$

Discussion of Results

By using the analysis described in the previous section, effective lateral thermal conductivities were computed for a system of concentric cylindrical radiation shields having a nominal radius of 3 in. and a length of 14 in. The gap spacing and associated packing densities for 0.00025 in. aluminized mylar are shown in Table 1. Aluminum film thickness was taken to be 270 Å. The thermophysical properties used in the calculations are listed in Table 2.[3]

Table 1 Packing Density for 1/4-Mil mylar radiation shields

Packing density layers/in.	Gap spacing in.
19.9	0.05
24.8	0.04
33.1	0.03
49.4	0.02
97.6	0.01
190.5	0.005

Table 2 Thermophysical properties

Property	Aluminum	Diffuse aluminum	Mylar	Tissu-glas
ε_d	0.030	0.030	0.273	0.0
ρ_s	0.024	0.970	0.065	0.330
ρ_d	0.946	0.0	0.662	0.0
τ	0.0	0.0	0.0	0.670

Two types of spacer materials were considered, these being silk net and tissuglas. It was assumed that the silk net could be approximated by a noninteracting spacer. The tissuglas was treated as a completely diffusing material with zero absorptance. That is, all energy absorbed was re-emitted. The transmission and reflection were increased to account for the effective re-emission.

The results of the calculations for a single aluminized mylar insulation system with a noninteracting spacer are shown in Fig. 3. This figure illustrates the variation in effective lateral thermal conductivity as a function of temperature for six gap spacings. As may be noted the smaller gap spacing (higher packing density) results in a decreased radiation contribution to the lateral thermal conductivity. This is because radiant energy leaving a given nodal area will be incident almost entirely upon the opposing nodal area as the gap spacing is decreased. Additionally, the higher the radiation shield temperature the greater the parallel radiation heat transfer. Fig. 4 illustrates this effect by presenting the ratio of the radiation effective lateral thermal conductivity to the combined radiative and conductive effective lateral thermal conductivity (radiative heat flux to the total heat flux). For gap spacings greater than 0.010 and temperatures above 500°R, more than 75% of the energy transport is by radiation. At the lower temperature most of the energy is transported by the conduction mechanism.

Similar results were obtained for the double aluminized mylar with a noninteracting spacer, Fig. 5. Effective lateral thermal conductivity calculations were completed for four gap spacings covering the same range of packing densities as for the single aluminized mylar. For the double aluminized case the limiting thermal conductivity of the radiation shield is nearly double that of the single aluminized shields, since there is twice the aluminum cross-sectional area. The increase in the lateral thermal conductivity with temperature is not as large as was determined for the single aluminized mylar. This is because the low emittance of both shields, results in a lower parallel radiative heat flux in comparison to the single aluminized mylar. Again, the radiation becomes the primary mode of heat transfer above 550°R and conduction the dominant mechanism below 200°R.

The effects of a diffusing spacer between double aluminized shields were calculated and the effective lateral thermal conductivity is shown in Fig. 6 for five gap spacings. These results are similar in form but reduced in magnitude when compared to the noninteracting spacer results illustrated in Fig. 6.

A comparison was made of the effective thermal conductivity as a function of temperature for the following four conditions: single aluminized mylar with a noninteracting spacer; double double aluminized mylar with a noninteracting spacer; double aluminized mylar with a tissuglas spacer; and double aluminized mylar with a noninteracting spacer and diffusely reflecting surfaces. The results of this comparison are illustrated in Fig. 7. At temperatures above 350°R the single aluminized shields exhibit a higher effective lateral thermal conductivity than do the double aluminized shields. This is a consequence of the larger radiation conductance occurring with the single aluminized shields than with the double aluminized mylar. However, below 350°R the effect of the conductance through the aluminized film becomes predominant and the double aluminized mylar has a larger conductivity than single aluminized shields.

In addition, Fig. 7 shows that by utilization of a diffusing spacer, tissuglas, the radiation heat flux is reduced by 64% at 500°R in comparison to the double aluminized mylar with a noninteracting spacer. Also, the use of a diffuse reflectance model for the radiation shields closely approximates the results computed for a diffusing spacer.

Figure 8 shows the ratio of the lateral effective radiative thermal conductivity to the combined lateral effective thermal conductivity for single aluminized mylar and double aluminized mylar with and without a tissuglas spacer. This figure further illustrates the reduction in the lateral radiation transfer that can be obtained by the use of a diffusing spacer.

The emittance of metalized films may vary substantially from roll to roll and is effected by handling. Therefore, the effects of variation in emittance were studied and the results given in Figs. 9 and 10. As is illustrated in Fig. 9, a change in total hemispherical emittance from 0.03 to 0.07 causes only a small change in the effective lateral thermal conductivity for single aluminized mylar radiation shields with a gap spacing of 0.02 in. and emittance of the mylar side of 0.273. This is because the emittance of the mylar side, $\varepsilon = 0.273$, is the controlling surface in so far as emission and absorption are concerned. A similar effect can be obtained by a relatively smaller percentage change in the emissivity on the mylar side of the shield. However, this is not the case with double aluminized mylar with a noninteracting spacer. A change in total hemispherical emittance of 0.03 to 0.07 results in a much larger change in lateral effective thermal conductivity, from 4.2 to 5.3 Btu/hr ft °R, at a temperature of 500°R. In both instances the effect of surface emittance increases with the temperature of the radiation shields, as does the lateral radiative heat flux.

The numerical technique was used to calculate effective conductivities for an NRC-2 insulation system and was compared with data obtained from the literature,[8-10]. The thermophysical property data was assumed to be that tabulated in Table 2. Fig. 11 illustrates a comparison of the calculated effective thermal conductivity values with experimental data measured with NRC-2 radiation shields. These results are shown as a function of an average of the two boundary temperatures since this was the manner in which the data was taken. As may be noted good agreement was obtained with the experimental data. The approximate method of Tien et al.[11] was also compared to the finite difference solution and the two curves are in agreement. A comparison of the numerical method with data taken with double aluminized radiation shields[10] separated with dexiglass spacers is contained in Table 3. Again the calculated values agree well with the data.

Table 3 Parallel conductivity of double-aluminized mylar with a dexiglas spacer

Boundary temperature, °R	Experimental K_{eff} BTU/hr, ft °R	Calculated K_{eff}
542-252	0.0301	0.0298
451-252	0.0265	0.0254
359-252	0.0257	0.0218
271-252	0.0251	0.0188
206-162	0.0165	0.0144
162-108	0.0154	0.0120
253-108	0.0175	0.0134
133-68	0.0120	0.0104
216-69	0.0153	0.0124
85.6-68	0.010	0.0078

Conclusions

The nodal model proposed permits the direct calculation of the lateral effective thermal conductivity between radiation shields. These calculations are dependent upon the thermophysical properties of the radiation shields and the radiation shield gap spacing.

In conclusion, this study of effective thermal conductivity along parallel radiation shields has shown that:

1) The radiative transport along the layers of a multilayer blanket is significant and can be the dominant mode of heat transfer.

2) A diffusing spacer will decrease the radiation transport along the layers by a significant amount.

3) The specular component of reflectance of the radiation shields must be considered in any numerical technique utilizing radiation view factors to determine lateral radiation heat fluxes. However, the use of a diffuse reflectance model will approximate the radiant heat flux parallel to radiation shields when a diffusing spacer is used.

References

[1]Cunnington, G. R., et al., "Performance of Multilayer Insulation System for Temperatures to 700°K," CR-907, Oct. 1967, NASA.

[2]Johnson, W. R. and Sprague, E. L., "Analytical Investigation of Thermal Degradation of High-Performance Multilayer Insulation in the Vicinity of a Penetration," TN D-4778, Sept. 1968, NASA.

[3]MacGregor, R. K. and Pogson, J. T., "Multilayer Insulation (Thermal Performance and Scale Modeling Studies)," Document D2-121307-1, May 1970, Boeing Co.

[4]MacGregor, R. K., Pogson, J. T. and Russell, D. J., "Radiation Interchange Interior to Multilayer Insulation Blankets," Journal of Spacecraft and Rockets, Vol. 7, No. 2, Feb. 1970, pp. 221-223.

[5]Drake, R. L., Lester, A. B., and MacGregor, R. K., "Thermal Radiative Interchange Factor Program," Document D2-114470-1, Jan. 1969, Boeing Co.

[6]Seban, R. A., "Discussion of an Enclosure Theory of Radiative Exchange Between Specularly and Diffusely Reflecting Surfaces," Journal of Heat Transfer, Transactions of the ASME, Series C, Vol. 84, 1962, pp. 299-300.

[7]Almond, J. C., "Thermal Analyzer II," Document AS 1917, June 1966, Boeing Co.

[8]Vliet, G. C. and Coston, R. M., "Thermal Energy Transport Parallel to the Laminations in Multilayer Insulation," Advances in Cryogenic Engineering, Vol. 13, 1968, pp. 671-679.

[9]Coston, R. M., "Handbook of Thermal Design Data for Multilayer Insulation Systems, Vol, II," CR-87485, June 1967, NASA.

[10]Coston, R. M. and Vliet, G. C., "Thermal Energy Transport Characteristics along the Laminations of Multilayer Insulation," AIAA Progress in Astronautics and Aeronautics: Thermophysics of Spacecraft and Planetary Bodies, Vol. 20, edited by G. B. Heller, Academic Press, New York, 1968, pp. 909-923.

[11]Tien, C. L., Jagannathan, P. S., and Armaly, B. F., "Analysis of Lateral Conduction and Radiation along Two Parallel Long Plates," AIAA Journal, Vol. 7, No. 9, Sept. 1969, pp. 1806-1808.

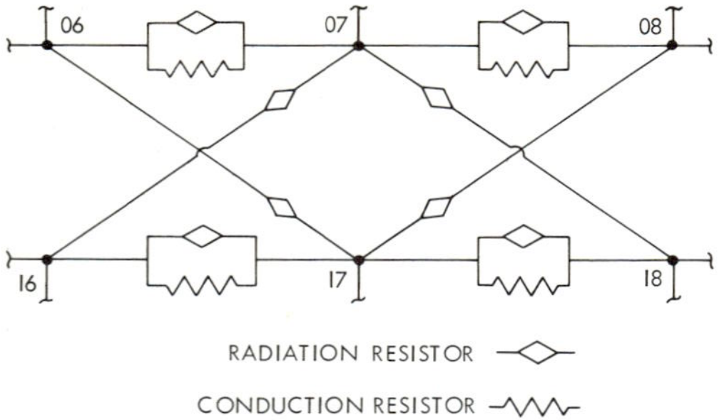

Fig. 1 Schematic of nodal network.

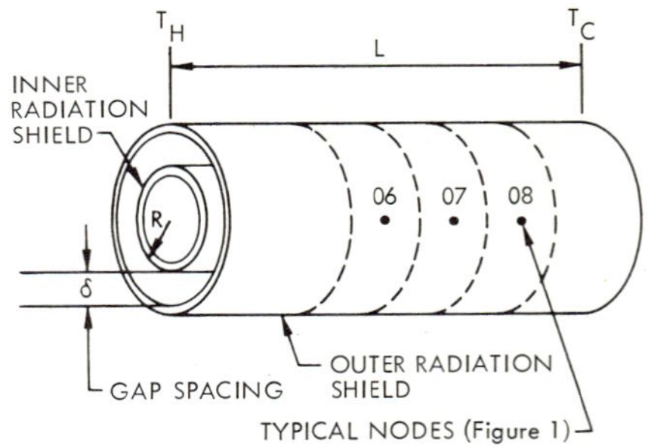

Fig. 2 Configuration and noding schematic.

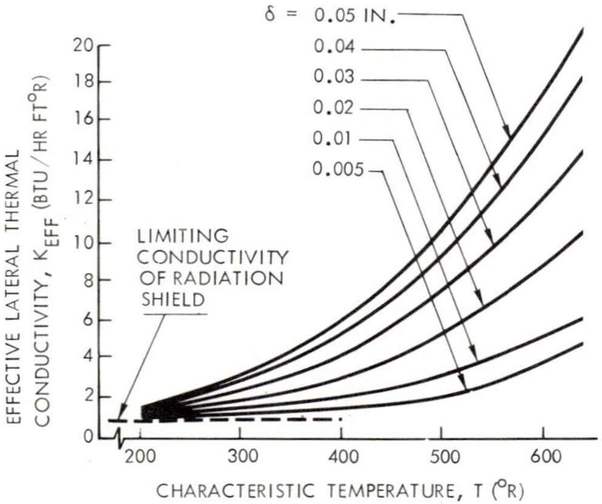

Fig. 3 Lateral thermal conductivity for single aluminized mylar-noninteracting spacer.

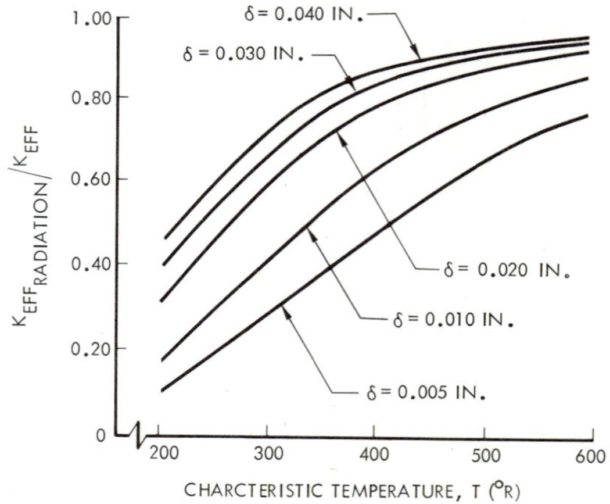

Fig. 4
Radiation contribution to lateral thermal conductivity for single aluminized mylar-non-

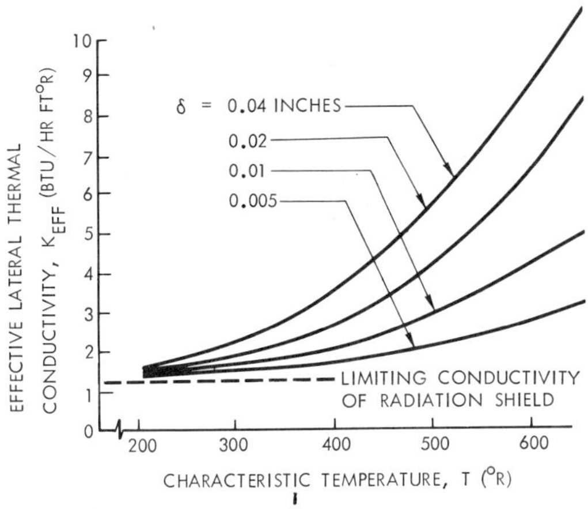

Fig. 5 Lateral thermal conductivity for double aluminized mylar-noninteracting spacer.

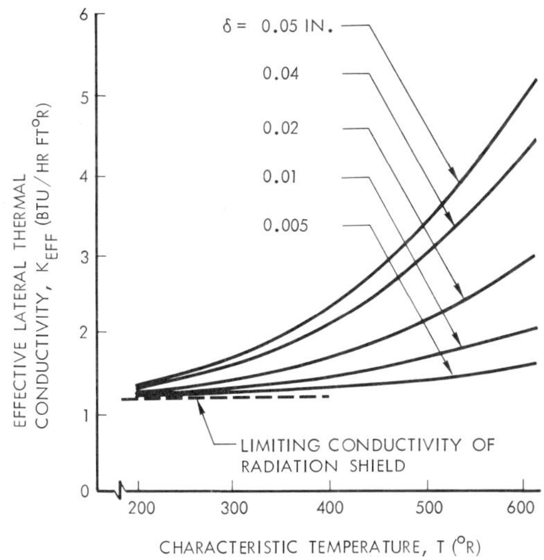

Fig. 6 Lateral thermal conductivity for double aluminized mylar-tissuglas spacer.

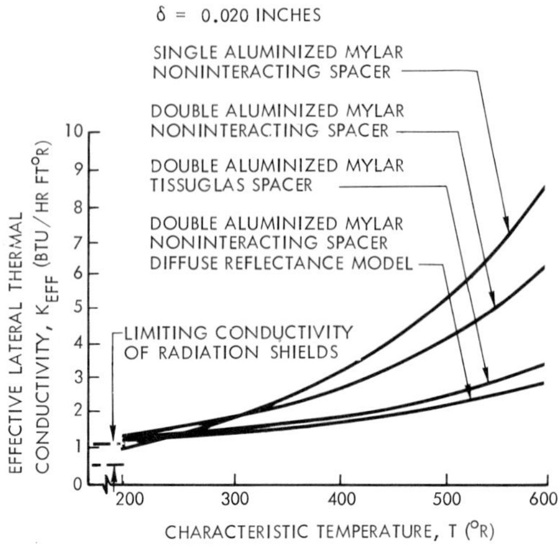

Fig. 7 Lateral thermal conductivity.

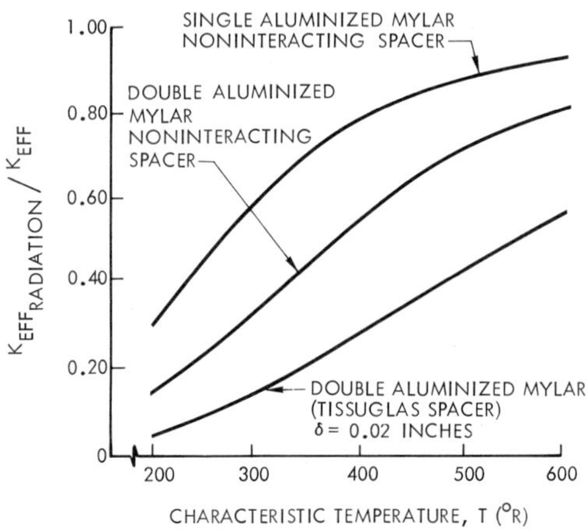

Fig. 8 Radiation contribution to lateral thermal conductivity.

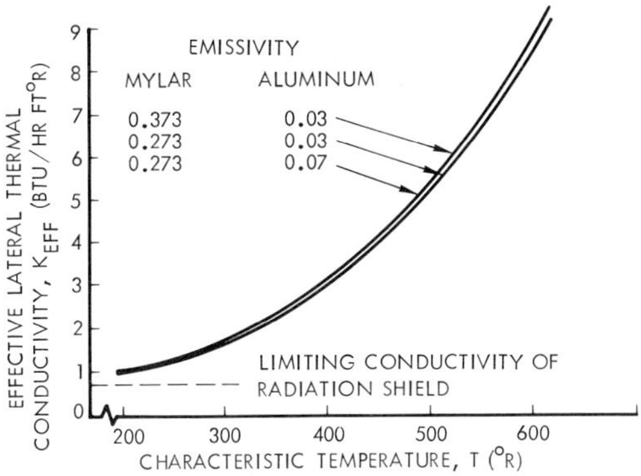

Fig. 9 Emissivity variation effect on lateral thermal conductivity - single aluminized mylar (no spacer; $\delta = 0.02$ in.).

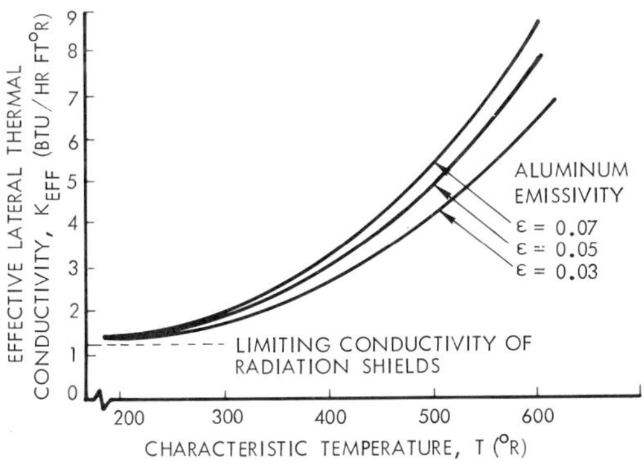

Fig. 10 Emissivity variation effect on lateral thermal conductivity - double aluminized mylar (noninteracting spacer; $\delta = 0.02$ in.).

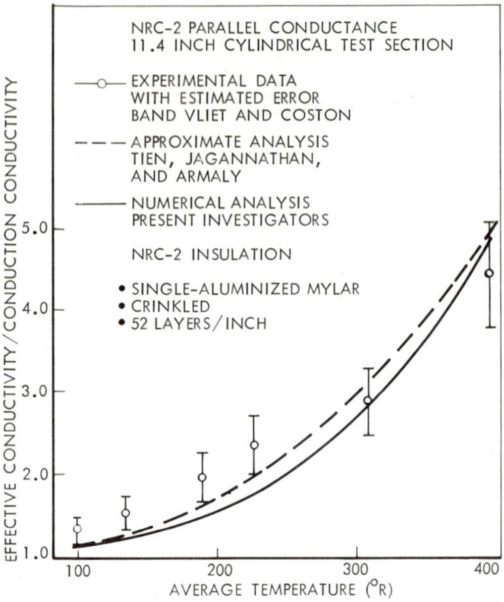

Fig. 11 Experimental comparison of lateral thermal conductivity.

A METHOD OF INCREASING THE LATERAL THERMAL RESISTANCE OF MULTILAYER INSULATION BLANKETS

John T. Pogson* and Robert K. MacGregor*
The Boeing Company, Seattle, Wash.

Abstract

Multilayer insulation provides a lightweight insulation system with a high thermal resistance. Consequently, multilayer insulation blankets are widely used in the thermal control of spacecraft and for the shrouding of cryogenic tankage. Although the heat transfer through the insulation is small, there are major energy losses at the seams and joints. Lateral conduction along the layers of the blanket to the joints can increase greatly the energy transfer through the blanket. Selective slitting of the individual layers of insulation normal to the direction of energy transport, in the region of the joints, will increase the lateral thermal resistance and hence reduce these energy losses. This study presents a solution for the increased resistance in sheetlike materials with slits normal to the direction of energy transfer. Closed form analytic solutions utilizing conformal transformations are developed for both the region between the rows of slits and the "entrance region" itself. The analytic solutions are subsequently compared to both finite difference and analog solutions. A further solution is presented to allow application of the results to either rectangular or axisymmetric blanket geometries.

Nomenclature

a, b, c = nominal slit dimensions in the physical plane (Fig. 2)

a, p, r, q = coordinates of the mapping points in the t plane (Fig. 3)

Presented as Paper 70-15 at the AIAA 8th Aerospace Sciences Meeting, New York, January 19-21, 1970.

*Research Specialist, Aerospace Systems Division.

A	=	area
B	=	blanket thickness
$F(\theta \backslash \phi)$	=	incomplete elliptic integral of the first kind
$\underline{K}(\theta)$	=	complete elliptic integral of the first kind
$\underline{H} = a/b$, $\underline{K} = c/b$	=	nondimensionalized slit dimensions in the physical plane (Fig. 3)
k	=	thermal conductivity
k_ℓ	=	conductivity along the radiation shield
k_s	=	conductivity of the spacer material
L	=	characteristic dimension of the system
n	=	number of layers
N	=	layer packing density
Q	=	energy transport
R	=	scale modeling ratio
R_{TH}	=	thermal resistance ratio
t	=	thickness of the radiation shield
T	=	temperature
θ	=	modular angle argument
ϕ	=	amplitude argument
λ	=	factor relating the cubic integral to the elliptic integral $(1/2\sqrt{1-p^2})$
Θ	=	nondimensional temperature
Φ	=	effective conduction contact area between layers
σ	=	Stefan-Boltzmann constant
ε	=	emissivity of the radiation shield

Subscripts

m	=	model
p	=	prototype
Z, t, Ω	=	denotes the plane in which the indicated lengths are measured

Introduction

An increasingly useful tool for the verification of spacecraft thermal design is thermal scale modeling. Thermal scale modeling allows the testing of a thermal model which can be scaled to fit in an existing space environment chamber where testing of the actual vehicle would not be possible.

One of the most difficult problem areas in thermal scale modeling is the effective scaling of multilayer insulation blankets. A multilayer blanket is made of alternate layers of a low emissivity radiation shield (e.g., aluminized mylar) and a spacer material (e.g., silk netting). These materials are usually selected from among the minimum thickness and minimum conductivity materials available to provide an effective low weight insulation system.

The normal three dimensional criteria for the temperature preservation thermal scale modeling of spacecraft require that all dimensions and material thermal conductivities be reduced in direct proportion to the scale ratio,[1-5]. As the insulation blankets are normally composed of minimum gage materials (e.g., 1/4-mil mylar with a 500 Å aluminum film) and fabricated for minimum normal thermal conductivity, the application of these scaling criteria is precluded.

It is therefore necessary to examine the scaling requirements for multilayer insulation blankets in detail to see if a criteria compromise might be effected which would allow the satisfactory scale modeling of multilayer insulation blankets.

Consider the segment of a multilayer blanket shown in Fig. 1. If this segment of an insulation blanket is treated as a three dimensional solid of volume L^2B, the energy equation may be written (for a steady state case with no internal generation) as

$$\frac{\partial}{\partial x}\left(k_x \frac{\partial T}{\partial x}\right) + \frac{\partial}{\partial y}\left(k_y \frac{\partial T}{\partial y}\right) + \frac{\partial}{\partial z}\left(k_z \frac{\partial T}{\partial z}\right) = 0 \quad (1)$$

The conductivities of the now anisotropic solid may be expressed as

$$k_x = k_y = k_\ell \, (nt/B) \quad (2)$$

$$k_z = \frac{QB}{A\,\Delta T} = \frac{B}{A\,\Delta T}\left[\frac{\sigma\varepsilon\,(1-\Phi)A\,\Delta T^4}{n(2-\varepsilon)} + \frac{k_s \Phi\, A\,\Delta T}{B}\right] \quad (3)$$

Substitution of Eqs. (2) and (3) into Eq. (1) with the change of variables

$$u = x/L,\; v = y/L,\; w = z/B,\; \Theta = T/T_o \quad (4)$$

results in the differential equation

$$\frac{\partial^2\Theta}{\partial u^2} + \frac{\partial^2\Theta}{\partial v^2} + \left[\frac{\sigma\varepsilon L^2(1-\Phi)\Delta T^3}{n^2\,k_\ell\; t(2-\varepsilon)} + \frac{k_s\;\Phi\;L^2}{k_\ell nBt}\right]\frac{\partial^2\Theta}{\partial w^2} = 0 \quad (5)$$

For the solution to Eq. (5) to apply both to a prototype and model insulation system the coefficient

$$\left[\frac{\sigma\varepsilon L^2}{n^2}\frac{(1-\Phi)\Delta T^3}{k_\ell t\ (2-\varepsilon)} + \frac{k_s \Phi\ L^2}{k_\ell nBt}\right] \qquad (6)$$

must be preserved between the model and the prototype.

If the following restrictions are placed upon the model: a) radiative properties are preserved, b) temperatures are preserved, and c) insulation material and fabrication techniques are preserved, then

$$\varepsilon_m = \varepsilon_p,\ \Delta T_m = \Delta T_p,\ \Phi_m = \Phi_p,\ k_{sm} = k_{sp},\ t_m = t_p \qquad (7)$$

With these restrictions the coefficient of Eq. (3) equated for model (subscript m) and prototype (subscript p) becomes

$$\frac{n_m^2}{n_p^2}\ \frac{k_{\ell m}}{k_{\ell p}} = R^2\ \frac{(1 + n_m/B_m)}{(1 + n_p/B_p)} \qquad (8)$$

If the model blanket is constructed of the same number of layers ($n_m = n_p$) and the same packing density is maintained Eq. (8) reduces to the requirement

$$k_{\ell m}/k_{\ell p} = R^2 \qquad (9)$$

Thus, it is shown that a multilayer insulation blanket may be modeled by an identical blanket if the effective conductivity along the layers can be reduced by the square of the modeling ratio.

A technique for selectively reducing the lateral conductivity of the individual material layers in the direction of the thermal flux has been suggested by Katzoff.[5] This technique consists of slitting the layers in a direction normal to the local heat flux. For many situations the direction of the local heat flux will vary with the spacecraft orientation. However, in regions of seams and penetrations through the blanket, the local fluxes will be directed toward the edges of the blanket. In these regions selective slitting of the layers will aid the scale modeling of multilayer insulation blankets.

The objective of this paper is to present an analytic solution for the thermal resistance increases resulting from this selective slitting of multilayer blanket materials. The solution considers lateral energy transport to be by conduction only.

Analysis

In the case of the insulation blanket with a constant temperature edge, as illustrated in Fig. 2, the direction of the heat flux is readily established and hence the orientation of the slits. In Fig. 2, is depicted a region of one-dimensional heat transfer in which the only significant heat flux is that normal to the plane of the paper. Near the boundary, however, energy is conducted towards the edge and the slits are placed in this region.

In the interior of the slit region of the sheet, the isotherms and heat flux lines occur in a symmetric pattern. A sketch of one such region is shown in Fig. 3a. This region has been denoted an interior region, and is representative of the interior heat flux paths. The region between the last slit and the constant temperature edge may also be represented by this sketch if point I is placed coincident with point G and H becomes identical to point F. The figure then represents a single slit placed between two constant temperature boundaries and has been called an end zone.

The solution of the thermal resistance for such a slit region was obtained by successive conformal mappings. A symmetric part of the slit region can be transformed into a rectangular region as shown in Fig. 3--that is, a normalized slit area, as illustrated in the Z plane, was mapped into the upper half of the t plane by a Schwarz-Christoffel transformation

$$Z = M \int \frac{t\,dt}{\sqrt{t+a}\,\sqrt{t-a}\,\sqrt{t+p}\,\sqrt{t-p}\,\sqrt{t+r}\,\sqrt{t-r}} + U \qquad (10)$$

The t plane was then transformed into the complex potential plane, by the transformation

$$\Omega = N \int \frac{dt}{\sqrt{q^2-t^2}\,\sqrt{t^2-r^2}} + V \qquad (11)$$

The rectangle in the Ω plane is recognized as an easily solved heat conduction problem. Due to the symmetry involved, the constants of integration U and V are zero. In addition Eqs. (10) and (11) can be rearranged in a more recognizable form. That is, by a change of variables $\tau = t^2$ Eq. (10) becomes,

$$Z = \frac{M}{2} \int \frac{d\tau}{\sqrt{\tau-a^2}\,\sqrt{\tau-r^2}\,\sqrt{\tau-p^2}} \tag{12}$$

A further change of variables,[6] gives

$$Z = (M/2)\ F\ (\theta_1 \setminus \phi_{1i}) \tag{13}$$

where $F(\theta/\phi)$ represents an elliptic integral of the 1st kind with amplitude ϕ_{1i} and a modular angle θ_1. In a similar manner Eq. (11) may be written in its canonical form by a change of variables $t = r \sin \phi_{2k}$ and $\sin \theta_2 = r/q$ resulting in

$$\Omega = N \int_{\phi_{2i}}^{\phi_{2j}} \frac{d\phi_2}{\sqrt{1-\sin^2 \theta_2 \sin^2 \phi_2}} \tag{14}$$

$$\Omega = N\ F(\theta_2 \setminus \phi_{2k}) \tag{15}$$

Since the region in the Z plane is symmetric the symmetry is retained in the t plane, Fig. 3b. Furthermore the magnitude of point a in the t plane may be set and was assigned a value of 1.0. The other mapping points of interest r and p, can be determined from Eq. (13)--that is, the following set of equations can be written for the lengths in the Z plane[6]:

$$\overline{(CD)}_Z = (M/2\lambda)\ F(\theta_1 \setminus \phi_{11}) \tag{16}$$

where

$$\phi_{11} = \cos^{-1} \sqrt{(r^2-p^2)/r^2} \tag{17}$$

$$\theta_1 = \sin^{-1} \sqrt{(r^2-p^2)/(1-p^2)} \tag{18}$$

Also

$$\overline{(DE)}_Z = (M/2\lambda)\ K(\theta_1) \tag{19}$$

where $K(\theta_1)$ is equal to $F(\theta_1 \setminus \pi/2)$, and

$$\overline{(EF)}_Z = (M/2\lambda)\ K(\theta_1') \tag{20}$$

where

$$\theta_1' = \pi/2 - \theta_1 \tag{21}$$

Equations (16, 19, and 20) can then be recombined to give

$$(\overline{CD}/\overline{DE})_Z = F(\theta_1 \setminus \phi_{11})/K(\theta_1) \tag{22}$$

and

$$(\overline{DE}/\overline{EF})_Z = K(\theta_1)/K(\theta_1') \tag{23}$$

Also, from Fig. 3 the ratios of the lengths can be determined from the slit dimensions as

$$(\overline{CD}/\overline{DE})_Z = (1-\overline{K})\ \overline{H} \tag{24}$$

$$(\overline{DE}/\overline{EF})_Z = 1/\overline{H} \tag{25}$$

Hence, the magnitude of r and p may now be determined, since there is only one value of θ_1 which will satisfy Eqs. (23) and (25)--that is,

$$1/\overline{H} = K(\theta_1)/K\ (\theta_1') \tag{26}$$

Once the modular angle, θ_1, is known then the amplitude ϕ_1 may be obtained from Eqs. (22) and (24).

$$(1-\overline{K})\ \overline{H} = F(\theta_1' \setminus \phi_{11})\ /\ K(\theta_1) \tag{27}$$

Thus, with the values of ϕ_{11} and θ_1 prescribed the value of r and p can be obtained.

$$r^2 = \sin^2 \theta_1/(1 - \sin^2 \phi_{11} \cos^2 \theta_1) \tag{28}$$

and

$$p = r \sin \theta_1 \tag{29}$$

The last mapping point of interest is q, whose value is obtained from the ratio of lengths $\overline{EH}$ to $\overline{EF}$--that is,

$$(\overline{EH})_Z = (M/2\lambda)\ F(\theta_1' \setminus \phi_{12}) \tag{30}$$

where

$$\sin^2 \phi_{12} = \frac{1-p^2}{1-r^2} \frac{q^2-r^2}{q^2-p^2} \tag{31}$$

Therefore

$$(\overline{EH}/\overline{EF})_Z = F(\theta_1' \backslash \phi_{12}) / K(\theta_1') \tag{32}$$

and

$$(\overline{EH}/\overline{EF})_Z = \overline{K} \tag{33}$$

The value of q can now be computed since $\overline{K}$ is a known quantity which sets the value of ϕ_{12} in Eq. (31).

The important mapping points representing the vertices and locations of the slits in the Z plane have now been determined in the t plane. In a similar manner relationships between the vertices in the complex potential plane Ω and the respective points in the t plane can be written. The ratio of the lengths $\overline{CE}$ to $\overline{HE}$ in the Ω plane are

$$(\overline{CE}/\overline{HE})_\Omega = K(\theta_2) / K(\theta_2') \tag{34}$$

Thus, the relative size of the rectangle in the complex potential plane has been determined. The solution to the conduction problem posed in the Ω plane is well known and can be expressed as

$$Q = (\overline{HE}/\overline{CE})_\Omega \ (tk_\ell/2) \ (T_2-T_1) \tag{35}$$

A solution in the Z plane can now be given from the mapping equations, substituting Eq. (34) into Eq. (35),

$$Q = [K(\theta_2')/K(\theta_2)] \ (tk_\ell/2) \ (T_2-T_1) \tag{36}$$

Thus, the ratio of the thermal resistance for a single slit region to that of an unslit region is,

$$R_{TH} = [K(\theta_2')/K(\theta_2)] \, \overline{H} \tag{37}$$

The above analysis also gives the thermal resistance ratio for an end region, i.e.,for a single slit next to a constant temperature boundary. In this case I and G become coincident points as H becomes identical to F. Hence, the modular angle θ_2 is defined as $\sin \theta_2 = r$.

If the region of interest in the multilayer blanket is in polar coordinates requiring circular arc slits a further transformation is required. This conformal map is shown in Fig. 4 and the mapping equation is

$$Z = \ln r + i\theta \tag{38}$$

defining

$$\theta_a = \pi/m \tag{39}$$

$$\theta_c = \pi/n \tag{40}$$

The ratio of a to b and c to a can now be determined as

$$a/b = (\pi/m) \ln (r_i/r_o) \tag{41}$$

$$c/a = \theta_c/\theta_a = m/n \tag{42}$$

Hence, the analysis of the thermal resistance for the rectangular region can be related to the circular slit region.

Discussion of Results

A digital computer program was written to facilitate the computational processes. The results of the analysis are presented in Fig. 5 in nondimensional form, for various values of the aspect ratio (a/b) as a function of the slit ratio (c/a). This curve is for an interior slit region as Fig. 2 indicates. As may be noted by preselection of the values for aspect and slit ratios, theoretically any resistance ratio from 1 to infinity may be obtained. Practically, resistance ratios between 1 and 20 may be obtained. At the larger values of aspect ratio and smaller values of slit ratio, small errors in cutting the slits will result in proportionately larger errors in the resistance ratio.

Figure 5 also shows the scaling requirement for multilayer insulation systems which was derived as Eq. (9). A scale model ratio of one-third requires an order of magnitude increase in the lateral thermal resistance of the radiation shields. This required increase in thermal resistance has become sixteen fold for a quarter scale model. One-quarter to one-fifth scale models of multilayer insulation blankets would appear to be a practical lower limit for this technique.

The results of this analysis are compared in Fig. 6 with the analog results presented by Katzoff.[5] Experimental values were measured by Katzoff utilizing Teledeltos' paper. As may be noted from the figure, at the lower values of aspect ratio, and higher values of slit ratio, the mapping analysis and that presented by Katzoff are nearly the same. However, at the larger values of aspect ratio and lower values of slit ratio a significant discrepancy exists. This difference is believed due to the anisotropic nature of the electrical conductivity for Teledeltos' paper[7] and also to the practical difficulties encountered in maintaining the slit spacing at the preselected values for aspect and slit ratios.

A further check on the analysis was performed by utilizing a finite difference technique. The values obtained using the finite difference numerical solution are compared to the mapping solution and are also shown in Fig. 6 for a single aspect ratio.

The analytic solution presented offers several advantages over the analog solution of Katzoff and the finite difference solution also presented in this paper. The closed form solution presents the results with an accuracy which cannot be matched by the analog solution and without the need for a digital computer.

The results of the end region are presented in Fig. 7. It should be noted, with reference to Fig. 2, that a multilayer blanket has both one dimensional regions (heat transfer normal to the insulation layers) and two dimensional regions (heat transfer along the layers). Slitting of the layers need only occur in the two dimensional regions near the seams and penetrations. However, a layer slit in this manner has both end zones and interim regions. The overall resistance of Eq. (9) relates to the resistances of both end zones and the sum of the individual interior regions.

Conclusions

In conclusion we note that practical resistance increases on the order of a factor of twenty are possible. This, however, places a lower limit on the scaling of multilayer insulation blankets using this technique. A lower limit of 1/4 to 1/5 scale ratios seems reasonably consistent with the difficulties inherent in slitting a sheet of material to the required accuracy.

The advantage gained by using this technique is that essentially identical blankets (number of layers, shield and spacer materials, packing density, etc.) can be used with only minor modifications in the regions of seams and penetrations to scale model the larger blankets of the actual flight vehicle.

References

[1]Vickers, J. M. F., "A Study of Thermal Scale Modeling Techniques," Technical Memorandum, Jet Propulsion Lab., Pasadena, Calif., Sept. 30, 1963, pp. 33-153.

[2]Chao, B. T. and Wedekind, G. L., "Similarity Criteria for Thermal Modeling of Spacecraft," Journal of Spacecraft and Rockets, Vol. 2, No. 2, March-April 1965, pp. 146-152.

[3]Thompson, R. K., "Thermal Scale Modeling Research," Document D2-113009-1, June 1966, The Boeing Company.

[4]Thompson, R. K., "Thermal Scale Modeling of Insulated Spacecraft," Document D2-114109-1, Nov. 1967, The Boeing Company.

[5]Katzoff, S., "Similitude in Thermal Models of Spacecraft," TN D-1631, April 1963, NASA.

[6]Abramowitz, M. and Stegun, I. A., Handbook of Mathematical Functions, U.S. Government Printing Office, Washington, D.C., 1964, p. 597.

[7]Schneider, P. J., Conduction Heat Transfer, Addison-Wesley, Reading, Mass., 1957, pp. 345-347.

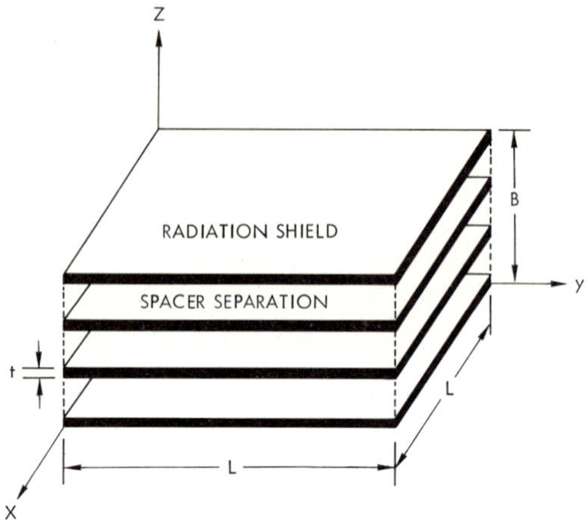

Fig. 1 Schematic of multilayer insulation blanket.

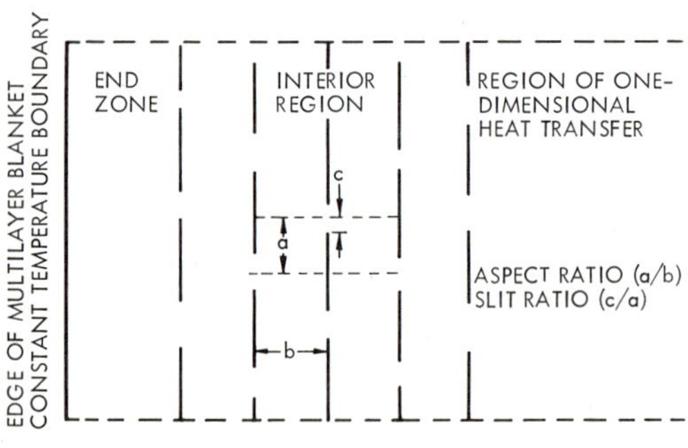

Fig. 2 Slit geometry in layers of multilayer insulation.

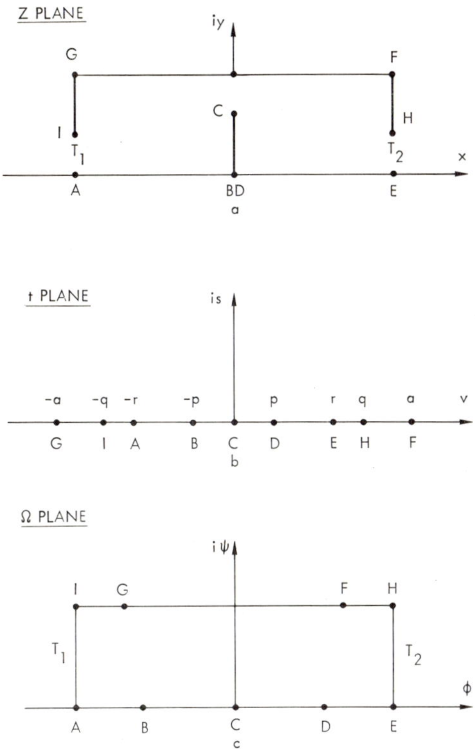

Fig. 3 Conformal mappings of a slit region.

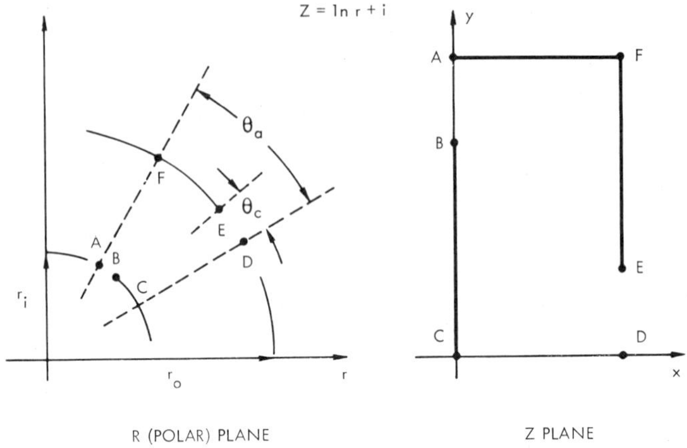

Fig. 4 Transformation relating polar to rectangular mappings.

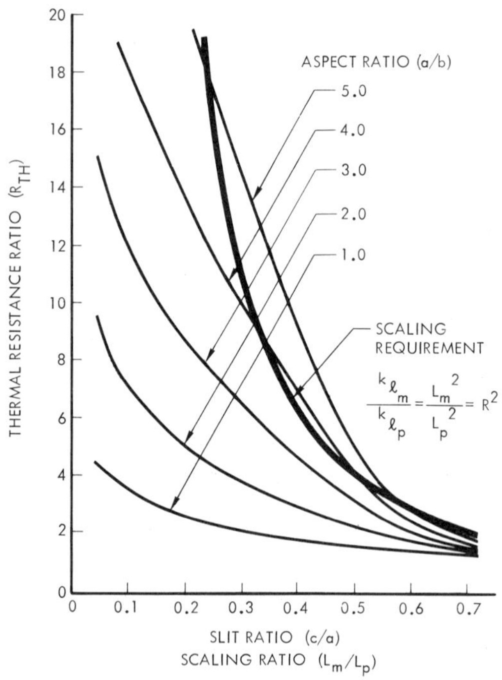

Fig. 5 Relative increase in thermal resistance for an interior region.

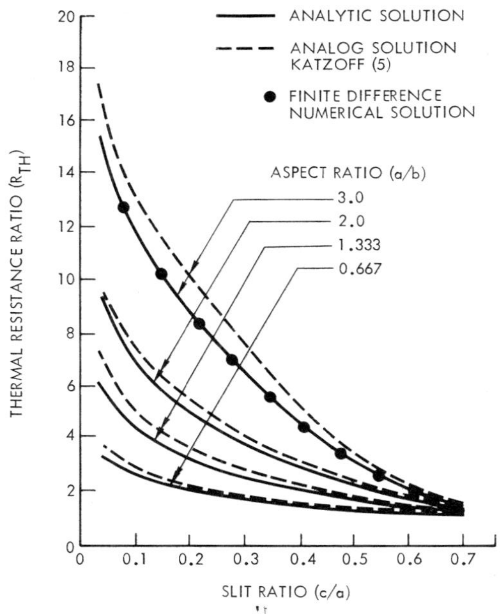

Fig. 6 Comparison of solutions for increase in thermal resistance for an interior region.

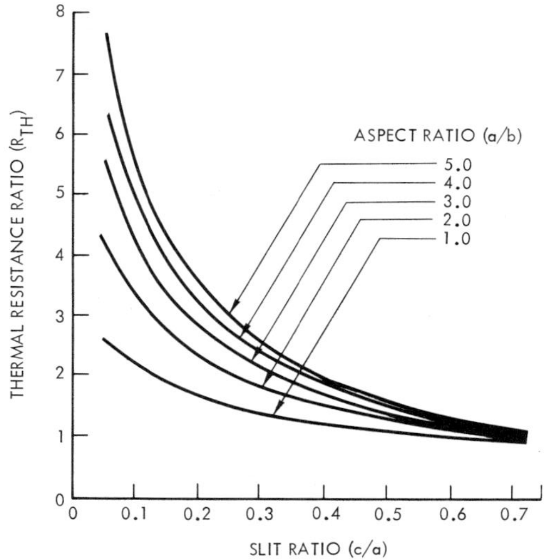

Fig. 7 Relative increase in thermal resistance for an end zone.

NUMERICAL EVALUATION OF MULTILAYER INSULATION SYSTEM PERFORMANCE

Robert K. MacGregor,* John T. Pogson,* and David J. Russell†

The Boeing Company, Seattle, Wash.

Abstract

The results of a study into the feasibility of computing multilayer insulation system performance through numerical techniques are presented. An insulation system incorporating a simple open joint between blanket sections has been treated both numerically and experimentally. Radiation and conduction components both along and between the layers were treated independently. The specular nature of the radiation shields has been accounted for in the calculation of radiation interchange factors between nodes. The results of the numerical study are compared with experimental data for three geometrically different insulation systems. It is concluded from these results that the performance of more realistic insulation systems can be predicted by numerical simulation if the effective contact area between layers can be determined.

Nomenclature

A	=	area
b	=	distance between layers of insulation plus thickness of one metalized radiation shield (Figure 1)
B	=	nominal distance between structure and outer layer of insulation (Figure 1)
k_s	=	thermal conductivity of spacer material
K_{eff}	=	effective insulation conductance
n	=	number of layers per inch
N	=	number of layers
q	=	heat transfer
R,θ,Z	=	cylindrical coordinate directions
t	=	thickness of metalized radiation shield

Presented as Paper 70-848 at the AIAA 5th Thermophysics Conference, Los Angeles, Calif., June 29-July 1, 1970.

*Specialist Engineer, Aerospace Systems Division.

†Engineer, Aerospace Systems Division.

T = temperature
x_s = distance between radiation shields
ε = emittance
ε_{eff} = effective insulation emittance
ρ^d = diffuse component of reflectance
ρ^s = specular component of reflectance
σ = Stefan-Boltzmann constant
ϕ = effective contact area between layers

Subscripts

c = refers to insulation cover sheet
i = refers to i^{th} layer of insulation
n = refers to outer layer of insulation
o = refers to structural substrate beneath insulation blanket
1 = refers to inner layer of insulation

Introduction

Multilayer insulation blankets[1] provide a lightweight insulation system with a high thermal resistance in vacuum applications. Consequently, multilayer insulation systems are widely used in the thermal control of spacecraft and for the shrouding of cryogenic tankage. Multilayer blankets consist of a number of highly reflecting radiation shields interspaced with a low thermal conductivity spacer material or separated by crinkling the radiation shields themselves. The radiation shields are generally a mylar or kapton film metalized on either one or both sides. Spacer materials range from very coarse silk nets to continuous materials such as borosilicate fiber sheets (i.e., tissuglas or dexiglas). Predicting the performance of a multilayer insulation system is a difficult task because of 1) the highly anisotropic nature of the system and 2) the indeterminacy of the local packing density.

Although heat transfer through the insulation normal to the layers is small, discontinuities such as seams or penetrations provide thermal "shorts" which degrade the over-all system performance. These relatively large energy losses at discontinuities are further increased by the anisotropic nature of the blanket. That is, conductance along the layers of the blanket may be two to three-orders of magnitude greater than conductance normal to the layers.[2] Thus, the energy losses at discontinuities in the blanket affect temperature distributions over substantial regions of the blanket surrounding these discontinuities.

Methods of controlling the blanket packing density (a critical performance factor) tend to degrade insulation performance and are undesirable. The very flexible and loosely packed nature of the blanket, combined with acceleration and depressurization forces, allows wide variations in local blanket density. The blanket packing density is generally determined as a mean or average value for the complete blanket. Thus, performance predictions using these mean packing densities are not locally exact and should be used only to establish trends for over-all system performance.

Several investigators[3,4] have utilized simplified nodal network models to examine local temperature distributions and heat fluxes within insulation blankets. These simplified nodal models combine the conduction and radiation transport normal to the layers into a single effective conductance which is evaluated from flat plate calorimeter data. These nodal models also combine the conduction and radiation transport parallel to the layers into a single effective conductance. This technique for performance prediction requires substantial experimental support. The experimental techniques evaluate an over-all effective conductance which is a function of the temperature boundary conditions. Local effective conductances, as a function of temperature, are required for a correct balance between radiative and conductive transport in the material. Several studies[2,5] have been conducted to determine experimentally the effective conductance along the layers of various blanket configurations.

Recent studies[6,7] have examined numerically the radiation and conduction transport along the layers and have shown good agreement with both the experimental results of Vliet and Coston[2] and the approximate analysis of Tien, Jagannethan, and Armaly.[8] One of these studies[6] has shown that consideration must be given to the specular nature of the radiation shields to obtain realistic radiative fluxes between adjacent nodal areas.

The objective of the present study, further documented in Ref. 9, is to integrate the numerical techniques developed for the calculation of heat transfer along the layers[6,7] into a realistic nodal network for the calculation of the over-all performance of a multilayer insulation system. The system examined contains a strong discontinuity in the form of an open seam, thus magnifying the effects of lateral conductance along the layers. The numerical calculations for insulation system performance will be compared with experimental data for three geometrically different insulation systems.

Experimental Investigation

The insulation systems utilized in this study consisted of two disc shaped multilayer blankets attached one to each side of a heater panel. A third blanket, in the form of a "belt," is wrapped around the perimeter of the heater panel. A typical sectioned view of this configuration is shown in Fig. 1. The heater panels served as test fixtures on which the mounted insulation systems were tested.

Insulation System Design

The insulation systems consisted of two coin shaped blankets and a single strip ("belt") blanket as shown in Fig. 2. Each blanket consisted of ten layers of single aluminized mylar alternated with single layers of silk net. Insulation blanket dimensions for each system, as shown in Fig. 2, are listed in Table 1. The mylar was 1/4 mil thick with 270 Angstroms of vacuum-deposited aluminum on one side. The aluminum film thickness was determined from electrical resistance measurements. The blankets were assembled with an inside layer of silk net and with the aluminized side of the mylar facing out.

The different system dimensions allowed three substantially different insulation configurations with regard to the relative blanket surface area/seam length parameter. These characteristics are presented in Table 2.

Table 1 Insulation system dimensions

	DIMENSIONS					
Model	D in.	P in.	G in.	L in.	A in.	B in.
30" model	30.2	1.0	1.1	95.6	1.0	6.0
15" model	15.1	0.5	0.6	48.1	0.5	3.0
8" model	8.3	0.25	0.35	26.8	0.25	1.5

Table 2 Insulation system characteristics

Model	Surface area ft^2	Seam length ft	Surface area/ seam length ft
30" model	10.44	7.90	0.756
15" model	2.76	3.95	1.430
8" model	0.895	2.17	2.43

The layers were joined together using a skip bonding technique. Each layer of mylar was bonded to the layer below it by drops of low temperature epoxy evenly spaced around the circumference of the blanket. These drops were staggered on alternate layers, as shown in Fig. 3, to eliminate a direct thermal short through the blanket.

Velcro tape was used for fastening the blankets to the heater panels. Patches of Velcro hook tape were bonded to the blankets as shown in Fig. 3 and located as shown in Fig. 2. Patches of pile were bonded to the heater plates in locations corresponding to those on the blankets.

Heater Panel Design

The heater panels consisted of two thin circular metal plates bolted together with an electrical heater in the center. A schematic of the heater plate construction is shown in Fig. 4, and the pertinent characteristics of the three models are summarized in Table 3.

Table 3 Heater panel dimensions

Model, in.	Heater panel material, in.	Diameter d in.	Thickness t in.	Diameter a in.
30	6061-T6 Aluminum	30.0	0.125	7.5
15	7075-T6 Aluminum	14.9	0.040	3.5
8	AISI 4130 Steel	8.1	0.040	1.5

Each heater panel assembly had a flat electrical heater centrally located between the two disks, as shown in Fig. 4. The heaters were wrapped with teflon tape to electrically insulate them from the metal disks. Additional protection was provided by a 25-mil brass ring which was placed around each heater. When the models were assembled, the heaters were placed at the center and all air spaces were filled with heat-conducting silicon grease. A regulated DC voltage supply controlled variations in the voltage drop across the heater to less than $\pm$ 0.03%. Voltage taps were connected to the power leads just outside the test specimen, and the voltage drop across the heater was measured. Heater current was determined by measuring the voltage drop across a known resistance in series with the heating element.

The disks were instrumented with three to five (depending on the model) AWG #36 chromel-constantan thermocouples, located along one diameter. The welded thermocouple beads were inserted and peened into small holes drilled into one side of each model. Since the disks were thin and highly conducting, temperature gradients perpendicular to the plane of the disks were assumed negligible, and the depth of the thermocouple bead in the disk was not critical. The wire used for all thermocouples was taken from a single spool. Forty-nine lengths of this wire were calibrated at temperatures near and above room temperature. For temperatures below 150°F (the maximum temperature of the thermal model tests), the maximum difference between the output of any two of the calibrated thermocouples was 1.5 μv (or 0.05°F), and the maximum absolute temperature error was ± 0.3°F. Thermocouple voltages were referenced to an ice bath junction and read manually on a potentiometer. Repeatability in the voltage measurements was found to be ± 0.3 μv or ± 0.01°F.

Test Equipment

The tests were conducted in the cylindrical vacuum chamber shown in Fig. 5 which is approximately 40 in. in diameter by 40 in. long. This chamber incorporates liquid nitrogen-cooled shrouded walls (nominally -320°F) coated with black paint. The test environment was maintained at a pressure below 10^{-8} torr throughout the tests. The thermal model was suspended in the center of the test chamber by a polyester chord, as shown in Fig. 5. The thermocouple and power leads also left the specimen at this point. Thermocouple, power and voltage measuring leads were routed through pass-through locations in the side of the chamber.

Three thin aluminum disks, approximately 1 1/4 in. in diameter, were suspended by thin threads in the chamber. Each disk was painted black and had a chromel-constantan thermocouple solidly staked in a hole drilled at its center. Since the disks were in thermal equilibrium with the model and the shroud, their temperatures indicated the effect of the various penetrations in the shroud on the effective sink temperature of the chamber. The temperature history of the reference disks also served to indicate fluctuations in the chamber operating conditions.

The electrical power input to the test model at steady state was dissipated by heat transfer through the multilayer blanket to the chamber walls, and heat transfer through the thermocouple and power leads and the cord supporting the specimen in the test chamber. It was necessary to evaluate the lead losses, to determine the power dissipated through the multilayer insulation for steady state conditions. For this purpose, a thermal analyzer computer program[10] was used to solve a nodal representation of the leads. The program was run for the thermocouple, power and voltage measurement leads at various disk temperatures, and the results used to correct the performance data for lead losses.

Numerical Analysis

The numerical analysis performed in support of this study[9] consisted of reducing the idealized representation (depicted in Fig. 1) of the experimental configuration into a discrete element nodal network model. Nodal elements were defined as shown (Fig. 1) in an axisymmetric pattern on each blanket layer. Due to azimuthal symmetry only a small sector of the total configuration required consideration in the nodal model. The individual components of radiation and conduction heat transfer, both along and normal to the layers, were evaluated to determine the overall system performance by a thermal analyzer computer program.[10]

Nodal Network

Typical elements of the nodal network utilized are shown in Fig. 6. The nodes prefixed with a "U" lie along one radiation shield while those nodes prefixed with an "L" lie along the opposed radiation shield. Conduction is considered along each layer and in a normal direction between the layers. Radiation transport is considered between a node and: 1) the two adjacent nodes on the same layer, and 2) the three nearest nodes on the two opposed layers. Thus, each node within the multilayer insulation has four conduction connections and eight radiation connections to the adjacent nodes. An earlier study[6] has shown that radiation transport along the layers is adequately treated by consideration of only the first adjacent nodal radiation connections.

Parallel Energy Transport

The techniques utilized to calculate heat transfer along the layers have been developed in previous studies,[6,7]. Radiation interchange factors between nodal elements are calculated using a Boeing developed Monte-Carlo computer program.[11] This program accounts for the highly specular nature of the radiation shields through the use of Seban's[12] specular-diffuse reflectance model. These radiation interchange factors (script-F's) are input to the nodal analysis as the coefficients of radiative heat transfer. The radiative properties of the shields used in this study, as determined from infrared reflectance data, are presented in Table 4.

Table 4 Material radiative properties

Property	Aluminized side	Mylar side
ε	0.030	0.273
ρ^d	0.024	0.065
ρ^s	0.946	0.662

Normal Energy Transport

The energy transfer between and normal to the adjacent layers is a combination of conductive and radiative transport. The temperature distribution and energy flux normal to the blanket is strongly dependent on the interlayer contact. Conduction between layers occurs at points of contact between the radiation shields themselves or, where spacer materials are present, through the spacer material between its points of contact with the radiation shields.

Consider the simple one-dimensional model, depicted in Fig. 7, of a multilayer insulation system composed of a subsurface, (o), alternate layers of spacer and radiation shields, (n), a cover sheet, (c), and a low-temperature space environment which is represented by the subscript s.

The thermal resistances pictured in Fig. 7 denote the parallel radiation and conduction paths between the layers of insulation. For a typical element of cross-sectional area A, ϕ will be the effective fractional contact area for conduction and $1-\phi$ will be the fractional radiation area. This model assumes that conduction occurs through the spacer material over a distance x_s which is equal to the spacing distance between the radiation shields. Note that x_s is thus related to the packing density, n, by the equation

$$x_s = (1/n) - t \qquad (1)$$

The energy transport across each of these pairs of thermal paths may be evaluated from the following expressions:

1) between the subsurface and the first radiation shield

$$\frac{q}{A} = \frac{\phi k_s}{x_s}(T_o - T_1) + \frac{\sigma(1-\phi)}{[(1/\varepsilon_o) + (1/\varepsilon) - 1]}(T_o^4 - T_1^4) \quad (2)$$

2) between the n layers of insulation

$$\frac{q}{A} = \frac{\phi k_s}{(n-1)\,x_s}(T_1 - T_n) + \frac{\sigma(1-\phi)}{(n-1)[(2/\varepsilon)-1)]}(T_1^4 - T_n^4) \quad (3)$$

3) between the nth layer and the cover sheet

$$\frac{q}{A} = \frac{\phi k_s}{x_s}(T_n - T_c) + \frac{\sigma(1-\phi)}{[(1/\varepsilon_c) + (1/\varepsilon) - 1]}(T_n^4 - T_c^4) \quad (4)$$

4) between the cover sheet and the space sink

$$\frac{q}{A} = \sigma\varepsilon_c T_c^4 \quad (5)$$

These equations may be solved simultaneously to determine the one-dimensional energy transfer through a multilayer system as a function of its thermal properties and boundary conditions, if the effective contact area (ϕ) is known. These equations have been used to examine performance sensitivity to effective contact area[13] with the following conclusions:

1) At large values of fractional contact area the heat transfer mechanism is conduction dominant. At minimal values of contact area, radiation transport is dominant.

2) Small changes in fractional contact area can radically change the dominant mode of energy transport through the insulation.

3) As the total flux level decreases, the insulation system becomes more sensitive to the effects of fractional contact area.

Equations (2 - 5) may be used in conjunction with experimental evaluations of one-dimensional insulation system performance (flat plate calorimetry) to evaluate the effective contact area between layers. With ϕ thus evaluated, the numerical analysis technique could be utilized to evaluate over-all insulation system performance. As the required flat plate calorimetry data was not available for the insulation systems tested in this study, one experimental point, as determined by this study, for each of the three systems was utilized to determine ϕ. Once determined, ϕ remained constant for each of the three insulation systems.

End Effects

The radiative interactions of the end nodes with space were treated in a manner suggested by Emslie,[14]. Radiosity nodes were established at the end of each layer and the open zone at the end of each layer was treated as a black body when viewed from space. In this fashion the radiosity node formed an interface between the blanket interior and conditions exterior to the blanket.

In the case of the radiosity nodes at the end of each "disk" blanket, radiative interactions with the backside of the belt blanket and space were calculated. The radiosity nodes at the end of each layer of the belt interacted only with the surrounding environment.

Discussion of Results

The over-all performance characteristics (dissipated power as a function of substrate temperature) of the three insulation systems are presented in Fig. 8. From the experimental data shown, the effective contact area, ϕ, of each insulation system was evaluated at that data point lying closest to 20°F. Those values of ϕ are shown on Fig. 8. The results of the numerical analysis for each insulation system are also shown over the range of experimental data. Excellent agreement over the entire range is shown and this is indicative of the ability of the numerical scheme to correctly apportion the interacting conductive and radiative fluxes both along the layers and normal to the layers. The performance degradation shown by the smaller models is both a function of increased seam length/ surface area and increased interlayer contact area (ϕ).

Although the curves of Fig. 8 are σT^4 in form, it should be noted that the effective emittance of the systems was a function of temperature. Figures 9 and 10 show the effective conductance and effective emittance determined for the insulation systems tested. A comparison of the two figures indicates the change in the interaction between conductive and radiation fluxes. At the higher temperatures, radiative transport dominates. At the lower temperatures, conductive transport strongly influences the overall performance. Effective emittance is indicated as the preferred parameter for the correlation and prediction of insulation system performance due to its more nearly constant value in the room temperature range of substrate temperatures.

During the fabrication of the insulation system, it was anticipated that the effective contact area would be a single constant for all three systems due to identical blanket lay-up procedures. During testing, it was determined that gravitational forces slumped the larger blanket more than the 15-inch specimen. Correspondingly, the 15-in. specimen slumped more than the 8-in. specimen. The "slumping" for each blanket configuration was shown to be repeatable for that configuration. The data for each blanket system shown in Figs. 8-10 were collected as the results of two separate tests on each configuration. As the blankets were skip bonded only around the outer perimeter, slumping caused significant variations in the effective contact area, and corresponding average packing density variations, as shown in Fig. 11.

The distance between support points shown in Fig. 11 is the diameter of each insulation blanket as the epoxy drops, used to fasten the blanket together, were spaced only around the outer perimeter of the blankets. A structured spacing of epoxy drops throughout the blankets would appear desirable in order to maintain the effective contact area (and packing density) at the pretest values.

The one-dimensional performance of each insulation system was altered according to the variation shown in effective contact area. Thus performance degradations for the smaller systems shown in Figs. 8-10 is partially the result of increased contact area between layers.

Conclusions

The results of this study have shown that simple multilayer insulation systems can be analyzed numerically with a high degree of accuracy if the effective contact area of the system can be determined. Insulation system performance could be predicted if design and fabrication techniques combine to maintain the system geometry through depressurization. If system geometry can be maintained, one-dimensional calorimeter tests will allow the evaluation of effective contact area and its relationship to packing density. If suitable one-dimensional calorimetry data is not available, then a single test of an insulation system could be performed to evaluate the effective contact area. The analysis technique could then be used to predict insulation system performance over a wide range of temperatures without the necessity of additional insulation testing.

The techniques presented in this study and in a previous study[7] should now be applied to a more realistic insulation system. An analysis of this type would allow an evaluation of the effect of seams, fasteners, and penetrations on insulation system performance. Design of an insulation system could be influenced by optimizing the number, location and size of fasteners and seams. An understanding of the local temperature distributions and an identification of those energy fluxes which dominate various regions of the blanket could result in design changes for improved performance with a minimal amount of verification testing.

References

[1]Glaser, P. E. et al., Thermal Insulation Systems, SP-5027, 1967, NASA.

[2]Vliet, G. C. and Coston, R. M. "Thermal Energy Transport Parallel to the Laminations in Multilayer Insulation," Advances in Cryogenic Engineering, Vol. 13, Plenum, New York, 1968, pp. 671-679.

[3]Cunnington, G. R. et al., "Performance of Multilayer Insulation Systems for Temperatures to 700°K," CR-907, Oct. 1967, NASA.

[4]Johnson, W. R. and Sprague, E. L., "Analytical Investigation of Thermal Degradation of High Performance Multilayer Insulation in the Vicinity of a Penetration," TN D-4778, Sept. 1968, NASA.

[5]Androulaskis, J. G. and Kosson, R. L., "Effective Thermal Conductivity Parallel to the Laminations, and Total Conductance of Multilayer Insulation," Journal of Spacecraft and Rockets, Vol. 6, No. 7, July 1969, pp. 841-845.

[6]MacGregor, R. K., Pogson, J. T. and Russell, D. J., "Radiation Interchange Interior to Multilayer Insulation Blankets," Journal of Spacecraft and Rockets, Vol. 7, No. 2, Feb. 1970, pp. 221-223.

[7]Pogson, J. T. and MacGregor, R. K., "Effective Conductance along Parallel Radiation Shields," AIAA paper 70-847 (this volume).

[8]Tien, C. L., Jagannathan, P. S. and Armaly, B. F., "Analysis of Lateral Conduction and Radiation along Two Parallel Long Plates," AIAA Journal, Vol. 7, No. 9, Sept. 1969, pp. 1806-1808.

[9]MacGregor, R. K. and Pogson, J. T., "Multilayer Insulation (Thermal Performance and Scale Modeling Studies)," Document D2-121307-1, May 1970, Boeing Co.

[10]Almond, J. C., "Boeing Engineering Thermal Analyzer," AS 1917, June 1966, Boeing Co.

[11]Drake, R. L., Lester, A. B. and MacGregor, R. K., "Thermal Radiative Interchange Factor Program," Document D2-114470-1, Jan. 1969, Boeing Co.

[12]Seban, R. A., "Discussion of An Enclosure Theory of Radiative Exchange Between Specularly and Diffusely Reflecting Surfaces," Journal of Heat Transfer, Series C, Vol. 84, 1962.

[13]MacGregor, R. K., "Limitations in Thermal Similitude," Document D2-121352-1, Final Report on Contract NAS 8-21422, Dec. 1969, Boeing Co.

[14]Emslie, A. G., "Radiation Heat Transfer Through Seams and Penetrations in Panels of Multilayer Metal Foil Insulation," Rep. ADL-63270-04-04, April 1962, A. D. Little.

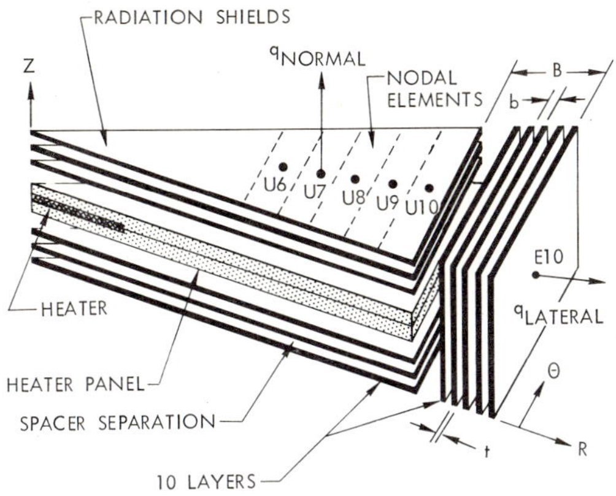

Fig. 1 Section view of multilayer insulation system installed on heater panel (typical).

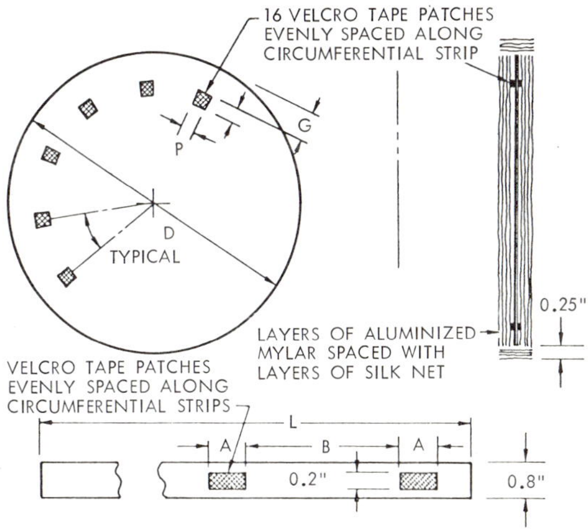

Fig. 2 Insulation system details.

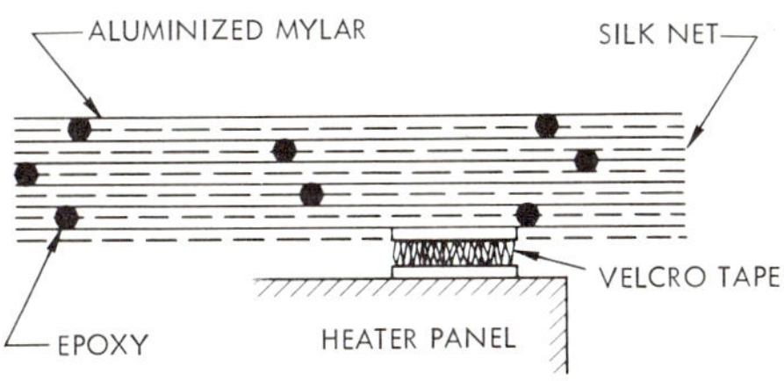

Fig. 3 Insulation system assembly and attachment.

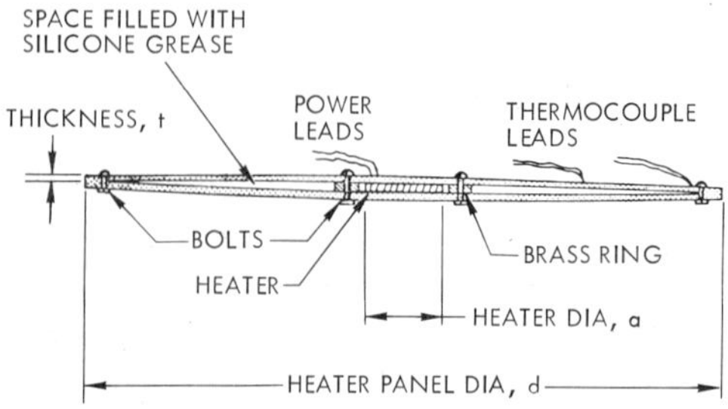

Fig. 4 Heater panel details.

Fig. 5 Thirty-inch insulation model mounted in space simulation chamber.

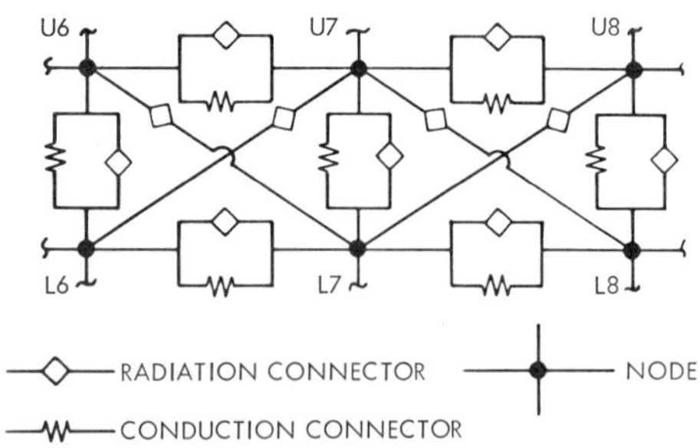

Fig. 6 Nodal network used for numerical analysis (showing parallel radiation and conduction connectors).

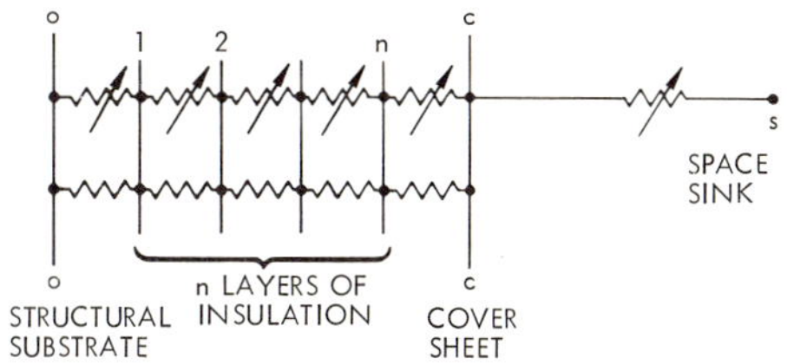

Fig. 7 One dimensional heat transfer model.

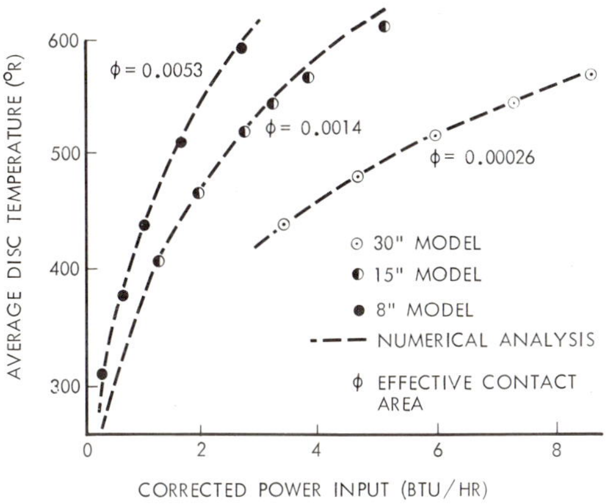

Fig. 8 Multilayer insulation system performance data.

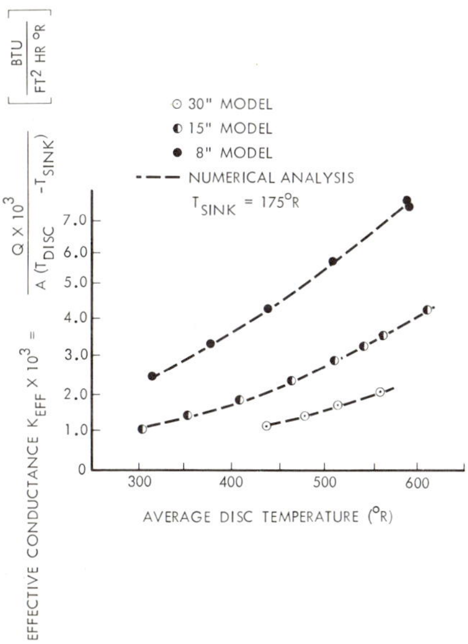

Fig. 9 Effective conductance of multilayer insulation systems.

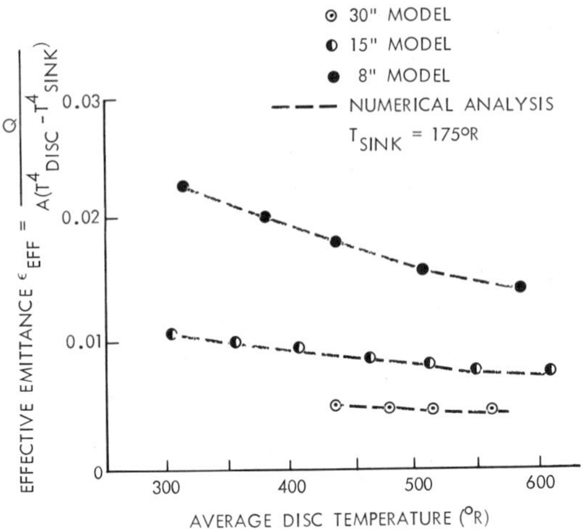

Fig. 10 Effective emittance of multilayer insulation systems.

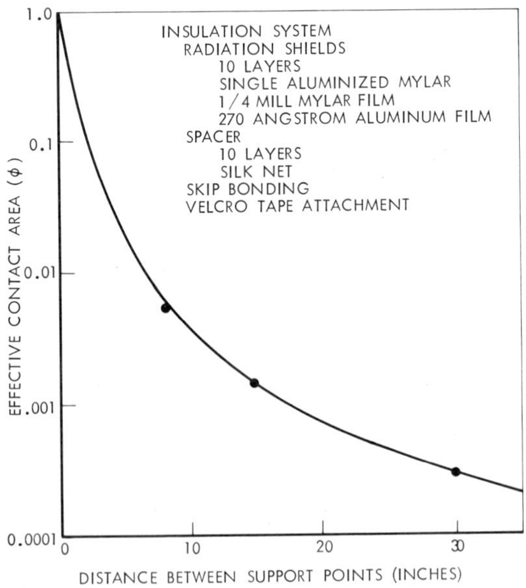

Fig. 11 Effective contact area for skip bonded insulation systems.

EXTRAVEHICULAR SPACE SUIT THERMAL INSULATIONS

David L. Richardson* and Frank E. Ruccia†
Arthur D. Little, Inc., Cambridge, Mass.

and

B. French‡
NASA Manned Spacecraft Center, Houston, Texas

Abstract

A space suit thermal garment whose insulation properties, wear durability, and nonflammability is substantially better than a Gemini garment was developed. Radiation shield and spacer materials were selected, screened, tested to determine their wear durability and flammability in 100% oxygen. Radiation shields were tested to determine their abrasion resistance and total hemispherical emittance. The abrasion resistance of aluminized radiation shields was substantially improved by overcoating with 500 Å of vapor-deposited germanium without degrading the low-emittance surfaces. Improvements were also achieved by applying gold to a polyimide film substrate by a standard commercial process. New insulations were wear-tested for 10,000 cycles. Then the thermal conductance of these insulations was measured under compressions ranging from no-load to 15 psi in simulated lunar nighttime and daytime. The best were those made of germanium-overcoated aluminized polyimide film radiation shields and open-mesh Beta fiberglass marquisette (Style 4897). Fabrication techniques for space suits were analyzed, and improvements reflecting cryogenic multilayer insulation fabricating techniques were developed. The greatest improvements in space garment integrity and thermal performance were achieved with insulation layers hung loosely (not load bearing) on the pressure-retaining layers of the space suit. A flexing elbow-calorimeter was developed to test the flexure durability and heat flow in the new garment sections.

Presented as Paper 70-851 at the AIAA 5th Thermophysics Conference, Los Angeles, Calif., June 21-July 1, 1970. Performed under NASA Contract NAS 9-7519.

*Program Director, Engineering Division.
†Project Engineer, Engineering Division
‡Physicist, Crew Systems Division

Introduction

A vital component of an astronaut's extravehicular space suit is the part which protects him from his external environment. For this the suit must have a thin, flexible, lightweight insulation that will not only shield him from thermal radiation but will also be impenetrable to micrometeoroids. In Gemini-type space suits, multilayer insulations consisting of alternate layers of heat-reflecting radiation shields and low-conductivity spacers provided the thermal and micrometeoroid protection.

Multilayer insulation is necessary for thermally isolating an astronaut from his external environment. The maximum permissible heat loss from a space suit should not exceed 10 Btu/sq ft hr.[1] On the lunar surface at night, this corresponds to a maximum space suit conductance of 0.03 Btu/sq ft hr°F. Multilayer insulation evacuated by exposure to the space environment is the only type which can provide this low conductance in lightweight thin construction. Other insulations, such as evacuated fibers or powders, can achieve this low conductance; however, they require up to 10 times more insulation thickness for the same thermal protection as multilayer insulation.

Multilayer insulations consist of alternate layers of heat-reflecting radiation shields separated by low-conductivity spacers. A thin, low-emittance radiation shield in each layer permits it to reflect a large percentage of the radiation it receives from a warmer surface. The spacers reduce the heat transferred from shield to shield by contact conduction. To minimize gas conductance the space between the shields is evacuated to less than 10^{-4} torr.

The predicted extravehicular temperature extremes on the surface of the Apollo Thermal-Micrometeoroid Garment range from 250° to -250°F.[2] To maintain an interior temperature of 70°F, several configurations of space suit thermal micrometeoroid garments were made in lightweight, thermally efficient, form-fitting configurations which were either an integral part of, or an overgarment for, the intravehicular space suit. The garments consisted of one or more outside protective layers, a multilayer insulation, and one or more inside bladder layers. The multilayer insulation was constructed of seven or more radiation shields interlayered with seven or more spacers. The radiation shields consisted of a metal vapor deposited on one or both sides of a flexible plastic film substrate. To reject thermal energy in a hot environment and retain body heat in a cold environment, the shield had to be highly reflective and have a low hemispherical emittance. To present a low heat conductance path from shield to shield, the spacer had to be a nonmetallic material which prevents the radiation shields from touching.

To date, the multilayer insulations used for extravehicular space suits have been adequate for extravehicular activities of short duration. The heat-transfer properties have been adequate; the weight and bulk characteristics have been optimum. However, those multilayer insulations are flammable; they tear, abrade, and degrade during normal use; and they are difficult to fabricate. A continuing responsibility of the NASA Manned Spacecraft Center (MSC) is to develop insulations without these disadvantages.

Our objective was to develop better multilayer insulations and fabrication techniques for extravehicular space suits, particularly the suits' insulation properties, wear durability, and nonflammability.

Our program was divided into four major phases: 1) development of better radiation shield and spacer materials, 2) evaluation of radiation shield and spacer combinations in thermal garments, 3) analysis of thermal garment construction techniques, and 4) system tests of a thermal garment section in simulated lunar environments. For flight qualification, the specimen materials, material combinations, and fabrication techniques will require additional testing.

Development and Testing of Radiation Shield and Spacer Materials

Candidate materials for radiation shields and spacers were surveyed, screened for applicability, and selected on the bases of flammability, mechanical properties, ease of fabrication, availability, and cost. All materials were experimentally tested in component form to determine their flammability in 100% oxygen at 16.5 and 6.2 psi and in air at ambient temperature and pressure, and measured to determine their physical properties such as tensile strength, tear resistance, fold endurance, and vacuum stability. In addition, radiation shields were tested to determine their abrasion resistance and the total hemispherical emittance of their surfaces.

Flammability

Flammability testing for the insulation program was conducted at NASA Manned Spacecraft Center (MSC). The test procedures for measuring the samples were those specified for flight materials. Each sample was cut and installed in a vertical sample holder so that a 2x5 in. rectangle of material was exposed

(Figure 1).[§] The sample was ignited by a 1-in. flame parallel to and touching the bottom edge of the sample. The flame was generated by passing a current through a nichrome wire that had a 1 x 2 in. rectangle of tissue paper wrapped around it. The wire, heated by the current flow, ignited the paper.

In our tests of radiation shield materials, all the samples made from polymeric film materials proved to be flammable in oxygen atmospheres (Table 1). All of the samples made of one-half-mil-thick polyimide film (Kapton) proved to be self-extinguishing in air. Polyester film, used in Gemini-type insulations, was flammable in air.

In our tests of spacer materials, fiberglass was the only material that was nonflammable in oxygen at 16.5 and 6.2 psi pressure and in air at ambient pressure (Table 1). Fiberglass plain weave Style 104, stabilized with 5 to 8% by weight of L-3306 Fluorel, proved to be self-extinguishing in an oxygen atmosphere of 6.2-psia pressure and in air at ambient pressure. Foam spacers and the non-woven Dacron batt spacers from Gemini-type insulations were flammable for all test conditions.

Physical Properties

In addition to flammability measurements, we made selected physical property measurements of all candidate shields and spacers. The measurements--conventional laboratory procedures--included tensile strength, propagating tear strength, flexure endurance, and vacuum stability (Tables 2 and 3).

The abrasion resistance of all the candidate radiation shield materials was determined by mounting each sample on a 2.5-in. diam. aluminum disc and placing the disc on an oscillatable (bottom) platen in a Stoll abrasion tester (Fig. 2). A fresh sample of Style 2530 Beta marquisette was attached to the fixed top platen to serve as the abrasive. A one-pound weight was added to this platen, and the sample was subjected to 500 oscillation cycles of wear. Its surface was then examined visually. If there was little degradation, an additional one-pound weight was added to the top platen and the sample was subjected to another 500 cycles of wear. This procedure was repeated until the sample either showed noticeable degradation of the low-emittance surface or until the load-limit of the abrasion tester was reached. The load was varied from a minimum of 0.20 psi to a maximum 1.83 psi in nine incremental steps (corresponding

§Figure 1 shows two flammability test chambers with identical sample holders. A flammability comparison test is being made.

Table 1 Flammability of candidate radiation shield and spacer materials in oxygen and air

Radiation Shields	Oxygen 16.5 psi	Oxygen 6.2 psi	Air 14.7 psi
Polyimide film - one-half-mil-thick Kapton with vapor-deposited aluminum and vapor-deposited germanium on both sides	F	F	SE
Polyimide film - one-half-mil-thick Kapton with vapor-deposited aluminum on both sides	F	F	SE
Polyimide film - one-mil-thick Kapton with liquid bright gold and vapor-deposited germanium on both sides	F	F	SE
Polyimide film - one-mil-thick Kapton with liquid bright gold on both sides	F	F	F
Polyester film - one-quarter-mil-thick mylar with vapor-deposited alumium on both sides	F	F	F
Spacers			
Fiberglass - Beta Marquisette, Style 2530	NF	NF	NF
Fiberglass - Plainweave Style 104 (unstabilized)	NF	NF	NF
Fiberglass - Plainweave Style 104, stabilized with 5 to 8 percent (by weight) of L-3306 Fluorel	SE	SE	NF
Dacron batt - nonwoven, 15 gm/yd^2, 0.0015 inch thick	F	F	F
Polyurethane foam - 100 pores per inch and 0.030 inch thick	F	F	F

F = flammable
NF = nonflammable
SE = self-extinguishing

to successive addition of one-pound weights to the upper platen).

The emittance (total hemispherical emittance) of each shield was measured before and after it was subjected to abrasion resistance tests, in order to determine the material's degradation. The emittance tests were performed in an emissometer (Fig. 3) specifically designed and constructed[3] by ADL in earlier NASA programs for the development of high-performance multilayer insulations for cryogenic storage vessels. Table 2 clearly shows that the abrasion resistance of germanium-overcoated aluminized Kapton and liquid-bright-gold-coated

Table 2 Physical properties of radiation shields

Material Description	Tensile Strength (Pound/inch)	Propagating Tear Strength (grams)	Flexure Endurance: Cycles	Flexure Endurance: Load (Kg)	Vacuum Stability (Weight Change In Percent)	Useful Temperature Range (°F)	Abrasion Resistance: Load (PSI) And Cycles	Abrasion Resistance: Emittance Before	Abrasion Resistance: Emittance After
Polyimide film – one-half-mil-thick Kapton with vapor-deposited aluminum and germanium on both sides	35	2.7	26,000	1.5	0.9	−450 to +750	0.2 to 1.8, 5,000	0.026	0.028
Polyimide film – one-half-mil-thick Kapton with vapor-deposited aluminum on both sides	35	2.7	11,000	1.5	1.1	−450 to +750	0.2, 1000	0.025	0.20
Polyimide film – one-mil-thick Kapton with liquid bright gold and vapor-deposited germanium on both sides	76	not tested; estimated to be ≥ 8	not tested estimated to be ≥16,000 for 1.5 kg		0.8	−450 to +750	0.2 to 1.8 1,500	0.020	0.020
Polyimide film – one-mil-thick Kapton with liquid gold on both sides	76	8	16,000	1.5	0.5	−450 to +750	0.2 to 1.8, 5,000	0.018	0.025
Polyester film – one-quarter-mil-thick Mylar with vapor-deposited aluminum on both sides	15	1.3	27,000	0.5	not tested	− 75 to +300	0.2, 500	0.023	destroyed

Table 3 Physical properties of spacer materials

Material Description	Tensile Strength (Pounds/inch)	Flexure Endurance: Cycles	Flexure Endurance: Load (Kg)	Vacuum Stability (Weight Change In Percent)	Useful Temperature Range (°F)
Fiberglass - Beta Marquisette Stevens Style 2530	36 warp 54 fill	20,000 warp 10,000 fill	0.5 0.5	0.9	–450 to +1000
Fiberglass - Plainweave Style 104 (unstabilized)	2.7 warp 2.5 fill	40,000 warp 10,000 fill	0.5 0.15	not tested	–450 to +1000
Fiberglass - Plainweave Style 104 stabilized with 5 to 8% (by weight) of L-3306 Fluorel	not tested	not tested	not tested	not tested	–100 to +900
Dacron batt - Nonwoven, 15 gm/yd^2 0.0015 inch thick	7.5 warp 1.6 fill	10,000 warp fill not tested	0.5	not tested	–75 to +350
Polyurethane foam - 100 pores per inch and 0.030 inch thick	1.5	not tested	not tested	2.0	– 30 to +275

Table 4 Experimental space suit insulations

Sample Number	Outer Layers	Radiation Shields	Spacers	Sample Weight (oz/sq yd)	Remarks (All samples made up with two Style 1807 inside layers)
ADL-05	(2)[a] Style 15035[b]	(7) 0.5-mil Kapton with 800 Å aluminum and 500 Å germanium both sides	(8) 0.030-inch flexible open-cell polyurethane foam	31.8	Sample subjected to 10,000 cycles of simulated wear before testing
ADL-06	(2) Style 15035	(7) 0.5-mil Kapton with 800 Å aluminum and 500 Å germanium both sides	(8) Style 2530 Beta marquisette	42.5	Sample subjected to 10,000 cycles of simulated wear before testing
ADL-08	(2) Style 15035	(7) 0.5-mil Kapton with 800 Å aluminum and 500 Å germanium both sides	(8) Style 2530 Beta marquisette	39.9	
ADL-09	(2) Style 15035	(7) 0.5-mil Kapton with 800 Å aluminum and 500 Å germanium both sides	(8) 0.030-inch flexible open-cell polyurethane foam	32.1	
MSC-04	(2) Style 15035	(7) 0.5-mil Kapton with 1200 Å liquid bright gold both sides	(24) Style 104 with 5% (by weight) L-3306 Fluorel stabilization		
MSC-05	(2) Style 15035	(7) 0.5-mil Kapton with 800 Å aluminum both sides	(8) Style 2530 Beta marquisette	42.3	
MSC-06	(2) Style 15035	(7) 0.25-mil Mylar with 360 Å aluminum both sides	(8) Style 2530 Beta marquisette	39.8	
MSC-08	(2) Style 15035	(7) 0.5-mil Kapton with 800 Å aluminum both sides	(24) Style 104 with 5% (by weight) L-3306 Fluorel stabilization	41.2	
ADL-02	(2) Style 15035	(7) Laminate of 0.5-mil Kapton with 300 Å aluminum both sides; adhered to Style 2530 Beta marquisette	no spacers	41.8	
MSC-02	(1) Style 4190B with carboxy nitroso coating both sides	(7) Laminated composite shield and spacer Style X993	no spacers	37.4	
MSC-03	(1) Style 4190B with carboxy nitroso coating both sides	(7) Laminated composite shield and spacer, Style X993	(7) Style 4897 Beta marquisette	47.1	

[a] Number of layers in sample. [b] Manufacturer's designation.

-320F, the temperature of liquid nitrogen. The heat flux through both insulation samples was determined from measurements of the liquid nitrogen boil-off.

The effectiveness of multilayer insulations in an astronaut's space suit (or in a cryogenic tank for space applications), is determined by the amount of heat which will flow through the insulation under any given boundary conditions. Thus, in comparing various space suit insulations, the parameter of importance is the heat flux per unit area for a given temperature driving force--often called the conductance--given by the equation,

$$C = Q/A(t_{warm} - t_{cold}) \qquad (1)$$

where C = conductance (Btu/sq ft hr°F), Q = heat flux (Btu/hr), A = area (sq ft), and t = temperature (°F).

The heat flux through the sample and the boundary temperatures were measured for each sample at zero-load conditions and for successively larger compressive loadings up to 15 psi. The sample conductance was calculated from the heat flux and the sample boundary temperatures. The use of a midplane temperature measurement between two identical samples of the space suit insulation allowed us to simultaneously measure the conductance of two samples of space suit insulation for each heat flux measurement. In most instances, the midplane temperature was higher than the 70°F required for the inside boundary condition of the space suit; therefore, extrapolations of the basic data were required to determine the insulation conductance for the two temperature ranges of interest. The measured conductance at each sample pressure loading was plotted as a function of the temperature parameter:

$$\tau = (T_1^4 - T_2^4) / (T_1 - T_2) \qquad (2)$$

where τ = temperature parameter $(°R)^3$ and T = absolute temperature (°R) and values of the conductance at the desired temperature differences (300 to 70°F and 70 to -250°F) were obtained by straight line interpolation.

Figures 5 and 6 show the conductance for the significant samples tested under this program as a function of the sample loading. The figures summarize the conductances of samples with separate shield-and-spacer insulations and samples with laminated radiation shield-and-spacer insulations.

At the no-load condition, the sample in the thermal conductance apparatus rests under its own weight. The compressive

Kapton are signficantly higher than the abrasion resistance of either aluminized polyimide or polyester film. The aluminized surface of polyester film was almost completely destroyed after 500 cycles of wear at 0.2 psi and emittance measurements were not made after wear. The emittance of the aluminized surface of polyimide film was degraded from 0.025 to 0.2 after only 1000 cycles of wear at 0.2 psi.

Evaluation of Radiation Shield and Spacer Combinations

The best radiation shield and spacer materials were made into multilayer insulations, incorporated into space suit layups (Table 4) and evaluated for their resistance to wear. These insulations (plus some others provided by NASA MSC) were tested in a 12-in. diam. guarded cold plate calorimeter under compressive loads ranging from "no-load" to 15 psi and in vacuum of less than 10^{-5} torr.

Simulated Wear Tests

Prior to being tested for thermal conductance, two candidate insulation combinations were subjected to 10,000 cycles of non-destructive, simulated wear in an insulation wear tester (Fig. 4), developed during the program specifically for this purpose. In this device, the relative position of the insulation layers changes as the layup passes over one roller through the pinch and out over the other roller. By tensioning the insulation with a weight, a compression of 0.8 psi was applied to the space suit insulation as it passed between the rollers. Germanium-overcoated aluminized shields showed no noticeable wear when used with either Beta fiberglass marquisette or thin flexible polyurethane foam spacers.

Thermal Conductance Tests

The thermal conductances of the samples were measured in a double-guarded cold plate calorimeter,[3,4] which used a bellows-mounted floating plate apparatus to develop compressive loads of less than 0.1 psi.

Two identical samples of each space suit layup were used for measurements of the thermal conductance of candidate space suit insulation systems in the temperature ranges of +300 to +70°F and 70°F to -250°F. The samples were separated by a mid-plane temperature measurement disc of 0.002-in.-thick aluminum to which were attached chromel-constantan thermocouples.

The warm-plate temperature of the apparatus was maintained at either 300°F or 200°F, and the cold-boundary temperature at loads within the sample itself, however, were accounted for. For the samples under lunar nighttime conditions (Fig. 5), the self-compression load range in the sample was determined as follows: the weight of the sample divided by its area was taken as the maximum; one-half of the weight of the sample divided by its area was taken as the minimum. For the sample under lunar daytime conditions (Fig. 6), the self-compression load range was determined as follows: the sum of the weights of the two samples and the midplane measuring disc divided by the sample area was taken as the maximum; the sum of one sample weight and the midplane temperature measuring disc divided by the sample area was taken as the minimum.

Conductance at No-Load

For the cold conditions (+70° to -250°F; Fig. 5) all of the samples with separate shields and spacers have thermal conductances ranging from 0.0012 to 0.003 Btu/sq ft hr°F irrespective of the type of spacer material used between radiation shields. The samples with laminated radiation shields and spacers have conductances ranging from 0.0026 to 0.0045 Btu/sq ft hr°F, which are higher than those for separate shield-and-spacer insulations.

For the warm condition (+300 to +70°F; Fig. 6) the conductances of the insulations with separate shields and spacers range from 0.0044 to 0.0082 Btu/sq ft hr°F. The conductances of the three laminated radiation shield-and-spacer samples range from 0.013 to 0.018 Btu/sq ft hr°F. Consequently, space suit insulations made from composite or laminates will have no-load conductances that are almost twice those of space suit insulations which use separate shields and spacers.

Conductance at Higher Compressive Loads

Figures 5 and 6 show that generally the insulations made of laminated radiation shields and spacers have lower conductances at higher compressions than do the insulations made of separate shields and spacers. At compressive loads of less than 1.0 psi, insulations with separate shields and spacers have lower conductances than those with laminated shields and spacers; above 1.0 psi, the composites have lower conductances. At a compressive load of 1.0 psi, both the laminated insulations and the insulations with separate shields and spacers had approximately the same conductance.

At 15 psi, of all the samples tested, Sample MSC-03, with a loose spacer between laminated shields and spacers, had the lowest conductance, attributable, we believe, to the added thermal resistance of two spacers between each pair of radiation shields (one spacer laminated to the shield, the second loose). Since this sample had more spacer material between the radiation shields than any other sample, it was the heaviest. The extra layers thus provided more contact resistance and more conduction resistance to heat flow at high compressive loads than in similar samples with single spacers.

Analysis of Gemini-Type Thermal Micrometeoroid Garment

Minor failures have occurred in the insulations of astronaut thermal micrometeoroid garments. Most of the reported failures have been mechanical, induced by the garment fabrication technique. Post-flight photographs of one insulation layup showed that the radiation shields had separated from a length of seam joining the overgarment to the space suit. The radiation shields and their spacers appeared to have pulled out at the seam.

In dismembering a Gemini thermal cover layer, we found that the radiation shields and spacers were all firmly anchored in all seams of the garment. Since the radiation shields (aluminized polyester film) have very low tear strength (particularly when pierced many times with a sewing machine needle), failure at seams was considered to be comparable to the action of a paper that is torn along a perforated line.

Although extremely tough, polyester film is notch-sensitive, thus allowing tears to propagate very easily. Because dynamic cycling disturbs the mechanical integrity of the insulation layer, the radiation shields naturally tear at the weakest points--near the seams.

The radiation shields and spacers extended into the seams and penetrations of the garment, where they were tightly stitched--an undesirable construction technique because of the resulting thermal short circuits through the insulation when the layers are tightly compressed. In addition, both the outer and inner layers of this overgarment were load-bearing members and, as such, the insulation was subject to compression and tearing at the seams.

Preferred Assembly Techniques

In applying multilayer insulations, the insulation must be kept as loose as possible, subjected to a minimum of compressive loads during application, and each radiation shield prevented from touching adjacent shields.

All of the conductance data on space suit insulations indicate that very low thermal conductance can be achieved in space suit insulations if compressive loads are kept to a minimum. Therefore, the insulation should hang as loosely as possible on the extravehicular garment and allow only the internal layer of the garment to be load-bearing. For example, as shown in Fig. 7, the load-bearing internal bladder-cloth layers are firmly stitched to one another. Through careful seaming techniques, it is possible to anchor the external layers in the bladder cloth layers and then to wrap the seam loosely over the outside of the insulation and anchor it with loose stitches. The radiation shields stop just short of the edge seam. The shield spacers are carried into the edge seam and are anchored by firm, tight stitches. The external Beta fiberglass layers are carried around the end of the insulation for flame protection in the event of fire. The seam stitches are made with double threads, one of Nomex and one of Beta fiberglass, the Nomex providing the strength at normal operating conditions and the Beta fiberglass providing backup protection in case the Nomex thread is destroyed by a fire.

Figure 8, showing a recommended overlap type of closure, embodies many of the design features that we have discussed. The inner bladder layers of the garment are load-bearing. The zipper used for structurally closing the garment is firmly anchored in the bladder layers. The outer double layer of Beta fiberglass is used to loosely hold the insulation layers in place on the outside of the garment. An overlap of insulation is necessary to minimize heat migration either in or out through the overlap. A Velcro (or equivalent) type of seal with a suitable edge for grasping during actuation completes the overlap closure.

System Test of a Thermal Micrometeoroid Garment in a Simulated Lunar Environment

Thermal and durability evaluations of experimental, development, and flight-quality thermal micrometeoroid garments are expensive and time-consuming when complete garments are used. Since a technique for rapidly evaluating garment subsystems in simulated space and lunar environments was desirable, we developed a flexing elbow calorimeter for rapidly evaluating sections of thermal micrometeoroid garments in a laboratory-sized space simulation chamber.

Elbow Calorimeter

The elbow calorimeter (Fig. 9) consists of a central calorimeter measuring section over which is placed a flexible pressurizable section of a space suit elbow. Guard sections attached at each end of the calorimeter minimize the heat leak into or out of the calorimeter measuring section. The experimental thermal micrometeoroid garment section covers the calorimeter measuring section and the two guard sections.

Figure 10 shows the assembled calorimeter with the flexible elbow section mounted in the test chamber. The left end of the calorimeter is anchored by low-conductance supports to a support structure in the space simulation chamber. The right end of the calorimeter is attached to a flexing mechanism with a yoke which guides the right end as the elbow section is flexed back and forth.

The requirement that the space suit insulation be tested in both lunar nighttime and lunar daytime conditions made it necessary to have a calorimeter that could measure heat flux outward from the calorimeter (during the lunar nighttime) and inward (during lunar daytime). Heat flux inward is determined by measuring the temperature rise and quantity of water which flows through the calorimeter. Heat flux outward is determined by measuring the water flow and its decrease in temperature as it flows through the calorimeter.

The elbow calorimeter can be pressurized with a small amount of dry nitrogen at 3.0 psi (typical extravehicular space suit pressure) to make it assume the shape it normally would have during space operations. The presence of dry nitrogen within the elbow calorimeter also couples the calorimeter thermally to the inside wall of the garment, ensuring good heat transfer.

The lowest heat flow into the calorimeter could be about 1 Btu/hr. At water flow rates of less than 5 lb of water/hr, however, the temperature rise across the calorimeter would be only about 0.2°F. To measure such a low temperature rise accurately--i.e., to $\pm$ 5%--we used a 10-junction copper-constantan thermopile so that we could measure the temperature rise across the calorimeter to within 0.01°F. To keep the error in the heat flux less than $\pm$ 2%, the temperature rise of the water entering the calorimeter was kept to less than 0.1°F per hour.

Thermal Micrometeoroid Garment Section

A section of a thermal micrometeoroid garment was made from the best of the candidate materials (Table 5) using the garment fabrication technique developed in the program. The section, 24 in. long, was fabricated on a mandrel 5.75 inches in diameter.

Table 5 Garment test section materials

Layer Sequence	
Outside	2 layers Style 15035 tightly woven Beta fiberglass cloth.
Insulation	7 radiation shields of 0.5-mil polyimide film with 800 Å vapor-deposited aluminum and 500 Å vapor-deposited germanium on both sides.
	8 spacers of Style 2530 Beta marquisette with 9362 finish.
Inside	2 bladder layers--Style 1807 with neoprene-coated surfaces together.

The thermal micrometeoroid garment section was then installed on the elbow calorimeter for testing. To minimize the heat loss from the ends of the insulation and the calorimeter section, cap insulations consisting of single radiation shields of double-aluminized polyester and double spacers of silk bridal veil were attached over the end seams. Figure 11 shows the complete layup with the end radiation shield guard. The elbow calorimeter is bent slightly to show how the space garment was flexed during the 10,000-cycle wear test.

Space Simulation Chamber

The garment section was tested on the elbow calorimeter in ADL's space simulation chamber. To simulate lunar nighttime, this chamber, three feet in diameter, has blackened shrouds through which liquid nitrogen was circulated. To simulate lunar daytime, the garment and calorimeter were enclosed in a heated baffle. The elbow calorimeter was held in the axially elongated position and a heater attached to the inside of the heated baffle was used to raise the temperature to +300°F. Liquid nitrogen was circulated in the test chamber shrouds to cryopump the chamber and achieve pressures below 10^{-4} torr.

The temperature of the calorimeter was measured with copper constantan thermocouples. Thermocouple millivolt outputs were measured with a digital voltmeter. The output of the 10-junction thermopile was continuously recorded so that trends in the temperature could be determined. Accurate measurements of the inlet temperature and the thermopile output were made with a potentiometer. The calorimeter water flow rate was determined by collecting the water at the end of each hour and carefully weighing it. Temperatures and water flow were measured every half hour after equilibrium was reached. For all tests, the equilibrium heat flux either into or out of the calorimeter varied less than ± 5% from the average. Equilibrium measurements were made for periods ranging from 6 to 9 hr.

Test Procedure

The first step in the test procedure was to measure the heat flow into the garment section prior to wear during simulated lunar daytime. The test chamber was evacuated, checked for leaks, and then the hot baffle was adjusted for an equilibrium temperature of approximately 300°F. Liquid nitrogen in the chamber baffles kept the pressure below 10^{-4} torr. After this test, the chamber was opened, the hot baffle removed, and the chamber again closed.

Lunar nighttime conditions were then simulated by cooling the shrouds in the chamber to liquid nitrogen temperature. After equilibrium was reached and measurements were completed, the elbow section was flexed for 10,000 cycles while the chamber was maintained at liquid nitrogen temperature.

After 10,000 cycles of wear, the performance of the garment section was measured under cold conditions with the elbow section in its axial position. The chamber was again opened and the hot baffle placed over the garment and elbow calorimeter. Then the performance for the warm condition was measured again.

Thus, the performance of the thermal micrometeoroid garment was measured for lunar daytime and nighttime conditions prior to wear and for lunar nighttime and daytime conditions after wear.

As indicated in Table 6, the insulation was degraded by 10,000 cycles of flexure wear. The measured heat loss from the garment during simulated lunar nighttime conditions was 0.54 Btu/sq ft hr before wear and 0.76 Btu/sq ft hr after 10,000 cycles of flexing. The heat flow into the garment during simulated lunar daytime conditions was 1.47 Btu/sq ft hr before wear and 1.92 Btu/sq ft hr after 10,000 cycles of wear.

After all the tests, the third radiation shield in the insulation layup was removed and the emittance of both sides measured. From these emittances, and the measured emittances of the unworn radiation shield, we calculated the heat flow to the insulation based on radiation heat transfer alone. As indicated in Table 6, the heat flux per unit area based on radiative heat transfer alone accounts for between one-third and one-half of the measured heat flux per unit area. The remainder of the heat flux is by solid conduction through the spacers.

Table 6 Thermal performance of a section of a thermal-micrometeoroid garment

	Unworn	After 10,000 cycles of lexure
Simulated Lunar Day		
Measured temperatures, inside/outside, °F	72/254	75/262
Measured heat flux per unit area, Btu/sq ft hr	1.47	1.92
Measured shield emittance	0.022	0.041
Heat flux per unit area based on the radiative heat transfer, Btu/sq ft hr	0.484	0.92
Simulated Lunar Night		
Measured temperatures, inside/outside, °F	78/-310	70/-310
Measured heat flux per unit area, Btu/sq ft hr	0.54	0.76
Measured shield emittance	0.022	0.041
Heat flux per unit area based on the radiative heat transfer, Btu/sq ft hr	0.23	0.39

Postexamination of the Thermal Micrometeoroid Garment Section

The thermal micrometeoroid garment section was removed from the elbow calorimeter and partially disassembled so we could examine the condition of the insulation and the other components. The third radiation shield from the outside showed signs of wear on the lateral side of the elbow section. We removed two pieces from this area of the radiation shield for emittance measurements, both sides. Where the emittances of the unworn germanium-overcoated radiation shields ranged from 0.021 to 0.023, the emittances of the shields after 10,000 wear cycles ranged from 0.038 to 0.045--a not unexpected degradation considering the condition of the radiation shield after flexure wear. Although the germanium adhered well to the aluminum, and the aluminum adhered well to the shield substrate, the substrate itself failed, as evidenced by several small cracks in the radiation shield. Several holes large enough to pass a pencil lead were observed in the vicinity of the lateral side of the elbow. In addition, several cracks approximately 1/2-in. long were observed in this area.

Examining the remaining shields in the insulation layup in situ, we found that the same substrate failures had occurred. There was no evidence on the germanium-overcoated radiation shields of abrasive wear against the Beta fiberglass spacers and no failure of the spacers or of the bladder layers. The external layers showed evidence of degradation by mutual abrasion in areas where wrinkles were induced during each cycle.

Summary

In summary, we found that

1) The abrasive wear resistance of low-emittance aluminized surfaces of radiation shields is substantially improved by overcoating the surfaces with 500 Å of vapor-deposited germanium, without degrading the low emittance surfaces.

2) Effective radiation shields can also be achieved by applying gold to polyimide film by a commercial liquid bright gold process.

3) All polymeric film materials available for use as radiation shields are flammable in oxygen at 16.2 and 6.2 psi.

4) The greatest improvement in integrity and thermal performance of the thermal micrometeoroid garment is achieved when the insulation layers hang loosely (are not load bearing) on the pressure-retaining layers of the space suit.

On the basis of our investigations, we recommend that

1) Future multilayer insulations for astronauts' extravehicular space garments have a) radiation shields of 0.5-mil polyimide film with 500 to 800 Å of vapor-deposited aluminum and an overcoating of 500 Å of vapor-deposited germanium on both sides, or radiation shields of 0.5-mil polyimide film with

1000 Å of gold applied to both sides by a liquid bright gold process; and b) shield spacers made from single layers of Beta fiberglass yarn in an open leno-weave similar to the marquisette materials--Style 2530, or shield spacers made from three layers of commercial fiberglass yarns in plain weave (similar to Style 104) with 5-8% (by weight) of nonflammable yarn stabilization (L-3306 Fluorel) to prevent unraveling.

2) Seven radiation shields and eight spacers in combinations with internal and external garment layers as required for normal wear and micrometeoroid protection, with a spacer between each radiation shield and the inner and outer garment layers to minimize contact conduction.

3) The garment be designed and fabricated so that a) the insulation hangs loosely, b) the inner layer of the garment is load-bearing, c) the spacer layers are firmly anchored in the seams and penetrations, and the radiation shields terminate no closer than 0.5-inch to seams or penetrations, and d) tight stitches never be made through the radiation shield layers.

4) That all developmental thermal micrometeoroid garments have a section tested for flexure endurance and heat loss (or gain) in simulated space thermal environment in order to prove their durability and thermal protection capability.

References

[1]Roth, E. M., "Bioenergetics of Space Suits for Lunar Exploration," SP-84, 1966, NASA.

[2]Adorjan, A. S., "Temperature Distribution in Shadowed Lunar Craters," Journal of Spacecraft and Rockets, Vol. 7, No. 3, March 1970, pp. 378-380.

[3]"Advanced Studies on Multi-layer Insulation Systems," Rept. NAS CR-54929, Contract NASA 3-6283, June 1966, Arthur D. Little, Inc.

[4]"Basic Investigations of Multi-Layer Insulation Systems," Rept. NASA CR-54191, Contract NAS3-4181, Oct. 1964, Arthur D. Little, Inc.

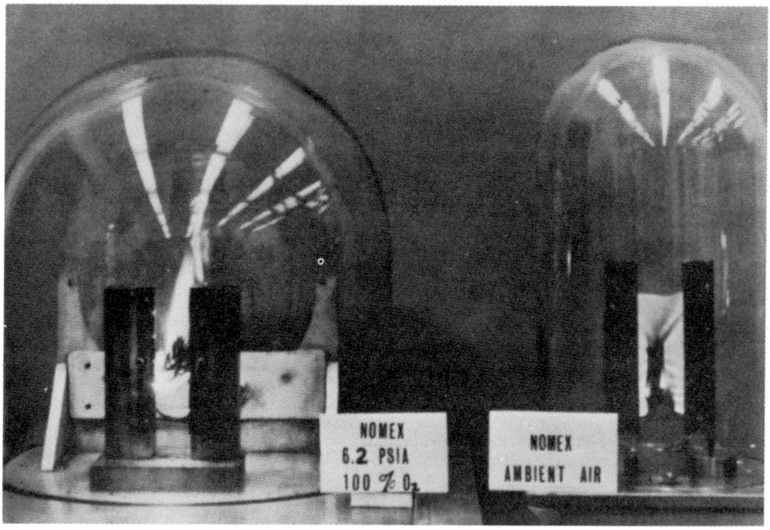

Fig. 1 Flammability test facility.

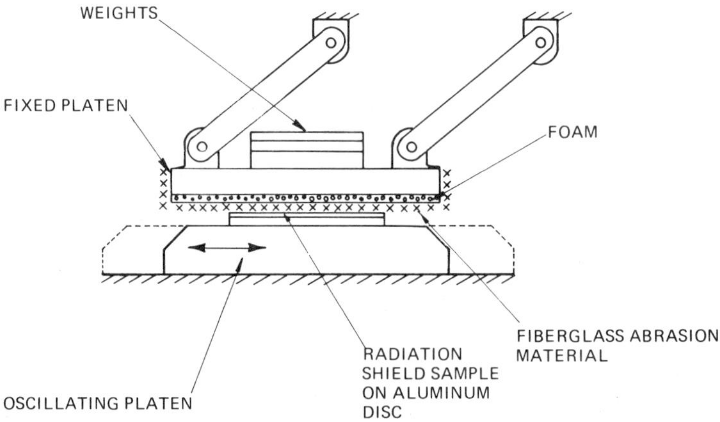

Fig. 2 Schematic of stoll abrasion test apparatus.

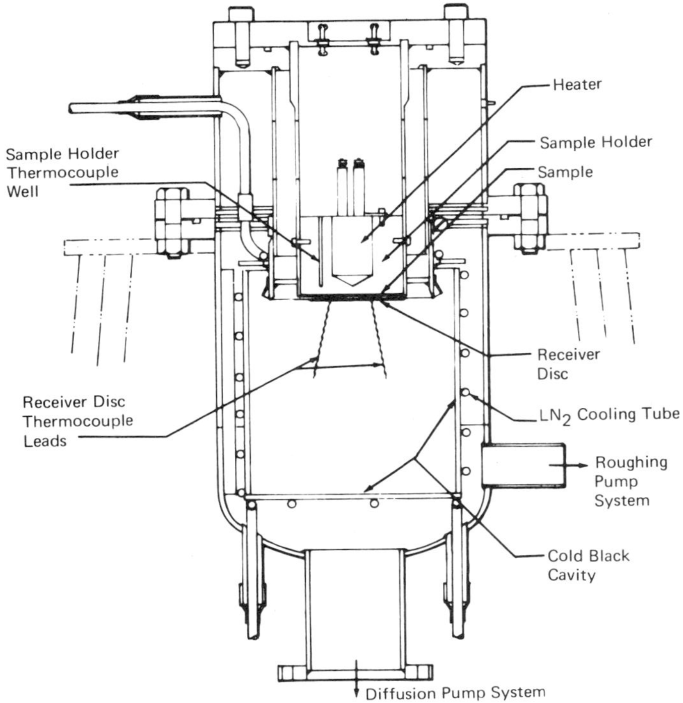

Fig. 3 Emissometer.

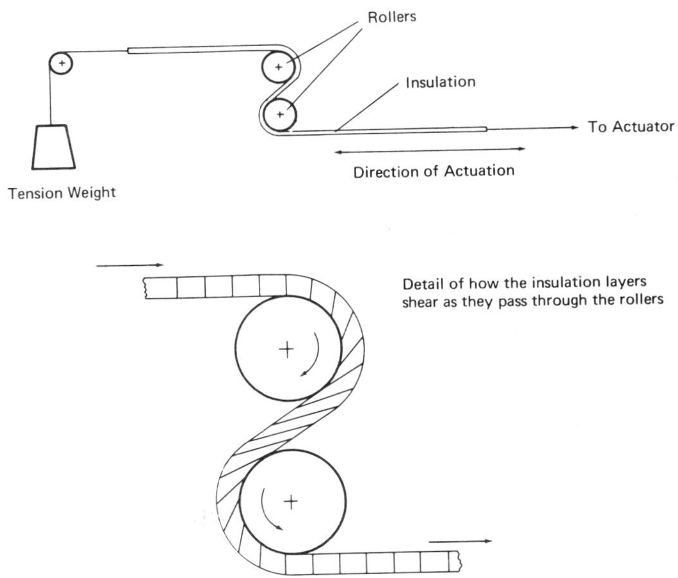

Fig. 4 Schematic of insulation wear test.

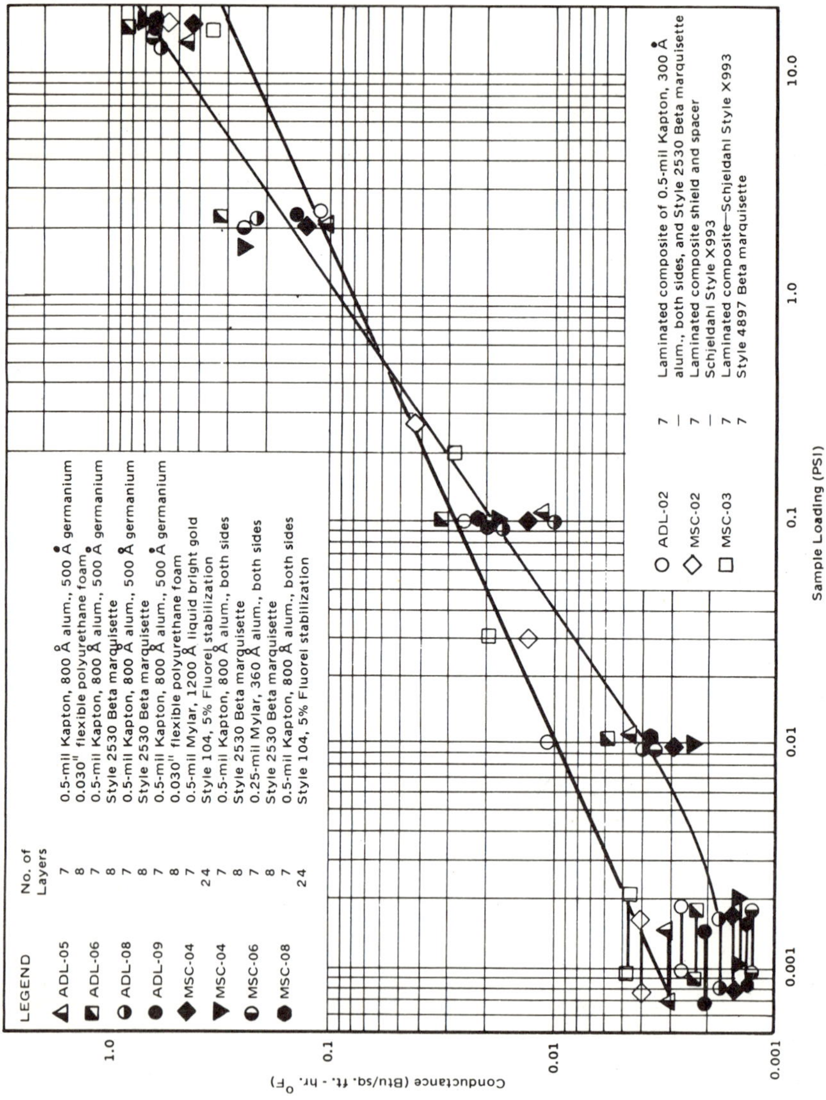

Fig. 5 Summary of insulation conductance: boundary temperature -250 to 70°F.

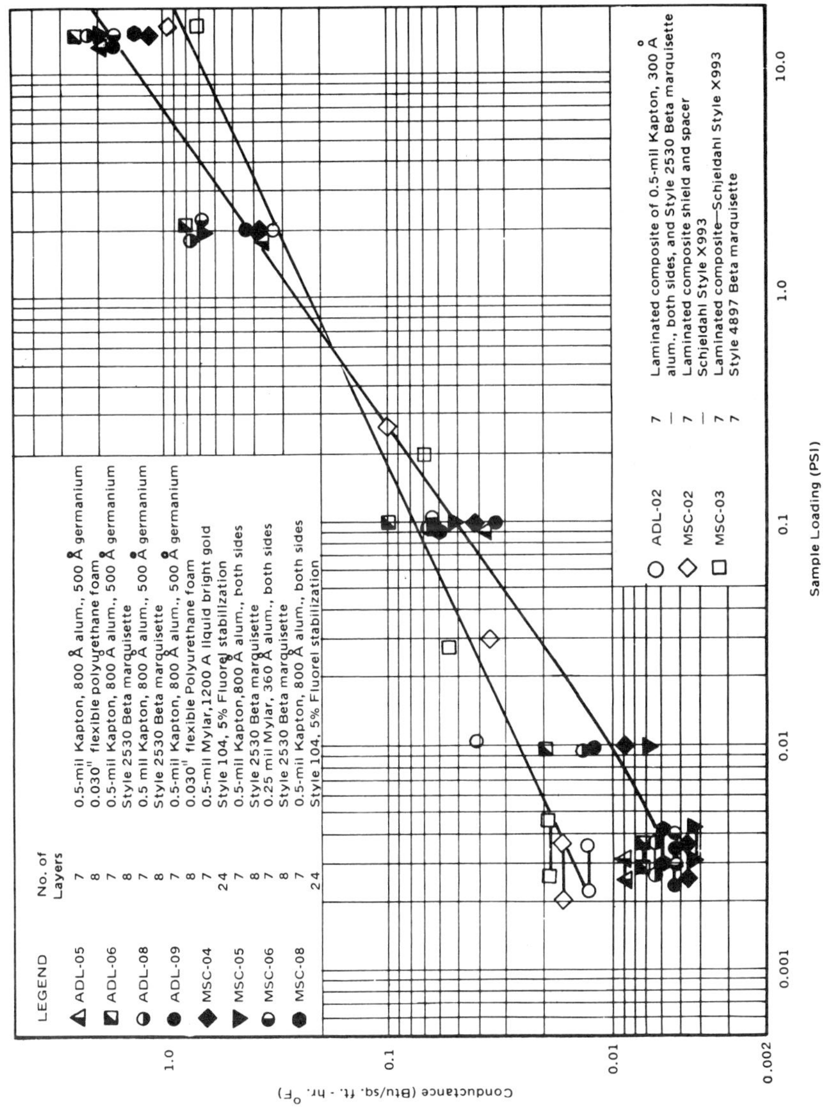

Fig. 6 Summary of insulation conductance: boundary temperatures 300 to 70°F.

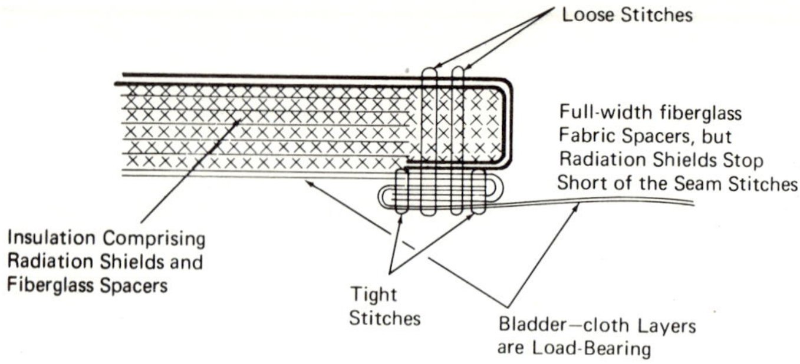

Source: Arthur D. Little, Inc.

Fig. 7 Typical load-bearing edge seam.

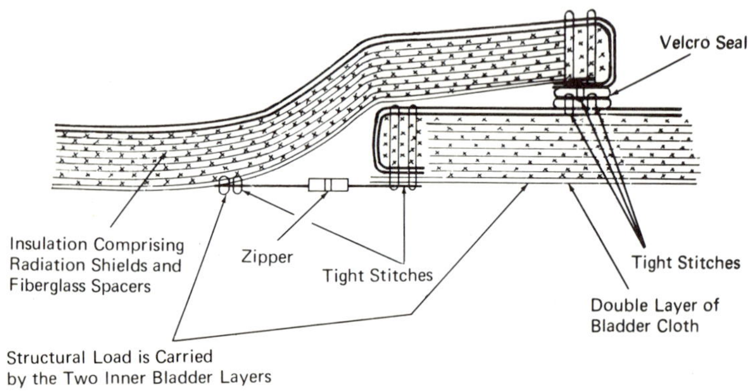

Source: Arthur D. Little, Inc.

Fig. 8 Recommended closure.

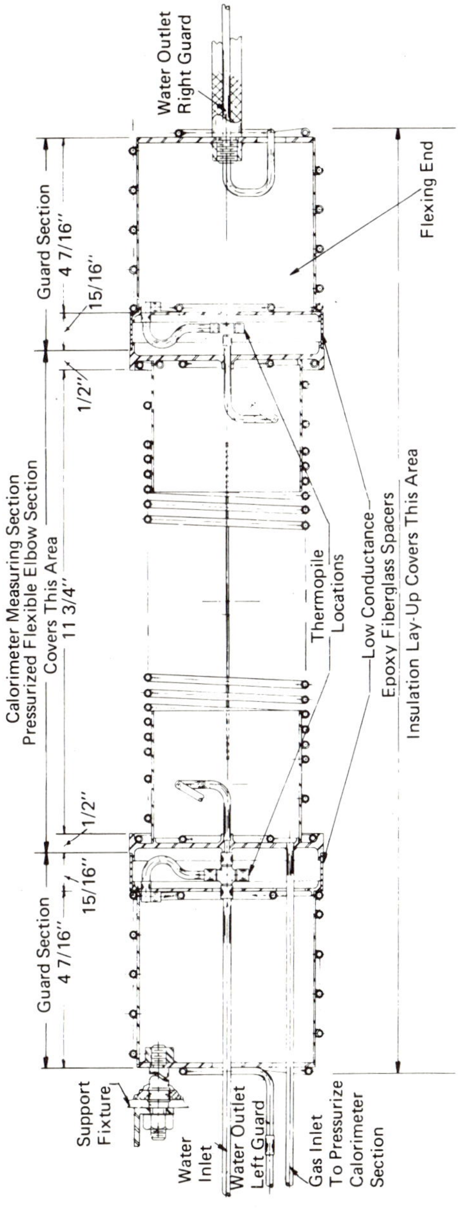

Fig. 9 Elbow calorimeter assembly.

Fig. 10 Guarded elbow calorimeter.

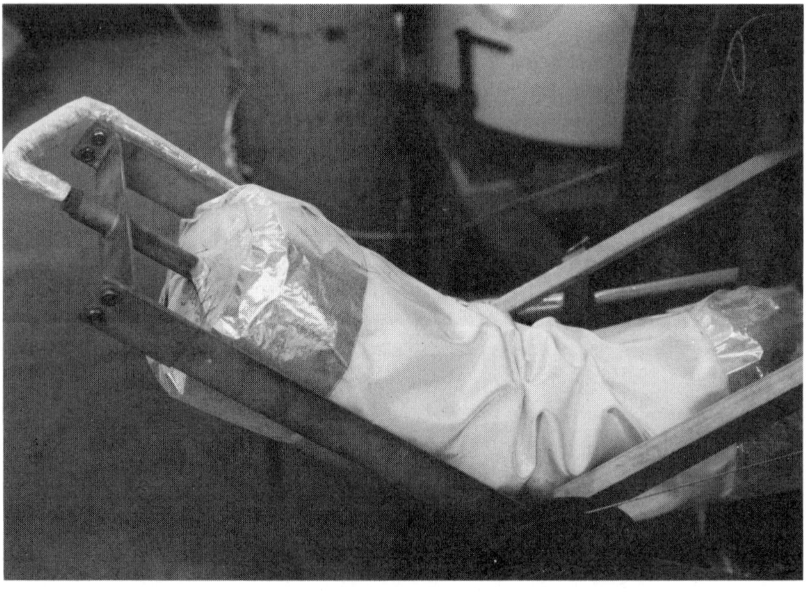

Fig. 11 Garment test section on the elbow calorimeter.

VI. THERMAL CONTROL DEVICES

SELECTION OF SOLID-LIQUID PHASE-CHANGE MATERIALS FOR SPACECRAFT THERMAL CONTROL

Philomena G. Grodzka*

Lockheed Missiles & Space Company, Huntsville, Ala.

Abstract

Performance criteria for phase-change materials (PCMs) for spacecraft thermal control are developed on a logically consistent basis. The definition of these criteria indicates the scope and nature of required testing programs. Two of the most important of the defined criteria, other than a high latent heat and appropriate melting point, are rapid nucleation and solidification kinetics. Simple freezing curves serve as a simple, initial experimental test of the fulfilment of these two primary criteria. Twenty-seven non-paraffinic PCM candidates were screened by means of freezing curves. Four of these candidates appear especially promising and are designated as prime candidates. They are lithium nitrate trihydrate with zinc-hydroxy nitrate catalyst (melting point, 30°C, fusion heat 71 calories per gram); acetamide (melting point, 81°C, fusion heat, 58 calories per gram); methyl fumarate (melting point, 102°C, fusion heat 58 calories per gram); and myristic acid (melting point, 58°C, fusion heat 48 calories per gram).

I. Introduction

The experience of the Apollo 12 astronauts with clinging moon dust gives a timely importance to thermal control devices not dependent primarily on optical properties for efficient operation. The idea of using materials with high heats of fusion for thermal control of spacecraft has been current for years.[1-3] Past research and development, however, has considered only limited aspects of the problems which might be

Presented as Paper 70-12 at the AIAA 8th Aerospace Sciences Meeting, New York, January 19-21, 1970. This work was sponsored by NASA Contract NAS8-21123.

*Staff member, Huntsville Research & Engineering Center.

encountered in a space environment and a limited number of material classes from which potential phase change materials (PCMs) might be chosen. Potential PCMs have been restricted in the past, for the most part, to the paraffin family. The search for new PCMs, therefore, requires performance criteria by which the materials can be tested experimentally. Failure to develop these criteria can result in needlessly extensive and costly or even totally unnecessary experimental programs. Once a logically consistent set of criteria is developed, moreover, an overall logical experimental program can be delineated for prime PCM candidates.

Definition of PCM performance criteria for space conditions necessitates that both the thermal performance requirements and the manner in which phase change is likely to be affected by space environments shall be considered. The defined criteria, furthermore, should be such that they can be translated into meaningful experimental tests. An extensive examination of the physical processes involved in melting and freezing was conducted in the course of the present study.[4,5] From this examination it became apparent that the roles of heat transfer and phase change kinetics in solid-liquid phase change phenomena usually are not clearly delineated. The present study considers the roles of heat transfer and phase change kinetics in efficient PCM operation. The knowledge so obtained is then utilized to define PCM performance criteria. As will be subsequently discussed two of the most important PCM performance criteria, other than a high latent heat and an appropriate melting temperature, are rapid nucleation and solidification kinetics. A simple experimental test is devised for ascertaining the degree of fulfilment of these two primary criteria by a potential PCM. The effectiveness of a potential PCM as measured against these defined criteria is checked experimentally by means of simple freezing and melting curves. The possible influences of space conditions, with particular emphasis on a zero-g gravity field, is also considered. The present study, in short, represents an initial screening of potential, non-paraffinic PCMs. Subsequent studies will further screen the prime candidates with respect to thermal stability, container compatibility, container material and configuration, and other similar criteria.

II. Definition of PCM Thermal Performance Criteria

A representative schematic for the operation of PCM thermal control devices is shown in Fig. 1. Ideally, when the heat fluxes from the space environment or the internal environment cause the RSM (Radiating Surface Material) - PCM or EC

(Equipment Component) - PCM interface temperature to reach the PCM melting point, the PCM will melt, absorbing an amount of heat equivalent to its latent heat of fusion. The RSM-PCM or EC-PCM interface temperature will not rise appreciably above the PCM melting point as long as both solid and liquid PCM phases are present. If the RSM-PCM or EC-PCM interface temperatures fall as the result of external or internal cooling, the PCM solidifies, liberating latent heat. Again, the RSM-PCM or EC-PCM interface temperatures do not appreciably depart from the PCM melting point as long as both solid and liquid PCM phases are present. After the PCM is completely converted to solid or liquid, further heat absorption or liberation is determined by heat capacity considerations.

The preceding description of PCM behavior is, of course, highly idealized. Any actual PCM will deviate from this behavior to varying degrees. One very important deviation which is likely to occur unless precautions are taken is that after PCM is completely liquid, initiation of solid phase formation by cooling may be extremely difficult. Some materials can be cooled far below their solid-liquid equilibrium temperature without any formation of solid phase. In this state the liquid is said to be supercooled. The initiation of a solid phase from a liquid phase is called nucleation. It is possible also that although a solid phase may have been initiated within the liquid phase, subsequent solid growth rate is so slow that it cannot keep up with the rate of heat extraction. Further cooling, therefore, only supercools the remaining liquid. As a result, the temperature of the solid-liquid interface drops until the rate of crystal growth matches the rate of heat removal. Supercooling in the presence of both solid and liquid is called two-phase supercooling in contradistinction to supercooling when only liquid phase is present. The situation in the latter case is called one-phase supercooling.

Another important deviation that is likely to occur with materials of low thermal diffusivity is that the EC-PCM or RSM-PCM interface temperatures will rise to unacceptable levels even when solid is still available but separated from the heated surface. Along with these major material limitations, a number of secondary factors may present problems. These include deleterious effects of thermal cycling, container-material incompatibility and undesirable volume changes.

Figure 2 presents a graphic representation of some possible material responses to a cooling environment. The possible responses shown depend on a large number of adjustable parameters and on the nature of the material itself. It is possible, therefore, to modify or alter material response in some cases by adjusting the independent or adjustable parameters. For example, almost any material will form a glass from the melt if it is cooled rapidly enough. Material response behavior that depends primarily on the nature of the material itself can sometimes be altered by suitable additives. For example, a nucleation catalyst, if one can be found, will overcome a large degree of supercooling.

Regarding efficient PCM operation in thermal control devices, consideration of the illustrated behaviors makes possible a listing of the most important attributes required of an ideal PCM. These would include appropriate melting temperature (0 - 100°C in present case); a large heat of fusion (in the range of 50 cal/g in the present case); small degree of one-phase supercooling; small degree of two-phase supercooling; large thermal diffusivity in both solid and liquid phases; composition of liquid and solid phases the same; perfect contact between PCM and container walls; no reaction between PCM and container; and small volume changes accompanying phase change. The requirements that the composition of liquid and solid phases be the same limits PCM to single-component materials (with no appreciable amounts of soluble impurities) and eutectic materials. Materials which do not freeze to a solid of the same composition as the liquid will in most cases show variable freezing and melting temperatures as the freezing or melting progresses.

III. PCM Performance in Space Environments

The gravity and radiation fields encountered in space orbit are radically different from those present on the surface of the earth. Of these two fields, the zero-*g* gravity field is of particular concern in the present study because of the possibility that the absence of convection may alter appreciably heat transfer through the melt in zero-*g* environments. In most cases, this would be an undesirable condition for PCM thermal control devices because rapid heat transfer rates are difficult enough to achieve with most non-metallic materials. An extensive literature search, however, revealed that gravity is not the only driving force for fluid flow. Surface tension, interfacial tension, and the volume change which accompanies solid-liquid phase change are also potential driving forces. Fluid flow caused by surface tension gradients is called the Marangoni effect. (Several excellent general reviews of surface tension flows are given in Ref. 6.) The various driving forces for fluid flow and phase change as well can couple in complex reinforcing or cancelling manners. Convection kind and magnitude in zero-*g* environments, therefore, cannot generally be predicted without actual flight data on any given system configuration.

Gravity can also affect phase change by its influence on the thermodynamic parameters of phase change. Its influence in this regard, however, is judged to be insignificant for the amount of PCM which would be used in any actual device. For example, the pressure of air is greater near the surface of the earth than at higher altitudes, but the temperature distribution is not affected. An equilibrium atmosphere, in other words, has a constant temperature.[7] Conceivably, therefore, a very long column of materials in solid-liquid equilibrium standing vertically in the earth gravity field will have different equilibrium temperatures for different levels of solid-liquid interface position. For the amount of PCM material that would be used in thermal control devices, however, no significant effect of gravity on the thermodynamic parameters of phase change would be expected.

The influence of electric, magnetic, and radiation fields is also judged to be insignificant or slight for the case of near earth orbit. Slight or negligible effects are predicted because of the small magnitude of these flights which would be ordinarily encountered in earth orbit.[5]

Although convection by its effect on heat transfer through the melt is recognized to be the primary factor capable of affecting phase change, a simple meaningful experimental test of its effect is not easily defined. The present study, therefore, does not consider convection as a specific criterion for PCM selection. The question of whether the kind and magnitude of natural convection occurring either in one-g or zero-g can appreciably alter phase change kinetics is the subject of studies now in progress.

IV. Experimental Design for Evaluation of New PCM Materials

Simple freezing and melting curves provide a convenient means for determining the potential of a candidate material as a PCM. Such curves are obtained by noting the temperature registered by a thermometer, or a thermocouple, located in the center of an unstirred sample while the PCM sample is being cooled or heated. The experimental arrangement is shown in Fig. 3. The curves obtained are plots of temperatures vs time. Ideally, the rate at which the sample is heated or cooled should be controlled so that linear rates are achieved. Linear heating and cooling rates aere generally not achieved with the simple apparatus shown in Fig. 3.

This condition, however, does not alter substantially the following interpretation of freezing-melting curves which assumes linear heat rates. The interpretation of these curves in regard to the fulfilment of the stated material criteria requires an explanation. Figure 4 can be deducted for an ideal PCM of properties listed previously. Lines A' B' represent the heating and cooling curves of a reference material which is heated or cooled under the same conditions as the PCM. The reference material, however, does not undergo phase change in the temperature region of interest. Curves ABCD are the melting and freezing curves of the PCM. At point B on the melting curve, ideal PCM begins to melt. On the freezing curve, ideal PCM begins to freeze at Point B. In both cases the temperature at Point B is the melting temperature, T_m. The melting temperature is held until Point C is reached. Point C on both curves indicates that the phase transformation is complete. The liquid or the solid temperature now rapidly adjusts to that of the surroundings.

Actual materials, of course, will deviate from this ideal behavior to varying extents depending on material nature, rates of heating and cooling, and degree and kinds of impurities. Actually the parameters of material nature and rates of heating and cooling are interdependent. All pure materials would show ideal behavior if the rates of cooling or heating were suitably slow. In many cases a suitably slow rate is so slow that it is impractical. With relatively rapid rates of cooling a pure, single-component material may show some of the type of curves as depicted on Fig. 5. These curves can be interpreted on the basis of extent of one-phase supercooling, two-phase supercooling, convection, and low thermal diffusivity. For example, Point G on Curves A, B and C represent the point in time of maximum one-phase supercooling. At this point nucleation occurs and the temperature thereafter rises and levels off at approximately T_m. Assuming that crystallization begins at the wall of the container, the rate of temperature rise from Point G can be taken as an indication of thermal diffusivity of the liquid phase. In Curve A it is low and in Curve B it is high.

Curve C illustrates both one- and two-phase supercooling.[8] After nucleation, the temperature does not rise to T_m but levels off at some value below T_m. This plateau represents a steady-state phenomenon. Heat is being withdrawn as fast as it is being liberated. The rate of heat generation, however, is slow compared to the rate of heat transport. The process may be said to be "kinetic-controlled." In the case in which the rate of heat transport is the slow process, compared to the rate of heat liberation, the term "heat-controlled" is applicable.

In Curve D, one-phase supercooling does not occur. At the beginning of the plateau the temperature is maintained at T_m, but begins to fall before solidification is complete. This temperature decrease before solidification is complete can be interpreted as the result either of two-phase supercooling (because of rapid rate of heat removal towards the end of the solidification) or of the small amount of crystallizing material. Toward the end of a solidification the amount of heat being liberated by crystallizing material is so small, compared to that which is being withdrawn, that it has little effect on the temperature sensor. The decision concerning the major effect would depend on a consideration on the size and location of the temperature sensor. Severe departure from T_m early in the plateau would, of course, indicate two-phase supercooling. Also convection can play a role in maintaining the temperature at T_m early on the plateau.[9] Convection, however, is less effective as solidification proceeds.

Distorted shapes such as those shown in Curves E and F are the result of low thermal diffusivity. In Curve E, T_m is reached later than in the ideal case because of low thermal diffusivity in the liquid phase (Convection is assumed as negligible.) In Curve F, resumption of steady-state cooling after complete solidification is delayed considerably because of low solid thermal diffusivity.

In the case of melting curves (pure, one-component materials), only low thermal diffusivities need be considered, for superheating is rarely encountered. It would be expected that low thermal diffusivities would lead to the distorted shape shown in Fig 6. At a higher heating rate the curve would be even more distorted.

Another source of deviant melting-freezing curves is impurities. Generally speaking, the concentration of impurities increases in the liquid phase as crystallization proceeds. Consequently, the melting temperature is progressively lowered. The effect of increasing amounts of impurity on freezing curves is generally as shown in Fig. 7. The effect of impurities on the melting curves is generally the same, i.e., the melting occurs at lower temperatures and the plateau becomes less level with increasing impurity content.

The experimental technique as used in the present study provides reliable information only on the degree of one-phase and two-phase supercooling on materials as received. A much more refined and complicated apparatus is required to untangle curve distortion occasioned either by low thermal diffusivity or impurities or both. A small degree of one-phase and two-phase supercooling, however, are two of the most important criteria that a candidate PCM must meet. Thus if some materials in a primary screening show good performance in regard to the supercooling criterion, it is more expeditious to designate these materials as candidates than to pursue materials which show large supercoolings perhaps because of impurities.

V. Results of Experimental Materials Evaluations

First Screening Tests

Table 1 presents the results of the evaluation tests on 27 candidate PCMs. A very qualitative rating scale as a PCM (0 to 10, 0 poorest to 10 best) was set up for the materials that were tested. A rating of seven or above was judged, again rather arbitrarily, as sufficient to qualify a material for further tests. By this standard, 16 of the materials tested rated as potential PCMs.

The shapes of the freezing curves obtained in the first screening ran the gamut of behaviors that were considered possible. In addition, some unexpected types of shapes were observed. Considering the crudity of the experimental apparatus employed in this first screening, the relatively small departures from ideality exhibited by many of the compounds is surprising. Especially impressive was the small amount of two-phase supercooling exhibited by such high temperature melting compounds as $Ba(OH)_2 \cdot 8\,H_2O$, Cerrolow 136 eutectic, stearic acid, and others. The magnitudes of the thermal heads during the freezing of these compounds under ice bath cooling was large (in the range of 58 to 78°C). The rate of cooling, therefore, was rather extreme.

The major class of materials eliminated was a number of low melting eutectic combinations between the various hydroquinones and quinone-hydroquinone. The one eutectic combination tried, 1-1 pyrocatechol and resorcinol, showed such extreme one- and two-phase supercooling that further combinations were not tried. Also eliminated were the majority of salt hydrates which have incongruent melting points. An incongruently melting material undergoes decomposition with formulation of another solid phase before its congruent or "true" melting point is reached. The material, in other words, melts to form a new solid phase and liquid of composition different

Table 1 Results of first screening of candidate PCMs

Candidate	Reported melting temp. (^{o}C)	Rating scale (0-10)[a]
Hydrated Salts		
Lithium nitrate trihydrate	29.8	9
Barium hydroxide octahydrate	78	7
Disodium phosphate dodecahydrate	36.1	8
Disodium arsenate dodecahydrate	28	1
Sodium carbonate decahydrate	33	6
Sodium sulfate decahydrate	31	5
Ferric nitrate enneahydrate	47.2	1
Lithium acetate dihydrate	58	0
Metals		
Gallium	29.8	6
Ga-GaSb eutectic	29.8	7
Cerrolow 136 eutectic	58	6
Cerrobend 158 eutectic	70	8
Organic materials		
Acetic acid	16	7
Choloracetic acid	α 61.2	8
	β 56.0	
	γ 50.0	
Methyl fumarate	102	7
Acetamide	81	9
	(69.4)	
Cyanamide	42.9	6
Elaidic acid	47	7
Myristic acid	58	8
Stearic acid	69.4	8
Tristearin	54.5	7
	70.8	
Glycerol	18	1
Polyethylene glycol 600 (carbowax)	20-25	7
1 methyl - 4 piperidone hydrochloride hydrate	94-95	2
Oxazoline wax TS970	74	8
Oxazoline wax ES254	37	8
Resorcinol-pyrocathecol eutectic	70	0

[a] Freezing behavior, poorest to best.

from the original solid phase. These included sodium arsenate dodecahyrate, sodium sulfate decahydrate, and sodium carbonate decahydrate. The chief reason for eliminating these compounds was the lack of easy reversibility of the transition reaction. For sodium sulfate and sodium carbonate decahydrates, the addition of fillers or gels presumably would keep insoluble, lower hydrates suspended and thereby improve the reversibility. In one experiment with sodium sulfate decahydrate, the addition of chopped glass fibers did improve the shape of the freezing curve. The improvement, however, was not marked. About the only advantage that sodium sulfate and sodium carbonate decahydrates offer over lithium nitrate is one of cost. Sodium sulfate decahydrate is commonly known as Glauber's salt and sodium carbonate decahydrate is common washing soda. For home heating purposes where large, tonnage quantities are being considered, it might be worthwhile to try to modify the properties of these two salts. Lithium nitrate is moderately expensive (presently $1.55 to $1.25 per pound, depending on amount bought) but not expensive enough to eliminate its consideration.

The lack of reversibility of the fusion reaction of sodium arsenate dodecahydrate is easily understood when the solubility data is consulted.[10] Sodium arsenate dodecahydrate has a transition point at 22°C where it transforms into liquid and the heptahydrate. Actually, the same condition should occur with sodium phosphate dodecahydrate. The transition in this case occurs at about 35°C. With the phosphate, however, the phase diagram shows that the compounds transition point is not far removed from being almost a congruent melting point.[11] In present observations of the melting and freezing of sodium phosphate dodecahydrate, no incongruent type behavior was noted, i.e., no separation of salt into liquid and lower hydrate was observed on melting or freezing. This observation confirms a similar observation reported by Telks.[12]

Nucleation Catalyst Search and Definition

The previously described evaluation tests indicated some materials that performed well in almost every respect except that they showed severe one-phase supercooling. As a result, nucleation catalysts were sought for a number of these materials. The discovery was soon made that a large element of luck is involved in finding a suitable nucleation catalyst for any given material. Only one theory is known that can somewhat guide the experimenter in his search for a suitable catalyst. This theory designates as potent nucleation catalysts materials that have low-index planes in which the atomic arrangement is similar to that in certain low-index planes of the forming crystals. The problem with applying this theory is that a rather specialized knowledge of X-ray crystallography and its complex nomenclature is required.

The present study settled for searching out materials that have the same crystal structure as candidate PCM and possess lattice constants which closely match those of candidate PCMs. Even this approach is not without problems. For one, lattice parameters for a number of potential PCMs are either not listed in readily available sources or are not available. Secondly, a great number of the materials turned up are quite exotic and are not readily available. Their preparation would require a major effort. Finally, a great number of materials must be rejected because of their solubility in PCM melt. A great number of materials also were tried as catalysts "on hunch." One hunch - stressed metals - proved profitable for it led to a class of compounds called basic oxides which appear to have great possibilities as nonreactive nucleation catalysts.

The actual testing of materials for catalytic activity was also beset with problems. Some materials would work well one day through any number of thermal cycles, only to show no activity the next day. With some, catalytic activity would reappear after a time. In spite of all these difficulties a couple of apparently reliable catalysts for two of the PCM candidate materials were found in the present study. The exact nature of the catalysts found has not been established. Their probable nature, however, is easily assessed. In the case of the PCM candidate lithium nitrate trihydrate, a pyrolysis product, probably $Zn(NO_3)_2 \cdot 2\ Zn(OH)_2$,[13] of $Zn(NO_3)_2 \cdot 6\ H_2O$ nucleated the melt with only 1 to 3°C supercooling. Furthermore, the catalytic activity appears to be retained through any number of thermal cycles provided that the melt is heated no higher than about 100°C. A reaction product between iron and the phosphate melt nucleated sodium phosphate dodecahydrate through a number of thermal cycles with about 7 to 8°C supercooling.

Designation of Four Prime PCM Candidates

The previously described evaluation tests, although adequate as a basis for selecting candidate PCMs, provided only very limited information on freezing-melting characteristics. As mentioned previously, an ice bath represents a rather extreme cooling environment for some of the higher melting temperature materials. Further information is also desirable on the effects of variables such as impurity kind and concentration, thermal cycling, container size and material, effect of fillers of various kinds, etc. A great deal of this kind of information was generated during further testing in connection with development of an apparatus in which melting and freezing curves could be obtained at known, linear heating and cooling rates. The details of the development of the apparatus and the results obtained with it are not presented here. A full discussion is given in Ref. 5. It may be mentioned, however, that the apparatus is an adaptation of a differential thermal analysis unit. Weighing the information generated by this means along with further information found in the literature makes possible a final designation of four prime PCM candidates. The specific considerations in selecting a final four candidates are now discussed.

In the first screening test, 16 materials received a rating of seven or above, thus qualifying them as candidate PCMs. If the criterion of non-reactivity with aluminum is set, the 16 materials are reduced to 12. Candidate compounds barium hydroxide octahydrate, sodium phosphate dodecahydrate, gallium and chloroacetic acid are all corrosive to aluminum. In the case of sodium phosphate dodecahydrate, aluminum metal can probably be protected from corrosion by adding sodium silicate (water glass) to the melt. However, because the melting point of $Na_2HPO_4 \cdot 12\ H_2O$, 36°C, is so near that of $LiNO_3 \cdot 3\ H_2O$, 30°C, there does not appear to be any compelling reason for further considering $Na_2HPO_4 \cdot 12\ H_2O$.

The next materials which came off second-best in an intra-comparison are the oxzaoline waxes, polyethylene glycol, stearic acid and tristearin. The PCM-container separation observed with the oxazoline waxes in quartz and pyrex containers might also present a problem in aluminum containers. The point was not investigated further because other puzzling and unresolved features of oxazoline wax phase change behavior would require further investigation before a final evaluation could be made. In particular, although the DTA indicates a high heat of fusion, the wax-container separation complicates the thermogram to the degree that an unequivocal statement about the value of heat of fusion is not possible. Also, the indicated low thermal diffusivity and presence of undesirable impurities make the oxazoline waxes a poor second in comparison to the other remaining materials.

The compound stearic acid is eliminated from consideration as a prime candidate for two reasons. First, compound-container separation during freezing was observed for the case of a pyrex container. Also, although no unexpected freezing behavior was noted in the present study, reference is made in the literature to an irreversible polymorphic conversion.[14] Unfortunately, the original Russian reference[15] is not readily available. What the irreversible polymorphism means in terms of PCM performance awaits further study.

Tristearin also is eliminated from further consideration as a prime candidate because of a reported polymorphism.[16] Under certain conditions the following sequence can occur:

$$\text{Solid I} \longrightarrow \text{Liquid I} \longrightarrow \text{Solid II} \longrightarrow \text{Liquid II.}$$

The transition Liquid I to Solid II is exorthermic, having a heat of transition of - 12.4 cal/g. Again, as with stearic acid, no unusual freezing behavior was actually noted.

Polyethylene glycol 600 exhibited good freezing characteristics. Its heat of fusion, however, is one of the lowest of the remaining compounds left in the list.

At this point the list of candidates for designation as prime candidates numbers seven. These seven materials are acetic acid, methyl fumarate, acetamide, elaidic acid, myristic acid, and Cerrobend alloy. Final selection of four prime candidates is based on rather subtle and intuitive considerations. Cerrobend, for example, has a very low heat of fusion on a gram basis - 7.78 cal/g. On a volume basis, however, the value is a respectable 73 cal/ml. Because the other compounds have a much higher heat of fusion, however, Cerrobend is eliminated, even though its thermal diffusivity is undoubtedly much higher than that of the other compounds. For some applications Cerrobend will probably be superior to the other six compounds when all tradeoffs are considered.

Acetic acid is eliminated as a prime candidate mainly because a nucleation catalyst was not found in the present study to overcome the exhibited 16°C one-phase supercooling. Although a catalyst was not found in the present study, further effort will undoubtedly be successful. A secondary reason for disqualifying acetic acid as a prime candidate is that its volume expansion on melting is quite large (approximately 16%).[17]

The final elimination is elaidic acid on the intuitive and therefore nebulous grounds that the longer chain organic acids are more prone to polymorphic transitions than the shorter chain acids. No report, however, has been found in the literature of any polymorphism.

The volume expansion on melting for two of the four prime candidates was checked crudely by noting the volume of melt, allowing it to crystallize, and then noting the volume of the solid. Control over temperature or crystal form was not undertaken. Lithium nitrate trihydrate and myristic acid show an approximate 8% increase in volume on melting. Acetamide is reported to also have an approximate 8% volume expansion on melting.[17] The liquid density of methyl fumarate was determined to be about one g/ml and the volume change on melting a rather high 18 to 20%. The volume change figure 18 to 20% is a very approximate because methyl fumarate wet a pyrex container so well that it remained clinging to the walls when it crystallized. It was extremely difficult to estimate, therefore, the volume of the solid.

The four prime PCM candidates designated in the present study are: lithium nitrate trihydrate with zinc-hydroxy nitrate catalyst (melting point 30°C, fusion heat, 71 cal/g); acetamide (melting point 81°C, fusion heat, 58 cal/g); methyl fumarate (melting point 102°C, fusion heat, 58 cal/g); and myristic acid (melting point 58°C, fusion heat, 48 cal/g). The order in which the compounds are presented represents a predicted effectiveness as PCMs. The order also represents to some degree the confidence with which the recommendation is made. Lithium nitrate trihydrate, because of its convenient melting point, was used as a test case in most of the development work on the adaptation of the DTA apparatus. More data, therefore, was generated on it than on the other compounds. Acetamide rates second place because of its relatively high density and low volume expansion on melting. On a volume basis the heat of fusion is a relatively high 67 cal/ml. Methyl fumarate is ranked third because of its high expansion on melting, 18 to 20%. This high volume expansion, however, may not be enough to disqualify methyl fumarate as a PCM especially when the properties of other PCM candidates are considered. For example, the paraffins have a low enough thermal diffusivity so that packing with metal foams has been considered. Packing with metal foams would increase not only the weight but also the volume of the PCM system. Very vigorous convection, on the other hand, has been observed in methyl fumarate melts. This convection may be largely the surface tension driven kind because the compound does appear to have unusual surface tension properties. If all tradeoffs are considered, therefore, it may turn out that methyl fumarate performs as well or better than a paraffin in a system which has the same internal container volume but with a considerable weight saving. Myristic acid is placed fourth because of its low density and a feeling that it may exhibit polymorphism.

References

[1]Leatherman, R.A., "Component Thermal Control Via Heat of Fusion Radiator," Paper 63-AHGT-12, 1963, American Society of Mechanical Engineers.

[2]Fixler, S.Z., "Satellite Thermal Control Using Phase-Change Materials," Journal of Spacecraft and Rockets, Vol. 3, No. 9, Sept. 1966, pp. 1362-1368.

[3]Bannister, T.C. and E.W. Bentilla, "Research and Development Study on Space Thermal Control by Use of Fusible Materials," Institute of Environmental Sciences' 1966 Annual Technical Meeting Proceedings, pp. 593-607.

[4]Grodzka, P.G., and C. Fan, "Thermal Control by Freezing and Melting: Interim Report on Space Thermal Control Study," LMSC/HREC A791342, Lockheed Missiles & Space Co., Huntsville, Ala., March 1968.

[5]Grodzka, P.G., "Space Thermal Control by Freezing and Melting Second Interim Report on Space Thermal Control Study," LMSC/HREC D148619, Lockheed Missiles & Space Co., Huntsville, Ala., May 1969.

[6](a) Scriven, L.E. and C.V. Sternling, "The Marangoni Effects," Nature, Vol. 187, July 1960, pp. 186-188; (b) Sternling, C.V., and L.E. Scriven, "Interfacial Turbulence: Hydrodynamic Instability and the Marangoni Effect," AIChE, Dec. 1959, pp. 514-523; (c) Kenning, D.B.R., "Two-Phase Flow with Nonuniform Surface Tension," Applied Mechanics Review, Vol. 21, No. 11, Nov. 1968, pp. 1101-1111.

[7]Allis, W.P. and M.A. Herlin, Thermodynamics and Statistical Mechanics, McGraw-Hill, New York, 1952, p. 171.

[8]Smit, W.Mn. "Errors Occurring in the Determination of Temperature-Heat Content Curves," Purity Control by Thermal Analysis, Elsevier, New York, 1957, pp. 23-35.

[9]Steiner, L.E. and J. Johnston, "Development of a Method of Radiation Calorimetry, and the Heat of Fusion or of Transition of certain Substances," Journal of Physical Chemistry, Vol. 32, 1928, pp. 912-939.

[10]Mellor, J.W., A Comprehensive Treatise on Inorganic and Theoretical Chemistry, Vol. IX, Wiley, New York, 1922, p. 152.

[11] Mellor, J.W., A Comprehensive Treatise on Inorganic and Theoretical Chemistry, Vol. II, Wiley, 1922, p. 855.

[12] Telkes, M., "Development of High Capacity Heat Storage Materials," M.I.T. Instrumentation Lab., R-380, July 1962, p. 50.

[13] Pannetier, G., and G. Djega-Mariadassoux, "Preparation and Characterization of the Hydroxynitrate of Zinc," Comptes Rendus Academie Sciences Paris, Vol. 261, Oct. 1965, pp. 3146-3149.

[14] Smothers, W.J., and Y. Chiang, Handbook of Differential Analysis, Chemical Pub. Co., New York, 1966, p. 179.

[15] Arutyunova, L.B., "A Study of the Polymorphism of Higher Normal Aliphatic Acids by Differential Thermal Analysis and Infrared Spectral Analysis," Zhurnal Strukturnoi Khimii, Vol. 1, 1960, pp. 145-150.

[16] Barrall, E.M., II, and J.C. Guffy, "The Polymorphism of Tristearin," Advances in Chemistry Series 63, American Chemical Society, Washington, D.C., 1967.

[17] International Critical Tables, Vol. IV, McGraw-Hill, New York, 1928.

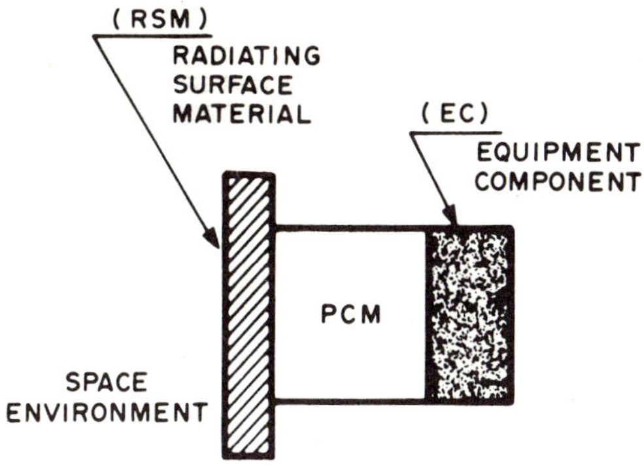

Fig. 1 Representative schematic for operation of PCM thermal control device.

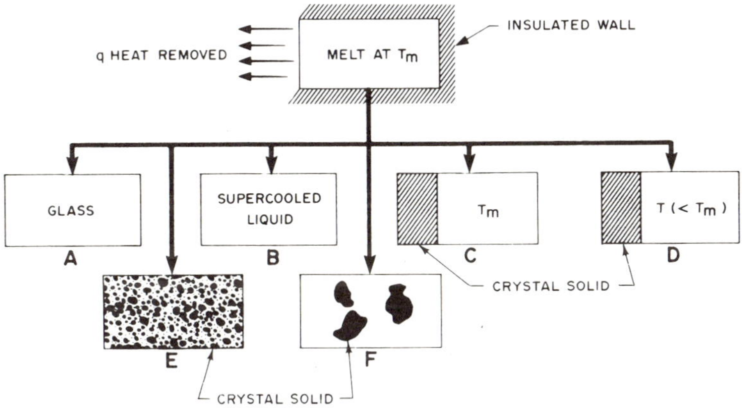

Fig. 2 Some possible responses of a material melt to cooling.

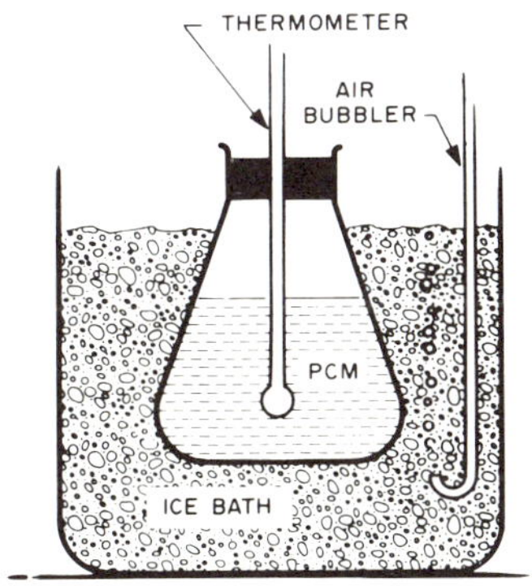

Fig. 3 Experimental apparatus for determining freezing-melting curves.

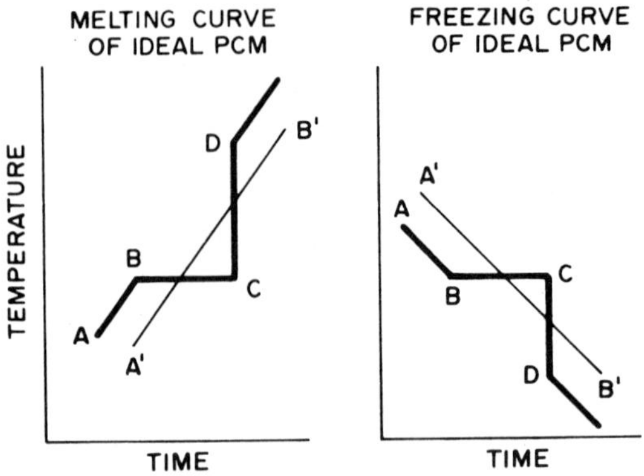

Fig. 4 Ideal PCM melting and freezing curves.

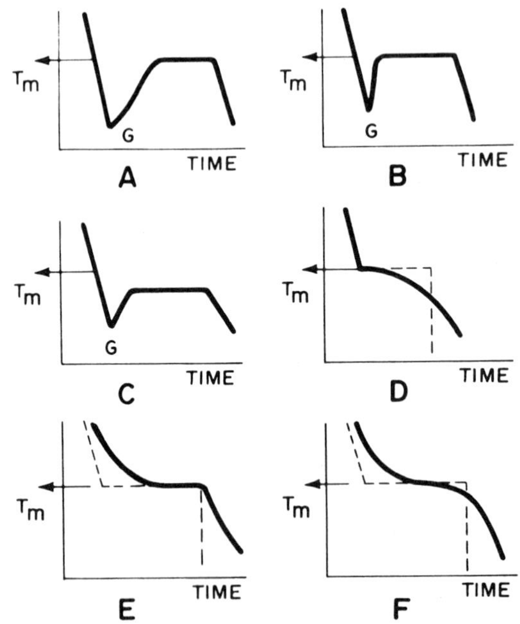

Fig. 5 Possible types of freezing curve (dashed curves = ideal freezing)

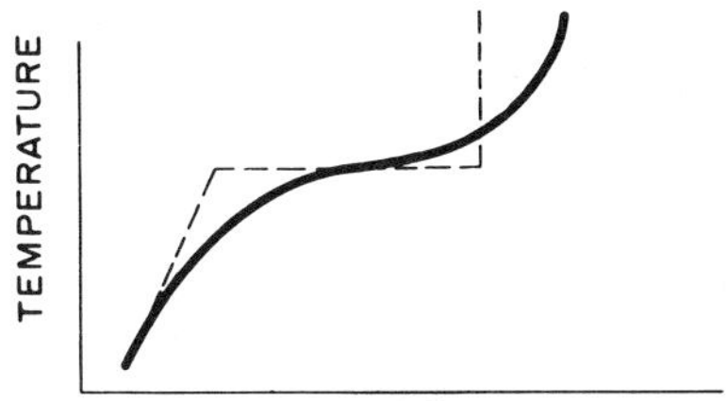

Fig. 6 Melting curve of material with low thermal diffusivity.

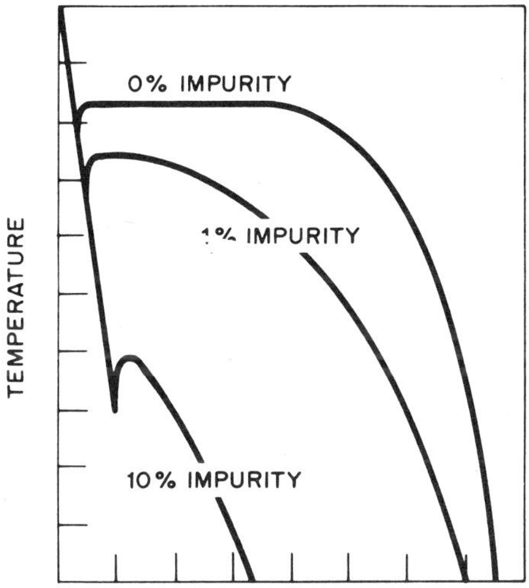

Fig. 7 Influence of increasing impurity on freezing curves.

TRANSIENT RESPONSE OF A SOLID WITH THERMAL-GALVANO-MAGNETIC EFFECTS

M. R. El-Saden*

California State College, Fullerton, Calif.

Abstract

Problems of thermal-galvano-magnetic effects involve the multual interaction between electric and thermal conduction processes in the presence of an applied magnetic field. Although several steady-state studies are available in the literature, no transient response analysis has been reported. The model employed in this study is that of a block initially at a uniform temperature T_1, subjected to a constant electric field E_y. The transient response is investigated when a constant external magnetic field B_o is turned on in the z direction while the side of the block at x = 0 is maintained thermally insulated and the side x = L is held at the temperature T_1. The results are presented in graphical form and show the temperature and the thermal and electric current densities in terms of position and time, as well as the degree of cooling at x = 0.

Nomenclature

$\vec{B}$ = magnetic field vector
b_z = induced magnetic field
c = specific heat
$\vec{E}$ = electric field vector
e = dimensionless electric field
$\vec{J}$ = electric current density vector
$\vec{J}_{st}$ = entropy current density vector
J_u = energy current density vector
j = dimensionless electric current
k = isothermal heat conductivity

Presented as Paper 70-17 at the AIAA 8th Aerospace Sciences Meeting, New York, New York, January 19-21, 1970. This work was sponsored by the Air Force Office of Scientific Research, Office of Aerospace Research, U.S. Air Force, Grant No. AF-AFOSR-1272-67.

*Professor and Chairman, Mechanical and Aerospace Engineering.

L = length in x direction
R = isothermal Hall coefficient
T = temperature
t = time
u = internal energy per unit volume
X = dimensionless length
α = absolute thermoelectric power
ϵ = thermomagnetic figure of merit
ϵ_o = permittivity
η^o = isothermal Nernst coefficient
θ = dimensionless temperature
μ_o = magnetic permeability
τ = Righi-Leduc coefficient
ρ = mass density
ρ_e = electric charge density
σ = isothermal electrical conductivity
τ = dimensionless time

Introduction

The model for the analysis is shown in Fig. 1. It is a uniform rectangular block of an isotropic material of length L in the x direction and unit length in the y and z directions. A constant electric field E_y is applied in the y direction. Initially, the block is at a temperature T_1 throughout. The side of the block at x = 0 is thermally insulated while the side at x = L is maintained at the temperature T_1. The problem is to determine the transient response when a magnetic field, B_o, of constant strength is turned on in the z direction.

The steady state solution shows that a temperature gradient will be set up and an energy current will flow up the gradient (from the cold to the hot end).[1] This results in a heat current to the surroundings at the hot end (x = L) of the block and a useful cooling effect at the other end (x = 0). Thus, this configuration serves as a model for a thermomagnetic cooler.

The following governing equations for the analysis are well established in the literature, 2-4:

Electromagnetic equations,

$$\vec{\nabla} \times \vec{E} = -\partial\vec{B}/\partial t \quad (1)$$

$$\vec{\nabla} \times (\vec{B}/\mu_o) = \vec{J} + \partial(\epsilon_o\vec{E})/\partial t \quad (2)$$

$$\vec{\nabla} \cdot \vec{B} = 0 \tag{3}$$

$$\vec{\nabla} \cdot \vec{J} = -\partial \rho_e / \partial t \tag{4}$$

$$\vec{\nabla} \cdot (\epsilon_o \vec{E}) = \rho_e \tag{5}$$

Generalized laws of Ohm and Fourier,

$$\vec{E} = (\vec{J}/\sigma) - R(\vec{B} \times \vec{J}) + \alpha\vec{\nabla}T + \eta(\vec{B} \times \vec{\nabla}T) \tag{6}$$

$$T\vec{J}_{st} = T\alpha\vec{J} + \eta T(\vec{B} \times \vec{J}) - k\vec{\nabla}T + \lambda k(\vec{B} \times \vec{\nabla}T) \tag{7}$$

First law of thermodynamics,

$$\vec{\nabla} \cdot (T\vec{J}_{st}) - \vec{J} \cdot \vec{E} = -\partial u/\partial t \tag{8}$$

and the second law of thermodynamics,

$$(\eta^2 B^2 \sigma T/k) \leq 1 \tag{9}$$

Equation (9) for the second law was employed in the analysis to define the region of validity of the results.

The analysis is restricted to one dimension and the following assumptions are employed:

$$\partial/\partial y = \partial/\partial z = 0 \tag{10}$$

All properties of the material are constant.

$$B_x = B_y = 0 \tag{11}$$

Furthermore the magnetic field in the z direction may be expressed as $B_z = B_o + bz$ where b_z is the induced component and is assumed to be much smaller than the applied field B_o.

$$J_x = J_z = 0 \tag{12}$$

The former requires that the electrodes for E_y be segmented. Hence, the only significant current in the system is J_y.

The thermal boundary conditions are

$$T(x,o) = T_1 \tag{13}$$

$$T(L,t) = T_1 \tag{14}$$

$$J_{ux}(o,t) = 0 \tag{15}$$

where

$$J_{ux} = TJ_{stx} \tag{16}$$

Analysis

Using the foregoing assumptions, Eqs. (1) through (9) give

$$\partial b_z/\partial t = -\partial Ey/\partial x \tag{17}$$

$$(1/\mu_o)(\partial b_z/\partial x) = - J_y - \epsilon_o(\partial E_y/\partial t) \tag{18}$$

$$E_x = RB_oJ_y + \alpha(\partial T/\partial x) \tag{19}$$

$$E_y = (J_y/\sigma) + \eta B_o(\partial T/\partial x) \tag{20}$$

$$TJ_{stx} = - \eta B_oTJ_y - k(\partial T/\partial x) \tag{21}$$

$$TJ_{sty} = T\alpha J_y + \lambda kB_o(\partial T/\partial x) \tag{22}$$

$$\partial(TJ_{stx})/\partial x - J_yE_y = - \rho\, c(\partial T/\partial t) \tag{23}$$

$$(\eta^2B_o^2\sigma T/k) \leq 1 \tag{24}$$

At this point the following dimensionless variables are introduced and used in the analysis

$$X = x/L \tag{25}$$

$$\epsilon = \eta^2B_o^2\sigma Ta/k \tag{26}$$

$$\theta = T/Ta, \quad \theta_o = T_o/T_a, \quad \theta_1 = T_1/T_a \tag{27}$$

$$T_a = (T_o + T_1)/2 \tag{28}$$

$$e = L(\sigma/kT_a)^{1/2}E_y \tag{29}$$

$$j = [L/(\sigma kT_a)^{1/2}]J_y \tag{30}$$

$$j_{ux} = (L/kT_a)J_{ux} \tag{31}$$

$$\tau = (k/\rho\, cL^2)t \tag{32}$$

Using Eqs. (20) and (21), Eq. (23) becomes

$$(1-\epsilon\theta)(\partial^2\theta/\partial X^2) - \epsilon(\partial\theta/\partial X)^2 + e^2 = \partial\theta/\partial\tau \tag{33}$$

and Eqs. (13-15) become

$$\theta(X,0) = \theta_1 \tag{34}$$

$$\theta(1,\tau) = \theta_1 \tag{35}$$

$$[1 - \epsilon\theta(0,\tau)](\partial\theta/\partial X)_{X=0} = -\epsilon^{1/2} e\theta(0,\tau) \tag{36}$$

Equation (36) results from Eqs. (16, 20, and 21). Equations (33-36) govern the temperature distribution in the block in terms of X and τ. Once this is established, the electric current may then be obtained from Eq. (20).

$$j = e - \epsilon^{1/2}(\partial\theta/\partial X) \tag{37}$$

The energy current is obtained from Eqs. (16, 20, and 21),

$$j_{ux} = - (1 - \epsilon\theta)(\partial\theta/\partial X) - \epsilon^{1/2} e\theta \tag{38}$$

At X = 1 the energy current is the same as the heat current exchanged by the block with the surroundings.

In the Eqs. (37) and (38) the electric field E_y is treated as a constant except in Eqs.((17) and (18), where the small variations in E_y and B_z may be determined once the electric current is established. Equations (19) and (22) give E_x and J_{sty} which are not relevant in this analysis and hence will not be considered.

Before the transient solution is attempted, the steady state solution and its region of validity must be established for guidance in the transient solution.

The Steady-State Solution

For this case

$$\partial\theta/\partial\tau = 0 \tag{39}$$

Equation (33) reduces to

$$(1-\epsilon\theta)(\partial^2\theta/\partial X^2) - \epsilon(\partial\theta/\partial X)^2 + e^2 = 0 \tag{40}$$

The closed form solution of this equation that satisfies the boundary condition of Eq. (36) is

$$(1 - \epsilon\theta)^2 = \epsilon e^2 X^2 + 2\epsilon^{3/2} e\theta_o X + (1 - \epsilon\theta_o)^2 \tag{41}$$

The boundary condition of Eq. (35) requires that

$$\theta_o = (1-\epsilon + 0.25e^2)/(1-\epsilon - 0.5\epsilon^{1/2} e) \tag{42}$$

where use is made of the relation

$$\theta_o + \theta_1 = 2 \tag{43}$$

Region of Validity

In terms of dimensionless variables Eq. (24) takes the form

$$0 < \epsilon\theta \leq 1 \tag{44}$$

Thus, the second law dictates that the product $\epsilon\theta$ must remain within the limit of unity everywhere in the block, otherwise this law would be violated. Hence,

$$\theta_o > 0, \qquad \theta_1 > 0$$

$$0 < \epsilon\theta_o \leq 1, \qquad 0 < \epsilon\theta_1 \leq 1 \tag{45}$$

In addition to these relations, the maximum value of θ in the block θ_m must be found and the product $\epsilon\theta_m$ must then be limited to unity.

To find θ_m, Eq. (41) is differentiated, the term $(\partial\theta/\partial X)$ is set equal to zero and the result is solved for X_m, the coordinate of the point where θ_m occurs. The result is

$$X_m = - \epsilon^{1/2}\theta_o/e \tag{46}$$

The existence of a maximum is ascertained by differentiating Eq. (41) twice,

$$\partial^2\theta/\partial X^2 = - 2e^2/(1-\epsilon\theta) \tag{47}$$

which, in view of Eq. (44), is always negative, and hence there is no minimum.

Equation (46) offers two possible outcomes. Either the point X_m lies within the block ($0 \leq X_m \leq 1$), or else it is outside the block. When $0 \leq X_m \leq 1$,

$$0 \leq (-\epsilon^{1/2}\theta_o/e) \leq 1 \tag{48}$$

Here, the applied electric field e must be negative. Substitution of X_m in Eq. (41) and the application of Eq. (44) results in the requirement that

$$0 < \epsilon\theta_o \leq 0.5 \tag{49}$$

Inequalities (45), (48), and (49), together with Eq. (42), are represented in Fig. 2 by the singly crosshatched region below the horizontal axis. When $X_m > 1$,

$$- \epsilon^{1/2}\theta_o/e > 1 \tag{50}$$

Here again e must be negative. The maximum temperature in the block is θ_1. This relation along with those of inequality (45) are represented in Fig. 2 by the doubly cross-hatched region. Finally, when $X_m < 0$

$$- \epsilon^{1/2}\theta_o/e < 0 \tag{51}$$

In this case, e is positive while the maximum temperature in the block is θ_o. This case is shown in Fig. 2 by the area above the horizontal axis. Thus, Fig. 2 shows the entire region of validity for the solution that is to be obtained in this analysis.

The Transient Solution

A digital computer was used for the numerical solution of Eqs. (33, 37, and 38) subject to the boundary conditions of Eqs. (34-36). The solution exists in a space-time plane such as the one shown in Fig. 3. The following difference expressions were employed:

$$\partial\theta/\partial\tau = (\theta_{i+1,j}-\theta_{i,j})/\Delta\tau \tag{52}$$

$$\partial\theta/\partial X = (\theta_{i,j+1}-\theta_{i,j-1})/2(\Delta X) \tag{53}$$

$$\partial^2\theta/\partial X^2 = (\theta_{i,j+1}-2\theta_{i,j} + \theta_{i,j-1})/(\Delta X)^2 \tag{54}$$

Substitution in Eq. (33) yields

$$\begin{aligned}\theta_{i+1,j} = {} & [\Delta\tau/(\Delta X)^2]\{[1-\epsilon(\theta_{i,j})][\theta_{i,j+1}-2\theta_{i,j}+\theta_{i,j-1}] \\ & -(\epsilon/4)[\theta^2_{i,j+1}-2(\theta_{i,j+1})(\theta_{i,j-1})+\theta^2_{i,j-1}]\} \\ & + (\Delta\tau)e^2 + \theta_{i,j}\end{aligned} \tag{55}$$

Equations (34) and (35) now may be expressed as

$$\theta(X,0) = \theta_{1,j} = \theta_1 \tag{56}$$

$$\theta(1,\tau) = \theta_{i,n} = \theta_1 \tag{57}$$

For the last boundary condition an imaginary column 1 is introduced such that

$$(\partial\theta/\partial X)_{X=0} = (\theta_{i,3} - \theta_{i,1})/(2\Delta X) \tag{58}$$

Thus, in difference form Eq. (36) becomes

$$\theta_{i,1} = [2(\Delta X)\epsilon^{1/2}e\theta_{i,2}]/(1-\epsilon\theta_{i,2}) + \theta_{i,3} \tag{59}$$

The solution was carried out starting with the first row (i = 1) where θ is given by Eq. (56) at j = 2, 3, ...,n. The value of $\theta_{1,1}$ at the imaginary column is determined from Eq. (59). This completes the values of the set of $\theta_{1,j}$ (j = 1,2,...,n). This set is used in Eq. (55) to yield the values of θ at the next row $\theta_{2,j}$ (j = 2,3,..., n-1). Equation (57) gives $\theta_{2,n}$ while Eq. (59) gives $\theta_{2,1}$, establishing the values of the set of $\theta_{2,j}$ (j = 1,2,...,n). The procedure is continued until the steady state solution is approximately reached. The electric and energy currents are determined from Eqs. (37) and (38) using Eq. (53) for all interior points. At X = 1, a backward difference expression was used for the first derivative

$$(\partial\theta/\partial X)_{X=1} = (\theta_{i,n} - \theta_{i,n-1})/\Delta X \tag{60}$$

Solution were obtained for ϵ = 0.4 and e = 0.3, -0.5, and -1.2 which are shown in Figs. 4-12. In all cases, the transient solution converged to the steady state solution which is given by Eqs. (37, 38, and 41). This provides a check on the validity of the numerical procedure. For additional results at other values of ϵ and e, see Ref. 5.

The results show that when e is positive the block heats up throughout. However, when e is negative, the end of the block at X = 0 cools off and the energy current is positive, i.e., toward the warm end at X = 1. It is interesting to note in Fig. 7 that when θ_1 is a maximum at steady state, θ increases above θ_1 and then falls during the transient phase.

Maximum Temperature Difference

The temperature difference $\Delta\theta$ between the ends of the block may be established from Eq. (42) and (43),

$$(\Delta\theta) = \theta_1 - \theta_o = 2-2\theta_o = -(\epsilon^{1/2}e+0.5e^2)/(1-\epsilon-0.5\epsilon^{1/2}e) \tag{61}$$

This expression can be maximized with respect to e giving

$$e = 2[1-\epsilon-(1-\epsilon)^{1/2}]/\epsilon^{1/2}, \qquad \epsilon \le 0.75 \tag{62}$$

This relation is represented in Fig. 2 by the solid curve separating the single and double cross-hatched regions below the horizontal axis. It follows that maximum $(\Delta\theta)$ occurs when θ_{max} is at X = 1. Substituting Eq. (62) in Eq. (61) yields

$$(\Delta\theta)_{max} = 2 + 4[1-\epsilon-(1-\epsilon)^{1/2}]/\epsilon, \qquad \epsilon \leq 0.75 \qquad (63)$$

The restriction on ϵ follows from the region of validity of Fig. 2. For $\epsilon > 0.75$, the relation between e and ϵ follows the remaining part of the solid boundary in Fig. 2. This boundary was obtained from Eq. (45), which gives

$$e = [1-2\epsilon + 2(\epsilon-0.75)^{1/2}]/\epsilon^{1/2}, \qquad \epsilon > 0.75 \qquad (64)$$

Substituting in Eq. (61)

$$(\Delta\theta)_{max} = 2(1-\epsilon)/\epsilon, \qquad \epsilon > 0.75 \qquad (65)$$

Figure 13 is a plot of Eqs. (63) and (65) and shows that the maximum $(\Delta\theta)$ obtainable is 0.667.

Conclusions

The results presented here are valid only for those values of e and ϵ that lie within the region of validity of Fig. 2. This is due to the assumptions introduced in the analysis and the restriction of the second law. Values of e and ϵ outside the region of validity are physically possible but they invalidate some of the assumptions of this analysis, such as the constancy of the material properties.

References

[1] El-Saden, M.R. and Arnas, O.A., "Transport Processes in Magnetosolidmechanics," Journal of Applied Mechanics, Vol. 33, Dec. 1966, pp. 770-776.

[2] El-Saden, M.R., "Theory of Nonequilibrium Thermodynamics with Applications to the Transport Processes in a Solid," Journal of Heat Transfer, Vol. 88, 1966, pp. 57-63.

[3] DeGroot, S.R. and Mazure, P., Nonequilibrium Thermodynamics, North-Holland, Amsterdam, The Netherlands, 1962, Chap. 13.

[4] Callen, H.B., Thermodynamics, Wiley, New York, 1960, Chap. 17.

[5] Reynolds, J.R., "Transport Processes in Magnetosolidmechanics," Ph.D. thesis, Mechanical Engineering Dept., North Carolina State University, Raleigh, N.C., 1966.

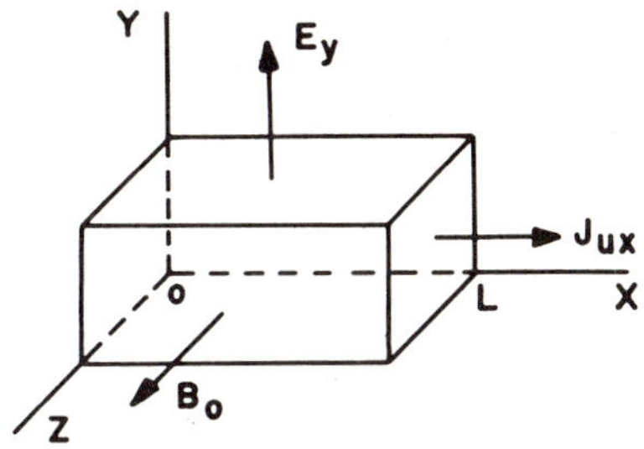

Fig. 1 Model employed in the analysis.

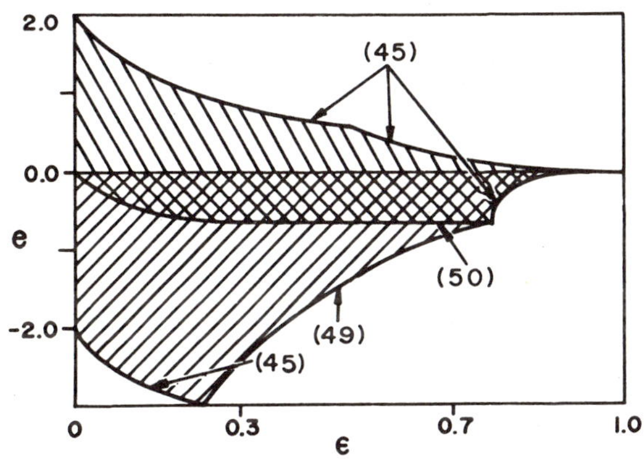

Fig. 2 Region of validity of the solution.

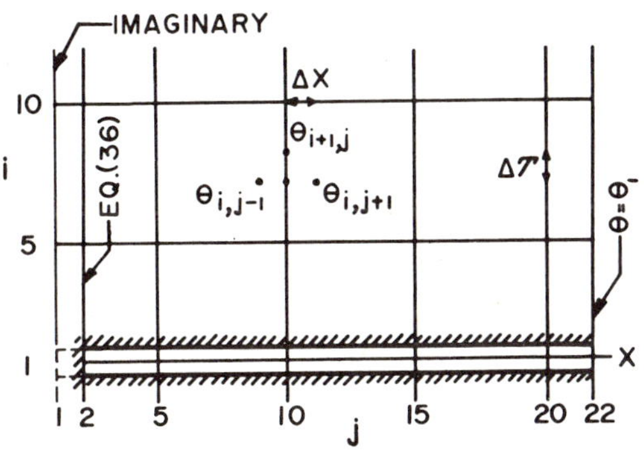

Fig. 3 Space-time plane of the solution.

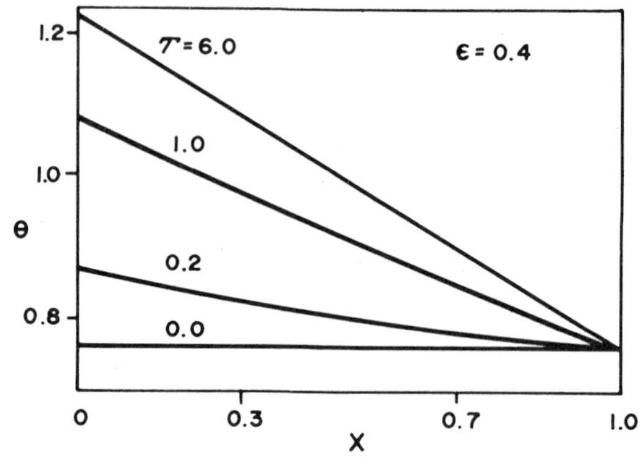

Fig. 4 Temperature distribution, e = 0.3.

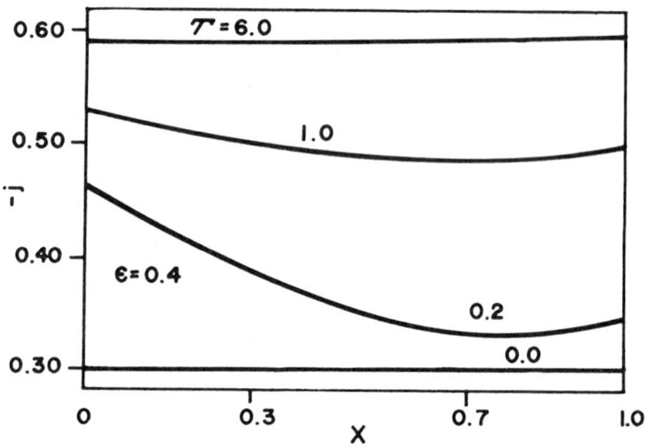

Fig. 5 Electric current distribution, e = 0.3.

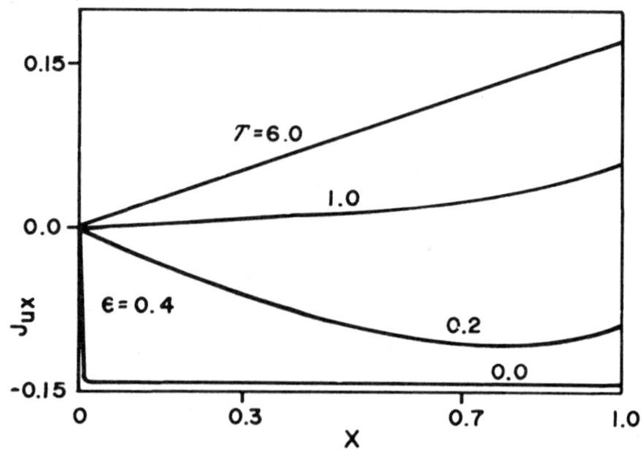

Fig. 6 Energy current distribution, e = 0.3.

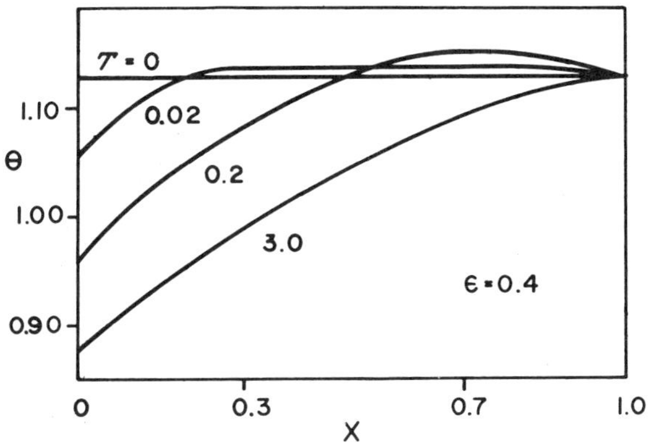

Fig. 7 Temperature distribution, e = -0.5.

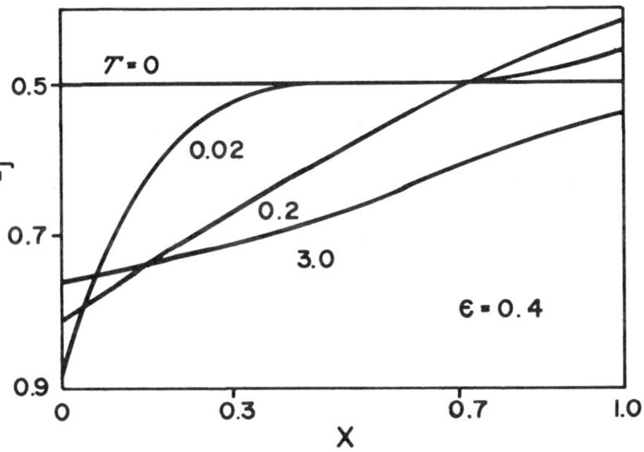

Fig. 8 Electric current distribution, e = -0.5.

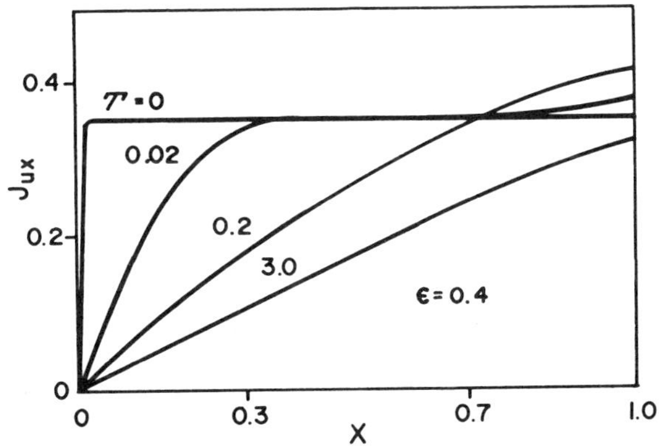

Fig. 9 Energy current distribution, e = -0.5.

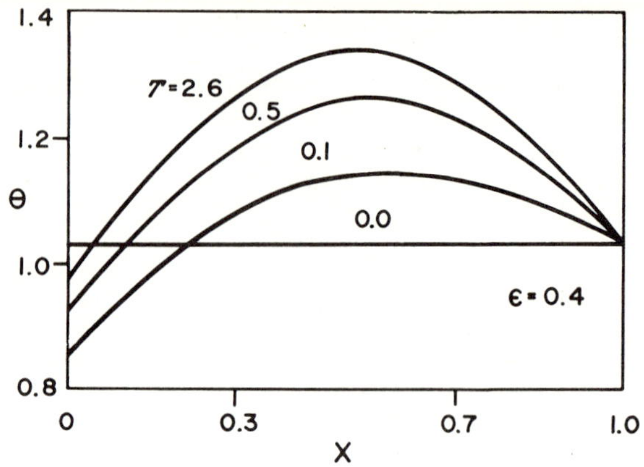

Fig. 10 Temperature distribution, e = -1.2.

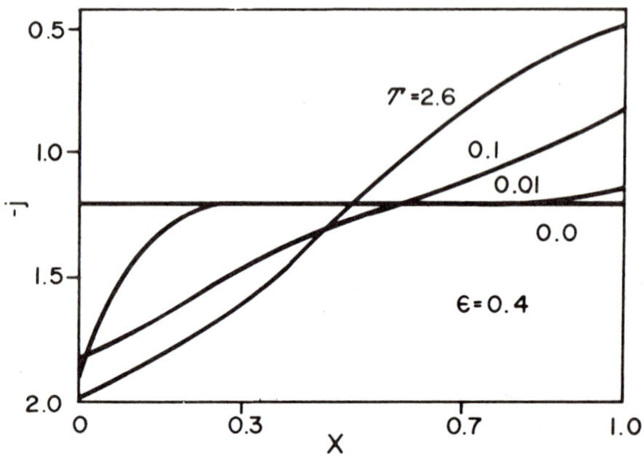

Fig. 11 Electric current distribution, e = -1.2.

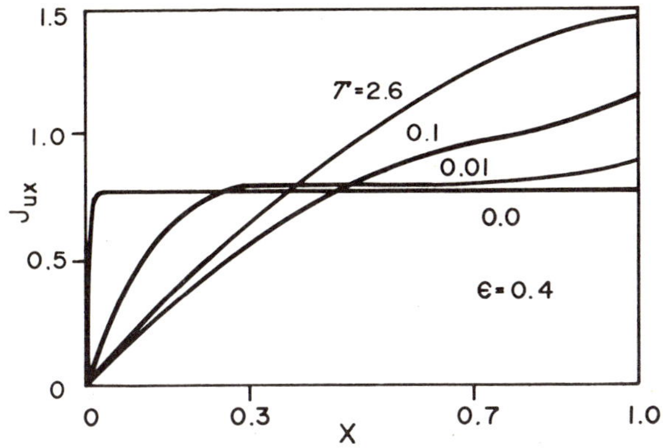

Fig. 12 Energy current distribution, e = -1.2.

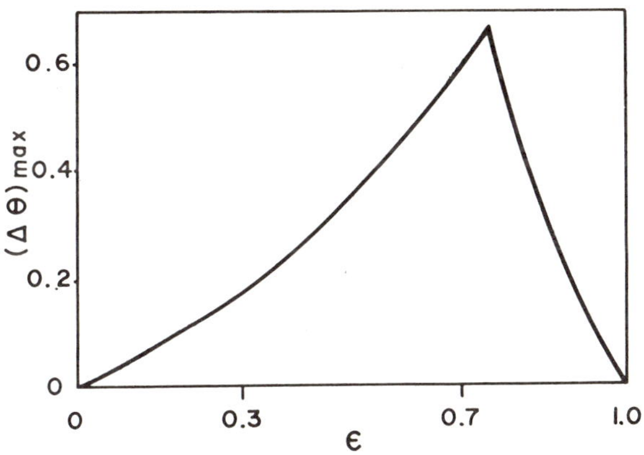

Fig. 13 Maximum temperature difference.

THERMAL TESTING OF INFLATABLE SOLAR SHIELDS FOR CRYOGENIC SPACE VEHICLES

R. O. Doughty* and L. R. Jones*

General Dynamics Corporation, Fort Worth, Texas

Abstract

The effectiveness of an inflatable spherical shield for protection of cryogenic propellants from direct solar radiation was determined experimentally. A high-emittance ($\epsilon \approx 0.90$) band over one-half of the backside (shaded) area of the shield reduced backside shield temperatures over 20%. Solar misalignment of the shield had little effect on propellant tank temperatures. For a staged, superinsulated Mars vehicle, the low backside shield temperatures (-175 to -200°F) could reduce the initial mass placed into Earth orbit by as much as 40,000 lbm compared to an unshielded, superinsulated vehicle. The effects on candidate thermal-control coatings from extended exposure to solar-wind protons and ultraviolet radiation were investigated. Tests with ¼-scale spherical shields under simulated space conditions verified the feasibility of the inflatable solar-shield concept.

Nomenclature

R^C_{ij} = resistance to heat transfer by conduction between nodes i and j

R^R_{ij} = resistance to heat transfer by radiation between nodes i and j

---

Presented as Paper 70-856 at the AIAA 5th Thermophysics Conference, Los Angeles, Calif., June 29-July 1, 1970. The research was sponsored by the National Aeronautics and Space Administration under Contract NAS8-21132 and was monitored by J. H. Pratt, Marshall Space Flight Center. The authors wish to acknowledge J. V. Clifton, Jr., J. K. Miles, and P. R. Cheever of General Dynamics Corp. for their contributions to this research effort.

*Project Aerothermodynamics Engineer, Research and Engineering Department.

T_j = temperature of node j in electrical network

T_m = temperature of solar shield model

T_s = effective sink temperature in thermal vacuum chamber

T_∞ = temperature of solar shield model with infinite skin conductance

Q_s = incident solar energy

α_s = solar absorptance

ϵ = surface emittance

θ = vertical-plane angle on thermal test model, deg (Fig. 8)

σ = Stephan-Boltzmann constant

ψ = horizontal-plane angle on thermal test model, deg (Fig. 8)

Introduction

Successful long-term storage of cryogenic propellants in space requires a highly effective thermal protection system. Present passive systems (multilayer insulation and surface coatings) do not prevent heat leaks through piping and propellant tank structural supports. An inflatable solar shield (Fig. 1) has been developed that can be folded and packaged in a small canister for integration with a vehicle and for launch into space. Once in space the canister cover is jettisoned and the shield is deployed, inflated, and chemically rigidized over the area of the shield that is subject to high inertial loadings from vehicle attitude-control thrusters. This provides a thermal shield that is external to the vehicle such that incident solar energy can be intercepted and then reflected and re-radiated back to space.

Inflatable-shield concepts for protection of cryogenically fueled space vehicles have been described previously.[1,2] Specific materials and structural development details for the rigidizable shield described in this paper are also available.[3] Recent studies[4] have shown that an external shield is very effective when used to complement other passive protection systems. In addition, the use of a solar shield can significantly reduce the need for active thermodynamic control systems such as partial recondensation, venting, and stratification reduction.

The objective of the work described in this paper was to establish thermal design criteria for inflatable solar shields and to conduct an analytical and experimental development program[5] leading to the fabrication and testing of scale-model shields. The primary study areas were 1) establishment of solar-shield configuration requirements, 2) investigation of solar wind and ultraviolet (uv) effects on candidate thermal-control-coating materials, 3) experimental determination of the thermal effectiveness of spherical shield models, 4) determination of the thermodynamic properties of the chemical accelerator used in the shield rigidization process, 5) demonstration of the resistance of the rigidizable material to various nuclear radiation environments, and 6) demonstration of the operational feasibility of the inflatable solar shield concept under simulated space conditions.

Thermal Design Considerations

Meaningful effectiveness of an off-standing solar shield is obtained when the temperature of the aft side (side facing the space vehicle) of the shield is minimized. Earlier studies[1] showed a spherical shield to be the most effective configuration for reducing aft-side shield temperatures. Shielding is further improved if internal radiation shields (membranes stretched across the inside of the sphere) and internal and external high-emittance bands over portions of the aft area of the shield are incorporated.

In a spherical shield the effect on aft-side shield temperature of the number of membranes and high-emittance bands is also quite pronounced. Typical variations in the aft-surface temperature of a spherical shield with changes in membrane and band coverage are shown in Fig. 2. These predictions are based on a spherical shell of GT-15 material (3/4-mil-thick aluminum/Mylar/aluminum tri-laminate fabricated by the GT-Schjeldahl Company with an emittance, ϵ, of 0.025) for various combinations of internal membranes ($\epsilon = 0.025$ each) and emittance bands. Skin conductance is assumed to be negligible for the data in Fig. 2. A high-emittance band ($\epsilon = 0.75$) is used on the inside and outside of the spherical shell as shown in the figure. An alodine coating (solar absorptance, α_s, = 0.3, and $\epsilon = 0.52$) is used on the outside hemispherical surface irradiated by the sun. The inside surface of the spherical shield between 120° and 180° from the subsolar point is covered with the rigidized-foam reinforcing cap ($\epsilon = 0.8$). This is one-fourth of the internal surface area of the sphere.

With a shield sized for an allowable ±5° solar misalignment of the vehicle longitudinal axis, the selection of the number of radiation shields and the high-emittance coating distribution is dependent upon the savings in liquid-hydrogen boiloff that results from reduction of the temperature on a portion of the aft section (i.e., vehicle side) of the shield. In the analysis described here, it was assumed that the area of the shield between the angles of 120° and 180° from the shield subsolar point is exposed to the vehicle liquid-hydrogen tank.

The total liquid-hydrogen boiloff resulting from this assumption was calculated for a nuclear vehicle Mars braking stage during the Mars transfer phase of a 1982 manned Mars stopover mission. For a shield that is 60 ft in diameter and separated by 42.2 ft from a 33-ft-diam vehicle liquid-hydrogen tank (minimum separation distance allowed by the modular-vehicle nuclear engine), there is no net heat transfer to the liquid hydrogen and, hence, no boiloff for a backside shield temperature of less than -272°F. With one internal radiation membrane and a high-emittance band, the backside shield temperature is decreased to -250°F and the sum of liquid-hydrogen boiloff and the mass of the membrane is 41 lb. Since the mass of two membranes alone is 43 lb, the desired configuration is a spherical shield with one radiation membrane across the center of the shield and a high-emittance band as shown in Fig. 2.

Thermal-Control-Coating Tests

Because the effectiveness of a solar shield is highly dependent upon the radiative properties of the shield surface, it is imperative that these properties remain relatively constant throughout the entire mission. In order to determine the effects of the space environment on candidate thermal-control-coating materials, radiative-properties tests were conducted under simulated space-radiation conditions. The test program consisted of subjecting various coating materials to simulated doses of solar-wind protons, ultraviolet radiation (uv), and combined solar wind and uv. The degradation of the coating materials was then determined *in situ* and selection of materials for the shield thermal-control coatings was made. All tests were based on radiation dose levels anticipated for an Earth-Mars mission.

The objective of the test program was to measure the effects of vacuum-irradiation on the spectral reflectance and solar absorptance of three candidate thermal-control-coating materials. The three coatings tested were silicon oxide, alodine, and a carbon-doped polyester deposited on a substrate of the aluminum/Mylar/aluminum tri-laminate, GT-15.[5] These were selected for testing because of their ability to maintain their surface properties after the shield is deployed and inflated from a tightly packed configuration in its storage canister.

A typical Earth-Mars mission (1982 Manned Mars Stopover, Opposition Class) consists of a 220-day outbound journey at an effective solar distance of 1.23 astronomical units (a.u.), a 40-day capture orbit about Mars averaging 1.47 a.u. from the sun, and a 211-day Mars-Earth journey at an effective solar distance of 0.791 a.u. The anticipated electromagnetic radiation and solar-wind proton doses for this mission are: 1) 145 equivalent solar days of uv (ESD) and 1.26×10^{15} p/cm^2 during the outbound journey, 2) 18.5 ESD and 1.60×10^{14} p/cm^2 during Mars orbit, and 3) 337 ESD and 2.91×10^{15} p/cm^2 during the inbound journey.

Experimental Investigation

The experimental technique used involves the measurement of the spectral reflectance of opaque thin films in vacuum. From the spectral reflectance, both the spectral absorptance α_λ and the solar absorptance α_s can be readily calculated.

The reflectance measurements were made using standard irradiation sources, vacuum chamber, pumping system, integrating sphere, transfer-optics system, and a Perkins-Elmer Model 13-U spectrophotometer utilized in the double-beam mode of operation. A sketch of the equipment setup is shown in Fig. 3.

A stainless-steel integrating sphere, with a 6-in.-diam spherical cavity, was internally coated with magnesium oxide. Apertures for illumination and viewing of specimen and reference coatings permitted total hemispherical spectral reflectance measurements to be accomplished using double-beam optics under conditions of diffused hemispherical illumination and 12.5° off-normal viewing. Illumination was by a hydrogen lamp in the 0.25- to 0.4-μ wavelength region, and by a tungsten lamp in the 0.3- to 2.6-μ region. The reference coating was magnesium oxide on a magnesium carbonate block.

Irradiation sources were a Texas Nuclear Model 150-1H accelerator, which supplied nominal 3-kev protons to simulate the solar wind, and General Electric B-H6 mercury lamps, which were used at two locations as ultraviolet sources. One source was for irradiation of up to four samples at a rate of six equivalent Suns, whereas the other was for single-sample uv irradiations at eight equivalent Suns.

Four specimens, including one control, of each of the three materials were mounted on the sample drum. The chamber was sealed and evacuated and initial reflectance measurements were made when the pressure had reached 1.1×10^{-6} torr. Pressures during proton irradiation were on the order of 8×10^{-6} torr because of the atomic hydrogen particles injected into the chamber from the accelerator. All irradiations and reflectance measurements were performed at room temperature.

Coating Test Results

Solar absorptance values obtained by integrating spectral absorptance data are given in Table 1. The corresponding plots of reflectance as a function of wavelength are presented in Figs. 4-6.

Table 1 Solar absorptance of irradiated thermal-control coatings

Irradiation Level	Solar Absorptance, α_s		
	Silicon Oxide	Alodine	Carbon-Doped Polyester
Preirradiation	0.21	0.33	0.96
18 ESD	0.34	...	0.95
36 ESD	0.36	0.34	...
72 ESD	0.37	...	...
155 ESD	0.41	0.34	0.93
6.3×10^{14} p/cm^2	...	...	0.96
3.2×10^{15} p/cm^2	0.20	0.32	0.95
8 ESD plus 3.7×10^{15} p.cm^2	0.27	0.34	0.95

All three materials were found to be very resistant to solar wind protons; changes in α_s of less than 5% were produced by 3.2×10^{15} p/cm^2. The alodine and carbon-doped polyester coatings were also found to be very resistant to ultraviolet radiation; changes in α_s of 3% were produced by 155 ESD of uv. The silicon oxide coating, however, was found to be highly susceptible to uv radiation. An increase in α_s of 62% was produced by 18 ESD. Further irradiation to 155 ESD resulted in a total increase in α_s of 95%. The change in solar absorptance of this coating as a function of uv dose is presented in Fig. 7. The change in solar absorptance observed for the silicon oxide coating in the simultaneous uv-proton irradiation corresponds to that expected for a sample irradiated with uv only.

Based on these results, the alodine and carbon-doped polyester coatings meet solar-shield mission requirements of radiation resistance, whereas the silicon oxide coating does not.

Scale-Model Thermal Tests

A series of thermal tests was conducted to determine the effects of variations in thermal-control-coating distributions on shield temperature and to measure the thermal effectiveness of spherical solar shield models under simulated operational conditions. Alodine was used to provide an acceptable ratio of solar absorptance to emittance ($\alpha_s/\epsilon \approx 0.67$) in the areas irradiated by the Sun, and carbon-doped polyester was used to provide areas of high emittance on the aft side of the shield.

Three tests were performed in which temperature measurements were taken on various combinations of a shield model and a model simulating the aft-end geometry of a liquid-hydrogen tank. All tests were performed under simulated space conditions: a vacuum of 10^{-5} torr or less, a liquid-nitrogen cold wall with a high-absorptance (ϵ=0.98) coating, and a source of simulated solar radiation.

In the first test, a tank model and a shield model without a high-emittance band of the black carbon-doped polyester paint were used. Temperature measurements were taken with various alignments of the shield and tank models relative to the solar source. A shield model with a black band was substituted in the second test to investigate, by comparison with the first test results, the effectiveness of the black coating in reducing the temperature of the backside of the shield model. In the third test, the tank model was removed to investigate its influence on the shield model temperature distribution.

Solar Shield Models

Two 30-in.-diam spherical shield models and one 16.5-in.-diam spherical model simulating the aft-end geometry of a hemispherical liquid-hydrogen tank were tested. A separation distance of 23.75 in. was used between the shield and tank models. These dimensions were scaled from a 60-ft-diam operational shield sized for the Mars braking stage of a nuclear vehicle for a 1982 manned Mars stopover mission.

Ordinary thermal model scaling techniques were not used since the 0.00075-in. skin of a full-scale shield is already approaching the limits of existing manufacturing, fabrication, and handling techniques. The alternative approach was to fabricate the models using materials and thicknesses identical to those of the full-scale shield, thereby retaining the desired optical properties of the model surfaces. The scaling of the effects of conduction were therefore neglected, resulting in slightly higher backside model temperatures than would occur on a full-size shield. Since only one shield model size (30-in.-diam) was used in the thermal tests, the specific influence of neglecting conduction scaling is not known.

All models were fabricated of the GT-15 aluminum/Mylar/aluminum tri-laminate material ($\epsilon = 0.05$) using gore-type construction with double-taped butt joints. The distribution of thermal control coatings used on the shield models is illustrated in Fig. 8. Radiative properties of the coatings are given in Table 2. The front hemisphere of the shield models (the side irradiated by the solar source) was coated with alodine. This coating provides a low ratio of solar absorptance to emittance with a negligible mass per unit surface area. Portions of the rear hemisphere were coated with a 2-mil-thick thermo-setting polyester resin filled with carbon black. Internal and external bands of this black coating (Fig. 8) were used on one of the shield models. Results of previous investigations[1,6] indicated that high-emittance bands of this type will increase the thermal effectiveness of the solar shield in terms of reducing the energy incident on the liquid-hydrogen tank. This effect was investigated experimentally by comparing data obtained with the model with the high-emittance coating with data obtained from the model with bare aluminum in this particular area instead of the black coating. Both of the shield models were coated on one-half the inside hemispherical surface toward the tank model with the same black coating to simulate the high emittance of rigidized foam reinforcing material on the inside surface of an operational shield. Each shield model also had an aluminized Mylar membrane stretched across the center (Fig. 8).

The same vehicle tank model was used in the tests of the shield model with and without a high-emittance band. The tank model was coated over its entire external surface with alodine, and the entire inside surface was bare aluminum.

Test Facility

The test facility (Fig. 9) consisted of a horizontal cylinder equipped with a black, 11-ft-diam, 21-ft-long liquid-nitrogen cold wall. The temperature of the cold wall during each test averaged -304°F. Chamber pressure was approximately 10^{-6} torr during all tests.

Table 2 Thermal model radiative properties[a]

Model	Material	Measurement location	Total emittance (75°F)	Solar absorptance (75°F)
Shield Model With Black Band	Alodine	Front Outside Surface	0.51	0.40
	Alodine	Front Outside Surface	0.62	0.40
	Black Coating	Outside Surface	0.87	...
	Black Coating	Inside Surface	0.93	...
	Aluminum	Internal Membrane	0.05	...
Shield Model Without Black Band	Alodine	Front Outside Surface	0.58	0.38
	Aluminum	Back Outside Surface	0.05	...
	Black Coating	Back Inside Surface	0.84	...

[a]Based on measurements of model surfaces taken after the tests with a Lion Research Corporation Model 25 reflectometer/emissometer.

Two Genarco Model ME-6 carbon-arc lamps were used as a simulated solar source. These lamps, positioned as shown in Fig. 9, had their beams projected through two closely-spaced quartz windows. Water-cooled "masks" were installed at the focal plane of each lamp to limit the solar pattern to a diameter slightly larger than the 30-in-diam models. This prevented any temperature errors from being induced on the back surface of the shield model because of reflected solar energy from the back chamber cold wall. The lamp optics were adjusted to produce a focal plane at the center of the shield model. With this arrangement, the collimation angle was 3° and the heat flux variation 9%/foot fore and aft of the focal plane.

Since it was necessary to adjust the diameter of the solar radiation pattern to as close to the diameter of the sphere as practical, no incident solar flux measurements were taken during the tests. Prior to the test, however, the two lamps used were adjusted such that each provided a 0.5 solar constant of incident radiation on a test plane perpendicular to the lamp and at the center of the shield model.

Model temperature distributions were measured with Rosemount Model 146BB platinum resistance sensors. These sensors, attached to the model as indicated in Specification 5 of Fig. 8, were connected in series with a precision 200-kΩ resistor excited by a 40-VDC source. The voltage drop across the sensors was measured with a DYMEC data acquisition system accurate to ± 0.02 mv, resulting in a temperature error to within $0.08^{\circ}F$. The temperature error due to I^2R heating was calculated to be a maximum of $1^{\circ}F$.

Analytical Predictions

Model temperature distributions were predicted by solving an electrical network representing the coupled radiation and conduction heat transfer. With analyses of this type, an energy balance is performed on isothermal zones or nodes of finite area; this results in a discontinuous temperature distribution. The accuracy of the resulting temperature distribution is a function of the number of nodes defined, with an infinitesimal area producing a continuous temperature distribution in the limiting case. The approach used here was to define a nodal size representative of an engineering-type analysis, with smaller, more closely spaced nodes used in the regions of maximum anticipated temperature gradients.

A typical energy balance performed on a model node and the part of the electrical network representing this energy balance are shown in Fig. 10. The network was constructed with temperature as the potential function, and the script-F method of Hottel was used to solve the net radiant exchange between the nodes. Heat transfer by conduction in the skin was also included in the analysis. The location of each zone relative to the focal plane was used to establish the incident flux, thus accounting for the variation in flux fore and aft of the focal plane.

An effective sink temperature was used to calculate the radiant exchange between the model and the chamber surroundings. This temperature was based on the equilibrium temperature reached between the models and the liquid-nitrogen cold-wall surroundings before the carbon-arc lamps were turned on.

A reference or scaling temperature for use in normalizing the shield surface temperature was determined by calculating the temperature to which the shield would come if it had infinite conductance, no high-emittance bands, and uniform emittance of alodine over the entire surface area of the shield at a given surface temperature. Equating the heat from the solar source that is absorbed by the shield model with the net heat lost to the chamber cold-walls (at the effective sink temperature) by radiation,

$$\alpha_s Q_s \, (\pi r^2) = \sigma \epsilon (4\pi r^2)(T_m^{\,4} - T_s^{\,4})$$

which yields a reference temperature of

$$T_\infty = T_{m_{ref}} = (\alpha_s Q_s / 4\sigma\epsilon + T_s^{\,4})^{1/4}$$

This reference temperature agrees essentially with the reference temperature chosen by McConnell and Campbell,[6] except that the reference used in the present work accounts for the effects of a finite sink (cold-wall) temperature whereas the McConnell and Campbell calculations are based on a sink (cold-wall) temperature of zero.

The radiative properties used in the analysis were based on measurements of the model surfaces taken after the tests with a Lion Research Corporation Model 25 reflectometer/emissometer. These results were shown earlier in Table 2. The measurements on the alodine coating indicate a significant variation in the emittance. The values used in the analyses were 0.51 for the model with a black band and 0.58 for the model without a black band. The solar absorptance of the alodine surfaces on the models was also noted to be slightly higher than that found in the thermal-control-coating tests (0.38-0.40 in Table 2 compared with 0.33 in Table 1). This was felt to be a result of handling and instrumenting the models prior to the test. No attempt was made to account for the effects of micrometeoroids on the coatings or overall thermal performance of the shield, although these effects were included in the structural analysis and testing.[3]

Model Thermal Test Results

The results of the thermal model tests revealed that a high-emittance band (both internal and external surfaces) over the area from 90° to 120° from the subsolar point is very effective in reducing backside shield temperatures. Figure 11 shows the predicted and experimental dimensionless temperature distributions over spherical shield models with and without black bands.

Examination of these results shows a reduction in aft-side shield temperature of over 20% for the black-banded model over the model without a black band. The existence of isotherms concentric about an axis passing through the subsolar point was verified by the close agreement between the temperature measurements in the vertical and horizontal planes. Because of the large temperature gradient in the model between $\theta = 50^o$ and 90° ($50^oF/in.$), it is difficult to make a quantitative comparison between predicted and measured results. Extremely accurate location of the temperature sensors and provision for more isothermal nodes in the analysis would be required before such a comparison could be made over this area of the shield model. This explanation, however, does not apply to the differences found in the temperature of the backside of the model where the temperature is relatively constant. In this area of the shield, radiation is the primary mode of heat transfer and the temperature gradients are small, resulting in a greater degree of confidence in the data obtained in this region.

A final point worthy of note is the slight increase in backside shield temperature in the region aft of the high-emittance band. This increase was noted in the predicted temperature distribution for a typical shield with zero skin conductance (Fig. 2). The present experimental results (Fig. 11) confirm that a temperature increase does occur between the black-banded and aft areas of the shield. This is true because of the unequal values of emittance of interior and exterior shield surfaces in the region aft of the high-emittance band.

Solar misalignment of the shield and vehicle models had no adverse effect on the net radiation exchange as long as direct solar energy was not incident on the vehicle model. In fact a slight reduction of vehicle model temperatures was observed with increasing misalignment angles up to 8°.

Analysis of the results with and without the vehicle model in place indicates that the shield model temperature distribution is unaffected by the presence of the vehicle model. This experimental result agrees with the analytical conclusion reached by McConnell and Campbell.[6] Future analyses using equivalent-scale shield diameter and vehicle separation distances may therefore be simplified by neglecting the effects of the vehicle.

Shield Operational Characteristics

Typical mission and vehicle requirements necessitate a solar shield that can be folded and stowed in a small volume for transport into space, subsequently deployed and inflated, and then rigidized in the space environment to form a structurally stable piece of space hardware after loss of pressurant gases. Because of the 0.00075-in. thickness of the GT-15 (aluminum/Mylar/aluminum tri-laminate) skin of the shield, some areas require reinforcement in order to withstand inertial loads imposed by vehicle attitude-control systems. The rigidizable material developed to provide this reinforcement is a flexible open-cell polyurethane foam that is bonded to the GT-15 and then impregnated with a monomeric compound. Rigidization of the foam may be accomplished by use of a polymerization accelerator, N,N-Diethylaniline (DEA). In order to utilize DEA as a polymerization accelerator, some thermodynamic properties (vapor pressure at low temperatures, specific heat of liquid DEA, and heat of vaporization) were required that are not available in the open literature. Results of these property determinations are presented in Ref. 7.

Since a solar shield is particularly well suited to protect nuclear cryogenic vehicles, the effects of nuclear radiation on the properties of the rigidizable material were investigated. Tests included both pure gamma and reactor-produced radiation. Results[3] indicate that the material will remain sufficiently flexible and that it can be successfully folded, stored, and then deployed after exposure in a vacuum at 30°F (required material storage temperature) to an accumulated gamma dose of up to about 4×10^6 ergs/g(C). This is well above the anticipated dose from a NERVA engine on a manned Mars flight.

Operational performance tests of large-scale (15-ft-diam) spherical shields were performed under simulated space conditions (10^{-5} torr, -320°F). These tests involved all of the operational phases required for a full-size shield — fabrication, storage, deployment, inflation, rigidization, venting, and loading. Results[3] show that the concept of a lightweight, inflatable solar shield that can be folded and packaged in a compact form for storage and transport into space, subsequently deployed and inflated to a large volume, and then rigidized in the space environment to form a structurally stable piece of hardware is feasible and practical.

Conclusions

The main conclusion derived from this research is that an inflatable solar shield can be utilized to provide cryogenically fueled space vehicles with effective protection from direct solar radiation. Shield temperature distributions obtained from scale-model thermal tests indicate that the spherical solar shield is capable of significantly reducing cryogen boiloff by maintaining low (-175 to -200°F) backside shield temperatures and by reducing penetration heat leaks to the propellant tanks. Thermal control coatings have been found which undergo very little degradation when exposed to extended solar wind and ultraviolet radiation, and which provide the optical properties required for proper control of shield temperatures. A lightweight, flexible material has been developed that can be chemically rigidized on command and that can be subjected to a gamma dose of up to 4×10^6 ergs/g(C) at the required material storage temperature of 30°F without undergoing premature polymerization. This means that a thermal control system utilizing an inflatable, rigidizable shield of the type described here is feasible for use on nuclear as well as chemically fueled vehicles.

References

[1]Jones, L. R., and Barry, D. G., "Lightweight Inflatable Shadow Shields for Cryogenic Space Vehicles," Journal of Spacecraft and Rockets, Vol. 3, No. 5, May 1966, pp. 722-727.

[2]Marshall, J. E., and Jones, L. R., "Inflatable Solar Shields for Cryogenic Space Vehicles," Proceedings of the 18th International Astronautical Congress, Vol. 2, Pergamon Press, London, 1968, pp. 229-236.

[3]Doughty, R. O., Jones, L. R., and Clifton, J. V., "Development and Testing of Expandable Rigidizable Solar Shields for Protection of Cryogenic Propellants in Space," AIAA/ASME 11th Structures, Structural Dynamics, and Materials Conference, AIAA, New York, 1970.

[4]Barry, D. G., and Hunter, T. S., "Parametric Study of Optimized Liquid Hydrogen Thermal Protection Systems for Nuclear Interplanetary Spacecraft," FZA-434 (NAS8-21080), 31 Aug. 1968, General Dynamics Corp., Fort Worth, Texas.

[5]Doughty, R. O., et al., "An Analytical and Experimental Development Program for Inflatable Solar Shields," FZA-441 (NAS8-21132), 22 May 1969, General Dynamics Corp., Fort Worth, Texas.

[6]McConnell, D. G., and Campbell, J. P., "Thermal Design of Inflatable Spherical Shadow Shields," TM X-52364, 1967, NASA.

[7]Clifton, J. V., et al., "Polymerization Characteristics of Tetraethylene Glycol Dimethacrylate," ERR-FW-828, 31 Dec. 1968, General Dynamics Corp., Fort Worth, Texas.

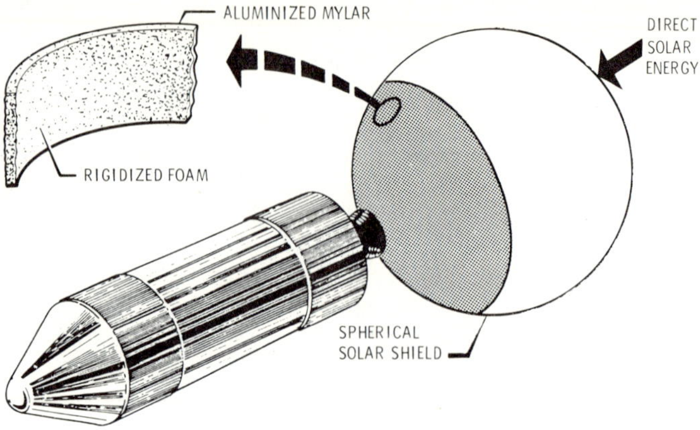

Fig. 1 Inflatable solar shield concept.

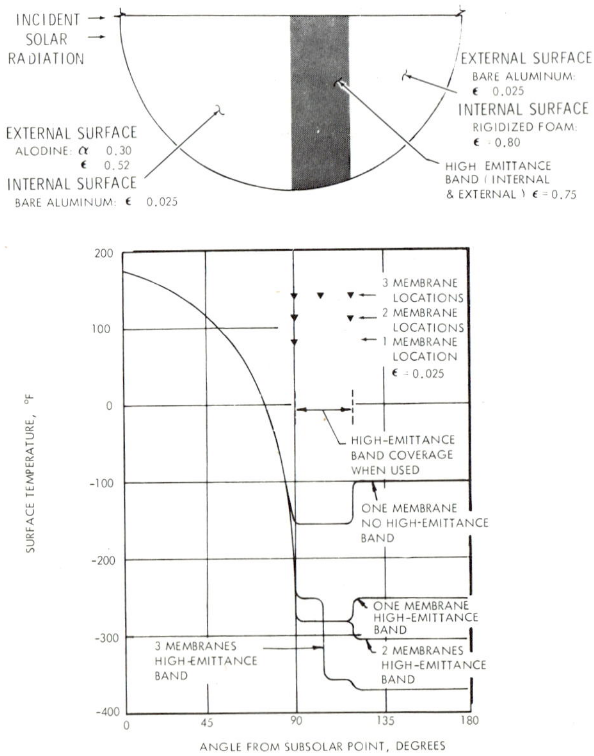

Fig. 2 Predicted spherical-shield temperature distributions for various internal-membrane/emittance-band combinations.

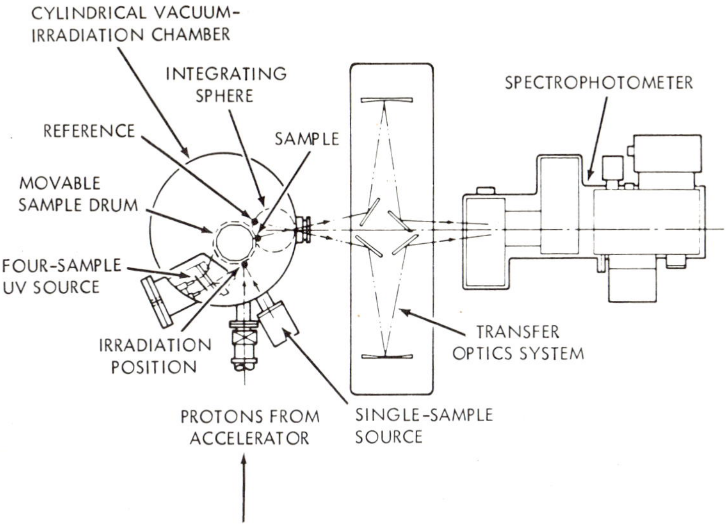

Fig. 3 Optical configuration for measurement of spectral reflectance.

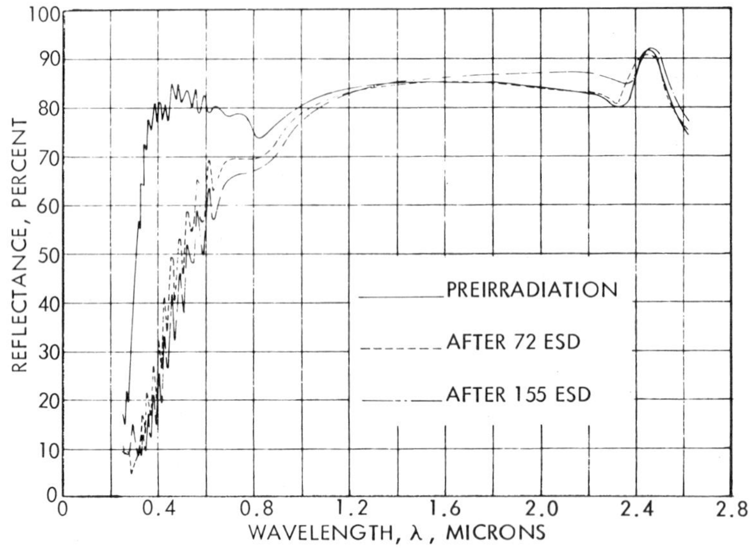

a) Ultraviolet irradiation only.

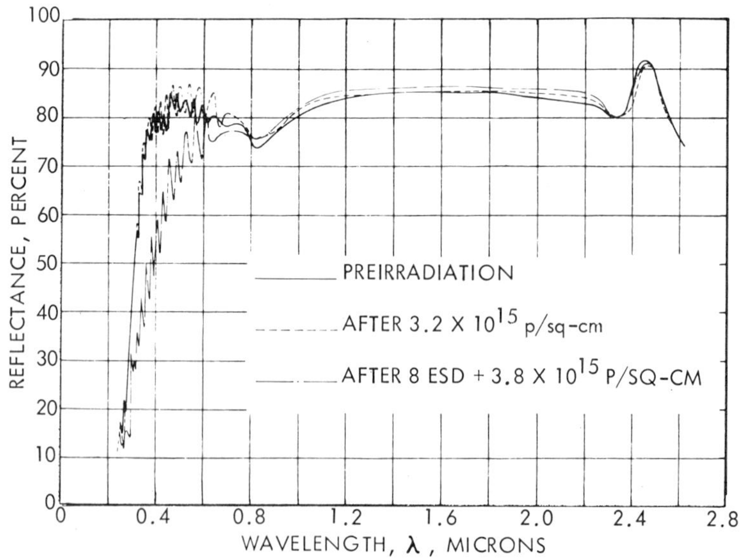

b) Combined ultraviolet and proton irradiation.

Fig. 4 Reflectance of silicon oxide on aluminum after ultraviolet and proton irradiation.

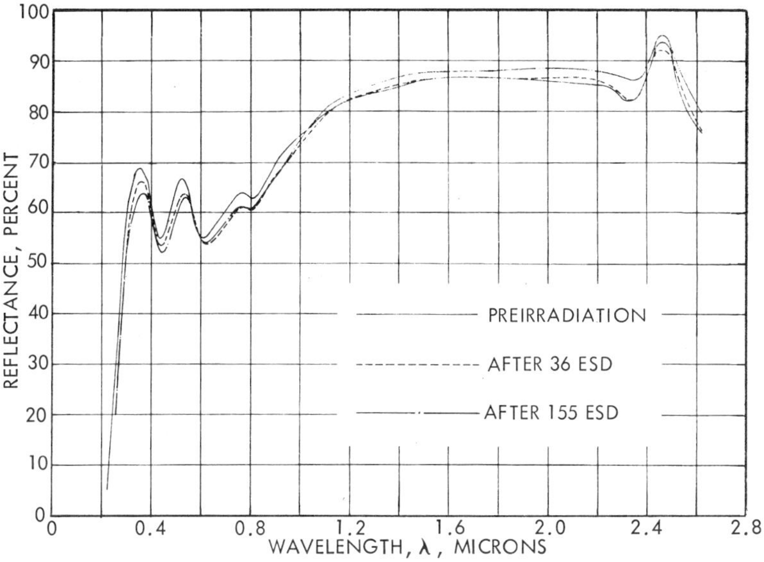

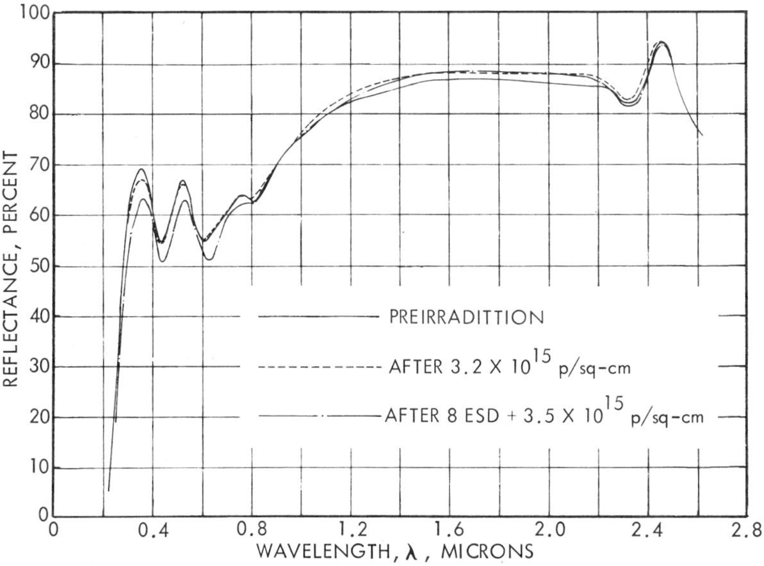

b) Combined ultraviolet and proton irradiation.

Fig. 5 Reflectance of alodized aluminum (alodine) after ultraviolet and proton irradiation.

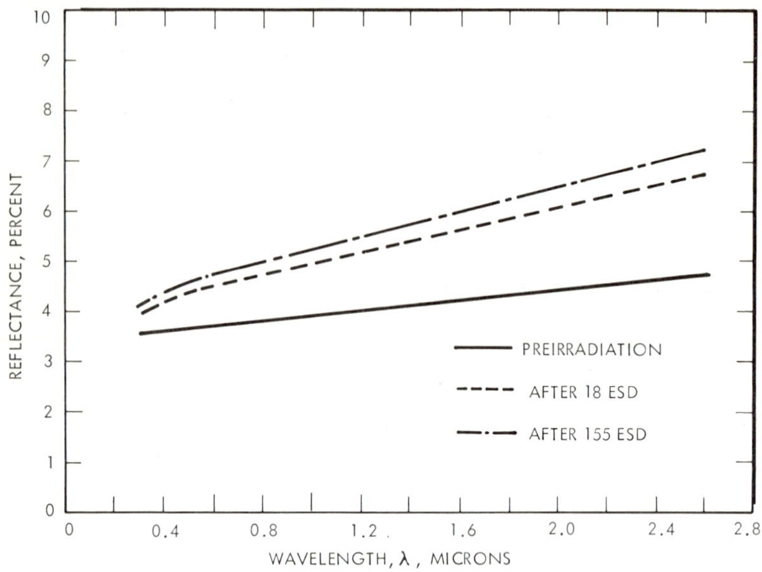

a) Ultraviolet irradiation only.

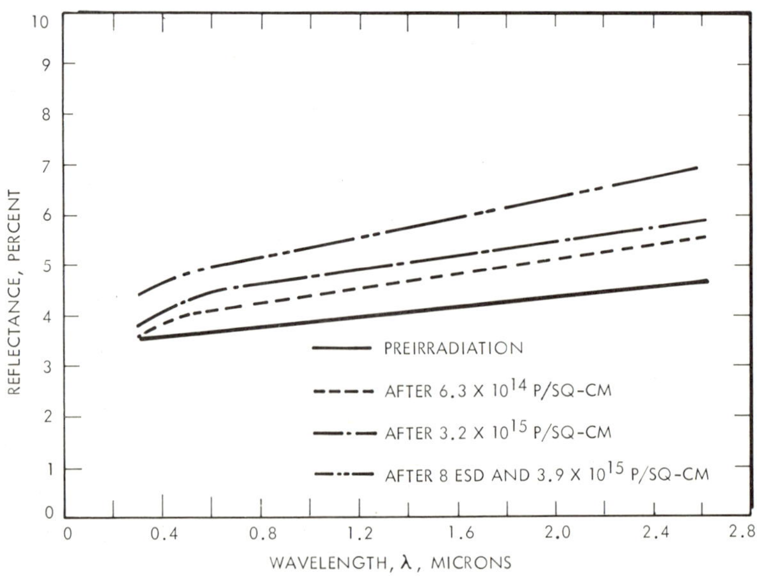

b) Combined ultraviolet and proton irradiation.

Fig. 6 Reflectance of carbon-doped polyester on aluminum after ultraviolet and proton irradiation.

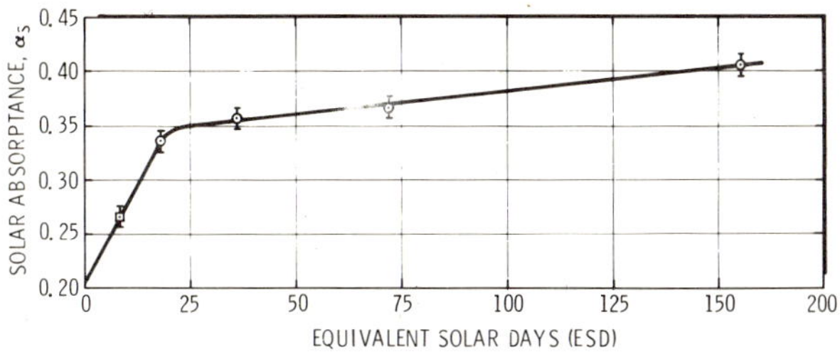

Fig. 7 Increase in solar absorptance of silicon oxide on aluminum with increasing ultraviolet dosage.

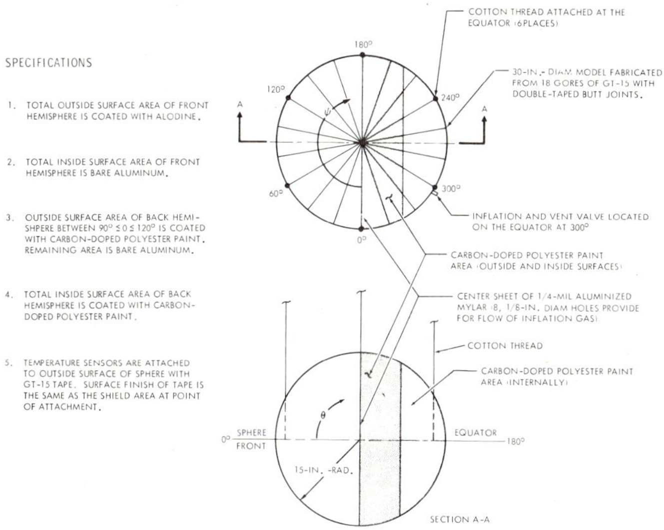

Fig. 8 Shield model with black band.

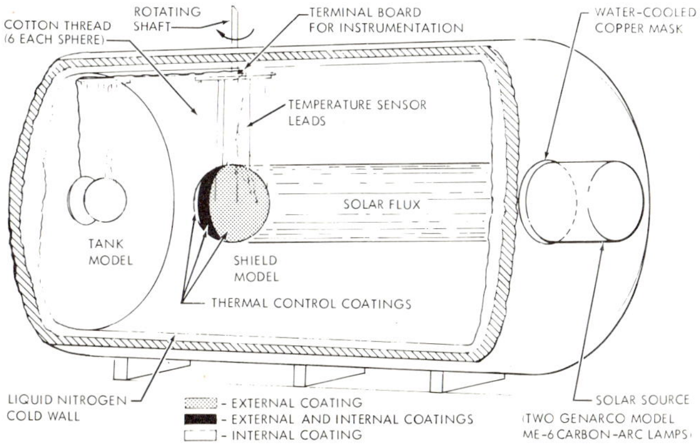

Fig. 9 Solar shield thermal test facility.

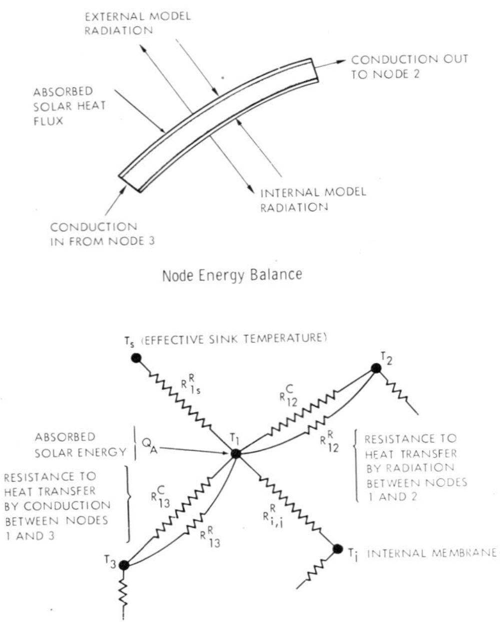

Fig. 10 Network representation of nodal energy balance.

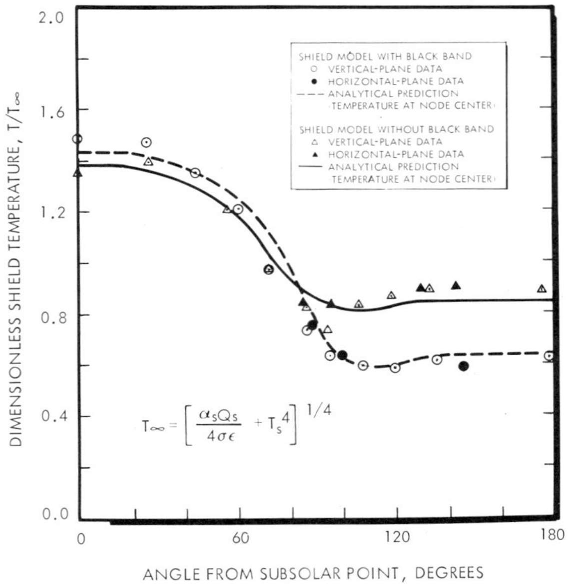

Fig. 11 Temperature distribution of spherical shield model with and without black high-emittance band.

TESTING OF AN ALUMINUM RADIATOR WITH LIQUID METAL COOLANT

Robert D. Cockfield* and Robert E. Killen†

General Electric Company, Philadelphia, Pa.

Abstract

An aluminum space radiator panel utilizing a liquid metal coolant was designed, fabricated, and tested in the 300° to 700°F temperature range. A minimum weight design with load bearing capability was achieved by use of a computer code which accounts for the thermal, hydraulic, and meteoroid protection requirements of an entire radiator system. Radiator flow tube sections were fabricated by a special casting process which allowed a metallurgical bond to be achieved between the aluminum armored section and a stainless steel tube liner. The conduction fins, which were coated with iron titanate, were made of 6061-0 aluminum sheet. Subsequent testing of the panel in air and in vacuum demonstrated the feasibility of an aluminum radiator utilizing liquid metal coolants and the ability of current analytical techniques to predict the performance of the radiator.

Nomenclature

Ap = radiating surface area of panel, ft^2

Cp = specific heat of NaK coolant, w-hr/lb-°F

n = number of measurements

Ti = temperature of a node along the fin, °R

T_{in} = coolant temperature at inlet header, °F

Presented as paper 70-855 at the AIAA 5th Thermophysics Conference, Los Angeles, California, June 29-July 1, 1970.

*Manager, Ocean Structures Engineering, Ocean Systems.

†Heat Transfer Engineer, Isotope Power Systems Operation.

T_{out} = coolant temperature at outlet header, ^{o}F

T_{root} = fin root temperature, ^{o}R

T_{sink} = sink temperature, ^{o}R

W = total coolant flow rate, lb/hr

P_{calc} = calculated radiator pressure drop, lb/in^2

P_{exp} = experimental radiator pressure drop, lb/in^2

Introduction

This paper describes the testing of an aluminum space radiator using liquid metal coolant at temperatures up to 700°F. Such a radiator would be suitable for the primary heat rejection system of a SNAP-8 Rankine cycle power system, or for power systems of advanced technology utilizing a Brayton cycle. For operating temperatures below 700°F, aluminum is the best choice for radiator construction because of its low density, high thermal conductivity, and ease of fabrication. Aluminum's strength and elastic modulus also make it effective as a meteoroid armor material, and suitable for use as primary structure during launch of the spacecraft. A disadvantage of aluminum, as well as the other light metals, is that it is not compatible with the liquid metal coolants that are required in the 700°F temperature range. However, by protecting the aluminum with a thin stainless steel tube liner the advantages of aluminum construction can be exploited with little additional weight penalty.

In order to demonstrate manufacturing feasibility, it was decided to fabricate a radiator panel that represented one of the redundant primary heat rejection loops of a typical SNAP-8 Rankine power system. To ensure that all of the problems of manufacturing were encompassed, a segment of a conical section, 4 ft x 6 ft, was chosen. This is consistent with the use of the radiator as a portion of the aerodynamic fairing and load bearing structure required of the spacecraft during launch. It has been shown that this approach results in a considerable weight saving over a nonload bearing radiator in a flat plate configuration.[1,2] Structural analysis of the radiator acting as a stiffened shell under the bending and axial loads encountered during a cold launch shows that the construction optimized for thermal performance requires only the addition of a few circumferential stiffening rings to enable it to serve as the primary spacecraft structure. A technique for minimizing the weight of a radiator used as primary structure is discussed in Ref. 3.

The objectives, then, in fabricating a radiator panel and performing the tests described in this paper, were to demonstrate the feasibility of a cone-cylinder construction and to verify the integrity of the critical stainless steel to aluminum bond under operating conditions. In addition, the tests were intended to verify the thermal performance of such a radiator as predicted by established computer codes and other methods of analysis.

Analysis and Design

A detailed, large scale, computer code was used to design a minimum weight radiator system. The geometric variables which were optimized included inside tube diameter, tube spacing, fin thickness and flow rate. The optimization was performed on a radiator "system weight" which included the effects of feed lines and a system weight penalty for pumping power, rather than minimization of only the panel weight. The over-all design included the effects of heat transfer, fluid flow and meteoroid protection requirements. The Kaufman-Lubarsky[4] heat transfer correlation and the familiar Fanning[5] friction factor were used to calculate the film temperature drop and pressure drop, respectively. Meteoroid protection requirements were based on the criteria and relationship provided in Ref. 1. Significant parameters of the resulting radiator design are given in Fig. 1.

The offset fin/tube configuration takes advantage of the self shielding effect of a cylindrical or conical radiator. This effect, illustrated in Fig. 2, allows a reduction in the armor thickness required for meteoroid protection on the back and sides of the tube because of the bumpering effect of the fin. In addition, this fin/tube arrangement provides an effective load bearing structure.

An important consideration in the design of the fin/tube element is the thickness of the stainless steel liner. The liner must have sufficient thickness to provide protection from the liquid metal coolant and to withstand the manufacturing processes. However, minimum weight is achieved when this liner is made as thin as possible. In addition, the thermal stresses in the interface between the stainless steel and the aluminum are minimized by a thin liner. This is because the aluminum has a higher coefficient of thermal expansion than the stainless steel and tends to put a tensile stress on the interface at elevated temperatures. If the liner is thin enough it will not provide an excessive radial restraint and despite the high elastic modulus of stainless steel, excessive tensile stresses in the interface will not occur.

Fabrication

The achievement of a bond of low thermal resistance between the aluminum armor and the stainless steel liner is the critical process in fabricating the radiator panel. Without a strong, reliable bond, the radiator will not maintain adequate thermal performance. Having fabricated the tube armor elements, all subsequent manufacturing processes must be closely controlled to avoid subjecting the bond to adverse thermal gradients or strains that might degrade its integrity. Because of the difference in coefficients of thermal expansion of aluminum and stainless steel, the bond is subjected to tensile forces that increase with temperature and are a maximum at operating temperature. However, if the bond is made by a process that induces high compressive stresses, subsequent exposure to elevated temperature tends to merely relieve these stresses.

The armored tube elements were produced by a casting process. The stainless steel tubes, 316 alloy, were given a surface treatment then placed under tension in a water-cooled mold. The aluminum was a special casting alloy of 99.5% purity whose composition was chosen to obtain a high thermal conductivity and to be compatible with welding to the fin material which was 6061-0. After casting, the armored tubes were straightened and machined to final dimensions. X-rays verified the soundness of the casting and the bond to the tube liner.

The fins were made from 6061-0 sheet, milled with a welding land for each tube element. The tubes and fins were welded using an electron beam process at a relatively high-power setting to obtain full penetration. Occasional pitting resulting from casting porosity was removed by subsequent wash passes. The fin/tube elements, which were limited in size to fit within the electron beam welding chamber, were then cut in trapezoidal segments and welded to form a panel. The weld joining the fins did not pose a threat to the tube-armor bond and so could be made by a conventional MIG process. After adding stainless steel headers and suitable framing members, construction of the radiator panel was complete.

The final fabrication process was to apply a high emissivity coating to the radiating surface. Iron titanate was applied using a plasma spray process. Because of the high temperature of the plasma spray, it was necessary to limit the area covered on each pass and to cool the radiator by flowing water through the tubes. Although the fins buckled noticeably due to the severe thermal gradients during application, no permanent distortion was evident. A tough , adherent coating with a uniform emissivity approaching 0.9 was achieved.

Test Procedure

Instrumentation and testing of the radiator panel was performed at the liquid metal test facilities of the General Electric Company, Nuclear Systems Programs, in Evendale, Ohio. After cleaning and leak testing, the panel was charged with NaK eutectic and purged by circulation for 12 hr. The radiator was then drained and refilled with fresh NaK; a picture of the panel is shown in Fig. 3.

The panel was instrumented with 170 thermocouples which were either flame sprayed or spot welded to the radiator. A calibration test was performed at 300°F by insulating the entire panel to achieve isothermal conditions. Total NaK flow rate was measured by a magnetic flow meter, located on a feed line to the panel, outside of the test chamber. In an effort to measure the flow through each tube, small magnetic flow meters were mounted on each pigtail section between the inlet header and the armored section of each flow tube. A pressure drop across the entire panel was measured with Taylor gauges utilizing NaK filled Bourdon tubes.

Preliminary tests were then conducted in air, outside the vacuum chamber, with the panel in a vertical position at temperatures up to 300°F. The panel was then placed horizontally in a 5-ft x 12-ft vacuum chamber shown in Fig. 4. The back side of the panel was insulated to minimize heat losses. Chamber wall temperature was maintained at approximately 70°F during all tests and a vacuum of 0.04 torr or less was maintained. Tests were conducted over a range of inlet temperatures from 300°F to 700°F and a range of flow rates from 11.9 to 39.0 lb/min. Total test time in the chamber was 550 hr, including a continuous run of 175 hr at design conditions during which no degradation in performance was observed.

Test Results

In general, the test data showed excellent agreement with predicted performance. The only sizable discrepancy was the overall panel pressure drop which was substantially higher than that calculated.

Heat Rejection Performance

Total panel heat rejection rate was calculated from temperature measurements at the inlet and outlet headers and from the total flow rate. The following expression yields the heat flux from the panel:

$$q = w\ C_p\ (T_{in} - T_{out})\ /\ A_p,\ \ w/ft^2$$

These results are shown in Fig. 5 where the heat rejection rate per square foot is plotted as a function of fluid ΔT. Also provided are the results from the radiator computer code. Over the entire range of inlet temperatures, 300°F to 700°F, the test observations were in close agreement with the analytical predictions.

Fin Efficiencies in Vacuo

Since the over-all thermal performance of the radiator compared closely to the analytical results, it is reasonable to assume that the fin efficiencies would also match the analytical data. Experimental fin efficiencies were estimated by calculating an effective fin radiating temperature, T_{eff}, by graphical integration using the fin profiles shown in Fig. 6. Typical examples of observed temperatures from which fin efficiencies were calculated are provided in Fig. 7. The effective temperature is defined as

$$T_{eff} = \left[\frac{1}{n} \sum_{i=1}^{i=n} T_i^4 \right]^{0.25}$$

Analytically derived fin efficiencies were calculated using an approach provided in Ref. 6 and contained within the radiator computer code. A comparison of predicted and experimentally determined fin efficiencies is shown in Table 1.

Table 1 Comparison of calculated and experimental fin efficiencies

Fin root temp. °F	Fin Efficiency	
	Calculated	Measured
610	78	80
563	80	82
471	84	84
389	87	87
265	91	91

Flow Distribution

Attempts to measure the flow rate in each tube using magnetic flow meters were unsuccessful. Observed measurements indicated variations in the flow rate from tube to tube by as much as 50 to 1. Results from the thermal analysis, and infrared photographs of the radiator during operation, however, indicated relatively little maldistribution. The incorrect readings given by the magnetic flow meters were attributed to their proximity to the header/pigtail joint which induced excessive turbulence at that location.

As shown in Fig. 8, the ΔT per tube is fairly constant for the representative cases shown except for tube five which exhibits a 21% higher ΔT than tube 1 at a total flow rate of 0.506 lb/sec, and a 15% higher ΔT at a total flow of 0.198 lb/sec. These differences can be attributed to manufacturing deviations at the pigtail to header joint.

Radiator Pressure Drop

Figure 9 shows that the experimental radiator pressure drops were substantially higher than that predicted; the expressions for each curve in terms of the radiator flow rate are $\Delta P_{exp} = 67.6\ w^2$ and $\Delta P_{calc} = 28.6\ w^2$. The reason for the discrepancy is believed to be in the geometry at both the inlet and outlet header to pigtail welds. Physical observation of these welds by means of a borescope revealed that the tube openings were constricted by weld beads surrounding the tube on the inlet header openings. Although the opening diameter could not be measured accurately, the constriction was sufficiently severe to account for the increased pressure drop. Tube five appeared to have the greatest constriction which agreed with the observation of a higher temperature drop for this tube.

Conclusions

The successful operation of the radiator demonstrated the feasibility of utilizing aluminum as a radiator material with liquid metal coolants. This concept offers a lightweight, high performance heat rejection system in the temperature range below 700°F which is not possible with more conventional radiator systems.

The uniform temperature distribution observed during the test and flawless thermal performance of the radiator demonstrated the integrity of the bimetallic bond. Any contact between aluminum and the NaK coolant at the temperatures experienced would have revealed itself by causing a leak within the system.

The predictable performance of the radiator during the entire test proved the ability of the iron titanate to remain unaffected by temperature cycling and air operation. Physical examination of the coating after test completion revealed no apparent degradation.

The over-all thermal performance of the radiator was approximated closely over the temperature range of operation by analytical relationships found in the literature. Therefore, existing analytical techniques can be used to design minimum weight radiators of this type.

References

[1]Cockfield, R. D., "Comparison of Load Bearing and Nonload Bearing Radiators for Nuclear Rankine Systems," CR-72307, May, 1967, NASA.

[2]Hagen, Kenneth G., "Integration of Large Radiators With Nuclear Electric Spacecraft Systems," presented at the American Society of Mechanical Engineers Air Transport and Space Meeting, New York, April, 1964.

[3]Cockfield, R.D., "Structural Optimization of a Space Radiator," Journal of Spacecraft and Rockets, Vol. 5, No. 10, Oct., 1968, pp. 1240-1241.

[4]Kreith, F., Principles of Heat Transfer, International Textbook Company, 1958, pp. 349-351.

[5]El-Wakil, M. M., Nuclear Power Engineering, McGraw-Hill Book Company, Inc., 1962, p. 243.

[6]Lieblein, S., "Analysis of Temperature Distribution and Radiant Heat Transfer Along a Fin of Constant Thickness," TN D-196, 1959, NASA.

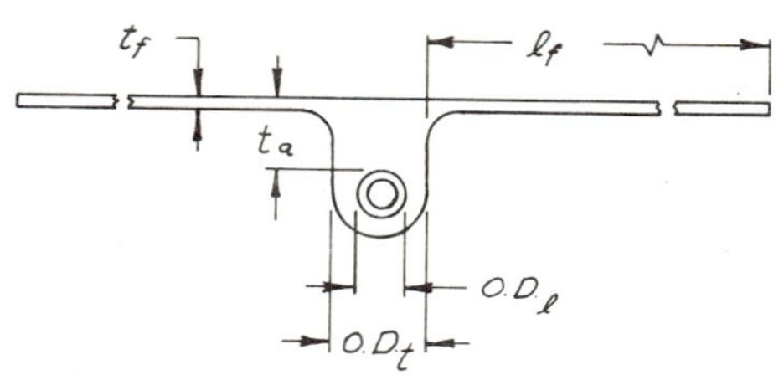

FIN THICKNESS, t_f = 0.050 in.
FIN LENGTH, ℓ_f = 3.320 in.
TUBE ARMOR, t_a = 0.450 in.
LINER OUTER DIAMETER, $O.D_\ell$ = 0.210 in.
LINER INNER DIAMETER, $I.D_\ell$ = 0.180 in.
TUBE OUTER DIAMETER, $O.D_t$ = 0.440 in.

Fig. 1 Offset tube configuration.

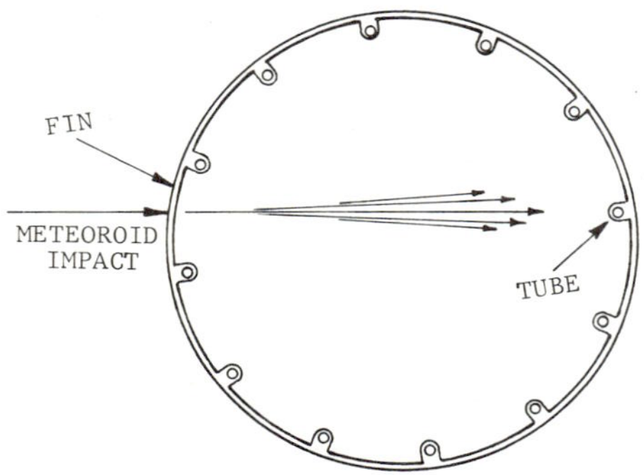

Fig. 2 Bumper effect of cylindrical radiator configuration.

Fig. 3 Radiator panel rear surface.

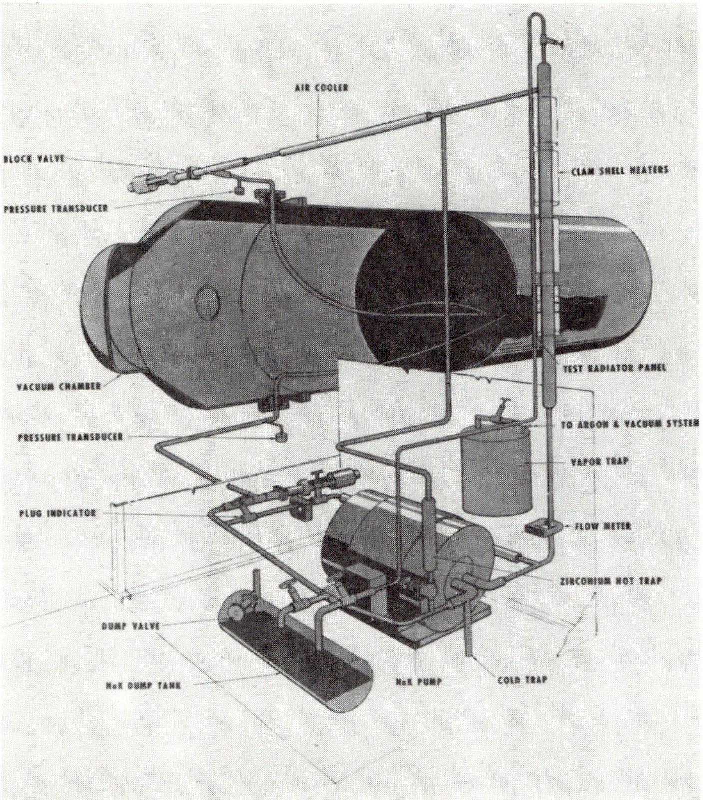

Fig. 4 Liquid metal vacuum chamber facility.

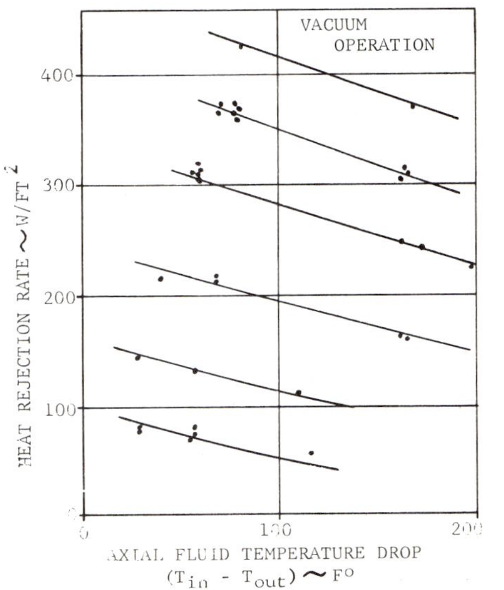

Fig. 5 Comparison of calculated and experimental heat rejection rates.

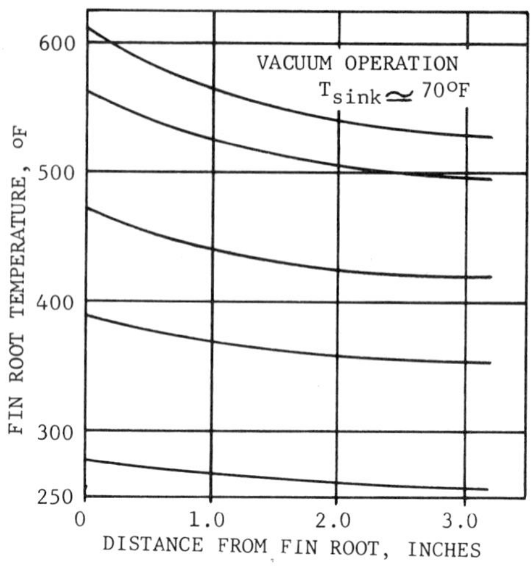

Fig. 6 Fin temperature profiles at various root temperatures.

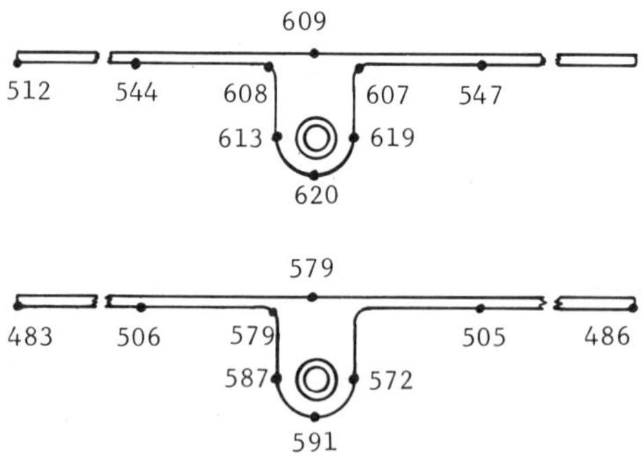

Fig. 7 Typical fin/tube temperature profiles.

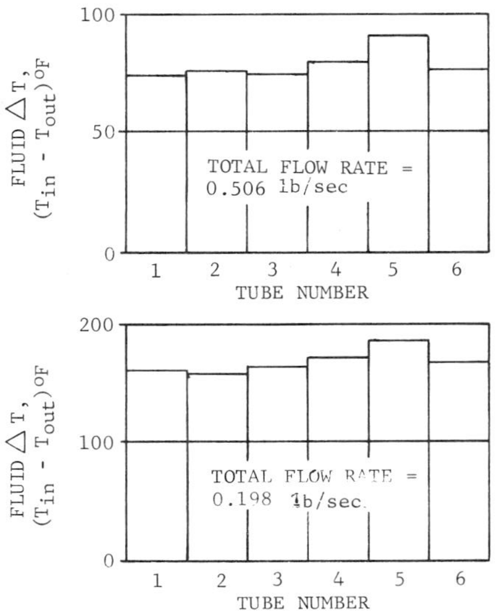

Fig. 8 Typical axial fluid temperature drop per tube at inlet temperature of 650°F.

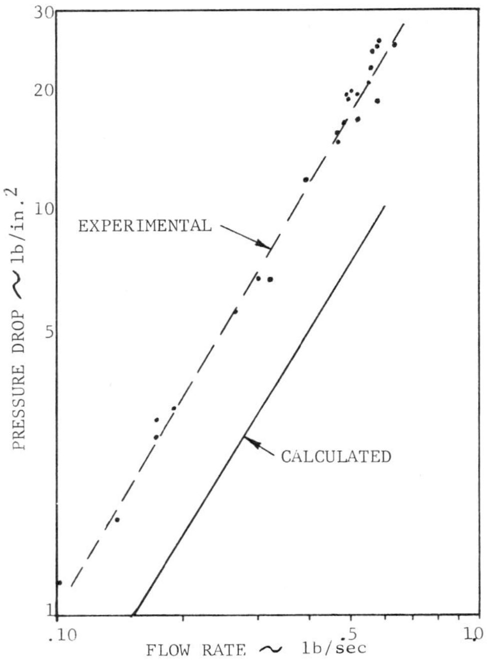

Fig. 9 Radiator pressure drop as a function of flow rate.

STAGED RADIATOR DESIGN FOR LOW TEMPERATURE SPACECRAFT APPLICATIONS

R. H. Hulett* and C. A. Zierman†

Philco-Ford Corporation, Electronics Group, Palo Alto, Calif.

Abstract

The development of passive radiative coolers is a key area of technology in the utilization of infrared detector systems for spacecraft applications. The staged radiator not only acts as an insulator by rejecting the undesired heat leak from the vehicle before it is transferred to the primary radiator but also has the added feature of acting as a structural member between the primary radiator and the vehicle. This paper describes the results of a parametric design analysis and both thermal and structural testing of a four-stage passive radiator. For a heat load of 50 mw, a radiator surface area of 0.23 sq ft, a surface emittance of 0.90, and a view factor of unity to deep space, a perfectly isolated radiator attains a temperature of 150°R. A four-stage passive radiator under similar conditions and with a spacecraft interface temperature of 530°R can achieve a temperature of 172°R.

Nomenclature

A = area, ft^2

U = effective unit conductance $Btu/hr\text{-}ft^2\ {}^\circ F$

d = diameter, ft

$\mathcal{F}$ = radiant interchange factor

h = height, ft

n = number of stages

P = heat dissipation, mw

Presented as Paper 70-854 at the AIAA 5th Thermophysics Conference, Los Angeles, Calif., June 29-July 1, 1970.

*Project Engineer, Thermal Technology Section.

†Supervisor, Thermal Technology Section.

q = heat rate, Btu/hr

T = temperature, $^{\circ}R$

ϵ = infrared emittance

β = cone half angle, deg

Subscripts

B = base

0 = sink

R = radiator

S = spacecraft

Introduction

The cooling of infrared sensors and optics is of primary importance to such spacecraft systems as military reconnaissance, Earth resource, communications and weather reporting. Recent advances in infrared detection systems have made it feasible to consider passive radiation coolers to reach operating temperatures of 140° to 200°R. In addition to cooling of the detectors to cryogenic temperatures, passive radiators have been proposed to cool primary and secondary infrared optics for detector systems that operate at ultra low temperatures and require backgrounds in the cryogenic temperature regime.

The major problems in achieving cryogenic temperatures passively in spacecraft systems is twofold: 1) isolation from the external environment, and 2) thermal decoupling of the system from the spacecraft which operates at temperatures of the order of 500°R and above. This report describes the thermal design of a radiator that by means of staging isolates the system from the high temperatures of the vehicle. The staged radiator concept was brought to the attention of the authors through work related to Refs. 1 and 2.

The radiator stages are rigidly supported to the spacecraft interface with low conductivity foam spacers. Structural analysis and vibration tests were performed and demonstrated that the alignment of the radiator (or detector) with respect to a fixed spacecraft reference could be held to within ±0.001 in. Thus, for the staged radiator concept developed it is not required to cage radiator assembly to survive launch and ascent loads typical of a spacecraft booster.

Concepts and Parametric Relations

Concept

A staged radiator consists of a number of disks or stages which are separated by low conductive supports. The surface areas are arranged in order of decreasing size with the largest stage adjacent to the spacecraft interface, as shown in Fig. 1. The portion of each stage exposed to space has a high emissivity while all interior surfaces have low emissivities. Ideally, the stages are isothermal and are made from a high conductivity material. The isolation effect of the staged radiator is achieved by minimizing radiation between the stages due to the low emittance surfaces, minimizing conduction through the supports by the use of low conductive materials, and by providing a radiative coupling to space (high emittance surface) on each stage for the minimal heat leak which occurs.

The main parameters governing the staged disk radiator include: 1) the desired operating temperature, 2) the heat dissipation on the radiator, 3) radiator and base (last stage) areas, 4) the conductive coupling of the support structure between the stages, and 5) the effective sink temperature. Secondary parameters associated with the staged radiator are 1) the number of stages, 2) the spacecraft interface temperature, and 3) the emittance values.

The range of operating temperatures for IR detectors extends from 140°R to 200°R. Heat dissipation requirements are usually between 20 to 200 mw. The resulting surface areas are in the neighborhood of one square foot. In most cases the effective sink temperature and spacecraft interface temperature are near 0°R and 530°R, respectively. The thermal conductivity of suitable support materials is 0.010 Btu/hr-ft-°R. High and low emittances of 0.90 and 0.05 are quite practical.

Parametric Relations

Assuming a perfectly isolated radiator (adiabatic back surface), the relation between desired operating temperature, effective sink temperature, and power density is

$$P/A \approx \sigma(T_R^4 - T_0^4). \qquad (1)$$

Fig. 2, which illustrates this relation, shows that as the operating temperature decreases from 140°R the power dissipation capability is sharply reduced and also that a relatively high effective sink temperature can dominate the operating temperature level making the influence power dissipation a secondary factor. Knowing the required operating temperature, heat dissipation, and approximate radiation interchange factor, the minimum radiator area can be determined from Eq. (1) or Fig. 2.

The following parametric relations were determined using thermal network models which were analyzed on a timesharing computer. The computer models are similar to that used for the development model thermal analysis (refer to analyses of development models section of text).

Neglecting conduction effects of the stage supports, stage temperature distributions and radiator temperature were determined as a function of the number of stages. The base to radiator area ratio, A_B/A_R, was assumed to be 2.25 (based on a diameter ratio of 1.5) with a linear variation for the intermediate stages. From the results shown on Fig. 3, it appears that four stages are the maximum number of stages needed to achieve a minimum radiator temperature. Additional stages produce no significant reduction in radiator temperature even though the intermediate stages may be at a lower temperature than the radiator.

In dealing with the conduction between stages, through the supports it is convenient to define an effective unit conductance,

$$U = KA_{support}/\ell A_{stage}$$

where K is the thermal conductivity of the support material and ℓ is the length between stages. The dimension ℓ, and area specifications for $A_{support}$ and A_{stage} are determined by structural analyses. Then the effective unit conductance, which is used for all stages, is established based on the thermal conductivity of the support material considered. For a given effective unit conductance for all stages the conduction thermal resistance varies inversely proportionately to the surface area of each stage. The radiator temperature was determined as a function of base to radiator area ratio, A_B/A_B, for effective unit conductance (U) values of 0.01, 0.05, and 0.10 Btu/hr-ft^2. The results, which are plotted in

Fig. 4, show that for each effective unit conductance value, a minimum radiator temperature exists. Figure 5 shows the optimum A_B/A_R (ratio A_B/A_R producing the minimum radiator temperature) as a function of effective unit conductance. This indicates that for a fixed base diameter there is a tradeoff between radiator area (for heat dissipation) and the remaining area of the following stages exposed to space (for heat leak from the spacecraft). For optimization, radiative coupling to space from intermediate stages must be increased at the expense of radiator area as effective unit conductance increases.

The radiator temperature dependence on effective unit conductance for spacecraft interface temperatures of 530° and 600°R are shown in Fig. 6. The radiator temperature appears to be relatively independent of effective unit conductance for values below 0.02 Btu/hr-ft^2 and significantly dependent for values above 0.04 Btu/hr-ft^2. The results presented in Fig. 6 also show that the radiator temperature is a rather mild function of spacecraft interface temperature.

Design Requirements of Development Model

The following requirements are considered to be typical of an infrared detector cooling application: 1) power dissipation of 35 mw, 2) detector sensitivity within a temperature range of 160° to 190°R, and 3) lateral and focal plane alignment of detector to optics within ±0.002 in.

The dynamic loads considered during ascent are presented in the Structural Test Section. The spacecraft interface temperature is assumed to be 530°R, and a maximum base diameter of 8 in. is arbitrarily defined.

Design of Development Model

To meet the operating temperature and heat dissipation criteria, the radiative coupling between the cooler and space should be maximized, isolation from the vehicle should be maximized, and solar irradiation impingement on the cooler and radiation exchange with the outer spacecraft surface must be minimized if not completely avoided. To accomplish these, the radiative cooler pertaining to this report consists of two basic components: 1) a staged radiator and 2) a conical shield. The staged radiator provides isolation from the spacecraft while the conical shield prevents radiation exchange with the spacecraft (outer surface). Operation of the shield was not the primary concern of this study; therefore, optimization of the shield design from a solar irradiation standpoint was not attempted.

Staged Radiator

Assuming perfect isolation from the spacecraft, a radiant interchange factor, $\mathcal{F}$, of 0.90, and a zero sink temperature, Fig. 2 shows that 0.20 ft^2 (5.5-in.-diam. disk) is sufficient area to radiate 50 mw≠ at a temperature of 160°R. This establishes the minimum radiator area. Thus, the optimum radiator diameter is bounded by a 5.5-in. minimum and an 8.0-in. maximum (design requirement). The optimum diameter, which is a strong function of conduction effects through the supports, is determined based on an effective unit conductance value of 0.03 Btu/hr-ft^2 (specified by structural requirements). Using Fig. 5 the optimum A_B/A_R ratio for an effective unit conductance of 0.03 Btu/hr-ft^2 is found to be approximately 2.3. Since the base diameter d_B is 8 in. the resulting optimum radiator diameter is 6 in. The staged radiator consists of four disks uniformly decreasing in diameter from 8 in. to 6 in.

The staged radiator is constructed by bonding foam rings between the disks with epoxy adhesive. The foam used for this design was 2.0 lb/ft^3 density polyurethane foam with a thermal conductivity of 0.010 Btu/hr-ft °F. The foam was fabricated in a ring approximately 3 in. in diameter and 3/8 in. wide. A constant conductance between the stages was maintained by changing the space between the stages to match their corresponding surface areas. Because conduction losses are many times greater than radiation losses between the radiator and its adjacent stage, a further optimization was made. The conductance between the radiator and its adjacent stage was decreased by a factor of three by increasing the length of the support ring proportionately. Then, instead of using a disk for the radiator, which would leave a sizeable opening between the two stages from a radiation standpoint, a conical-shaped radiator was used. This opening is undesirable because it allows radiation exchange between the high emittance surface of the lower stage to the bottom side (low emittance) of the radiator. The conical shape not only eliminates the exposure problem but also increases the available radiator surface area without an increase in diameter (base of cone). Similarly, the remaining stages could be made into frustums of a cone, however, the advantage is diminished due to the smaller support length, and the remaining stages were made in the shape of a disk to keep fabrication simple. Radiation from the high emittance surface of a stage to the low emittance surface of the stage above it was prevented by using

≠Estimated heat dissipation plus 15 mw for heat leak from electrical leads.

a strip of aluminized mylar around the edge of the upper shield as shown in Fig. 7.

Treating the disks as radiating fins, it was determined by analysis that a thickness of 0.020 in. (aluminum) provides the desired isothermal characteristics for each stage. Because of the radiator's relatively low emissive power ($\sigma T^4 \approx 1.0$ Btu/hr-ft^2), the radiator is expected to be isothermal within 2°R. For the base stage the analysis indicated that the temperature difference may increase by an order of magnitude (20°R), however, this is not considered high enough to merit the extra weight for a thicker fin. All the low emittance regions between the stages are highly polished aluminum ($\epsilon = 0.05$) while the high emittance surfaces on each stage are obtained with black paint ($\epsilon = 0.85$).

The detector assembly is bonded to the radiator, and the electrical leads to the detector are routed individually through small holes in each disk. Alignment of the detector assembly is performed optically and adjustment is provided by shimming at the support ring/spacecraft interface.

Shield

The shield used is a frustum of a right cone with a height of 3 in. and a cone angle of 60°. With these dimensions the spacecraft outer surface is completely blocked. No other external radiation direct or indirect is considered in this shield design. To maximize the radiative coupling between the radiator and space the shield surface should be specular with a low infrared emittance. The highly polished aluminum surface used satisfies both of these goals.

The complete radiative cooler assembly is shown in Fig. 8. The weight of the radiative cooler assembly (less insulation) is 1.3 lb.

Analyses of Development Models

Thermal Analysis

Thermal analyses on the radiative cooler system were performed using the 8-node computer model shown in Fig. 9. Each of the four stages is represented by a node, the shield is one node, and the multilayer insulation surrounding the system is represented with four nodes. Space and the spacecraft interface are constant temperature nodes at 0° and 530°R respectively. The external surface of the multilayer insulation surrounding the shield is assumed to be coated with black paint ($\epsilon = 0.85$). The other main assumptions applied to the model were 1) the stages and shield are isothermal, and 2) the shield surface is totally specular.

The predicted temperatures for analytical model are presented in Table 1. The temperatures are based on space conditions where the radiative cooler assembly is located on a surface away from the sun. The thermal analysis indicates that the radiator temperature is within the required range of 160° to 190°R with the maximum power dissipation level anticipated.

Table 1 Predicted temperatures for analytical model based on deep space conditions

Node	Description	Temperature, °R
1	Radiator (1st stage)	172
2	2nd stage	189
3	3rd stage	205
4	4th stage (base)	228
5	Shield (base)	262
6	Shield (side)	296
7	Spacecraft	(530) [a]
8	Space	(0) [a]

[a]Parentheses indicate fixed boundary temperatures.

Structural Analysis

Preliminary design considerations showed that the primary load carrying member in the radiation cooler is the foam cylinder used to support the stages. Consequently, structural analyses were performed to verify the integrity of this member under a worst case dynamic loading condition. Maximum shear loads and maximum bending moments imposed at the base of the cylinder were calculated based on a 12g vibration environment. An amplification factor of 10 was assumed; vibration tests performed on the unit showed that this amplification factor was conservative. A minimum factor of safety of 1.25 was calculated using the loads determined as described above, based on a shear strength of 35 psi for the foam. A factor of safety of 1.65 was calculated for the worst case normal stresses. These analyses indicate that the foam material possesses adequate strength. No other structural problems are expected within the radiation cooler.

Thermal Development Tests

Thermal/vacuum tests were conducted on the thermal development model in order to demonstrate the thermal isolation of the radiator from the spacecraft environment (70°F ± 20°F) and to verify the predicted temperature performance of the proposed radiator design. To meet the test objectives the thermal development model was tested in an 18-in.-diam bell jar. The spacecraft temperature environment was simulated by a heated enclosure controlled to room temperature and above. The heat sink consisted of a circular aluminum plate which fits in the aperture of the conical shield, with coolant lines on the opposite side. The surface of the aluminum heat sink was coated with black paint (ϵ = 0.85) and controlled to a temperature of 170° ± 4°R.

The thermal development model was instrumented with copper-constantan thermocouples and a carbon resistor was attached to the radiator detector mount to simulate detector heat dissipation.

Four steady-stage tests were run: 1) P = 0 mw, 2) P = 100 mw, 3) P = 183 mw, and 4) P = 0 mw with the vehicle boundary at 585°R instead of 530°R. The basic results are listed in Table 2.

Table 2 Thermal development test conditions and results

Heat dissipation, mw		Temperature, °R		Temperature, °R	
Test		Sink	Vehicle	Measured	Predicted[a]
1	0	172	530	216	213
2	100	172	532	236	233
3	183	168	530	247	249
4	0	173	585	218	216

[a]Predicted radiator temperatures are based on test boundary conditions as input to the analytical model.

The predicted and measured radiator temperatures show excellent agreement which indicates that the analytical model can predict the radiator temperature with great certainty. In comparing results of Tests #1 and #4, the high degree of isolation between the radiator and the spacecraft is evident.

Figure 10 shows the predicted and measured temperatures for the entire radiative cooler assembly. There is good agreement between predicted and measured temperatures for all regions except the 3rd and 4th stages (9° and 16°R differences, respectively). The radiant interchange factors from the high emittance stage surfaces to space were based on perfectly specular surfaces for the shield and aluminized mylar strips between the stages. This assumption will tend to underpredict the stage temperatures for surfaces that are not perfectly specular. Since the measured stage temperatures are higher than those predicted, the corresponding temperature discrepancies are attributed to this assumption.

Structural Development Tests

A development model radiator was subjected to a series of sinusoidal and random vibration tests to 1) verify its structural integrity and 2) demonstrate that the position of the detector can be maintained within allowable tolerances (±0.002 in.) following exposure to a worst case dynamic environment. The test procedure and results are discussed below.

The radiator was attached to a test fixture using three bolts at each of the gusset locations. The position of the detector relative to a point on the test fixture was determined optically prior to start of testing. An accelerometer was mounted on the top stage of the radiator, parallel to the axis of vibration. A low level (1g) sinusoidal vibration test was then performed parallel to the thrust axis (Z-Z) and one lateral axis (X-X), to obtain resonant frequencies and amplification factors. The accelerometer was then removed to avoid unrealistic inertia loads and qualification level sinusoidal and random vibration tests were performed as per Tables 3 and 4.

The location of the detector, relative to the selected point on the test fixture, was checked following each vibration test. No structural failures or inelastic deformations occurred in the radiator due to exposure to the vibration environment, thus its structural integrity was demonstrated.

Table 3 Random Vibration Test

Axis	Frequency, Hz	Test	PSD Level, G^2/Hz	G-RMS, O.A.
	20-550		Increasing at the rate of 3 db/octave to 0.2 at 550 Hz	
(Z-Z X-X)	550-1000	4 min ea.axis	0.2	15.7
	1000-2000		Decreasing at the rate of 6 db/octave	

Total test time: 9 min

Table 4 Sinusoidal Vibration Test

Frequency, Hz	Axis	Sweep Rate	Input Level
5-22			
22-200	Thrust (Z-Z	2 octaves/ min	0.5 in. double amplitude 12.0g (0-peak)
200-2000			5.0g (0-peak)
5-20			0.5 in. double amplitude
20-200	Lateral (X-X)	2 octaves/ min	5.0g (0-peak)

Total test time: 9 min

Conclusions

The following conclusions are supported by parametric design analyses and the thermal and structural tests of the passive radiative cooler design described in this report:

1) The temperature difference of less than $4^{\circ}R$ between the actual radiator performance and predicted performance at test conditions verifies the analytical model of the radiator.

2) The verification of the math model of the radiator by the results of the development thermal tests has increased the confidence in obtaining the predicted temperature performance, namely, $190^{\circ}R$ or less with 50 mw heat dissipation in the space environment.

3) A high degree of thermal isolation between the primary radiator and the spacecraft temperature environment has been attained with the staged radiator as demonstrated in comparing the results for test conditions Number 1 and 4.

4) The structural integrity and detector alignment of the staged radiator design is validated by the structural test.

References

[1]Phase B Study Report, "10.6 Micron Laser Communication System Experiment for ATS-F and G," Preprint X-52-68-208, NASA/GSFC, May 1968.

[2]Annable, R. V., "Radiant Cooling," Applied Optics, Vol. 9, No. 1, Jan. 1970, pp. 185-193.

Bibliography

"Materials Selector Issue," Materials Engineering, Vol. 70, No. 5, Oct. 1969.

"CPR Rigid Urethane (9000 Series)," CPR-2605, The Upjohn Co., 1969.

Handbook of Thermal Design Data for Multilayer Insulation Systems, Lockheed Missiles and Space Co., Vol. II, June 1967.

Handbook of Optical Properties for Thermal Control Surfaces, Lockheed Missiles and Space Co., Vol. III, June 1967.

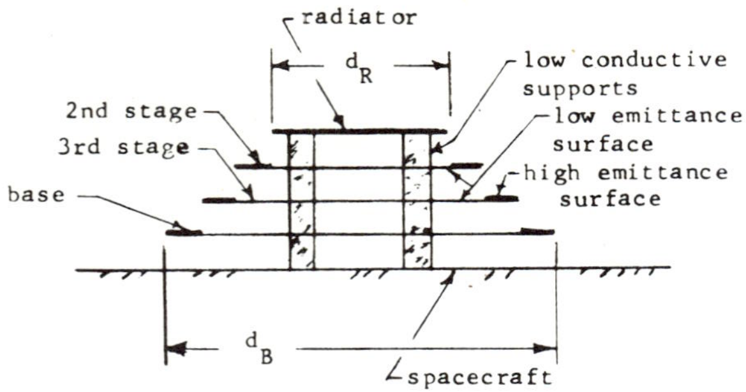

Fig. 1 Basic staged radiator.

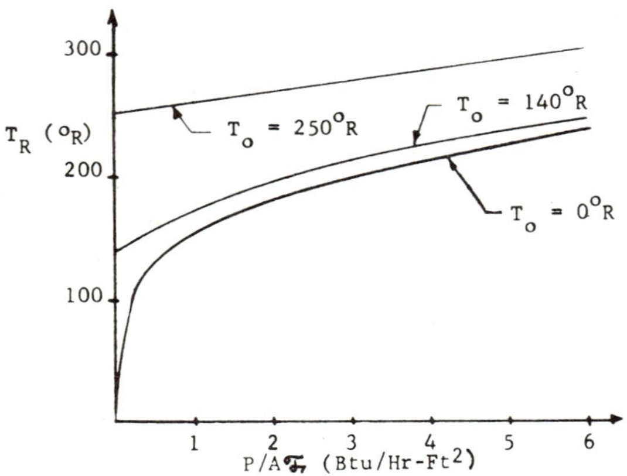

Fig. 2 Effect of power dissipation on radiator temperature.

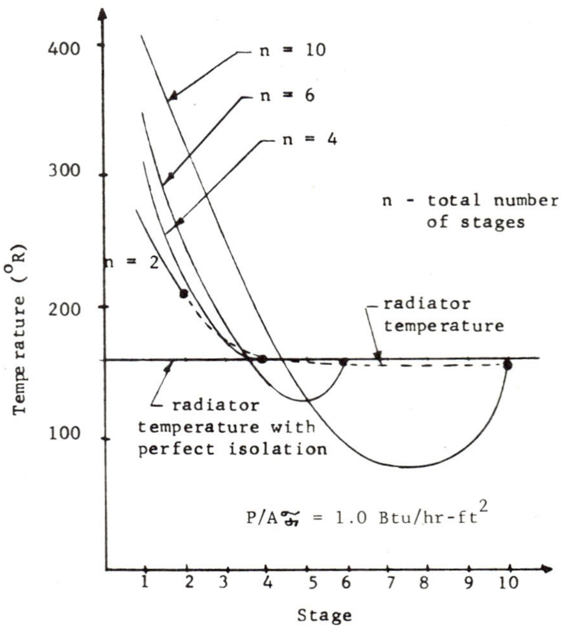

Fig. 3 Radiative cooler temperature distribution for various total numbers of stages.

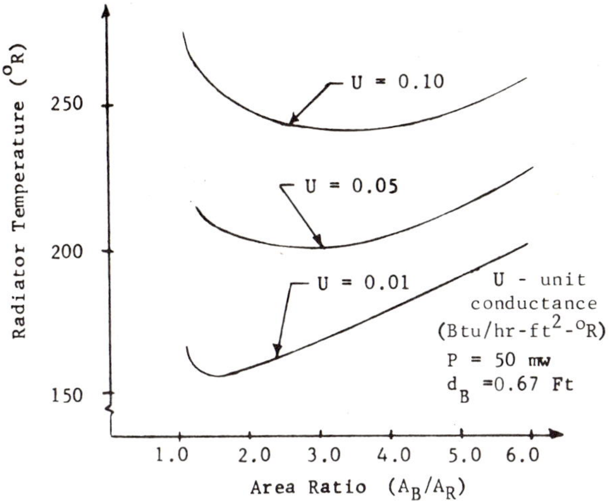

Fig. 4 Effect of area ratio and effective unit conductance on radiator temperature.

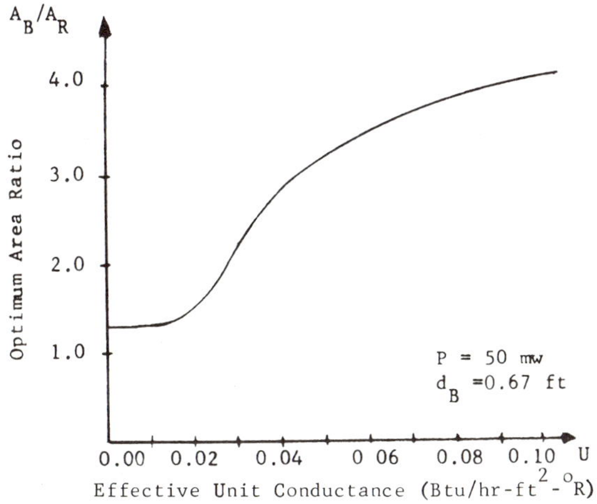

Fig. 5 Optimum diameter ratio as a function of effective unit conductance.

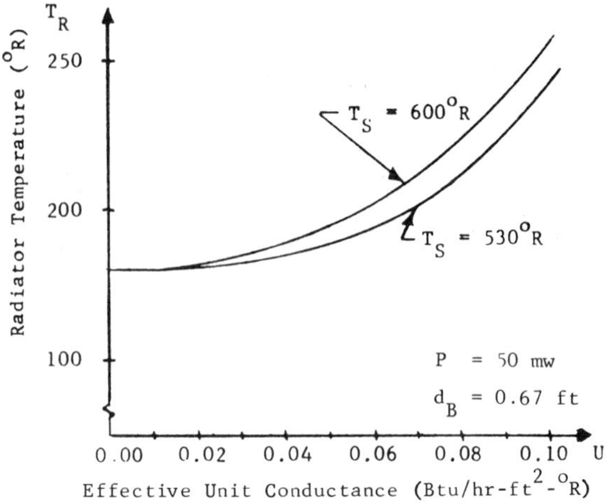

Fig. 6 Effect of unit conductance on radiator temperature.

Fig. 7 Staged radiator development model.

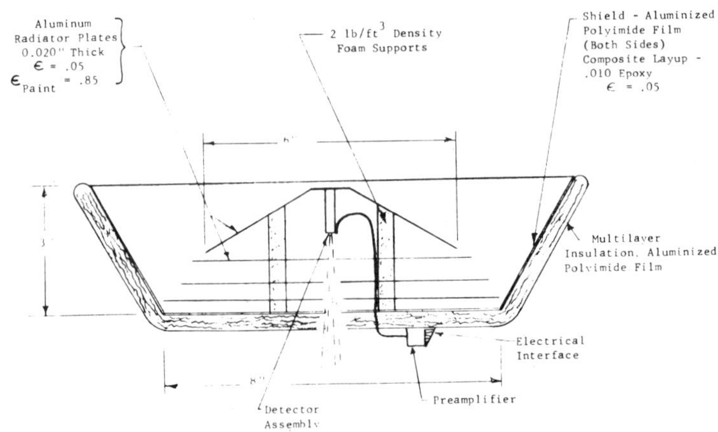

Fig. 8 Radiative cooler assembly.

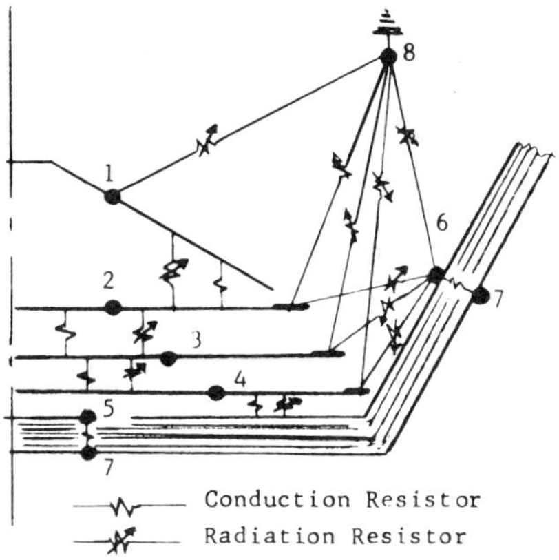

Fig. 9 Analytical model.

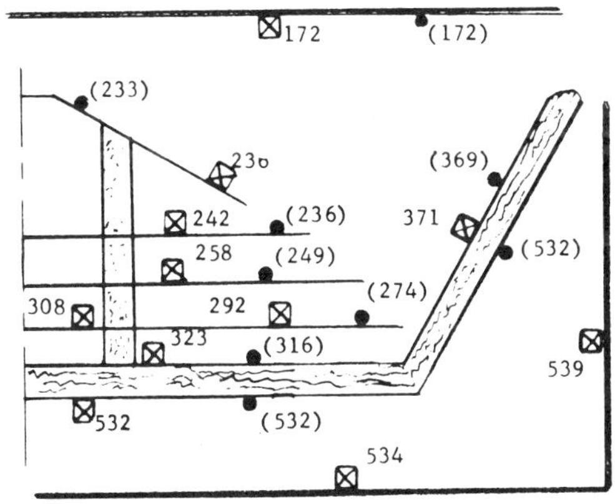

NOTES:

1. Temperatures in ^{o}R
2. Temperatures in parantheses are from Analytical Model
3. Test #2 conditions used for Analytical Model
   a. P = 100 mw
   b. $T_o = 172^{o}R$
   c. $T_S = 532^{o}R$
4. Code
   ⊠ T/C Location
   • Analytical Model Node

Fig. 10 Comparison of test and predicted temperatures.

MARTIAN SOFT LANDER INSULATION STUDY

O. J. Wilbers*

McDonnell Douglas Astronautics Company, St. Louis, Mo.

B. J. Schelden†

Jet Propulsion Laboratory, Pasadena, Calif.

and

J. C. Conti‡

McDonnell Douglas Astronautics Company, St. Louis, Mo.

Abstract

Early studies for a Martian soft lander indicated that the key uncertainty in the thermal control subsystem was the performance of the insulation after the Martian landing. Two unique environmental conditions required evaluation: heat sterilization and the effects of the Martian atmospheric gas. A test program was thus initiated using requirements for the Viking soft lander mission to Mars. Twenty-six candidate materials were screened down to two by a rapid transient heating technique for evaluating thermal conductivity and by heat sterilization compatibility tests. The two materials, isocyanurate foam and silicone bonded "AA" fiberglass, were tested further both on a sample basis and as installed in test panels identified as Insulation System Modules (ISM's). The ISM's were subjected to heat sterilization, launch venting and vibration, landing shock, and the Martian surface environment. The panel utilizing fiberglass insulation passed all tests success-

---

Presented as Paper 70-11 at the AIAA 8th Aerospace Sciences Meeting, New York, January 19-21, 1970. This paper presents the results of one phase of research carried out for the Jet Propulsion Laboratory, California Institute of Technology, under NASA contract NAS7-100.

*Group Engineer, Thermodynamics, Eastern Division.

†Cognizant Engineer.

‡Group Engineer, Materials and Processes, Eastern Division.

fully with no indicated change in performance, thus demonstrating a lightweight, low conductivity insulation configuration.

Introduction

The Viking program will send a modified Mariner spacecraft to orbit Mars and eject a soft landing capsule to land on the planet surface. The thermal control problem is complicated by two mission requirements: unknown insulation performance after exposure to heat sterilization and performance in the Martian environment. It was known that the presence of an interstitial gas would significantly degrade the performance of most insulations but not to what degree. Previous test programs had shown that heat sterilization and ethylene oxide decontamination environments had a severe effect on some insulation materials (foams and multilayers particularly), but the data was limited. In addition, very little literature data were available for any insulation materials in the Martian environment since most of the previous applications involved use in air or vacuum conditions. The Martian atmosphere, however, is different in composition than air and has a pressure of 0.005 to 0.02 atm. These differences result in thermal conductivity values intermediate between those at one atmosphere and hard vacuum, and for some materials this is the region where conductivity is highly sensitive to both pressure and gas composition. Because of this lack of data and resulting uncertainties in the insulation performance, the insulation weight estimates for a Martian lander varied by approximately 100%, causing a variance in the overall thermal control subsystem weight of approximately 50%.

A test program was initiated, based on Viking requirements.[1] The study was defined to include only the insulation component of the lander thermal control system and was limited to commercially available materials (no material development). The program was implemented as shown in Fig. 1 and outlined below.

Environmental parameters definition. All of the environmental combinations from each of the mission phases were assessed to determine the most severe test criteria. The Mariner '69 data were not yet available, so the Langley model atmospheres[2] were used. Most of the environmental parameters used for design and test are identified in this paper and are discussed in more detail in Ref. 1.

Candidate materials selection. An industry and literature survey was conducted to select approximately 25 materials for initial screening tests.

Heat sterilization screening test. A very short series of tests was conducted to determine which materials probably would not pass the long term heat sterilization test. Tests were conducted to determine weight loss, chemical reactions or phase change, and gas evolution.

Thermal diffusivity screening test. Fifty samples were evaluated simultaneously, both in air and in the Martian atmosphere. The test involved applying a heat pulse to the center of the insulation samples and deriving the thermal conductivities from analysis of the resulting temperature rise per unit heat input.

Insulation system design studies. The attachment/installation/fabrication techniques corresponding to the different types of insulation were studied. This allowed weight tradeoffs to be made, in addition to fabrication of the test panels.

Materials/design selection. The results of the screening tests and the design studies were integrated into the fabrication of two ISM's for further testing. One ISM, utilizing the foam insulation to carry part of the load, was considerably lighter than the other module which utilized fiberglass.

Heat sterilization test. The insulation ISM's, thermal conductivity samples and thermal diffusivity samples were sterilized in dry nitrogen at 275°F for 384 hours. All of these items underwent further testing to evaluate the effects of sterilization.

Thermal performance test. The two ISM's were to be compared for relative heat loss in vacuum and in the Martian atmosphere, before and after exposure to the mission dynamic environments.

Launch vibration test. The vibration levels for the ISM's were taken from Titan III data.

Landing shock test. The landing shock was estimated from the expected surface model, terminal velocity, and lander design.

Thermal conductivity test. A guarded hot plate was designed and the fiberglass sample was evaluated at six test conditions over the predicted range of Martian atmospheric pressure, temperature, and gas compositions.

Candidate Material Selection

Material selections for the initial screening tests were made by comparing available data with selection criteria developed for the Mars mission. Data were sought through a literature and vendor survey. The survey also helped identify those material parameters which would be most significant for parametric study. Twenty-six material configurations, consisting of four classes, fibers, foams, powders and multilayers were selected for initial evaluation. The primary criteria used to select the candidate materials are itemized in Table 1.

Table 1 Materials selection criteria

1) Thermal conductivity
   - Temperature
   - Pressure
   - Gas composition
2) Density
3) Conductivity - density product
4) Temperature resistance
   - Heat sterilization compatibility
5) Specific heat
6) Vibration and shock resistance
7) Long term vacuum compatibility
8) Handling characteristics
   - Penetration effects
   - Packaging and attachment techniques
9) Availability
10) Cost

Table 2 lists the candidate materials selected for the overall program. The rationale employed for the selection of these materials was as follows:

Samples 1-4 were unbonded fiberglass batt materials. The variable for these samples was fiber diameter. The objective was to learn how gaseous conduction and radiation is affected by varying the fiber diameter. Similarly, it was hoped to determine whether the change in fiber diameter could affect the apparent opacity. Fiberglass ("E" glass) was the fiber chemistry selected since previous data from numerous sources had already indicated that for the temperature region of interest, fiberglass had low conductivity and cost and was readily available.

Table 2 Candidate insulation material selection and thermal diffusivity test results before heat sterilization

Sample No.	Material Designation	Test Density pcf	Ambient 760 Torr (AIR) k Btu/Hr.Ft°F	Vacuum 1.5 10-6 Torr k Btu/Hr.Ft°F	Mars 20 mb k Btu/Hr.Ft°F	Evaluation Parameter
	Fiber:					
1A	JM unbonded	1.20	.023	.005	.018	Fiber diameter
1B	"A"	1.20	.023	.006	.019	
2A	JM unbonded	1.20	.022	.005	.018	
2B	"AA"	1.20	.020	.005	.018	
3A	JM unbonded	1.20	.021	.005	.017	
3B	"AAA"	1.20	.025	.005	.018	
4A	JM unbonded	1.20	.022	.005	.018	Density
4B	"AAAA"	1.20	.020	.005	.018	
5A	JM unbonded	0.70	.024	.014	.022	
5B	"AAAA"	0.70	.023	.012	.020	
6A	JM unbonded	2.20	.020	.005	.017	
6B	"AAAA"	2.20	.020	.005	.017	
7A	JM unbonded	1.70	.020	.005	.017	
7B	"AAAA"	1.70	.017	.005	.016	
8A	JM silicone	1.10	.021	.005	.018	Variations in binder chemistry from vendor
8B	bonded "AA"	1.05	.020	.005	.018	
9A	Hitco silicone	1.35	.019	.005	.017	
9B	bonded "AA"	1.34	.020	.005	.018	
10	JM phenolic bonded "AA"	1.12	.020	.004	.018	Vendor and binder variations
11	Hitco phenolic bonded "AA"	1.42	.020	.005	.018	
	Foams:					
12A	Upjohn HTF-200	2.03	.017	.010	.015	Basic chemistry differences
12B		2.05	.017	.010	.014	
13A	GE PPO	2.27	.026	.010	.024	
13B		2.63	.025	.010	.022	

Table 2 (Continued)

14A	Upjohn CPR 385D	3.21	.019	.011	.018	Vendor chemistry for same generic type
14B		3.29	.019	.010	.018	
15A	Diamond sham-	2.41	.020	.011	.018	
15B	rock G302	2.28	.020	.012	.019	
16A	Stafoam AA-1802	2.13	.017	.011	.015	
16B		2.14	.018	.011	.016	
17A	Scott foam	1.90	.024	.008	.021	Effective pore size
17B	80 cell/in.	1.88	.024	.009	.021	
18A	Scott foam	1.80	.027	.010	.023	
18B	60 cell/in.	1.83	.027	.010	.024	
19A	Scott foam	1.83	.028	.011	.025	
19B	45 cell/in.	1.81	.028	.011	.025	
	Powders:					
20A	Colloidal	3.74	.022	.007	.014	Particle chemistry
20B	alumina	3.54	.023	.007	.014	
21A	Colloidal	3.15	.021	.006	.012	
21B	silica	3.14	.022	.007	.013	
	Multilayer:					
22A	Goldized Kapton	2.12	.020	----	.020	Layer density
22B	40 layers/in.	2.13	.027	----	.022	
23A	Goldized Kapton	3.18	.027	----	.018	
23B	60 layers/in.	3.19	.024	----	.023	
24A	Goldized Kapton	4.25	.035	----	.020	
24B	80 layers/in.	4.26	.039	----	.020	
25A	Goldized Kapton	1.30	.020	.005	.019	Spacer thickness
25B	with 1/2" "AA" spacers	1.30	.020	.004	.019	
26A	Goldized Kapton	1.26	.020	.005	.019	
26B	with 1" "AA" spacers	1.26	.021	.005	.020	

Samples 5-7 were also unbonded fiberglass batt with constant fiber diameter (AAAA) and variable bulk density. The "AAAA" fiber was the smallest commercially available fiber size. Literature data indicated that minimum "k" would occur within the selected density range of 0.7 to 2.2 psf.

Samples 8 and 9 were similar to sample 2, the only difference being that samples 8 and 9 had a silicone bonding agent to stabilize the "AA" fibers. Selection of a bonded fiberglass permitted determination of the ability to reduce radiation by increasing the apparent opacity. Reduced radiation, however, would be traded off by the increased solid conduction caused by the bonding of the fibers rather than just point contact.

The rationale for samples 10 and 11 was identical to that for 8 and 9 except the binder material was phenolic.

Samples 12-16 were all organic foams selected to determine the effect of basic foam chemistry (samples 12-14) and formulation variations from different sources. With the exception of the polyphenylene oxide foam, all of these foams were "Freon blown" materials. Since the conductivity of the Freon gas is lower than that of air, the effective gas conduction component should also be less.

Samples 17-19 are also organic foam materials. These samples were flexible, whereas 12-16 are all rigid. Bulk density of this series of samples was essentially constant with the only variable being effective cell size.

Samples 20 and 21 consisted of two representative loose powders. Differences in the two were basic chemistry and particle size. It was realized that opacified powders may have lower conductivity than those tested, but rather than attempt to optimize, comparison of test data for these materials with the literature would reveal whether additional work including opacifiers was warranted.

The remainder of the samples, nos. 22-26, were multilayer type, using goldized 1/2 mil Kapton. By concentrating on this material and then comparing with other literature values, an appraisal could be made of any possible warranted additional work. Multilayer materials by their nature being reflective lamina very effectively reduce radiation. Samples 22-24 were therefore selected to evaluate the effect of layer density on solid and gaseous conduction.

The last two samples, 25 and 26, consisted of a combination of fibrous material and the goldized Kapton multilayer. Since radiation is a large contributor to overall heat transfer in fiberglass, it was reasoned that incorporation of radiation foils in a fibrous material would be beneficial. Layers of goldized Kapton at two different spacings were included to determine whether the radiation component could be reduced.

Candidate Materials Screening Tests

Two types of tests were conducted concurrently to rapidly identify the best two materials from the 26 candidates initially selected for consideration. Materials which probably would not pass heat sterilization were identified in the Heat Sterilization Screening Test. Thermal conductivity was derived from the Thermal Diffusivity Screening Test.

Heat Sterilization Screening

This test was employed to provide a rapid means of verifying probable heat sterilization compatibility of materials. For those materials that failed the test, quantitative reasons for rejection were provided by the test data. The tests conducted were as follows:

1) Thermogravimetric Analysis (TGA) in which sample weight is continuously monitored during heatup to 235°C (455°F) at a preselected rate;
2) Differential Thermal Analysis (DTA) in which temperature pulses denoting reactions are monitored during heatup.

The peak temperature of 235°C was selected to identify those materials which have marginal temperature stability above the 135°C required in the actual heat sterilization. The excess temperature was also chosen to accelerate long term changes which might occur during the 384 hr. of actual heat sterilization. The material samples for these tests were fabricated into small cylinders, each weighing approximately 100 mg. The DTA detected no reactions for the materials tested.

Based on the testing completed under this phase, the materials listed in Table 3 were judged to be acceptable for further consideration. It should be emphasized that this testing was used strictly as a screening tool and did not verify that either the actual materials tested or the class of materials that they represented would pass the actual heat sterilization cycle.

Thermal Diffusivity Screening

The purpose of this test was to rapidly obtain comparative thermal conductivity data on candidate insulation materials. A transient heating technique was selected to reduce the number of candidate materials to a more manageable number yet allow confident selection of those materials which best fulfill the planetary mission objectives. The basic objective of the test was to measure the temperature rise per unit heat input of a heater in the center of each specimen. From this data thermal conductivity could be derived. Including duplicates, a total of 50 specimens from all 26 candidate materials were tested.

Table 3 Materials considered to have passed sterilization screening

Material	Vendor
Microlite, silicone bonded "AA" fiberglass fibers	Johns-Manville Sales Corp.
Isocyanurate foam HTF-200, closed cell	The Upjohn Company
Polyphenylene oxide foam, closed cell	General Electric Company
Polyurethane foam, G302, closed cell	Diamond Shamrock Chemical Co.

The test specimen size was determined such that the energy dissipated by the central heater was retained within the insulation during the heating period. Desired temperature rise of the heaters was selected to be 40 to 70°F.

Based on analysis which indicated that essentially all the heater energy would be contained within the samples, and that initial transients associated with heater mass would be damped out, a heating time of 12 min was selected for all samples. Specimen size for the fibers, foam and multilayer material was a cube measuring 6 in. along each edge. The powder materials were placed in a standard one gallon can (6.45 in. diam). Each thermal diffusivity specimen, contained a centrally located Thermofoil heater, 1 in. in diameter by 0.08 in. thick. Each heater was painted black and instrumented with a 40-gage thermocouple. A wire mesh box retained the specimens in a cube configuration for handling protection and density control. The test samples were installed in an 8-ft chamber and heated simultaneously by application of power to each specimen heater.

A total of six thermal diffusivity test runs were made as shown in Table 4.

All 50 samples were subjected to the three test conditions before heat sterilization. Nine samples of materials that had successfully passed the heat sterilization cycle were subjected to the second test.

Table 4 Thermal diffusivity test conditions

Test environment	1st test (before sterilization)	2nd test (after sterilization)
Ambient air	X	X
Vacuum (5 x 10^{-6} torr)	X	X
20 mb simulated Mars atmosphere[a]	X	X

[a]10% CO_2, 60% N_2, and 21%A by volume

Data reduction. Two techniques were selected to calculate thermal conductivity, a closed form analytical solution and a finite difference computer model. The closed form solution[3] predicts the temperature rise of a uniformly heated disk in a semi-infite solid,neglecting the thermal mass of the heater. These constraints were essentially fulfilled by selection of specimen size, heater size, power level and duration of heating. Based on the good agreement during a checkout test, Fig. 2, the closed form analytical solution was selected to reduce the thermal diffusivity test data for all isotropic materials (fibers, foams, and powders). This solution was used because it was faster and less costly than the computer model technique. The computer model was used to reduce the data for the multilayer materials because different thermal conductivity values could be assumed in two directions to simulate heat transfer both parallel and perpendicular to the film layers. These data reduction techniques use the temperature rise and heater power from the test, along with the insulation properties of density and specific heat, to determine the insulation thermal conductivity. The resulting thermal conductivity values are shown in Tables 2 and 5.

The test technique employed for rapid thermal diffusivity screening purposes yielded data that permitted material comparisons. The accuracy of the technique for the simulated Martian condition was sufficient within the scope of this program. However, the accuracy for multilayers at vacuum conditions was poor, but this was expected.

Examination of the data enabled the following conclusions to be made for the Martian environmental conditions.

1) Minimum conductivity for fiberglass batt is at about 1.7 pcf density.

2) At the mean temperature and test density tested, the effect of fiber diameter for the fiberglass materials is negligible. Literal interpretation of the test data reveals that the "AAA" material has a lower "k" than the "AAAA", a conclusion not borne out by theory.

Table 5 Thermal diffusivity test results after heat sterilization

Sample No.	Material Designation	Test Density pcf	Ambient 760 Torr (AIR)	Vacuum $<1.5\times10^{-6}$ Torr	Simulated Mars 20 mb(15 Torr)
			k	k	k
3A	JM unbonded "AA"	1.20	.021	.005	.018
4B	JM unbonded "AAAA"	1.20	.021	.006	.018
8A	JM silicone bonded"AA"	1.10	.021	.005	.018
9A	Hitco silicone bonded "AA"	1.35	.019	.005	.017
10	JM phenolic bonded "AA"	1.12	.021	.005	.018
12A	Upjohn HTF-200 (preconditioned)	2.03	.018	.011	.016
12B	Upjohn HTF-200	2.05	.017	.011	.016
13A	GE - PPO	2.27	.026	.010	.023
21A	Colloidal silica	3.15	.022	.007	.012

3) Binder content or chemistry does not influence fiberglass performance for these test conditions.
4) For the same density, urethane foams have a lower "kρ" than fiberglass batts.
5) Isocyanurate foam belongs to the same family of "kρ" values as the polyurethane foam.
6) At simulated Martian condition, the multilayer materials have the family of highest "kρ" values.
7) Powders had the lowest thermal conductivities and belong to the family of lowest "kρ" curves.
8) Use of radiation foils as separators in fiberglass batt reduces "kρ" effectiveness rather than increacing it.

Insulation System Module Design and Selection

Design studies were performed to identify means of installing the insulation on the lander. The entire surface of the lander equipment package (about 35 ft^2 was assumed) required insulation to protect it from the extremes of the Martian environment. The equipment package was assumed to be of sheet metal construction consisting of axial members and thin aluminum shear skins. The insulation would be attached to these aluminum shear skins and enclosed in a fiberglass laminate cover. These considerations plus insulation installation studies and heat transfer evaluations allowed definition of the basic ISM designs for testing. Selected dimensions were 18 x 18 in.with insulation thickness of 3 in.

ISM Insulation Selection

In order to compare the materials on an equitable basis, predicted ISM weights were determined for insulation candidates having the same heat loss characteristics. The required thickness for constant heat loss was determined using thermal conductivity data from the Thermal Diffusivity Test. From Fig. 3 the comparable ISM weights were computed and are presented in Table 6 for representative materials of each class. These data were normalized such that the lowest weight configuration has a thickness of 3 in., and other materials have the thickness required for equal heat loss. Predicted edge effects were not included in this comparison since they would not alter the materials selection, i.e., edge losses for nonfoam materials are higher than for foam because the fiberglass laminate support structure was thicker (0.040 in. vs 0.024 in. for the foam).

Based on all of the evaluation results, two materials were selected to be incorporated into the ISM test panels: 1) Upjohn Company HTF-200 foam, and 2) J-M Microlite, silicone bonded "AA" fiberglass.

The foam material was selected based on its weight as installed. A fibrous material wassselected to avoid committing the remaining portion of the study to foam materials only (the G302 foam was nearly as light as the HTF-200 material). Powder and multilayer materials were removed from consideration because of their high weight, fabrication cost, and expected uncertainties in performance as installed.

Environmental Tests

The tests discussed in this section are heat sterilization, thermal performance, launch depressurization, launch vibration, landing shock and thermal conductivity. The samples tested included the foam and fiberglass ISM's, material samples of these two insulations, and selected material samples which were considered to be backup candidates. With the exception of the Thermal Conductivity Test, the objective was to demonstrate the applicability of the two ISM materials to the Martian lander mission, and to identify which ISM had the better characteristics.

Table 6 Insulation system module weight comparison

Insulation Material	Conductivity	Density lb/ft^3	Normalized[a] Thickness Inch	Insulation System Module Weight Lb	% Increment
Foam HTF-200	0.015	2.03	3.00	2.31[c]	0
Foam G302	0.019	2.35	3.80	2.64	14
Fibrous-unbonded "AAAA":					
ρ=0.7	0.021	0.7	4.2	3.34	45
ρ=1.2	0.018	1.2	3.6	3.44	49
ρ=1.7	0.016	1.7	3.37	3.59	56
ρ=2.2	0.017	2.2	3.49	3.91	69
Fibrous-unbonded "AAA"	0.017	1.2	3.58	3.41	48
Fibrous-silicone bonded "AA":					
(JM)	0.018	1.08	3.72	3.34[c]	45
(Hitco)	0.017	1.35	3.58	3.50	52
Fibrous-phenolic(JM)	0.018	1.12	3.68	3.40	47
bonded "AA" (Hitco)	0.018	1.42	3.60	3.53	53
Powder-colloidal silica	0.012	3.15	2.55	3.78	64
Multilayer-gold on Kapton:					
40 layers/in	.021[b]	2.12	4.28	4.25	84
60 layers/in	.021[b]	3.18	4.27	4.95	114
80 layers/in	.020[b]	4.29	4.04	5.5	138

[a]Thickness for HTF-200 foam selected at 3.00 in. - all others are for equal heat loss (neglecting edge effects).
[b]Assumes gold thickness of 350 A.
[c]Selected materials for ISM's.

Heat Sterilization Test

The purpose of this test was to evaluate the effects of dry heat sterilization on the candidate thermal insulation materials. In addition to the foam and fiberglass ISM's, two foam specimens which had been preconditioned for 20 hr at 275°F were included. Preconditioning was used to provide a basis for determining how rapidly changes occur during heat sterilization, and to evaluate whether preconditioning was an effective means of limiting these changes.

The heat sterilization samples included in the test (Table 7) were: 12 Thermal Diffusivity Test Specimens; the 2 ISM's; 2 Thermal Conductivity Test Specimens and spare materials. Sealed canisters were used to isolate samples of each material to prevent cross-contamination during the test. The canisters were provided with inlet and outlet T-connectors to allow continuous nitrogen purging plus a means of installing thermocouples to measure the nitrogen purge gas temperature.

To start the test, the specimen canisters were flushed with nitrogen gas. Heat-up for all ovens was at a rate of 35°F/hr. The 384-hr heat soak period was begun when the oven reference temperature reached 275± 10°F. Nitrogen purge was continued throughout the test with continuous venting and preheating when necessary to maintain the required temperature. A 6-hr cool-down to room temperature concluded the test exposure.

Weight and appearance changes of the candidate materials after testing are noted in Table 7, together with average canister temperatures. All but two of the specimens experienced a weight loss after the 384-hr heating. One specimen, goldized Kapton material, decrinkled, causing the layers to collapse; another specimen, a closed-cell polyurethane foam showed appreciable shrinkage. The closed cell isocyanurate foam specimen showed slight warpage and discoloration, and the phenolic-bonded fiberglass specimen darkened. The preconditioned samples, HTF-200 and G-302, suffered weight loss during preconditioning, but lost no additional weight during the 384-hr exposure, indicating that weight loss in these materials occurs during the first 20 hr of exposure.

The foam ISM peeled at three of the four corners at the bond between the fiberglass laminate and the aluminum cover. The foam thermal conductivity specimens and the spares were slightly discolored. The silicone bonded "AA" fiberglass materials in the other large canister showed no visible changes. Weight loss in this ISM was attributed to the fiberglass laminate casing since the bare Thermal Diffusivity and Thermal Conductivity samples exhibited negligible weight loss.

<u>Heat Sterilization Test Conclusions</u>

The test conclusions were as follows:

1) Multilayer goldized-Kapton, G-302 closed-cell polyurethane foam, phenolic-bonded fiberglass, and HTF-200 closed-cell

isocyanurate foam materials were affected by prolonged heating at 275°F.

2) The multilayer and G-302 foam materials were sufficiently affected to preclude their further consideration as candidate materials.

Table 7 Heat sterilization test results

Material Class	Designation	Visual Change	Average Temp °F	Weight Change %
Multilayer	Gold-on-Kapton	Decrinkled Layers collapsed	275	-0.58
Powder	Colloidal silica	None	276	-0.98
Fibrous	Unbonded "AAAA" J-M	None	270	-0.77
	Unbonded "AAA" J-M	None	270	-0.67
	Phenolic bonded "AA" J-M	Darkened	293	-1.88
	Silicone bonded "AA" Hitco	None	257	-0.80
	Silicone bonded "AA" J-M	None	269	-0.95
	Silicone bonded "AA" J-M therm. cond. & spare samples	None	274	-0.02
Fibrous ISM[a]	Silicone bonded "AA" J-M	None	274	-3.74
Foam	Polyphenylene oxide (PPO)	None	261	-1.58
	Polyurethane G-302	Shrank	281	-0.85
	Polyurethane G-302, pre-conditioned	Slight shrinkage	283	0.0
	Isocyanurate HTF-200	Discolored	273	-2.52
	Isocyanurate HTF-200, pre-conditioned	Discolored	276	0.0
	Isocyanurate, HTF-200 therm cond & spare samples	Discolored	280	-3.96
Foam ISM[a]	Isocyanurate HTF-200	Peeled at bonded flange	280	-3.10

[a]Sealed in a large canister, all others in individual small canisters.

3) Further consideration of the phenolic-bonded fiberglass and HTF-200 foam, which were somewhat affected, should be based on additional post sterilization test results.

4) Preconditioning of the two foam materials indicated that weight changes in these materials occur within the first 20 hr of heating, and that preconditioning of materials before fabrication can be used to limit further changes during heat sterilization.

Thermal Performance Tests (1 and 2)

The thermal performance tests were planned to provide a comparison of the heat loss characteristics of heat sterilized ISM panels before and after exposure to the launch pressure profile, and to the launch vibration and landing shock loads. This test was performed in vacuum and in the 20 mb maximum model Mars atmosphere conditions. In the initial test, prior to vibration and shock, the two ISM panels were to be evaluated. The foam panel failed during the initial chamber pump down, and a mock ISM panel made from HTF-200 foam was substituted. The second test, following vibration and shock, utilized the same mock ISM and the original fiberglass ISM.

The test apparatus developed to evaluate the thermal performance of these panels, Fig. 4 was a guarded assembly comprised of a central heater, core heaters for each ISM panel, closed cell foam guard insulation, and cold panels. The apparatus was installed in a 5.5-ft diam vacuum chamber. The three heaters were 16 x 16 in., consisting of nichrome wire interwoven in fiberglass cloth and impregnated with silicone rubber. The cold plates were of stainless steel fabricated into a single embossed panel with parallel flow paths. Cooling was provided by a continuously controlled gaseous nitrogen supply.

During testing the heaters were maintained at an average temperature of 50 ±5°F and the cold panels were chilled to -150 ±10°F. Thermal equilibrium was established for the system during each test under conditions of (a) vacuum, and (b) simulated Mars atmosphere. The criteria for thermal equilibrium was that system temperature changes be less than 2°F per hour.

Thermal performance test 1. The two ISM test panels were placed in the test apparatus with the fiberglass panel in the upper test zone and the foam panel in the lower test zone. During pretest checks, the vacuum chamber was evacuated at the normal pumpdown rate shown in Fig. 5. This rate is less than that required to simulate the launch pressure profile. When observation through the chamber viewport revealed a change in alignment of the test apparatus components, the chamber was back-filled and opened. The sealed foam panel was found to have failed by forcefully breaking open along the cover plate edge bond area. The foam within broke into many small fragments which were dispersed throughout the chamber volume.

Testing was continued on the fiberglass ISM, with one of the mock panels substituted for the foam ISM.

The chamber was evacuated to the rate shown in Fig. 5. Equilibrium temperatures were then attained for both vacuum and simulated Mars atmosphere (20 mb, 19% CO_2, 60% N_2, 21% A) conditions. These temperatures and the associated data are given in Figs. 6 and 7. At the completion of this test, the ISM panel was removed from the apparatus and subjected to vibration and shock testing.

Thermal performance test 2. The ISM panel and heater temperature profiles were better defined for this second test by installing two additional thermocouples on each surface. Data from this test are compared with test 1 in Figs. 6 and 7.

Thermal performance test conclusions. The silicone bonded fiberglass ISM successfully survived the Mars mission requirements to which it was exposed with no change in thermal performance of loss of structural integrity. The foam panel failed from unknown causes during a normal chamber pumpdown.

Launch Vibration Test

This test was performed following Thermal Performance Test 1. The test specimen was the fiberglass ISM panel which had previously undergone heat sterilization and the initial thermal performance testing. The ISM was mounted in a vibration test fixture to simulate a typical spacecraft installation. This assembly was then fastened to an electromagnetic exciter. A vibration pickup and two piezoelectric accelerometers located at the mounting edge of the specimen were used to monitor the vibration environment applied to the panel. Structural response acceleration levels were also monitored by piezoelectric accelerometers mounted on the cover plate of the ISM.

A frequency response survey was conducted on the ISM, over a range from 5 to 2000 Hz. Following the frequency response survey, random vibration testing was conducted. The ISM was then mounted on the fixture and subjected to 4.5 min of vibration testing to 16.1 g rms. Two test axes (lateral and vertical) were exposed to this test condition.

The frequency response of the test specimen was analyzed in terms of power spectral density (PSD). These PSD curves indicated the energy distribution of the vibration environment applied to the specimen, and demonstrated that these tests were conducted within the tolerances specified in MIL-STD-810B. No structural damage was observed during or after the vibration test.

Vibration test conclusions. Based on the visual examination and subsequent thermal performance test data, the ISM panel satisfactorily sustained the simulated launch vibration environment.

Landing Shock Test

The test specimen was the fiberglass ISM panel which had previously undergone heat sterilization, initial thermal performance and launch vibration testing.

The cover plate of the ISM was instrumented with three piezoelectric accelerometers to monitor the structural response of the ISM to the applied environment. The ISM was installed in the test fixture previously used for vibration testing and mounted on a Hyge shock tester.

The specified input shock spectrum was an idealized 120 g, 12.8 m sec half sine wave pulse. Three shock pulses were applied in each direction of the lateral and vertical axes of the ISM. Acceleration-time history data generated by the test were converted by computer analysis into the shock spectrum for comparison with that specified. Although the maximum response of the center of the aluminum cover to the applied shock indicated appreciable magnification of the input shock level, no structural damage was observed.

Landing shock test conclusions. Based on visual examination and the subsequent Thermal Performance Test results, the ISM panel satisfactorily sustained the Mars soft landing shock environment applied during this test.

Thermal Conductivity Test

The purpose of this test was to measure the absolute thermal conductivity of heat sterilized insulation material. Thermal conductivity measurements were made on the pure insulation material used in the fiberglass ISM. This test provided data for comparison with the Thermal Diffusivity test results.

The insulation material, Silicone-Bonded "AA" Microlite, manufactured by Johns-Manville, was assembled into a final configuration 14 in. in diameter and 4 in. thick. It consisted of two 0.008 in. flat Thermofoil heaters sandwiched between eight layers of 1 in. thick Microlite compressed to a total specimen height of 4 in. The sample density was 1.29 lb/ft^3.

The specimen was mounted in a specially developed guarded hot-plate type apparatus shown schematically in Fig. 8. Pressure variations from vacuum to one atmosphere can be provided in the apparatus, with specified gas composition and cold face temperature variations from below -150 to + 32°F. The hot face heaters consisted of a 6-in. diam Thermofoil heater (central test section) completely surrounded by a 4-in. wide Thermofoil guard heater. The heaters have very small heat capacity providing for quick response to power changes. The temperature difference between the guard and central heaters was automatically controlled to maintain an interface temperature difference within $\pm$1°F. Power was supplied to the center heater from a stabilized d.c. power supply.

The cold face heaters, a nichrome grid type embedded in phenolic prepreg, were bonded to the aluminum end plates. The cold temperature was monitored by a reference thermocouple attached to the inner surface of the end plate. Cold face heater power was adjusted to produce the desired cold face temperature.

Test data were obtained as follows: The test conditions, Table 8, were stabilized for a minimum of 2 hr before recording the temperatures or the heater power. Steady state conditions were then assumed to exist when a temperature change of less than 1.5°F/hr occurred. Power was determined by measuring the heater voltage and the voltage drop across a precision resistor connected in series with the heater. Corrections were made for lead losses. The power input to the central heater, temperature difference across the insulation, pressure, and the thermal conductivity of the insulation are given in Table 9 for each test condition. The thermal conductivity was calculated using the temperatures of the heater and cold plates respectively. Thermal conductivity versus mean temperature for each of the three gas models is plotted in Fig. 9, together with comparisons of other test data. As shown at -50°F mean temperature, thermal conductivity increases 50% from 6 mb in the minimum model atmosphere composition to 20 mb pressure in the maximum model.

Table 8 Thermal conductivity test conditions

Number	Hot Face/Cold Face Temperature (°F)	Pressure (mb)	Gas Composition CO_2 (% by weight)	N_2	A_r
1	50/-150	6	100	0	0
2	50/-150	9	74.4	12.8	12.8
3	50/-150	20	25.0	50.0	25.0
4	70/-100	6	100	0	0
5	70/-100	9	74.4	12.8	12.8
6	70/-100	20	25.0	50.0	25.0

Table 9 Thermal conductivity test results

Number	Heater Hot Face (°F)	Insulation Hot Face (°F)	Cold Plate (°F)	Insulation Cold Face (°F)	Pressure (mb)	Power (Watts)	Thermal Conductivity $\left(\frac{BTU}{Hr\text{-}Ft - °F}\right)$
1	50.0	46.2 48.0	-150.0 -150.0	-147.6 -149.1	6.0	1.09627	0.0078
2	49.2	45.5 47.4	-151.0 -151.5	-148.0 -151.0	8.9	1.31270	0.0093
3	49.2	46.0 47.3	-148.9 -150.4	-151.1 -151.1	19.6	1.65061	0.0118
4	69.6	66.3 67.8	-100.6 -100.6	- 98.6 -100.0	6.3	1.05400	0.0088
5	72.0	68.6 70.0	-100.1 - 99.7	- 98.3 - 99.1	9.1	1.20825	0.0100
6	71.3	67.4 69.5	-100.0 - 99.4	- 98.4 - 99.0	19.7	1.50624	0.0126

Thermal conductivity test conclusions. These results shown the sensitivity of this material to the Martian surface environment range. The test uncertainty has been estimated at ±6%, including an accounting for the small edge losses to the guard heater. The measured variation of thermal conductivity would require flexibility in the lander thermal control system.

Investigation of Foam Panel Failure

Small scale tests were performed to assess potential causes for failure of the foam ISM panel during chamber evacuation.

Several possible failure modes could be postulated, including a tension failure, poor or incomplete bonding and residual internal gas pressure resulting from outgassing during heat sterilization. The tensile strength of the foam was found to decrease during heat sterilization, but the value should have been sufficient to prevent failure of the ISM. The specimens also demonstrated no structural instability when subjected to a launch environment which was more severe than the original test panel had experienced. The failure mechanism is thus presently unknown and cannot be determined adequately without further investigation.

Extrapolation of ISM Weights to the Lander

It is anticipated that structural weight can be reduced on any future ISM designs. This conclusion is based on the successful shock testing of a fiberglass filled ISM to a level of 120 g as described previously. In addition, another ISM identical to that described, except not heat sterilized, was successfully shock tested to a level of 250 g to evaluate design margin. Since no failure was detected after either test, improved design configurations and weight should be considered for any subsequent evaluation.

Improved Panel Design

Lower weights are considered achievable by reducing the number of fiberglass plys in the casing, reducing the thickness of each ply, and using thinner aluminum for the cover. Minimum weight ISM panels, Table 10, were obtained by changing only the number of fiberglass laminate plys, i.e., from 3 to 1 ply for the foam design and from 4 to 2 for the fiberglass design. Local reinforcing at stress concentration points was included, and use of foam-in-place was assumed for the improved foam design.

Comparing the ISM designs at constant heat loss the difference between the foam and fiberglass concepts would be increased since the fiberglass material requires 3.72 in. thickness for heat loss equal to that from 3.0 in. to foam (Table 6). For these thicknesses the minimum predicted panel weights would be 1.47 lb for foam and 2.52 lb for fiberglass.

Effect on Lander Weight

A typical Martian lander would use the equivalent of about 20 of the ISM panels. Thus, the total predicted insulation system weights (minimum weight design) would be 29.4 lb for foam and 50.4 lb for fiberglass. The weight difference of

21.0 lb is significant in that it is about one third of the total weight originally assigned for science equipment and when appropriate weight ratios are applied, a difference in flight vehicle weight of about 92 lb could be necessary, thereby accounting for a major portion of the allowable weight contingency available. From these comparisons it is clear that even though the initial evaluation with the foam panel was unsuccessful, considerable gains are projected if this concept were explored further.

Table 10 Comparison of minimum weight designs (3-in. thickness)

Panel Type	Minimum Predicted Weight - Lb	Comparable Present Design Weight - Lb
Foam insulation		
Case (fiberglass)	0.20	0.59
Cover (aluminum)	0.39	0.38
Adhesive	----	0.66
Foam insulation (2.0 pcf)	0.88	0.88[a]
	1.47	2.51
Fibrous insulation		
Case (fiberglass)	0.65	1.28
Cover (aluminum)	1.04	1.05
Screen (stainless)	0.06	0.07
Bonding	0.03	0.03
Fibrous insulation (1.2 pcf)	0.53	0.50
	2.31	2.93

[a]Adhesive weight between layers deleted.

Conclusions

The objective of the study was to prove which of the readily available insulation materials would perform best in a Martian lander. In the broad sense, this objective was completed because the silicone bonded "AA" fiberglass survived the significant mission environments with no structural damage or change in performance and has a relatively low value of density-conductivity product. A lighter design was identified, but unfortunately the program constraints did not allow further effort to determine whether the foam panel could be modified to survive the launch venting phase. If the fiberglass is selected for use on the Viking lander, the guarded hot plate conductivity data will be necessary to analyze the lander heat transfer, including the effects of penetrations and joints.

The study also demonstrated several useful techniques by which high value insulation materials can be rapidly selected.

References

[1]Wilbers, O. J. et al, "Martian Soft Lander Insulation Study," Rept. MDC E0018, Sept. 1969, McDonnell Douglas Corp.

[2]Viking Project, "Mars Engineering Model," Rept. M73-106-0, NASA.

[3]Carslaw, H. S. and Jaeger, J. C., Conduction of Heat in Solids, Clarendon Press, Oxford, England, 1947.

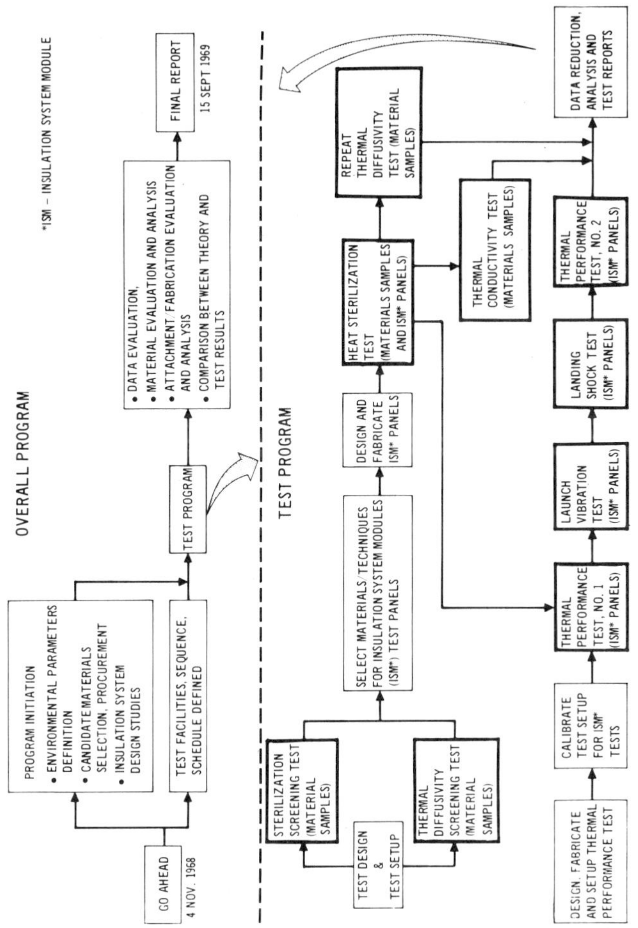

Fig. 1 Study flow plan.

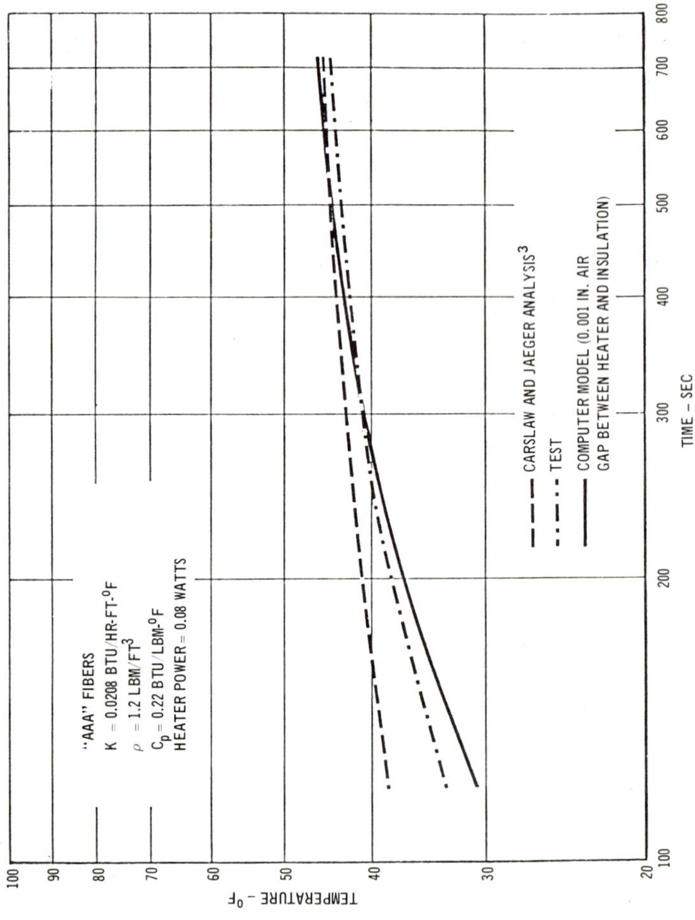

Fig. 2 Comparison of thermal diffusivity test and analysis results.

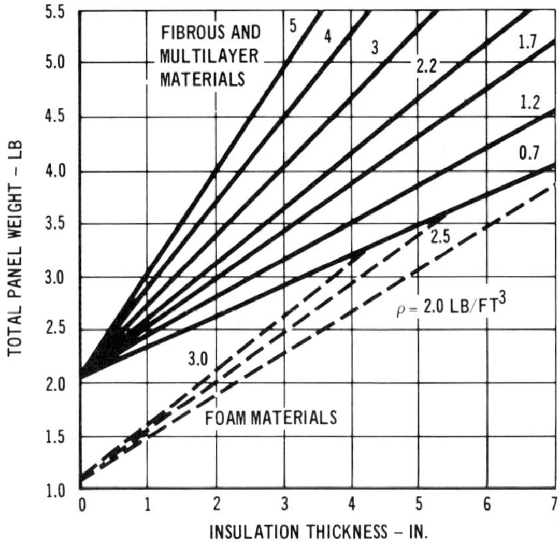

Fig. 3 Insulation system module weights for candidate materials.

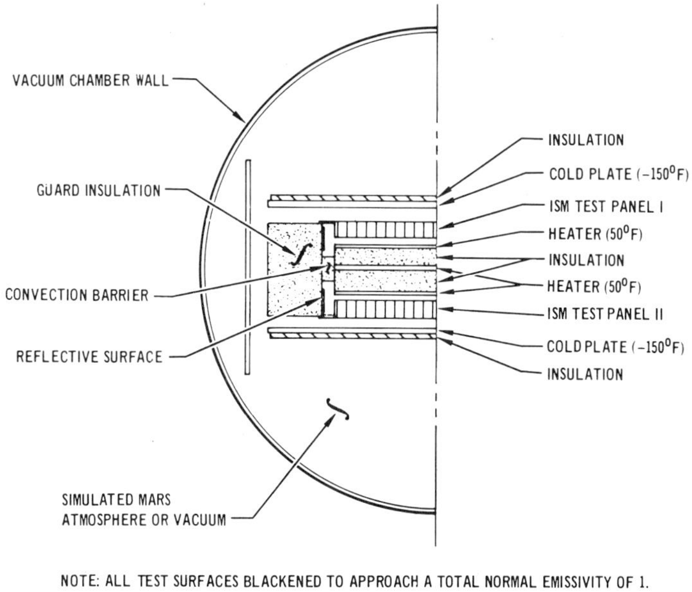

Fig. 4 Thermal performance test apparatus.

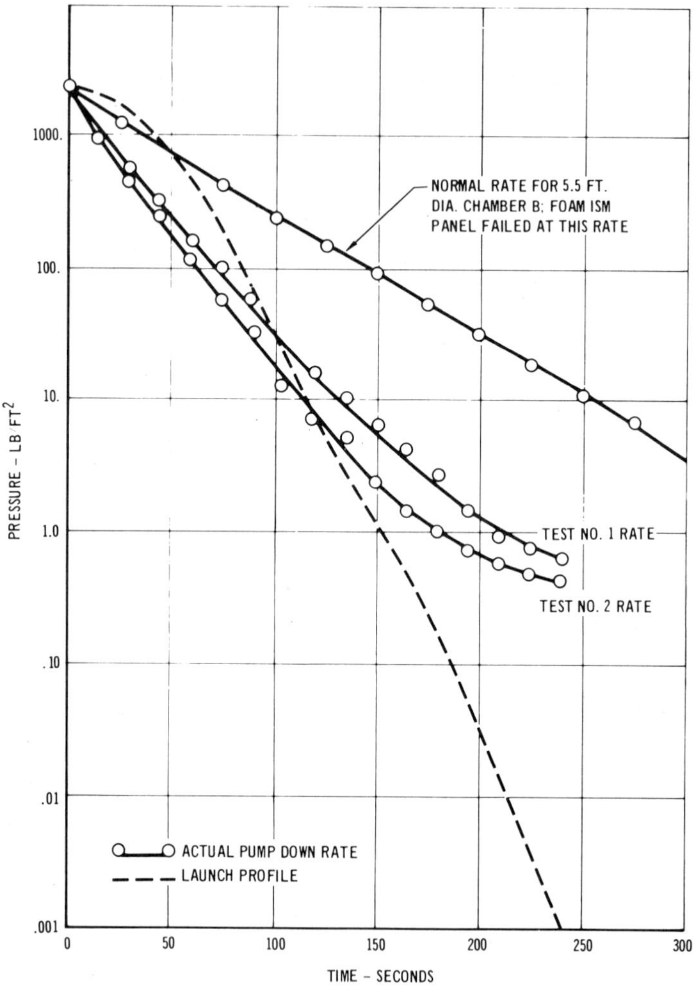

Fig. 5 Vacuum chamber pumping rates.

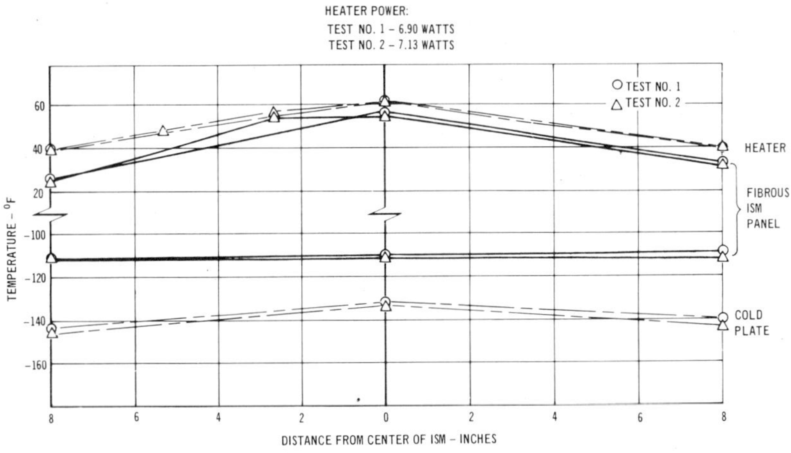

Fig. 6 Equilibrium temperature data - Thermal performance tests 1 and 2 - Vacuum phase.

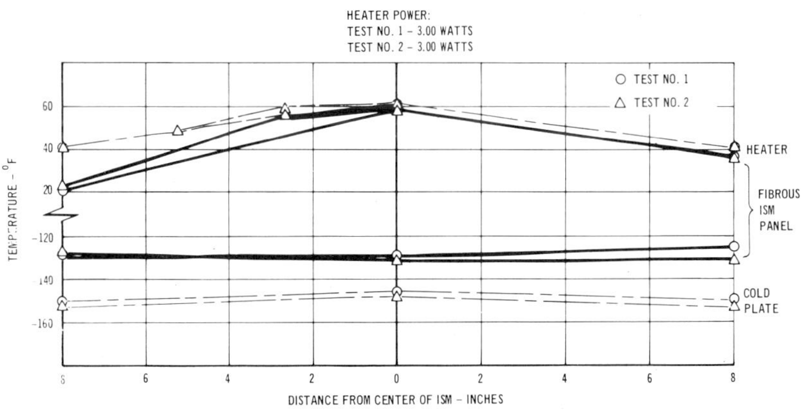

Fig. 7 Equilibrium temperature data - Thermal performance tests 1 and 2 - Mars atmosphere phase.

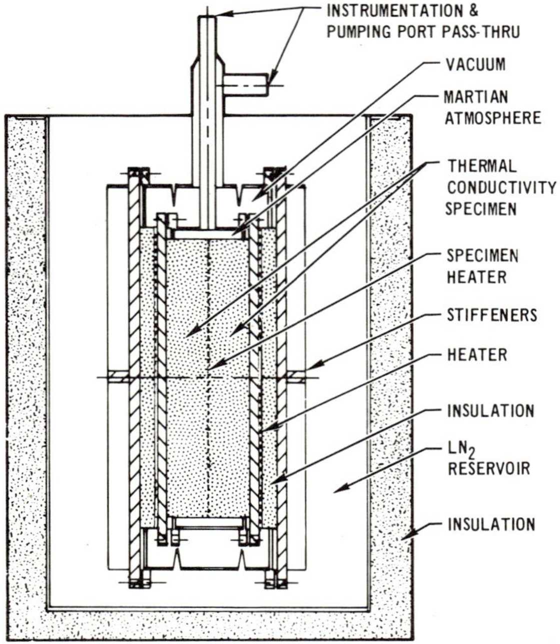

Fig. 8 Thermal conductivity test apparatus.

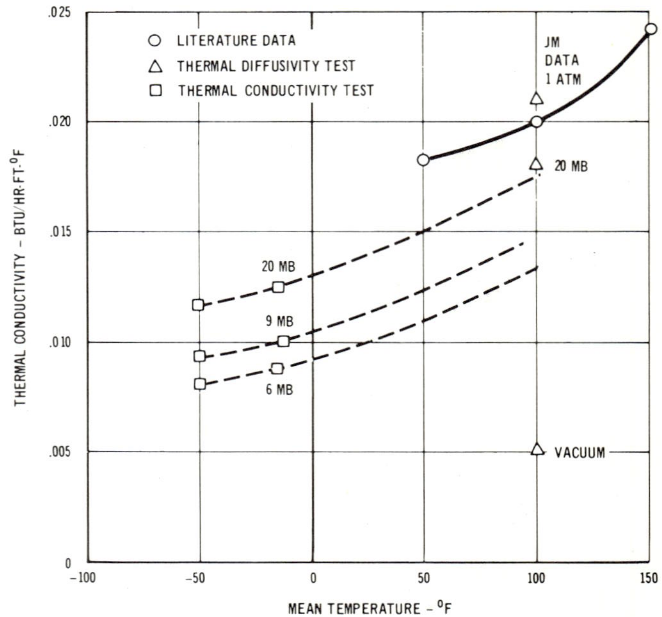

Fig. 9 Test data comparison - Silicone bonded "AA" fiberglass.

Index to Contributors to Volume 24

[1] Now at General Electric Company, Schenectady, N.Y.
[2] Now at North Dakota State University, Fargo.

[3] Now with Clark Brothers Company, Olean, N.Y.